# Solar and Wind Energy:
# An Environmental Approach

# Solar and Wind Energy: An Environmental Approach

Edited by Catherine Waltz

SYRAWOOD
PUBLISHING HOUSE

New York

Published by Syrawood Publishing House,
750 Third Avenue, 9th Floor,
New York, NY 10017, USA
www.syrawoodpublishinghouse.com

**Solar and Wind Energy: An Environmental Approach**
Edited by Catherine Waltz

International Standard Book Number: 978-1-68286-666-5 (Hardback)

#### Cataloging-in-Publication Data

Solar and wind energy : an environmental approach / edited by Catherine Waltz.
   p. cm.
Includes bibliographical references and index.
ISBN 978-1-68286-666-5
1. Solar energy--Environmental aspects. 2. Wind power--Environmental aspects.
3. Renewable energy sources. I. Waltz, Catherine.
TD195.S64 S65 2019
621.47--dc23

# TABLE OF CONTENTS

Preface......................................................................................................................................IX

Chapter 1    **Experimental Analysis of effect of Vegetation under PV Solar Panel on Performance of Polycrystalline Solar Panel**..............................................1
Sagar M Kande, Wagh MM, Ghane SG, Shinde NN and Patil PS

Chapter 2    **Solar Geyser using Spot Fresnel Lens**........................................................7
Aadesh Rajkrishna

Chapter 3    **A Review of** *in-situ* **Loading Conditions for Mathematical Modeling of Asymmetric Wind Turbine Blades**........................................................15
Bardsley A, Whitty JPM, Howe J and Francis J

Chapter 4    **Direct Current and Alternative Current based Solar Micro-Grid in Rural Energy Infrastructure**........................................................22
Vivek Kumar Singh, Lakshman Ravi Teja and Jitendra Tiwari

Chapter 5    **Relative Humidity effect on the Extracted Wind Power for Electricity Production in Nassiriyah City**........................................................27
Abdul-Kareem Mahdi Salih and Abdullh Saiwan Majli

Chapter 6    **Life Cycle Exergy Analysis of Solar Energy Systems**........................................33
Mei Gong and Göran Wall

Chapter 7    **Fuzzy Logic Control of a SEPIC Converter for a Photovoltaic System**..............41
Meryem Oudda and Abdeldjebar Hazzab

Chapter 8    **Potential of Wind Energy Development for Water Pumping in Ngaoundere**..............48
Ruben M. Mouangue, Myrin Y. Kazet, Daniel Lissouck, Alexis Kuitche and J.M. Ndjaka

Chapter 9    **Using Angstrom-Prescott (A-P) Method for Estimating Monthly Global Solar Radiation in Kashan**........................................................55
Ali Razmjoo, S. Mohammadreza Heibati, Mohammad Ghadimi, Mojtaba Qolipour and Javad Rezaei Nasab

Chapter 10    **Numerical Investigations of Coupling a Vacuum Membrane Desalination System with a Salt Gradient Solar Pond**........................................................59
Samira Ben Abdallah, Nader Frikha and Slimane Gabsi

Chapter 11    **Enhanced Biosolids Drying with a Solar Thermal Application**........................67
Domènec Jolis and Natalie Sierra

Chapter 12    **Understanding Various Technologies with Perspective of Energy and Environmental Audit**........................................................72
Harwinder Singh, Aftab Anjum, Mohit Gupta, Aadish Jain and Amrik Singh

Chapter 13    **Radical Urban Development in the Egyptian Desert**.............................................78
Abouelfadl S, Ouda K, Atia A, Al-Amir N, Ali M, Mahmoud S, Said H and
Ahmed A

Chapter 14    **Transmission Network Enhancement with Renewable Energy**.............................86
Oyedepo SO, Agbetuyi AF and Odunfa MK

Chapter 15    **Thermodynamic Analysis of a Combined Brayton and Rankine Cycle based on
Wind Turbine**....................................................................................................................97
Hossein Sheykhlou

Chapter 16    **Loss of Load Probability of a Power System**......................................................104
Vijayamohanan Pillai N

Chapter 17    **An Appropriate Extreme Value Distribution for the Annual Extreme Gust
Winds Speed**....................................................................................................................113
Banafsheh Abolpour, Bahador Abolpour, Hosein Bakhshi and Mohsen Yaghobi

Chapter 18    **Assessment of Wind Energy Potential for Small Communities in South- South
Nigeria: Case Study of Koluama, Bayelsa State**..................................................117
Akintomide Afolayan Akinsanola, Kehinde Olufunso Ogunjobi, Akintayo
T Abolude, Stefano C Sarris and Kehinde O Ladipo

Chapter 19    **An Experimental Appraisal on the Efficacy of MWCNT-H2O Nanofluid on the
Performance of Solar Parabolic Trough Collector**.............................................123
Harwinder Singh and Pushpendra Singh

Chapter 20    **Building Integrated Solar Thermal (BIST) Technologies and their
Applications: A Review of Structural Design and Architectural Integration**......129
Xingxing Zhang, Jingchun Shen, Llewellyn Tang, Tong Yang, Liang Xia,
Zehui Hong, Luying Wang, Yupeng Wu, Yong Shi, Peng Xu and Shengchun Liu

Chapter 21    **Performance Analysis and Comparison of Different Photovoltaic Modules
Technologies under Different Climatic Conditions in Casablanca**....................150
Elmehdi Karami, Mohamed Rafi, Amine Haibaoui, Abderraouf Ridah, Bouchaib
Hartiti and Philippe Thevenin

Chapter 22    **Energy Demand based Procedure for Tilt Angle Optimization of Solar
Collectors in Developing Countries**........................................................................156
Samer Yassin Alsadi and Yasser Fathi Nassar

Chapter 23    **Stored Heat Evaluation in Geothermal Systems: A Case of a Mexican Field**......160
Aragón-Aguilar Alfonso, Izquierdo-Montalvo Georgina, López-Blanco
Siomara and Gómez-Mendoza Rafael

Chapter 24    **Computational Examination of Utility Scale Wind Turbine Wake Interactions**......169
Tyamo Okosun and Chenn Q Zhou

Chapter 25    **Temporal Assessment of Wind Energy Resource in Algerian Desert
Sites: Calculation and Modelling of Wind Noise**..................................................178
Miloud Benmedjahed and Lahouaria Boudaoud

Chapter 26    **Application of Solar Energy Heating System in Some Oil Industry Units and
its Economy**.....................................................................................................................182
A.M. Abd El Rahman, A.S. Nafey and M.H.M. Hassanien

Chapter 27    **Exergetic Analysis of La Rumorosa-I Wind Farm**..................................................................................188
Rafael Carlos Reynaga-López, Alejandro Lambert, Oscar Jaramillo, Marlene
Zamora and Elia Leyva

Chapter 28    **Tilted Wick Solar Still with Flat Plate Bottom Reflector: Numerical Analysis for a
Case with a Gap Between Them**..................................................................................195
Hiroshi Tanaka

Chapter 29    **Application on Solar, Wind and Hydrogen Energy - A Feasibility Review for an
Optimised Hybrid Renewable Energy System** ..................................................................203
Subhashish Banerjee, Md. Nor. Musa, Dato' IR Abu. Bakar Jaafar and Azrin Arrifin

Chapter 30    **Small Wind Power Energy Output Prediction in a Complex Zone upon Five Years
Experimental Data**..................................................................................212
Ba MM, Ramenah Harry and Tanougast C

**Permissions**

**List of Contributors**

**Index**

# PREFACE

The main aim of this book is to educate learners and enhance their research focus by presenting diverse topics covering this vast field. This is an advanced book which compiles significant studies by distinguished experts in the area of analysis. This book addresses successive solutions to the challenges arising in the area of application, along with it; the book provides scope for future developments.

Solar and wind energy are renewable sources of energy. The field of renewable energy has shown remarkable growth over the years due to the improvement in the study of energy conservation and management. While solar energy is obtained by converting sunlight into usable energy, wind energy is harnessed by utilizing the power of air flow through wind turbines. This book is a valuable compilation of topics, ranging from the basic to the most complex advancements in the field of solar and wind energy. From theories to research to practical applications, studies related to all contemporary topics of relevance to this field have been included herein. It aims to equip students and experts with advanced concepts and upcoming topics in this area. Those in search of information to further their knowledge will also be greatly assisted by it.

It was a great honour to edit this book, though there were challenges, as it involved a lot of communication and networking between me and the editorial team. However, the end result was this all-inclusive book covering diverse themes in the field.

Finally, it is important to acknowledge the efforts of the contributors for their excellent chapters, through which a wide variety of issues have been addressed. I would also like to thank my colleagues for their valuable feedback during the making of this book.

**Editor**

# Experimental Analysis of Effect of Vegetation under PV Solar Panel on Performance of Polycrystalline Solar Panel

**Sagar M Kande[1*], Wagh MM[2], Ghane SG[3], Shinde NN[4] and Patil PS[5]**

[1]Energy Technology, Department of Technology, Shivaji University, Kolhapur, India
[2]Energy Technology, Shivaji University, India
[3]Department of Botany, Shivaji University, India
[4]Energy Technology, Shivaji University, India
[5]School of Nano Science and Technology, Shivaji University, India

## Abstract

The polycrystalline photovoltaic cell has an efficiency around 11-14%. The efficiency is low because of different factor, out of which the temperature is one of affecting factor on efficiency. The solar cell efficiency decreases with increase in temperature. So it is necessary to cool the PV panel to improve its efficiency.

Cooling of PV panel is one of critical issue during the planning of installation of PV plant. In the present work cooling of photovoltaic panel via different vegetation and water tray is carried out. The aim of this project is to optimize the panel efficiency by controlling the panel surface temperature by cultivating different vegetation below the panel. The experiment is done for polycrystalline silicon cell. The plants selected for the experimentation has a good evapotranspiration effect except aloe vera. The numerical value of increase in instantaneous efficiency is 3-4%, 1.8-2.2%, 1.2 -2%, 0.2 -0.5% for water tray, peppermint, tulsi and aloe vera respectively. The economical benefits due to cultivation of peppermint, tulsi aloe vera and water tray and also due to increase in power production from 1MW solar plant per year is forecasted has Rs. /- 455250, 436012, 219150 and 778850 respectively.

**Keywords:** Polycrystalline silicon; Vegetation cover; Evapotranspiration

## Introduction

Photovoltaic cells are devices which convert solar radiation directly into electricity. However, solar radiation increases the photovoltaic cell temperature [1-3]. The temperature has an influence on the degradation of the cell efficiency and the lifetime of a PV cell. But solar cells perform better in cold rather than in hot climate and as things stand, panels are at 25˚C which can be significantly different from the real outdoor situation [4-6]. For each degree rise in temperature above 25˚C the panel output decays by about 0.25% for amorphous cells and about 0.4-0.5% for crystalline cells. Thus, in hot summer days panel temperature can easily reach 70˚C or more. It means that the panels will put out up to 25% less power compared to what they are rated for at 25˚C [7,8]. Thus a 100W panel will produce only 75W in May/June in most parts of India where temperatures reach 42-44˚C and beyond during summer season and also the electricity demand is high. However, only a fraction of the incoming sunlight striking the cell is converted into electrical energy [9-11]. The remainder of the absorbed energy will be converted into thermal energy in the cell and may cause the junction temperature to rise unless the heat is efficiently dissipated to the environment [12]. Solar cell performance decreases with increasing temperature, shown in Figure 1, fundamentally owing to increased internal carrier recombination rates, caused by increased carrier concentrations. The operating temperature plays a key role in the photovoltaic conversion process [13-15].

The installation of MW scale power plant will leads to keep land to be barren and it leads to soil erosion in rainy season or in windy atmosphere. So to avoid such phenomenon we can use this barren land for cultivating the specific plant which provide good cooling effect by extracting heat from PV panel [16,17]. Hence due to the cooling of panel the performance of PV plant is enhanced also by cultivating specific plant certain economics benefit can be easily achieved. By this experimental method we can improve the performance of panel and ultimately the electricity production rate will increase [18].

## Effect of Temperature on Polycrystalline

For polycrystalline PV panels, if the temperature decreases by one degree Celsius, the voltage increases by 0.12 V so the temperature coefficient is 0.12 V/C. The general equation for estimating the voltage of a given material at a given temperature is [19]:

$$Voc\ module = Temp\ coefficient * (Tstc - T\ ambient) + Voc\ rated$$

Where:

Voc, mod = open circuit voltage at module temperature

T stc [°C] = temperature at standard test conditions, 25°C, 1000 W/m² solar irradiance.

$$Voc\ new = 0.12 * (25 - T\ ambient) + Voc\ rated$$

The on field experimentation is needs to study actual effect of vegetation on PV panel. So due to this reason an experimental prototype is made on the college terrace.

The experimental analysis gives the on field result with consideration of all environmental factors, the methodology used during experimentation is has mention below [20],

• The selection of plant which have high transpiration rate for providing good cooling effect.

• Design of experimental setup with vegetation

*Corresponding author: Sagar Maruti Kande, Energy Technology, Department of Technology, Shivaji University, Vidya Nagar, Kolhapur, Maharashtra 416004, India, E-mail: sagar.kande01@gmail.com

- • Installation of Experimental Setup

- • Testing and data collection

- • Data Analysis

- • Results

- • Evaluating the economic benefits of vegetation as per present market value

- • Conclusion, Limitations and Recommendations.

## Plant Selection

There are many plants available for vegetation which will grow under the shadow (under the panel). But during the selection of plant need to consider some limitation of prototype setup (Figure 2) [21].

The plant selected (medicinal plants) as per it's economical as well as health benefits.

Main criteria consider during the selections of plant [22]:

1. The experimental setup prepared has the limitation of height from ground. Therefore the plant selects must have height less than panel height from ground surface.

2. Root length - As I have prepared the artificial bed for cultivation of plants.

3. Water requirement for plant growth is most considerable factors during selection of plants.

4. Temperature Requirement of specific plant

5. Soil properties. The cooling media can be classified into two categories

a)    Medicinal Plant.

b)    Water tray (keeping tray filled with water at the below of PV panel).

## Experimental Setup

The 3- D model of experimental setup is design on Uni-Graphics. The actual setup may look like this as shown below in different view (Figure 3).

The experimental setup made on field for finding out the results is shown in Figure 4.

## Experimental Testing Procedure

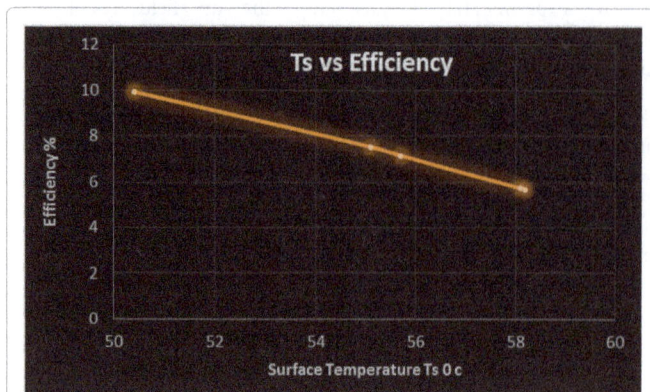

**Figure 1:** Ts vs Efficiency Graphs.

**Figure 2:** Selected Plant; A. Oscimum Sanctum (Tulsi); B. Aloe Vera (Korpad); C. Peppermint (Pudina); D. Water Tray.

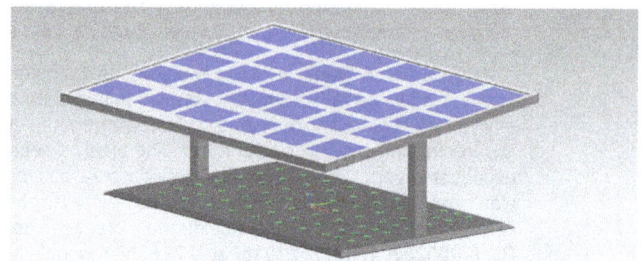

**Figure 3:** 3-D model on Uni-Graphics.

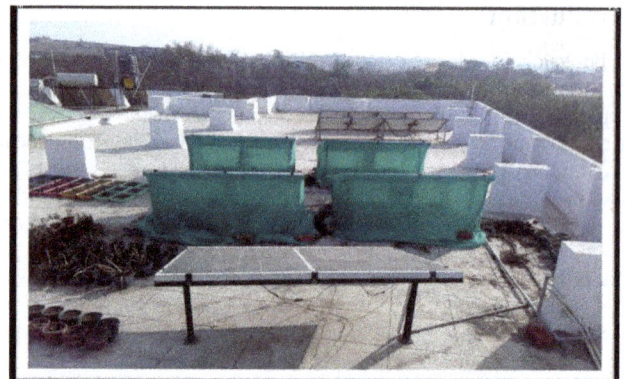

**Figure 4:** Experimental Setup.

Experimental setup is tested for two days and the reading on each day is mentioned in sheet. The different parameter are measured like voltage, current, radiation, panel surface temperature, chamber temperature, soil temperature, wind speed, atmospheric temperature etc. Various instrument are used to measure this parameter mainly data logger for recording temperature for half hour interval time, anemometer, multimeter, digital temperature gun, pyranometer for radiation measurement.

Procedure followed during testing is,

- • All sensor of temperature data logger is fixed at a particular point to record the temperature of that point continually.

- • The setting is set into data logger in such way that it record temperature with half hours of interval.

- • For the same instant of time the radiation are measured with the help of radiation meter.

- • The multimeter is used to measure the voltage and current at the same instant of time.

- All the parameter is taken at same time but there might be a 1-2 min time difference.

The load power and efficiency is calculated as,

Power At load = $I_m * V_m$

Efficiency = $\eta = \dfrac{I_m \times V_m}{A \times Radiat}$

Therefore, the sample calculation,

Power = Im × Vm = 2.7 ×74.8 = 201.96 W

Efficiency = (2.7 × 74.8) ÷ (3.87 × 632) = 8.25%

Similarly the all calculation is done and the following table is prepared (Table 1) [23].

## Results

From experiment it is seen that cooling effect from vegetation enhances the panel performance. The result obtains due to cooling by different vegetation and water tray are tabulated and plotted. The efficiency increased by water cooling is more as compare with other setup (Figure 5).

The increase in efficiency among the all setup is more for water tray setup and which will be in the range of 3-4%. Such increase in 3-4% efficiency causes the falling in the panel backside temperature upto 5-6°C.The water tray prototype gives very good cooling due to the high evaporation rate from the tray as it kept open. Peppermint also gives very good cooling due to its greenish nature. It has increase in efficiency in the range of 1.8-2.2 %. Tulsi also gives good cooling effect. The aloe vera has lowest increase in efficiency among the setup prepared for examining the cooling effect. The difference between cooling obtain

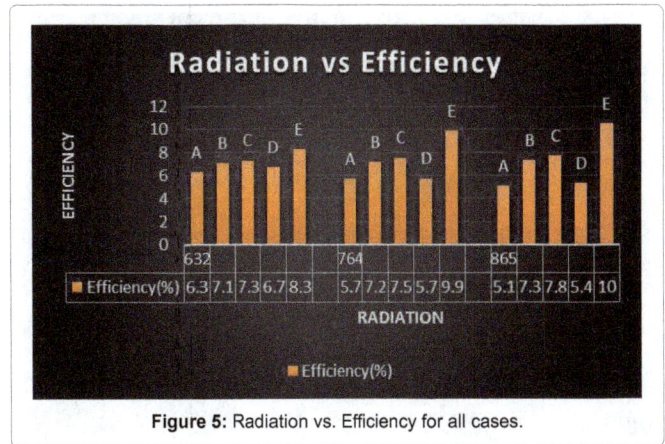

**Figure 5:** Radiation vs. Efficiency for all cases.

from various setup is clearly indicated in the graph drawn below (Figure 6).

## Analysis

The result obtains due to cooling by different vegetation and water tray plotted in graphs shown above. The efficiency is increased due to different type of vegetation and water tray, among of them water tray has highest efficiency. These results are analyzed by plotting the graphs between Radiation v/s Efficiency for each case, (Reading for analysis graphs are taken at no load on PV panel) [4-6].

1. The PV panel performance is improved due to the vegetation cooling.

2. Among the all cases efficiency increased for water cooling is more and which is 3-4%.

3. Aloe Vera does not give the good cooling has the efficiency improved for aloe vera is only 0.2-0.5 % because, aloe vera mainly grown in desert area it has characteristics of holding water for their survival.

4. As seeing the panel surface temperature (Ts where it is recorded) of panel if cooling is obtaining up to 1°C the panel efficiency will increase up to 0.33 -0.45%.

The graph plotted below show that there is increase in efficiency due to cooling via vegetation at the same radiation as comparing with empty model (A) [7,8] (Figure 7).

From above graphs plotted for empty case A show that the maximum efficiency obtain is near about 15 %.The panel surface is increased with increase in radiation level, as it starts increase above its rated temperature the efficiency goes on decreasing (Figure 8) [9-11].

The panel surface is increased with increase in radiation level, as it starts increase above its rated temperature the efficiency goes on decreasing. But due to the vegetation cover of tulsi under the panel it leads to a near about constant temperature inside the chamber. The graphs plotted for case B is clearly indicate that there is increase in efficiency doe to tulsi cover below the panel at same radiation as for empty case A. The maximum efficiency obtain is 22% (Figure 9) [12].

The panel surface temperature is increased with increase in radiation level, as it starts increase above its rated temperature the efficiency goes on decreasing. But due to the vegetation cover of peppermint under the panel it leads to a near about constant temperature inside the chamber as obtain in surface temperature vs. radiation graph. The efficiency is

| Model | Radiation (W/m²) | Im (A) | Vm (V) | Power (W) | Efficiency (%) | Ts |
|-------|------------------|--------|--------|-----------|----------------|------|
| A | 632 | 2.47 | 62.5 | 154.375 | 6.31 | 57.8 |
| B | | 2.63 | 66 | 173.58 | 7.096 | 56.2 |
| C | | 2.66 | 67 | 178.22 | 7.28 | 55.1 |
| D | | 2.59 | 63.7 | 164.983 | 6.74 | 57 |
| E | | 2.7 | 74.8 | 201.96 | 8.25 | 53.1 |
| A | 764 | 2.6 | 63 | 163.8 | 5.67 | 58.2 |
| B | | 3.1 | 67.5 | 209.25 | 7.16 | 55.7 |
| C | | 3.21 | 68.5 | 219.885 | 7.53 | 55.1 |
| D | | 2.62 | 64 | 167.68 | 5.745 | 58.1 |
| E | | 3.9 | 74.2 | 289.38 | 9.91 | 50.4 |
| A | 865 | 2.62 | 64 | 167.68 | 5.07 | 59 |
| B | | 3.4 | 71.1 | 241.74 | 7.31 | 55.3 |
| C | | 3.0 | 71.2 | 256.32 | 7.75 | 54.4 |
| D | | 2.69 | 66 | 177.54 | 5.37 | 58.1 |
| E | | 4.5 | 77 | 346.5 | 10.48 | 49.4 |

**ABCDE color codes specifications:** Model A: Empty setup; Model B: Tulsi setup; Model C: Peppermint setup; Model D: Aloe Vera setup; Model E: Water tray setup

**Table 1:** Reading at different Radiation.

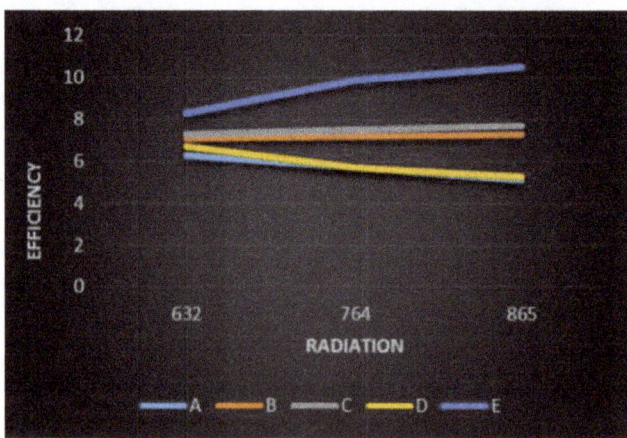

**Figure 6:** Increase in order of Efficiency for designed prototype.

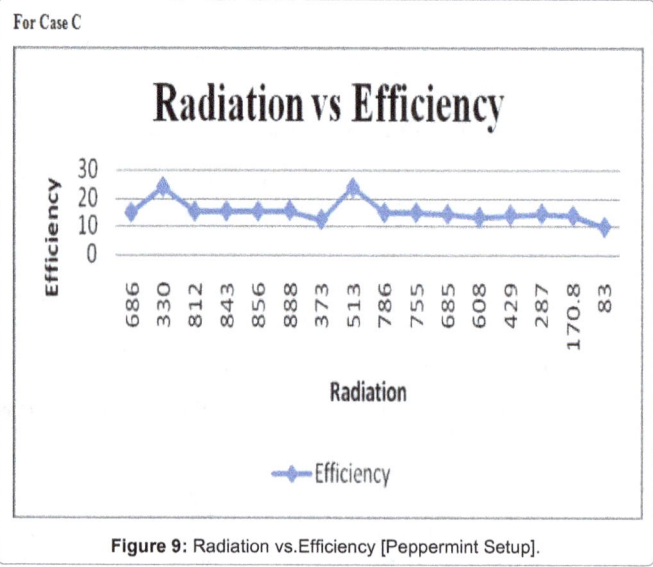

**Figure 7:** Radiation vs. Efficiency [Empty setup].

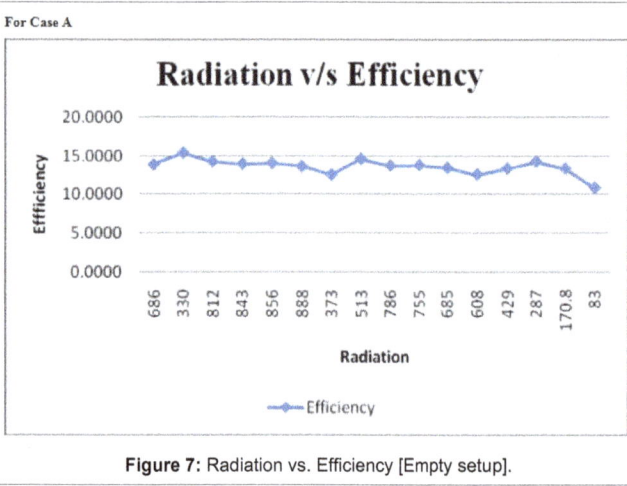

**Figure 8:** Radiation vs. Efficiency [Tulsi setup].

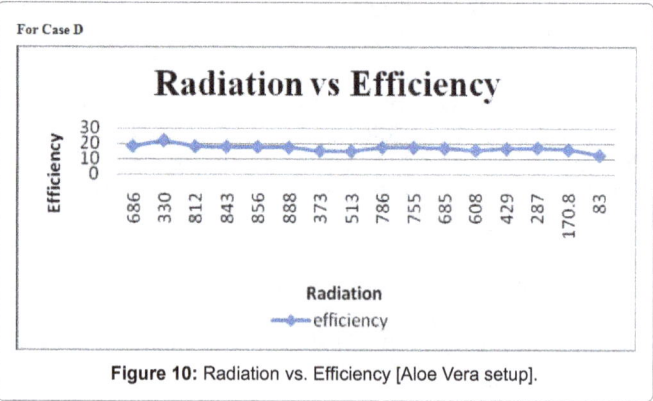

**Figure 9:** Radiation vs.Efficiency [Peppermint Setup].

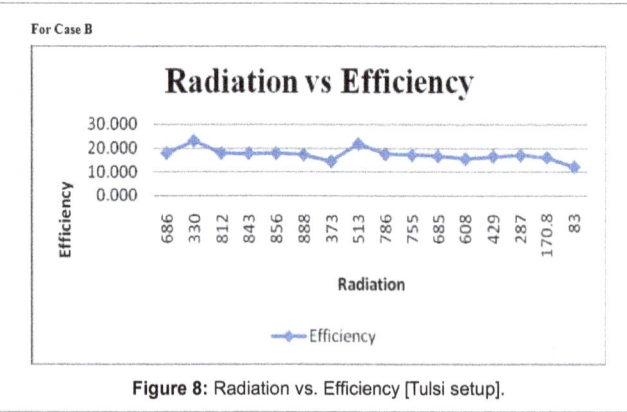

**Figure 10:** Radiation vs. Efficiency [Aloe Vera setup].

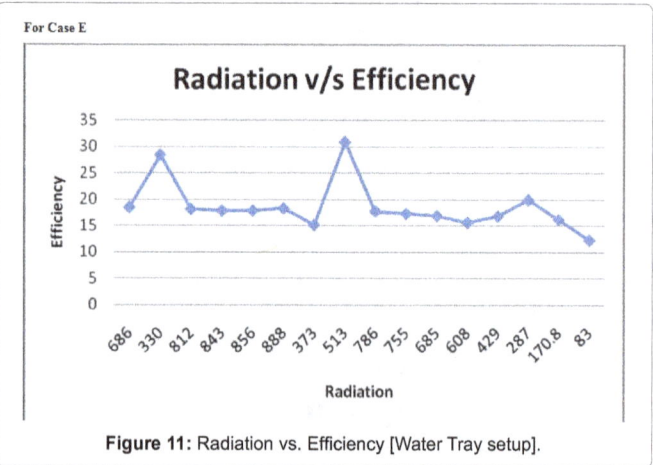

**Figure 11:** Radiation vs. Efficiency [Water Tray setup].

increased more than the tulsi setup due to the more greenish nature of peppermint. As the peppermint provide good cooling, the maximum efficiency obtain is 24% [13-15] (Figures 10 and 11).

The panel surface temperature is increased with increase in radiation level, as it starts increase above its rated temperature the efficiency goes on decreasing. But due the water tray kept under the panel it maintain the near about constant temperature inside the chamber as shown in surface temperature vs. radiation graph. The evaporation of water makes the panel back surface cool which leads to increase in panel efficiency due to decreasing panel temperature.

Among the all experimental prototype the water tray setup gives more increase in efficiency as shown in above graphs (Figure 12) [16].

The graph between Ts vs. Radiation clearly indicates that if there is increase in radiation will increases surface temperature of panel. Also it is possible to enhance the efficiency of panel due to cooling them by various cooling media likes vegetation and water tray as experimented in a project. The graph between radiations vs. efficiency show the increase in order of efficiency for various cases is Aloe Vera, Tuli,

Peppermint and water tray.

From all the above graphs and discussion it is concluded that the vegetation can made very good impact on panel cooling. The cooling effect due to vegetation enhances the panel efficiency as well as it will help to maintain surrounding temperature cool. For the same radiation the water tray case gives more increased in efficiency. On the graphs there are some pick out due to cloudy environment [17].

### Combined result analysis graph: for surface temperature

For different radiation the surface temperature versus efficiency is plotted, the graphs show that with increase in temperature the panel efficiency goes on decreasing. The temperature profile as a constant nature during 48 to 53°C and is slightly drop during next zone as shown in below graph (Figure 13) [24].

### Economical Analysis

It has been proved by practical experimentation that the vegetation gives the cooling effect under the PV panel, which leads to increase panel output power. The increase in power increases the unit generated per year which add the additional income into the existing income coming from power production also the by selling the medicinal plants parts (leaves, flower etc) will also add into the total income. The economical analysis is carried for the experimental setup as well as for MW scale power plant [25] (Table 2).

As the bar chart (Figure 14) clearly indicates that the benefits obtain in terms of money is more from the set of PV panel below which water tray is kept, as the water evaporation rate from open tray is more so it will cool panel at optimistic value which leads increase more number of k-Wh [26]. Also the peppermint and tulsi gives appropriate benefits

comparing with their initial capital investment. So when we cultivate such vegetation under PV panel for MW scale plant it will give income in both terms, also it will help to maintain environment cool (global warming effect) [27].

### Conclusion

1. The increase in temperature will decrease the efficiency; it has linear in nature with efficiency.

2. The efficiency is increased due to different type of vegetation and water tray, among of them water tray gives the highest efficiency.

3. The increase in efficiency due to water tray, peppermint, tulsi and aloe vera are 3-4%, 1.8-2.2%, 1.2-2% and 0.2-0.5%.

4. As the decreasing panel surface temperature by 1°C the panel efficiency will increase up to 0.33-0.45%.

### Limitations

1. The temperature data recorded as error of 1-2 degree Celsius due to sensor cable resistance.

2. The panel height from ground surface creates problem of maintenance so it is necessary to increase the panel height from ground surface.

3. The vegetation needs water so such cooling phenomenon creates the problem in the area of low water zone.

4. Need to select the plant having long life span.

### Suggestions

- If the water bodies are available then install the PV panel above the water body as they provide very good cooling.

- The area where we want to setup PV plant; if the water quantity

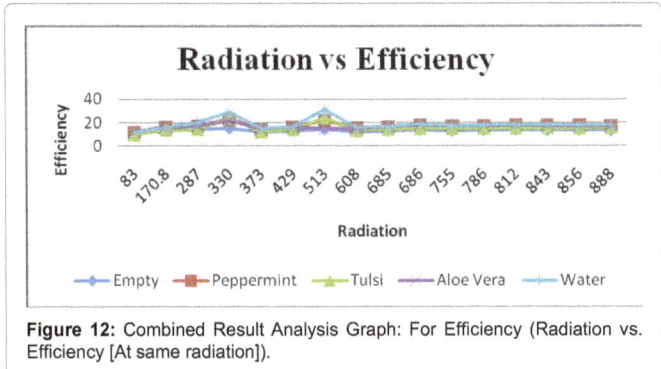

**Figure 12:** Combined Result Analysis Graph: For Efficiency (Radiation vs. Efficiency [At same radiation]).

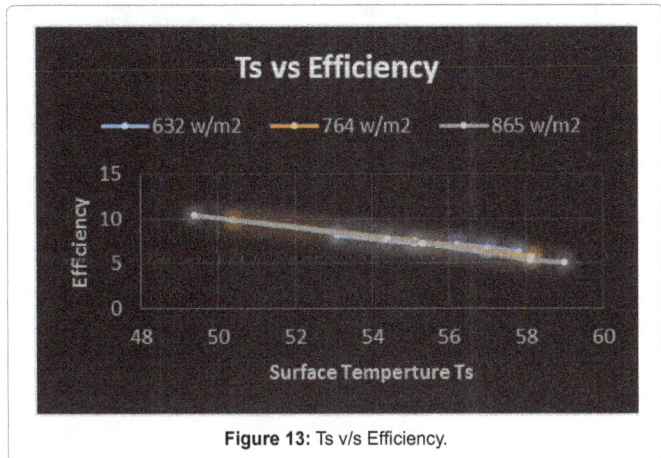

**Figure 13:** Ts v/s Efficiency.

| Type of Vegetation | Benefits Due to vegetation/ Hectare in Rs./- | Benefits due to vegetation obtain Rs./- per year | Benefits due to power increased from 0.5 kW in Rs./- | Total Benefit obtain from experimental setup/year | Benefit obtain/ MW Plant/ year |
|---|---|---|---|---|---|
| Tulsi | 84,575 | 169.15 | 559.97 | 729.12 | 436012 |
| Pudina | 55,000 | 110 | 678.27 | 788.27 | 455250 |
| Aloe Vera | 30,000 | 60 | 101.79 | 161.79 | 219150 |
| Water | - | 0 | 1339.05 | 1339.05 | 778850 |

**Table 2:** Benefits obtain in Rs. /-per year from experimental setup.

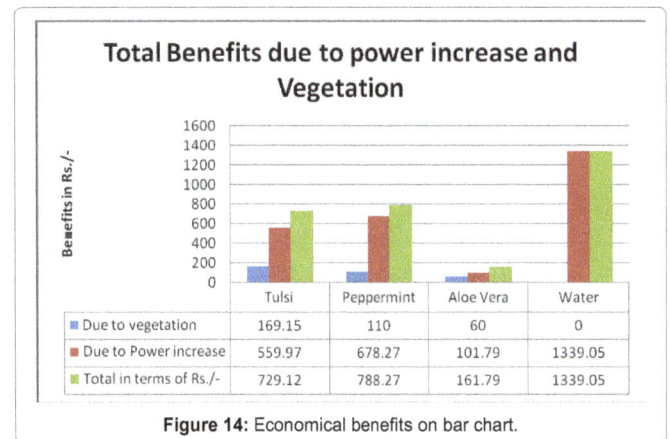

**Figure 14:** Economical benefits on bar chart.

available up to appropriate level then cultivate different type of vegetation suitable for that land. It will increase the annual income.

- But the area like Latur, Beed, etc. they have the problem of drinking water so for such area need to find out some alternatives for panel cooling.

- The way of cooling due to vegetation is make the surrounding temperature low also due to vegetation soil erosion reduces.

## Acknowledgment

The authors want to thank the department of Energy Technology, Shivaji University, Kolhapur, for providing lab support and testing equipment.

## References

1. Moharram KA, Abd S, Kandil H, El-Sherif H (2013) Enhancing the performance of photovoltaic panels by water cooling. Ain Shams Engineering Journal 4: 869-877.

2. El-Shae A, Tadros MTY, Khalifa MA (2014) Effect of light intensity and temperature on crystalline silicon solar modules parameters. IJETAE 4: 311-318.

3. Abdulgafar SA, Omar OS, Yousif KM (2014) Improving the efficiency of polycrystalline solar panel via water immersion method. IJIRSET 3: 8127-8132.

4. Cuce E, Bali T, Sekucoglu SA (2011) Effect of passive cooling on performance of photovoltaic cell. Int J Low-Carbon Technol.

5. Smith MK, Selbak H, Wamser CC, Day NU, Krieske M, et al. (2014) Water Cooling Method to Improve the Performance of Field-Mounted, Insulated, and Concentrating Photovoltaic Modules. J Sol Energy Eng 136: 1-4.

6. Mehrotra S, Rawat P, Debbarma M, Sudhakar K (2014) Performance of a solar panel with water immersion cooling techniques. Int J Sci Environ Technol 3: 1161-1172.

7. Tiwari GN, Mishra RK (2012) Advanced Renewable Energy Sources. Royal Society of Chemistry.

8. Royne A, Dey CJ, Mills DR (2005) Cooling of photovoltaic cell under concentrated Illumination: A Critical Review. Solar Energy Materials & Solar Cells 86: 451-483.

9. Dubey S, Sarvaiya JN, Seshadri B (2013) Temperature Dependent Photovoltaic (PV) Efficiency and Effect on Production in the world: Review. Energy Procedia 33: 311-321.

10. Hajjaj C, Rhassouli AE, Amiry H, Hasnaoui AE, Bounouar S, et al. (2015) Design, Implementation and Characterization of a cooling photovoltaic solar panel device.

11. Borkar DS, Prayagi SV, Gotmare J (2014) Performance evaluation of photovoltaic solar panel using Thermoelectric cooling. Int J Engineering Res 3: 536-539.

12. Rajaram R, Sivakumar DB (2015) Experimental investigation of solar panel cooling by the use of Phase Change Material. J Chem Pharm Sci 6: 238-239.

13. Karthick SP, Helen Catherine RL (2014) Modelling and analysis of solar photovoltaic Thermal Collector. IJAREEIE 3: 4.

14. Green Roofs Boost Photovoltaic Panels.

15. Lin BS, Lin YJ (2010) Cooling effect of shade trees with different characteristics in subtropical urban park. Hortscience 45: 83-86.

16. Jones AD, Underwood CP (2001) A thermal model for photovoltaic systems. Solar Energy 70: 349-359.

17. Tiwari GN, Mishra RK (2012) Advanced Renewable Energy Sources.

18. Nvis Technologies Pvt Ltd (2015) Hybrid solar and wind energy trainer.

19. Handbook for solar photovoltaic (PV) systems.

20. Mertens K (2014) Photovoltaics Fundamentals, Technology and Practice. Münster University of Applied Sciences, Germany 1st edition.

21. Bridge LJ, Franklin KA, Homer ME (2013) Impact of plant shoot architecture on leaf cooling: a coupled heat and mass transfer model. J Royal Soc Interface 10: 1-18.

22. Majewski W, Miller DC (1979) Predicting effects of power plant once-through cooling on aquatic systems. UNESCO 20: 1-170.

23. Haq F (2012) The Ethno Botanical Uses of Medicinal Plants of Allai Valley, Western Himalaya Pakistan. Int J Plant Res 2: 21-34.

24. Rajeswari R, Umadevi M, Rahale CS, Pushpa R, Selvavenkadesh S, et al. (2012) Aloe vera: The Miracle Plant Its Medicinal and Traditional Uses in India. J Pharmacogn Phytochem 1: 118-124.

25. Indian council of Agriculture Research.

26. Harvesting Crops along with Sunshine. Headway Solar Article.

27. Kumar S, Suresh R, Singh V, Singh AK (2011) Economical analysis of menthol mint cultivation in Uttarpradesh: A case study. Agricult Econom Res Rev 24: 345-350.

# 2

# Solar Geyser Using Spot Fresnel Lens

**Aadesh Rajkrishna\***

*Department of Engineering, University of Petroleum and Energy Studies, India*

**Abstract**

Fresnel lens based concentrating devices are not only gaining scope in the Photovoltaic industry, but also in solar thermal based industries. Fresnel lens is an optic device which concentrates the incoming light onto a spot or onto a line. This means that the temperature at that point will be significantly high. Utilizing this temperature for solar thermal applications will definitely be helpful for solar thermal power plants as the water can directly be converted into steam. This paper discusses about the various applications of Fresnel lens and how advantageous it can be to be utilized in solar thermal applications. Also better improvement on the collector design is projected that shows improvement in the overall performance of the system.

**Keywords:** Fresnel lens; Concentrated solar power; Photovoltaics; Thermal

## Introduction

Photo-voltaic (PV) concentrator systems and thermal concentrator systems are known to be gaining much scope over the past years. The PV concentrating system is an upcoming field where concentrating systems are used to direct the radiation onto a photovoltaic absorber that is specially manufactured for concentrating purpose. Parabolic trough collectors are another technology of advanced solar thermal heating systems. This technology uses significant amount of collector area [1]. Fresnel lens and Fresnel reflector technology reduces this size to only the aperture area and focal length of the lens. There are a lot of advantages of using Fresnel lens both in thermal and photovoltaic systems [2].

## Technologies based on Fresnel Reflectors and Fresnel Lens

### Fresnel lens PV concentrator systems

Fresnel lens is a component which focuses light onto a single point or a line. There are basically two types of Fresnel lenses. One is the Spot Fresnel lens and the other is the Linear Fresnel lens. The spot Fresnel lens focuses the light onto a spot on the object and the linear Fresnel lens focuses light onto an entire line on the object (Figures 1 and 2) [3].

The lens consists of concentric groves in spot Fresnel lens and parallel groves in linear Fresnel lens.

The groves are the main part of the lens. It focuses light according to its design based on concentric or parallel nature of the groves. Concentric groves focus light onto a single point and parallel lens focuses onto a line.

Major development in the field of Fresnel lens PV concentrator systems are being made due to the fact that radiation on the cell increases with increase in temperature.

Dual axis tracking, point focus Fresnel lens is used for concentrating the radiation on to the center. Currently there is a plant operating in Riyadh, Saudi Arabia. This plant produced 505Wh of Electricity in 12 hours.

The absorber is an Insulated Metal Substrate (IMS) board specially made for concentrating PV systems. The maximum temperature achieved by this technology is 92.4°C.

Figure 3 shows the graph that shows the maximum achieved temperature by the technology according to [1].

This proves that the Fresnel lens technology is definitely helpful for Solar Thermal Technology since the temperature achieved is high.

Although there is a significant temperature rise over the solar cell, there might be difficulties when operating at higher temperatures. Since the maximum temperature achieved is 92.4°C under STC, it might go high on field conditions making the IMS vulnerable to higher temperatures.

For a PV system, this might not be a good option since there is a risk of losing the circuit board due to high temperatures even though special protection measures are taken for high temperature operation.

### Flat plate collector (FPC) with fresnel cavity receiver

According to Iuliana [4], the Fresnel lens is a flat stretched glass with groves to focus the light onto the absorber plate.

The absorber is a U shaped glass tube made of copper.

The arrangement and the effect of sunlight on the absorber are shown in the Figure 4 [4].

The variation of thermal efficiency with difference in temperature of fluid mean temperature and ambient temperature with solar radiance divided. This is as shown in Figure 5 [4].

This study was observed over 4 seasons, summer sunny, summer cloudy, winter sunny and winter cloudy.

The insolation over the 4 seasons are,

Summer sunny - 905W/m$^2$

Summer cloudy - 462W/m$^2$

---

**\*Corresponding author:** Aadesh Rajkrishna, Department of Engineering, University of Petroleum and Energy Studies, College of Engineering Studies, Dehradun, India, E-mail: aadesh.rajkrishna@gmail.com

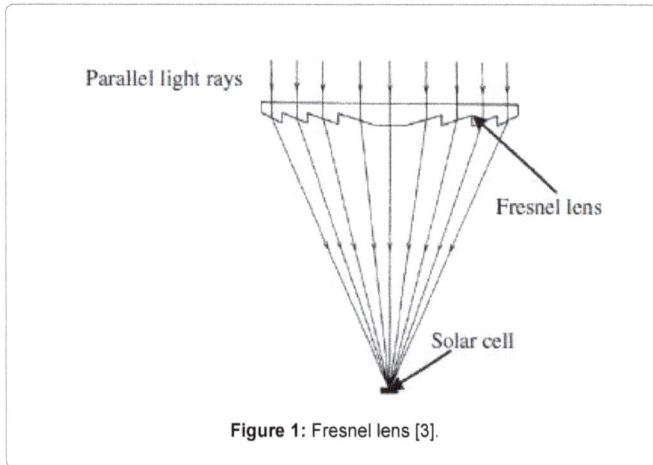

**Figure 1:** Fresnel lens [3].

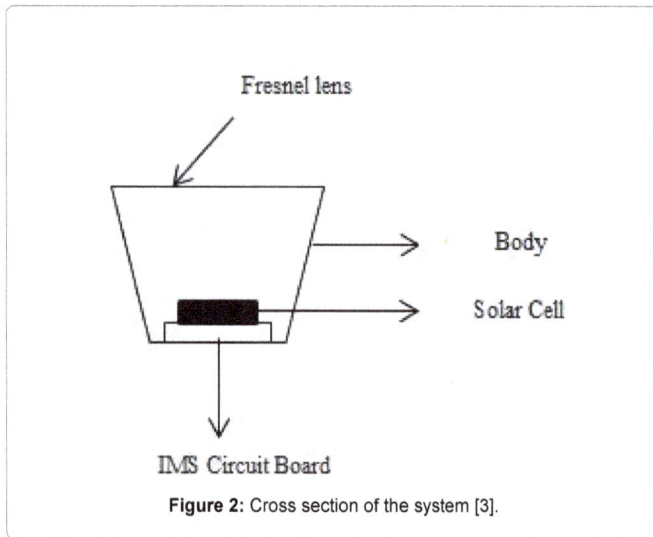

**Figure 2:** Cross section of the system [3].

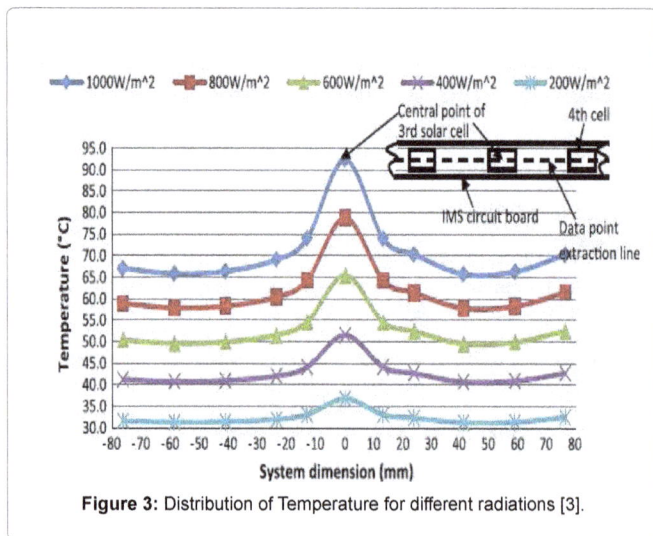

**Figure 3:** Distribution of Temperature for different radiations [3].

Winter sunny - 140W/m²

Winter cloudy - 68W/m²

The plot of tube length v/s fluid temperature during summer sunny

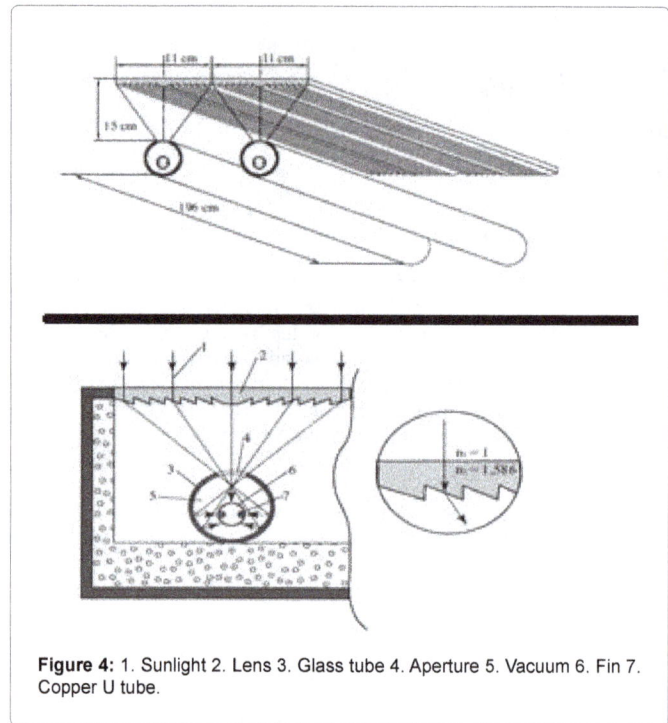

**Figure 4:** 1. Sunlight 2. Lens 3. Glass tube 4. Aperture 5. Vacuum 6. Fin 7. Copper U tube.

and winter sunny is as shown in Figure 6 [4].

According to Iuliana [4], there was significant increase in the Fresnel lens based collector efficiency.

This technology makes a good impact on the solar thermal energy concept. Bringing this technology into large scale production can be a difficult task since custom made Fresnel lens tends to be more expensive.

Also the absorber tubes may undergo chemical changes due to such high temperatures leading to corrosion.

**Flat linear Fresnel lens with sun tracking system**

This technology is much similar to that of Iuliana, but the major difference is that there is a sun tracking system which gives in more efficiency.

This study was done in Iraq which involved two lenses in series and two aluminum fin absorber tubes with black coating. Single lens and a single tube produce an output temperature of water of 37°C.

The maximum thermal efficiency achieved was 65% with a mass flow rate of 0.0070815 Kg/s.

This technology was a prominent and a promising technology according to [5].

A single lens and a single absorber produce 37°C output water. This means that more lenses and more absorbers in a single system produce a significant output temperature of water that can be used for both domestic and industrial uses (Figure 7).

This is a plot on all day efficiency where in the maximum obtained efficiency was 58% which was an acceptable value for a solar thermal system.

As the technology involves tracking, it is natural that the economic aspect of installing this system in developing countries is futile.

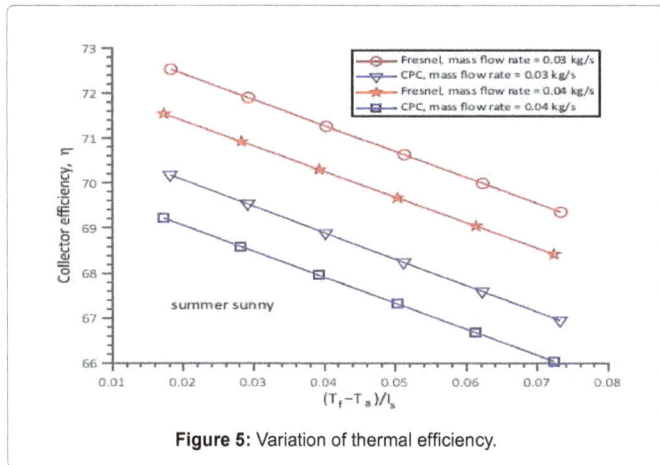

Figure 5: Variation of thermal efficiency.

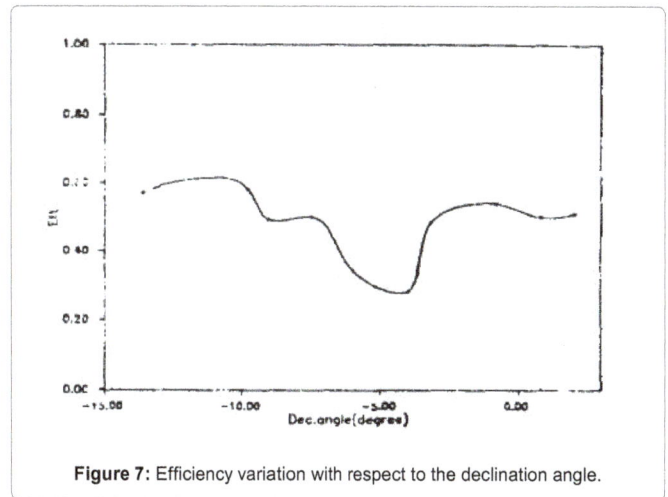

Figure 7: Efficiency variation with respect to the declination angle.

## Fresnel lens temperature and energy generation of concentrator PV system

According to Thorsten [6], there are basically two types of Fresnel lenses that are commercially available. One is the lens made up of PMMA (Poly-methylmethacrylate) and the other made up of a transparent silicone rubber on a glass. This is also called as SOG (silicone on glass).

Experiments were made and the two lens technologies was tested which resulted in proving that silicone based Fresnel lens are better than the conventional PMMA lenses.

Again, the cost of using silicone lens is high making the developing countries to look for another technology that is affordable.

## Fresnel reflector array technology

According to Abbas [1], this technology consists of basically reflectors and not lenses. This is another unique technology that is used for solar thermal applications that provides high temperature water as output.

The reflector used in this technology is called as the linear Fresnel reflector (LFR) technology. This is an important concentrated solar power (CSP) application.

Another technology which is very close to this technology is the Parabolic trough collector (PTC) technology which consists of a parabolic trough where at its focal point lays an absorber tube used for carrying water or any working fluid.

Abbas has stated that the Linear Fresnel reflector technology is better than the PTC technology. This is because there are practically no joints or welds in the LFR technology.

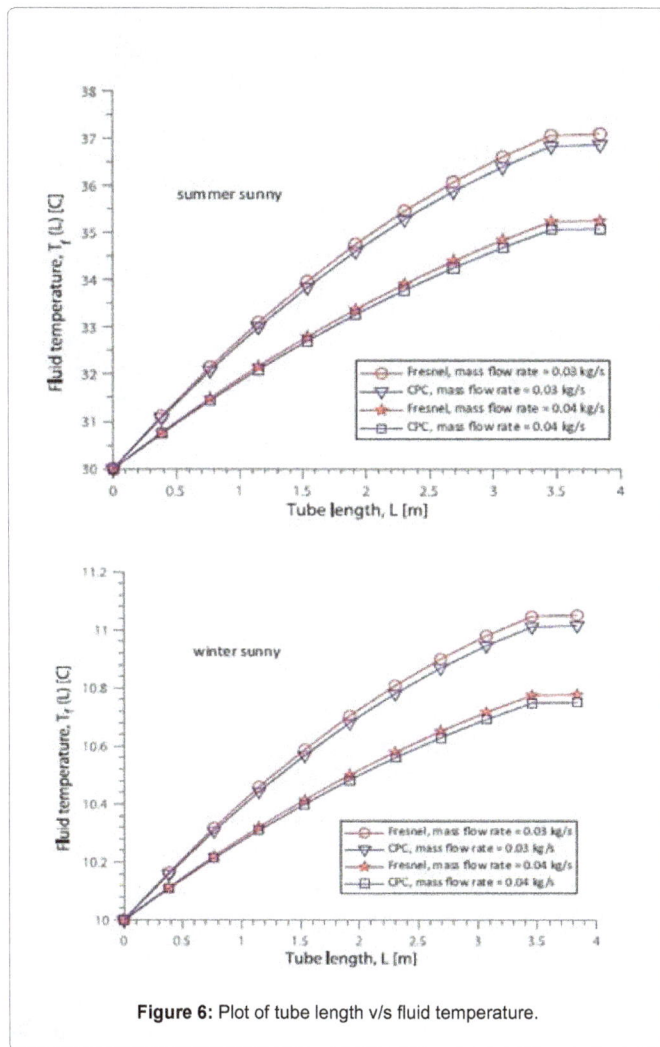

Figure 6: Plot of tube length v/s fluid temperature.

The setup is called as the Fresnel solar field. This basically consists of a series of Fresnel reflectors that are placed just like heliostats that focuses the radiation onto two absorbers. The absorbers consisting of a bundle of tubes, carries water or any working fluid. This is as shown in Figure 8.

Since there is absence of all moving parts, an important option would be direct steam generation.

Abbas has also stated that the efficiency of this technology is close to that of the PTC technology and the only weakness of this solar field is the concentration factor which was found to be lower than that of the PTCs.

Although there are a lot of systems already installed with tracking system, additional cost of running a thermal plant along with it makes it more expensive.

Technology wise, many losses are incurred with extra wiring for making way for the tracking system. Separate power consumption is also seen from the motors driving the tracking unit.

A solution also has been given which says that instead of using the energy directly, thermal storage can be done to achieve better efficiencies.

This technology deems itself to be very expensive and applied only at a large scale making it unaffordable to developing countries like India.

Large area is also required which affects the lives of the rural people. This is not a feasible solution in developing countries since the main occupation of the villagers is actually farming. Acquiring large area of lands is a major issue in countries like India.

Also, this system must be integrated with thermal power plants for power generation, again more land, more investment.

### Fresnel lens over LFPC

This technology was a test to determine any changes in the water temperature after just placing the Fresnel lens over the conventional LFPC (Figure 9).

This technology showed an increase of 20% instantaneous efficiency of that of an LFPC [7].

Figure 10 shows the variation of instantaneous efficiency along the day at different mass flow rates and both with and without the Fresnel lens.

We can easily see that the efficiency is significantly higher with the Fresnel lens.

Figure 11 shows the variation of glass temperature with respect to time at different mass flow rates.

Since the whole system is just a retrofit, all the losses that occur by incorporating the new system along with the losses of the previous system must be considered.

So, it can be seen that the glass temperature is very high at peak concentration temperatures.

This affects the efficiency of the system very much leading to undesired output and degradation.

The effect of output temperature throughout the day with different mass flow rates and both with and without lens is shown in Figure 12 [7].

Again, we can see the significant rise in output temperature of water with the lens retrofitted to the conventional system.

Just by retrofitting a conventional LFPC, we could see the rise in instantaneous efficiency and output temperature which gives more scope of new design of Fresnel lens based collector system for heating applications.

### Advantages of using fresnel lenses

Looking at all the technologies that are using Fresnel lens, much improvements and designs can be made for various applications bearing lesser cost and higher efficiencies.

Technologies can be so designed for solar thermal applications such that the primary optical element will be the Fresnel lens itself [8].

Fresnel lens based technologies produce ultra-high temperature of the working fluid.

The Fresnel lens technology is almost maintenance free.

Thermal efficiencies can be achieved at a higher level.

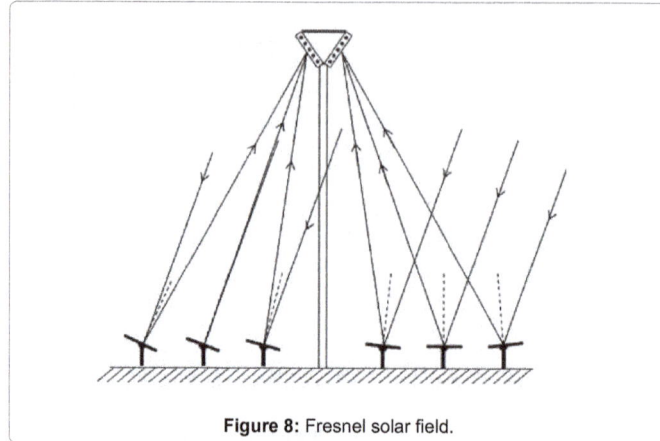

**Figure 8:** Fresnel solar field.

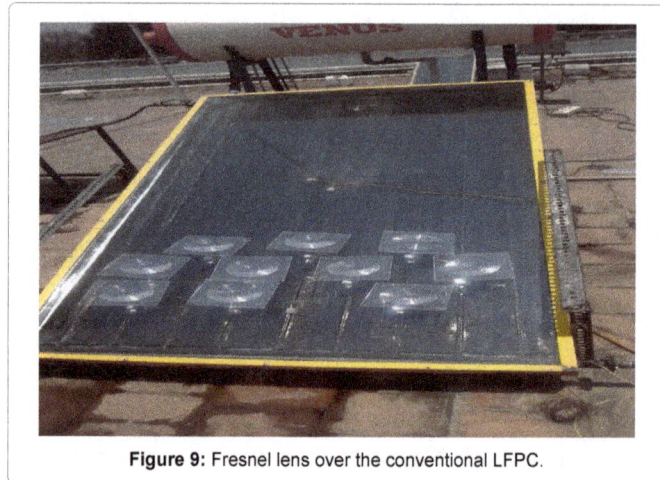

**Figure 9:** Fresnel lens over the conventional LFPC.

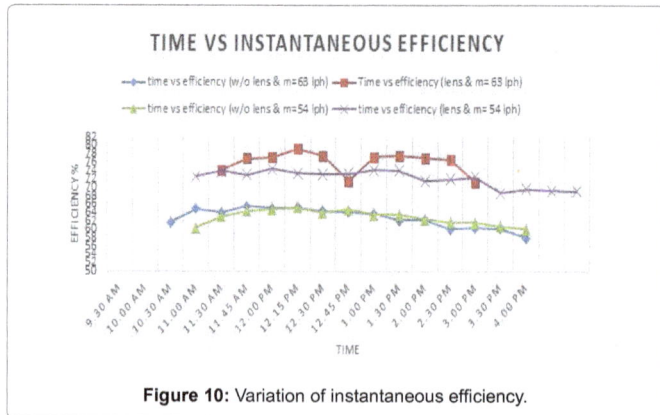

**Figure 10:** Variation of instantaneous efficiency.

Anti-reflection coating is not needed and also there is no need of reflecting surfaces [2].

By looking into all the technologies, Fresnel lens technology can further be improved by making innovative changes in the collector design.

The changes must be simpler, cheaper and more reliable so that it can be used by most people for domestic heating purposes.

### Review Conclusion

A general review of the technologies used using Fresnel lens is been

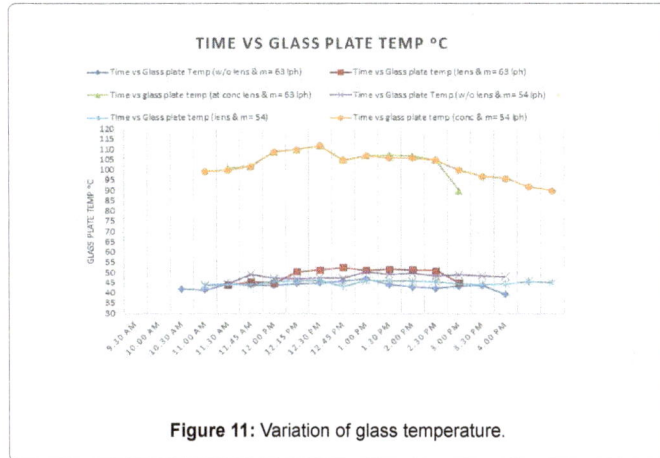

**Figure 11:** Variation of glass temperature.

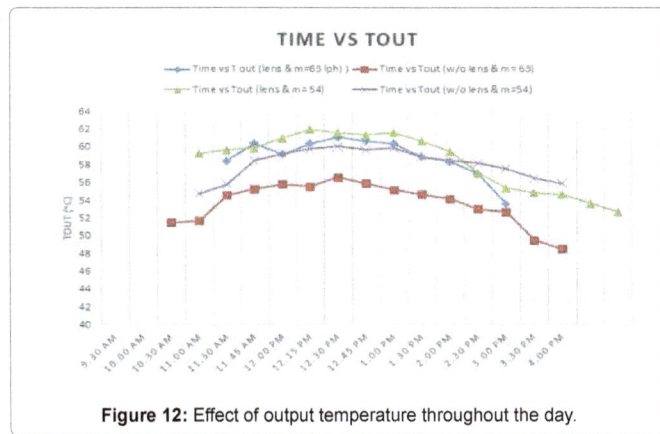

**Figure 12:** Effect of output temperature throughout the day.

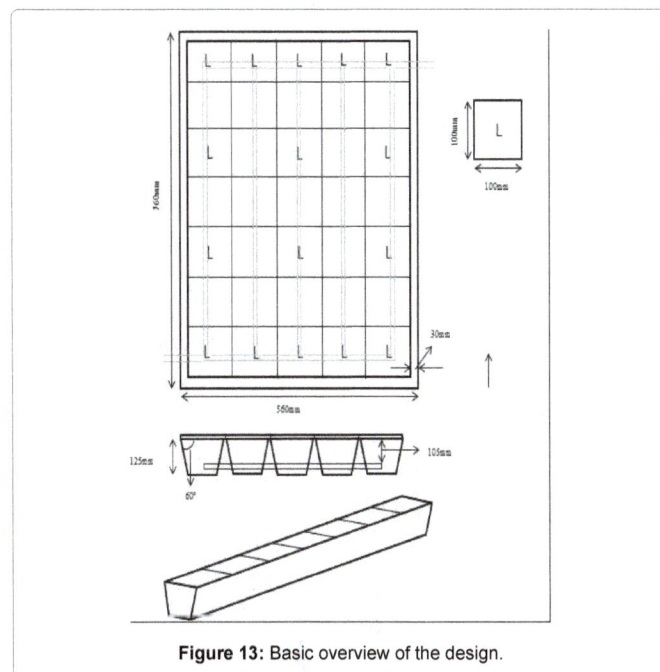

**Figure 13:** Basic overview of the design.

presented in this paper. Fresnel lens is a promising optical element that can be used to achieve high output temperatures of water in the solar thermal sector that can be used both for domestic and industrial usages. Cheaper and smaller water heaters for domestic purposes can be

designed where electricity is scarce in developing and underdeveloped countries. Advanced technologies of manufacturing Fresnel lens like the FK concentrator and the F-RXI concentrator [8] can be used for industrial and power plant sectors coupled with solar thermal systems.

The temperature achieved by Fresnel lens based technology and parabolic trough technology is comparable and is proved by [2] that Fresnel lens temperature is faster which means that more quantity of water can be heated throughout the day.

Fresnel lens is the most promising technology for the future designs of advanced water heaters.

## Design Proposed

By looking into the basic overview and all the advantages of using Fresnel lens, an innovative design is been presented in this paper that shows significant improvement in efficiency of the system and reduction in cost.

The design is as shown in Figure 13.

## Construction

The construction of this design is as follows.

Only the collector is designed and the rest of the system follows the conventional design. This system is designed for a tank capacity ranging from 50 to 200 liters. The dimension of the collector is 760 mm X 560 mm. The design is tapered at an angle of $60^0$ so as to follow the concentration pattern of the lens. L represents the Fresnel lens being used for concentrating the radiation onto a spot on the absorber. The lighter lines under the lens represent the absorber tubes. The cross sectional view of one segment is as shown in Figure 13.

The lens specifications are given below.

- Material: Poly Methoxy Methyl Acrylate (PMMA)

- Refractive Index: 1.49 (taken from catalogs)

- Type: Radial and not linear.

- Purpose: Solar thermal

- Maximum temperature at the focal spot: 400°Celsius at bright sun.

- Shape: Circular inside a square. The corners of the square have four holes for fixing in frames (R&D purpose)

- Size: 100 mm × 100 mm × 3 mm

- Groove density: Low for the purpose of long life (for avoiding scratches)

- Focal length: 105 mm

The piping system is 650 mm × 400 mm with an Outer diameter (OD) of 13.7 mm and Inner Diameter (ID) of 12.5 mm. The material used is Galvanized iron (GI)/steel with a thermal conductivity of 50 W/m²k.

## Working

The lens is placed at the top of the frame without any glass or plastic cover since the lens itself is very robust.

Sunlight passes through the lens and strikes the absorber tube, which heats up, changing solar energy into heat energy. The heat is transferred to liquid passing through pipes attached to the absorber tube. Absorber tubes are usually made of metal-typically copper or aluminum-because the metal is a good heat conductor. Copper is more expensive, but is a better conductor and less prone to corrosion than aluminum. In locations with average available solar energy, flat plate collectors are sized approximately one-half- to one-square foot per gallon of one-day's hot water use.

In this case, the sunlight passes through the lens and strikes the absorber tube. The Fresnel lens focusses the sunlight onto the absorber tube with a focal length of 100 mm.

The temperature achieved at the focal point is 400°C at peak sunshine. When water enters the collector, it is heated to a very high temperature. The water flow may or may not be via forced circulation.

The analysis of this design was done in MATLAB. This analysis shows the output temperature and the instantaneous efficiency achieved by the new design.

## Analysis

MATLAB analysis was done using the below mentioned inputs and successful outputs was obtained.

The location is considered as Dehradun, India on 15th May, 135th day of the year. The maximum radiation is considered to be 680 W/m² and the minimum to be 300W/m².

The latitude angle is 30°. The tilt of the panel is also taken to be the same as 30°. For the following inputs, we obtain outputs as mentioned.

## Inputs

1. Length of collector = $L_a$ = 0.76 m
2. Width of collector = $W_a$ = 0.56 m
3. Outer diameter of tube = $D_o$ = 0.0137 m
4. Inner diameter of tube = $D_i$ = 0.0125 m
5. Tube center to center distance = W = 0.1 m
6. Thickness of lens (or cover plate) = 0.003 m
7. Refractive index = 1.49
8. Reflectivity = 0.2
9. Location = Dehradun
10. Latitude = $\varphi$ = 30° for Dehradun
11. Day = n = May 15th = 135
12. Tilt = $\beta$ = Same as latitude
13. Beam radiation = $I_b$ = 680 W/m²
14. Diffuse radiation = $I_d$ = 300 W/m²
15. Water flow rate = m = 10 kg/hr.
16. Inlet temperature = $T_{fi}$ = 60°C
17. Ambient temperature = $T_a$ = 25°C
18. Wind speed = $W_s$ = 3.1 m/s
19. Back insulation thickness = 0.05 m
20. Reflectivity of surrounding surfaces = 0.2
21. Side loss = 10% of bottom loss

22. Thermal conductivity of steel = $I_t$ = 50W/m²-K
23. Permittivity of steel = $\varepsilon_p$ = 0.23

### Declination angle

$$\delta = 23.45 * Sin\left(\frac{360}{365}*(284+n)\right) = 18.79°. \ n = 135 \text{ on May } 15^{th}$$

### Local apparent time

LAT = 12h ± [4° (Std. Longitude – Longitude of location) + EAT] = 11 h 48 min

### Time correction

EAT = 9.87° Sin (2B) - 7.53° Cos (B) - 1.5° Sin (B) = 12.04 mins, where B = (n-1)° (360/364) $\omega$ = 3.01°.

### Angle of incidence of beam radiation

$$Cos\theta = Sin(\delta)Sin(\varnothing - \beta) + Cos(\delta)Cos(\omega)Cos(\varnothing)$$
$$= 0.9453 \text{ and } \theta_z = 19.02°.$$

$$Cos\theta_z = Sin(\delta)Sin(\varnothing) + Cos(\delta)Cos(\varnothing)Cos(\omega)$$
$$= 0.9798 \text{ and } \theta_z = 11.53°.$$

Solar flux incident on collector

$$r_b = \frac{\cos\theta}{\cos\theta_z} = 0.9658$$

$$r_d = \frac{1+\cos\varnothing}{2} = 0.9316$$

$$r_r = 0.2*\left(\frac{1-\cos\varnothing}{2}\right) = 0.0136$$

$$I_T = I_b\, r_b + I_d\, r_d + I_g\, r_r = 949.15 W/m^2$$

### $(\tau\alpha)_b$ and $(\tau\alpha)_d$

For beam radiation,

Angle of refraction = $\theta_1 = \sin^{-1}\dfrac{\sin\theta}{Refractive\, index} = 12.7670°$

$$\rho = 0.5*(\rho_1 + \rho_2)$$

$$\rho_1 = \frac{sin^2(\theta_2 - \theta_1)}{sin^2(\theta_2 + \theta_1)} = 0.0428$$

$$\rho_2 = \frac{tan^2(\theta_2 - \theta_1)}{tan^2(\theta_2 + \theta_1)} = 0.0313$$

$$\tau_{r1} = \frac{1-\rho_1}{1+\rho_1} = 0.9179$$

$$\tau_{r2} = \frac{1-\rho_2}{1+\rho_2} = 0.9393$$

$$\tau_r = \frac{\tau_{r1}+\tau_{r2}}{2} = 0.9286$$

$$\tau_a = e^{\frac{-k*\delta_c}{\cos\theta_2}} = 1$$

This value is 1 since $\delta_c$, the cover thickness is 0. This is because the lens itself acts as the cover.

$$\tau = \tau_r * \tau_a = 0.9286$$

For diffuse radiation, Angle of refraction = $\theta_1 =$
$\sin^{-1}\dfrac{\sin\theta_2}{Refractive\ index} = 40.2685°$, $\theta_2 = 60°$ for diffuse radiation.

$$\rho = 0.5*(\rho_1+\rho_2)$$

$$\rho_1 = \frac{\sin^2(\theta_2-\theta_1)}{\sin^2(\theta_2+\theta_1)} = 0.1177$$

$$\rho_2 = \frac{\tan^2(\theta_2-\theta_1)}{\tan^2(\theta_2+\theta_1)} = 0.0042$$

$$\tau_{r1} = \frac{1-\rho_1}{1+\rho_1} = 0.7894$$

$$\tau_{r2} = \frac{1-\rho_2}{1+\rho_2} = 0.9916$$

$$\tau_r = \frac{\tau_{r1}+\tau_{r2}}{2} = 0.8905$$

$$\tau_a = e^{\frac{-k*\delta_c}{\cos\theta_2}} = 1$$

This value is 1 since $\delta_c$, the cover thickness is 0. This is because the lens itself acts as the cover.

$$\tau = \tau_r * \tau_a = 0.8905$$

$$\rho_d = \tau_a - \tau = 0.1095$$

$$(\tau\alpha)_b = \frac{\tau\alpha}{1-(1-\alpha)*\rho_d} = 0.8787, \text{ Consider beam radiation}$$
values of $\tau$ and $\alpha$.

$$(\tau\alpha)_d = \frac{\tau\alpha}{1-(1-\alpha)*\rho_d} = 0.8426, \text{ Consider diffuse radiation}$$
values of $\tau$ and $\alpha$.

**Incident flux absorbed by absorber plate**

$$S = [I_b\ r_b\ {}^{'}(\tau\alpha)_b] + [(I_d\ r_d + I_g\ r_g)^{'}(\tau\alpha)_d]$$

**Collector heat removal factor and overall loss coefficient**

Assume $U_l = 2\ /m^2\text{-K}$.

$$m = \sqrt{\frac{U_l}{I_t * Plate\ thickness}} = 4.4721\ m^{-1}$$

$$m\left(\frac{W-D_o}{2}\right) = 0.1930$$

$$\Phi = \frac{\tanh(0.1930)}{0.1930} = 0.9878$$

$$F' = \frac{1}{U_l*W*\left[\frac{1}{U_l*[(W-D_o)*\varnothing_1+D_o]} + \frac{1}{\pi*D_i*h_f}\right]} = 0.9765,$$

$h_f$ is the heat transfer coefficient of the inner walls of the tube and is assumed to be 380 W/m²-K for water as fluid and steel as the absorber.

$A_p$ = Total area of tubes = $\dfrac{\pi*D_o*L_a*5}{2} + \dfrac{\pi*D_o*L_a*2}{2} + (L_a*W_a) = 0.3579$.

The area of the collector is taken as the area of the tubes running throughout the collector area. Vertically there are 5 pipes and so, in the 1st equation, 5 is used. Similarly horizontally there are 2 pipes. Although this may not be the exact area, it can be assumed that the taken area is correct.

$$F_r = \frac{\dot{m}*C_p}{U_l*A_p}*\left[1-e^{\frac{-U_l*A_p*F'}{\dot{m}*C_p}}\right] = 0.9477, \ \dot{m} = 10\ kg/hr\ and$$

$C_p = 1.1611$, specific heat of air.

$$q_u = F_r * A_p[S - U_l*(T_{fi}-T_a)] = 255.5569W$$

$$q_l = S * A_p - q_u = 39.1621W$$

Also, $q_l = U_l * A_p * (T_{pm}-T_a)$, From this, we can find the value of $T_{pm}$. So, $T_{pm} = 352.7086$ K = 79.7086°C. $T_{sky} = T_a - 6 = 292$ K = 19°C.

$h_{p\text{-}c}$ = heat transfer coefficient between plate and cover is 0 since there is no cover involved.

$$\text{So, } \frac{q_l}{A_p} = h_{p-sky}*(T_{pm}-T_{sky}) + \frac{\sigma*(T_{pm}4-T_{sky}4)}{\frac{1}{\varepsilon_p}-1}$$

$$h_{p-sky} = h_w = 8.55 + 2.56*W_s = 16.486\ W/m^2\text{-K}.$$

$$\text{So, } \frac{q_l}{A_p} = 1001\ W/m^2$$

$$U_t = \frac{q_l}{A_p*(T_{pm}-T_a)} = 18.2964 W/m^2\text{-K}$$

$$U_b = \frac{k_i}{\delta_b} = \frac{0.04}{0.05} = 0.8 \text{W/m}^2\text{-K}$$

$$U_s = 0.1 * U_b = 0.08 \text{W/m}^2\text{-K}$$

$$U_l = U_t + U_b + U_s = 19.1764 \text{W/m}^2\text{-K}$$

**Water outlet temperature**

$$q_u = \dot{m} * C_p * (T_{f_0} - T_{f_i})$$ Substituting the values of $\dot{m}$ $C_p$, $T_{f_i}$ and $q_u$, we get $T_{fo}$. $T_{fo} = 355.0099\text{K} = 82.0099°\text{C}$

**Instantaneous efficiency**

$$n_i = \frac{q_u}{I_t * A_c} * 100 = 63.263\%$$

**Summary**

The analysis is done according to the inputs given. Here, the declination angle is calculated to 135th day of the year. Accordingly, the local apparent time is calculated.

Solar flux incident on the collector is calculated with the geographical data assumed.

The main factors, heat removal factor and the overall loss coefficient are calculated for loss analysis for the given size of the collector.

The output temperature and the instantaneous efficiency are calculated for the obtained values of the loss coefficients and heat transfer coefficients.

**Results and Discussion**

$$F_r = \frac{\dot{m} * C_p}{U_l * A_p} * \left[ 1 - e^{\frac{-U_l * A_p * F'}{\dot{m} * C_p}} \right]$$

This is the main equation guiding the output of the system. This

equation guides the overall heat lost from the collector, $q_u$. This further guides the efficiency of the system.

The output of the conventional system is 65°C and 43%. This design gives a higher output temperature and higher instantaneous efficiency.

Assumption of a few parameters in the design analysis gave us higher temperature and higher efficiency. This may vary on field. Also, this system is a basic design on which further improvements can be made by incorporating a motor for forced circulation so as to get better efficiency.

The system is made from less expensive materials such a ply wood and cheap glue materials. Industrial standard grade materials can be

used which increases the cost slightly without affecting the performance of the system.

This design analysis gave us 63.263% efficiency with an outlet temperature of 82°C. This is still under the theoretical calculations involving assumptions. Practically this is going to be more efficient and the temperature is going to be high considering a lot of practical parameters.

This system is cost effective and smaller in size making it easier for transportation and installing it in rural areas for the welfare of the poor and where electricity is scarce.

This output water which is of a higher temperature can be used for cooking in big institutions.

Further, this system can be integrated with water driven Vapor Absorption Machine (VAM) for building cooling systems. VAM can operate easily with hot water as the input. Since the hot water obtained in this application is about 85 to 90°C, water driven VAM can be operated without any issues.

**Conclusion**

The thermal analysis and cost analysis was done and it showed that the new design is more efficient, less expensive and smaller in size when compared to conventional flat plate collector.

The same analysis is done for a conventional Flat plate collector and the efficiency and the outlet temperature is 43% and 65°C respectively.

This design can also be installed at places where solar radiation is low since the lens focusses the light onto the absorber at a temperature of 400°C making the water to be heated even at low radiation.

**References**

1. Abbas R, Montes MJ, Piera M, Martínez-Val JM (2012) Solar radiation concentration features in Linear Fresnel Reflector arrays. Energ Convers Manag 54: 133-144.

2. Kumar V, Shrivastava RL, Untawale SP (2015) Fresnel lens: A promising alternative of reflectors in concentrated solar power. Renew Sustain Energ Rev 44: 376-390.

3. Wu Y, Eames P, Mallick T, Sabry M (2012) Experimental characterisation of a Fresnel lens photovoltaic concentrating system. Solar Energy 86: 430-440.

4. Soriga I, Neaga C (2012) Thermal analysis of a linear fresnel lens solar collector with black body cavity receiver. U P B Sci Bull Series D 74: 106-116.

5. Al-Jumaily KEJ, Al-Kaysi MKA (1998) The study of the performance and efficiency of flat linear fresnel lens collector with sun tracking system in Iraq. Renew Energy: 41-48.

6. Hornung T, Steiner M, Nitz P (2012) Estimation of the influence of Fresnel lens temperature on energy generation of a concentrator photovoltaic system. Solar Energy Mater Solar Cells 99: 333-338.

7. Performance Enhancement of Solar Flat Plate Collector by using Fresnel lens.

8. Miñano JC, Benítez P, Zamora P, Buljan M, Mohedano R, et al. (2013) Free-form optics for Fresnel-lens-based photovoltaic concentrators. Opt Express 3: A494-A502.

# 3

# A Review of *in-situ* Loading Conditions for Mathematical Modeling of Asymmetric Wind Turbine Blades

**Bardsley A\*, Whitty JPM\*, Howe J and Francis J**

[1]Energy and Power Management Research Group, School of Computing, Engineering and Physical Sciences, University of Central Lancashire, Preston, PR1 2HE, UK

## Abstract

This paper reviews generalized solutions to the classical beam moment equation for solving the deflexion and strain fields of composite wind turbine blades. A generalized moment functional is presented to effectively model the moment at any point on a blade/beam utilizing in-situ load cases. Models assume that the components are constructed from in-plane quasi-isotropic composite materials of an overall elastic modulus of 42 GPa. Exact solutions for the displacement and strains for an adjusted aerofoil to that presented in the literature and compared with another defined by the Joukowski transform. Models without stiffening ribs resulted in deflexions of the blades which exceeded the generally acceptable design code criteria. Each of the models developed were rigorously validated via numerical (Runge-Kutta) solutions of an identical differential equation used to derive the analytical models presented. The results obtained from the robust design codes, written in the open source Computer Aided Software (CAS) Maxima, are shown to be congruent with simulations using the ANSYS commercial finite element (FE) codes as well as experimental data. One major implication of the theoretical treatment is that these solutions can now be used in design codes to maximize the strength of analogues components, used in aerospace and most notably renewable energy sectors, while significantly reducing their weight and hence cost. The most realistic in-situ loading conditions for a dynamic blade and stationary blade are presented which are shown to be unique to the blade optimal tip speed ratio, blade dimensions and wind speed.

**Keywords:** Wind turbine blades; Mathematical modeling

## Introduction

In the wind turbine industry it is paramount that components are designed to obtain the greatest obtainable efficiency. As time has progressed selecting the correct materials and aerofoil shape has become predominant in the determination of the optimum blade tip-speed ratio, thus increasing a wind turbines power coefficient, $(C_p)$ and overall turbine efficiency $(\eta_{total})$. It is in the industry's best interest to produce high performance turbine components at the lowest costs, without compromising the turbines structural stability. Solving problems of this ilk drive wind turbine design, research and development forward and one such problem is the subject of this paper.

There is currently a surprising lack of computational models in literature, which accurately predict the deflexion and strain fields of wind turbine blades. However, it has proven from relatively simple physical tests on the blades that relatively straightforward and well developed underpinning theory can be used to predict the deflexion of wind turbine blades remarkably well [1]. This said, testing protocols are quite restrictive, especially for the larger expensive components. For such components therefore, effective computational predictive modeling is required to ascertain specific structural static and dynamic design limits; such models can be employed to virtually test any component in a safe and cost free environment.

The work described throughout this paper considers small scale wind turbines (5m blades) which are representative of those used on wind-turbines producing outputs of 1.5 - 50 kW; additionally these size of blades are ideal for validation experimentally [1,2]. Moreover validation of larger components becomes technically and financially prohibitive [3]. Whence, modeling the performance and the actual working limits of the blades (i.e. blades angular velocity and tip speed ratio) will be taken from real case studies evident in the literature [4].

## Previous work

Recently [1] researchers have used the classical beam elasticity

differential equation in order to effectively model the displacement of standard aerodynamic geometries. Though this work was informative in predicting the required displacement and hence strains fields remarkably well given the simplicity modeling method, these researchers do submit that further work is required particularly with respect to the use of composite materials under salient loading conditions. Moreover, conducting such investigations better predictions can be made of blade behavior when attached to the turbine. In practice it is generally accepted that the geometry of a blade is determined by the aerodynamic properties which the designer requires. The ruling characteristics are aerodynamic coefficients: chord, twist and thickness distribution. When designed blade materials are selected with the aim of minimizing tip deflexion, which reduces the probability of a blade/tower collision [5]. It is also important that these components last the desired lifetime, around 20 years, while under accepted loads [6].

The aforementioned Whitty et al paper [1] presented closed form solutions to the classical beam elasticity differential equation. The developed analytical method was utilized to effectively model the displacement of standard aerodynamic geometries used for wind turbine blades. These models assume that the components are constructed

**\*Corresponding author:** Bardsley A, Energy and Power Management Research Group, School of Computing, Engineering and Physical Sciences, University of Central Lancashire, Preston, PR1 2HE, UK, E-mail: ambardsley@uclan.ac.uk

Whitty JPM, Energy and Power Management Research Group, School of Computing, Engineering and Physical Sciences, University of Central Lancashire, Preston, PR1 2HE, UK, E-mail: jwhitty@uclan.ac.uk

from in-plane quasi-isotropic composite materials, forming a blade of shell thickness with 5 mm, with or without a stiffening web. In these cases, a composite comprising of 40 layers of laminated carbon and glass substrate with an epoxy matrix was used, the effective modulus of the material being 42 GPa. The authors [1] derive an explicit solution for the strain and displacement fields for an elliptical shell formulation which is presented for the first time in the literature. This is then expanded to encompass the FX66-S-196 and NACA 63-621 symmetric aerofoil approximations. However, it has been shown that the aerofoils stated above are not symmetric (Figure 1). Thus this work investigates the effects of the asymmetry on the displacement and strain fields using comparable methodologies.

The experimental static testing of a composite turbine blade has also been carried out by a Jordanian group [2]. While focusing mainly on the manufacturing of the blade, it promotes a link between their local academic work and local industries. The works in depth look into the design and construction of the blade as well as a thorough application of static testing protocols provided valuable data in which to verify modeling work [1] and ultimately design codes [2].

## Aerofoil selection

Since it was demonstrated that small changes in the geometry had negligible effect on the structural response, the original Whitty et al [1] aerofoil (Figure 1) used in their simulations was significantly simplified. The aerofoil is effectively half an ellipse forming the leading edge and a triangle forming the trailing edge. Following a survey of literature it was found that there is normally no symmetry between the top half and the lower half of the aerofoil, thus the Whitty et al. (adjusted) aerofoil was designed. These are then compared with the FX66-S-196 and NACA 63-621 aerofoils which are plotted using discrete points obtained from "Airfoil Tools" [7]. Finally the Continuous NACA aerofoil is aerofoils constructed from the following equations, derived from the Joukowski transform [8]:

$$y_{top}(x, d_{top}, l_c) = d_{top} \cdot \left( A \cdot \sqrt{\frac{\infty}{l_c}} - B \cdot (\frac{\infty}{l_c}) - C \cdot \left(\frac{\infty}{l_c}\right)^2 + D \cdot \left(\frac{\infty}{l_c}\right)^3 - E \cdot \left(\frac{\infty}{l_c}\right)^4 \right) \quad (1)$$

$$y_{bottom}(x, d_{bottom}, l_c) = -d_{bottom} \cdot \left( A \cdot \sqrt{\frac{\infty}{l_c}} - B \cdot (\frac{\infty}{l_c}) - C \cdot \left(\frac{\infty}{l_c}\right)^2 + D \cdot \left(\frac{\infty}{l_c}\right)^3 - E \cdot \left(\frac{\infty}{l_c}\right)^4 \right) \quad (2)$$

Both equations depend on the length of chord, x, the diameter of the aerofoil at the furthest point, $d_{top}$ and $d_{bottom}$, respectively and the total chord length, lc. The constants A, B, C, D and E are all specific to the aerofoils aerodynamic properties [9]. Equations (1) and (2) are effectively the same equation manipulated to produce two connected curves at (0,0) and (0,600). These sections were plotted using the CAS Maxima (Figure 1).

## Scope

This paper provides a review of asymmetry of particular aerofoil sections for the modeling of displacement and strain fields of a wind turbine blade in service. A generalized blade moment functional which can incorporate any loading condition on the blade is also reviewed. These are employed to define the so called generalized aerodynamic load equation and different benchmarking procedures are described which should be routinely used for verification and validation of the computational models used across the industry.

## Methods

In this section we review mathematical and simulation methods which can be employed to describe standard industrial testing protocols. These take the form of a generalized moment functional

aerodynamic loading conditions, tip-speed optimization for the model blades considered and salient benchmarking procedures.

## Generalized moment equation derivation

The most important equation which is needed to model these loading cases is the second order differential equation for the displacement of a beam (equation (3)) [10].

$$\frac{d^2 u_y(z)}{dz^2} = \frac{M(X, z)}{E_l I} \quad (3)$$

The differential equation comprises of two other function which vary depending on, z, the distance along the blade. M (X, z), is the generalized moment functional, this equation will depend upon the loading case being modeled. These will be discussed and explained in the sections to follow. I (z), is the second moment of area, this function also depends on, z.

To our knowledge no derivation of a generalized moment functional for a wind-turbine blade is evident in the literature. This derivation requires the consideration of Figure 2 which shows a blade fixed at one end with general load intensity applied. Taking moments about the root renders:

$$M (X, z) = W (\zeta) \cdot \zeta - M (\zeta) \cdot \bar{\zeta} \cdot A \quad (4)$$

From the diagram (2), there are four variables, M, the moment, W, the force, q, the load intensity and ζ, the centroid of the section of load intensity to the left of z. Additional there are spacial coordinates z and ζ. These being respectively the distance from the root and an arbitrary distance to the centroid to the left of z. consider an infinitely small section of the load distribution, dz; the force at this point is given by:

$$dW = q (z) \cdot dz$$

The total force on the component is therefore found from the integral of the distribution over the whole length:

$$W(z) = \int_0^L q(z) \cdot dz \quad (5)$$

The moment equation can be derived, considering the load distribution acting over the infinitely small distance dz, whence the moment is:

$$dM = q(z)dz \cdot \left( z + + \frac{dz}{z} \right) = q(z)dz \cdot z + q(z)\frac{(dz)^2}{z}$$

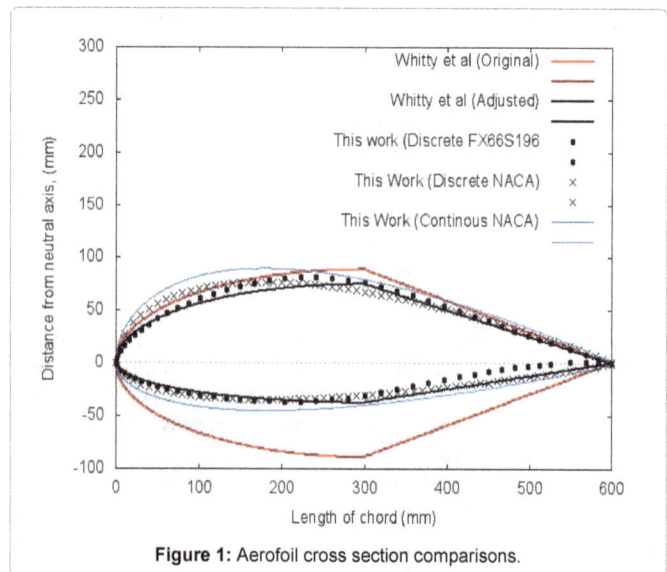

**Figure 1:** Aerofoil cross section comparisons.

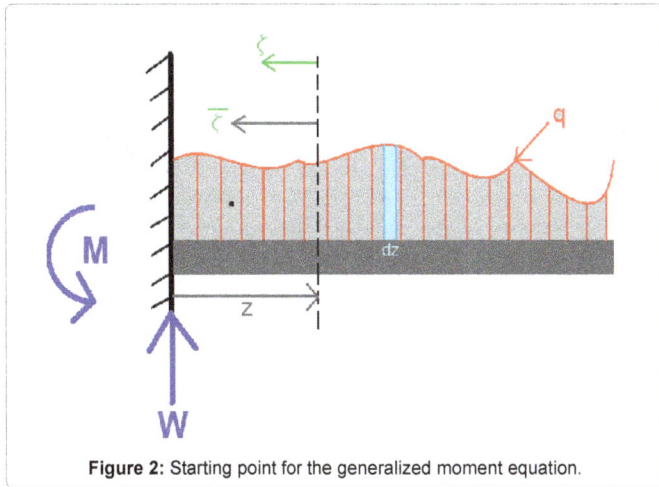

**Figure 2:** Starting point for the generalized moment equation.

If higher order infinitesimals are neglected:

dM = q (z). z dz

Hence, the moment at the root is found by integrating over the length of the blade:

$$M(z) = \int_0^L q(z).z dz$$

To make a general solution the moment equation must consider a changing centroid as moments are taken at points further from the root. It is important therefore that the moment equation be considered in the $\zeta$ direction. As the moment equation varies with length this creates a unique problem, when any load distribution is modelled the centroid of said load distribution will continually change the further along the blade moments are take. Hence, the well known expression for the centroid of area is employed [11].

$$\bar{y} = \frac{\int_0^L q(\zeta).\zeta d\zeta}{\int_0^z q(\zeta)d\zeta}$$

It is now possible to substitute all the known parameters into equation (4) which gives the required generalized form:

$$M(x,\zeta) = \left\{\int_0^z q(\zeta)d\zeta\right\}.\zeta - \int_0^L q(\zeta).\zeta d\zeta - \int_0^L q(\zeta).\zeta d\zeta \quad (6)$$

Where z, is the distance along the blade length from the root, the function X($\zeta$) is the load distribution function at $\zeta$, which is the distance along z, from the root, at which the centroid of the load distribution occurs. With the displacement field now solvable, the global strain field is required. This takes a much simpler algebraic relationship:

$$\varepsilon_z = \frac{M(z).y(x,z)}{E.I(x,z)} \quad (7)$$

Equation (7) is employed in the models of this work and it is important to note that these models will only consider cross-sectional variations in the strain with material properties remaining constant.

## Force Distribution Derivations

In reality there are a number of forces on a turbine blade that can cause deflexion. The purpose of designing an aerofoil for a wind turbine application is so that at relatively low wind speeds the lift force, produced by the aerodynamic properties of the blade, is large enough to overcome the drag force and the force on the blade due to gravity. This said, in this work the models will not consider the effect of gravity

on the system. It is important to simulate the worst possible scenario when testing the blades after manufacturing; this ensures that the blade meets the design standards and is to for purpose. When considering tip deflexion the worst case scenario is to assert a point load on the blade at the end (i.e. furthest distance from the rotor axis). The magnitude of this force would be the maximum possible dictated by the particular standard under consideration [1,12,13], although this is usually unrealistic, though informative as it imposes a further factor of safety into the blade design process. When the blade is modeled, determining a suitable load is very important. The lift force, $F_L$, can be determined by utilizing equation (8):

$$F_L = \frac{C_L.\rho.L.L_C.V_R^2}{2} \quad (8)$$

Here $C_L$, is the Lift coefficient, $\rho$, is the density of air, 1.225 kg/m³, L, is the Blade length, $L_C$, is the chord length in meters which, due to the tapering, depends on the blade length thus:

$$L_C = b_r - \frac{(b_r - b_t).z}{L} \quad (9)$$

Where $b_r$, is the breadth at the root, and $b_t$, is the breadth at the tip of the blade. The parameter z here is the changing distance along the blade. Finally the important function $V_R^2$, must be explained, this is the relative velocity along the surface of the blade, depending upon the tangential wind velocity, $V_L$ m/s, the angular velocity of the blade, $\omega$, the radius of the turbine, r, and the axial velocity, $V_\theta$, equation (10):

$$v_R^2 = v_t^2 + (\omega.r - v_\theta)^2 \quad (10)$$

Equation (10), can be determined from the Bernoulli momentum equation for a free flow stream. From equations (8) and (10) a worst case lift force can be determined and used in the models, the worst case occurring at the highest possible tangential velocity that the turbine could experience in its lifetime [5]. As previously indicated (Figure 2 and equation (6) an important parameter which must be determined for all the models is load intensity, q(z), that is, the force per unit length; which is relent on equation (8).

$$F_L = \frac{C_L.\rho.L_c.v_R^2}{2} \Rightarrow \frac{F_L}{L} = \omega(z)$$

$$\omega(z) := \frac{C_L.\rho.L_c.v_R^2}{2}$$

where, $\omega(z)$ is the load intensity for the particular loading case (e.g. PL, UDL, etc.), as indicated in Table 1.

These load distributions were developed after some key findings during modeling. Originally the UDL distribution was considered to be analogous to the stationary distribution on the blade, along with the RTDL which was an approximation to the dynamic load distribution. However, further investigation reveled that the most realistic loading case was that of the so called Bernoulli Dynamic Distributed Load (BDDL) and the special case[1] of this the BDDL termed the Bernoulli Stationary Distributed Load BSDL. It should be noted that the Point Load (PL) is modeled here using the theory of distributions. That is, multiplication of the Dirac distribution[2] thus concentrating the force at a point, rendering what is referred to in Table 1, as the Dirac Distributed Load (DDL). Figure 3, depicts the comparisons between each of the dierent loading cases, here application of equation (5) shows that the total load in each case corresponds to the, required Det Norske Veritas standard DNV-DS-J102, test load of 7600N.

Whence, in that case of the DDL, equation (12), the resulting distribution is that of an impulse showing a zero value until the tip

| Loading Case | Equation | | Peak Load intensity, (N/mm) |
|---|---|---|---|
| DDL (PL) | $\omega_{PL}(z)=P.\delta\,(z-L)$ | (12) | 7600 |
| UDL | $\omega_{UDL}(z)=q_e$ | (13) | 1900 |
| RTDL | $\omega_{RTDL}(z)=2.q_e.\left(\dfrac{z}{L}\right)$ | (14) | 3600 |
| QDL | $\omega_{QDL}(z)=3.q_e.\left(\dfrac{z}{L}\right)^2$ | (15) | 5400 |
| BDDL | $\omega_B(z)=\dfrac{C_L.\rho.b(z).(v_n^2+(\omega.L-v_\theta)^2)}{2}$ | (16) | 2000 |
| BSDL | $\omega_s(z)=\dfrac{C_L.\rho.b(z).(v_n^2-v_\theta^2)}{2}$ | (17) | 2750 |

**Table 1:** Load distributions for specific cases.

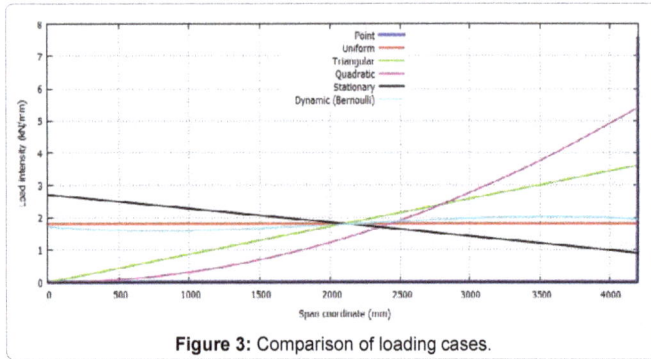

**Figure 3:** Comparison of loading cases.

of the blade. These distributions (Figure 3) were all obtained using the parameters given in Table 2. We note here for completeness that unrealistically high speeds of 68 m/s and 110:4 m/s must be applied in order to maintain the required factor of safety as dictated by the standards [12,13].

## Tip speed optimization

When the initial blade selection took place, the blade dimensions were taken directly from reference [1] and then modified to give the asymmetry. The blade is then designed for the optimization of the tip speed ratio which is the most important factor. The tip speed ratio is defined by:

$$\lambda=\frac{\omega.L}{U_\infty} \qquad (18)$$

where $\omega$, is the rotational speed in radians per second and $U_\infty$, is the free stream velocity. Typically accepted values for the tip speed ratio lie between 4 - 8 [5,3]. This is the classical mean of blade design (e.g as used in industry). From reference [4], it is shown that the optimal tip speed ratio for a wind turbine depends on the number of blade n, and can be calculated:

$$\lambda=1.3.\frac{4\Pi}{n}$$

In the case utilized in this work, the turbine has 3 blades; as such application of this expression renders an optimum tip speed ratio of 5.45. This tip speed ratio should be maintained at rated wind speed, which for the vast majority of small scale turbines is 10 m/s [3]. Rearrangement of equation (18) and the substitution of this expression, the rotational speed of the turbine at 10 m/s is 10.9 rad/s. As this work

considers the worst case scenarios which occurs at wind speeds of greater than 24 m/s, when the optimal tip speed ratio is maintained the rotational speed of the turbine is increased by a factor of 2.6.[1]

## Blade testing:

In static testing, the blade should be loaded to its most severe loading. In accordance to the Det Norske Veritas standard DNV-DS-J102, deflexion of the blade should always be less than a specified upper limit in order to avoid blade contact with the tower or other components. The design criterion for deflexion of the blade is given by the following inequality [12]:

$$\Delta_d(\gamma fF_k)\le\frac{\Delta_i}{\gamma_n-\gamma_n} \qquad (19)$$

Here, $F_k$ is the characteristic load, $\Delta d$ is the largest tip deflection when passing the tower, $\Delta i$ is the smallest distance from blade tip to the tower or other obstacle in the unloaded condition, f is the load factor, $\Upsilon_m$ is the material factor and $\Upsilon_n$ is the consequence of failure factor. Typically, the load factor takes a value between 0.9 and 1.5 depending upon the desired situation. In the case where the blade could experience the most load (transportation and installation) the suggested value is 1.5 [13]; when under favorable loads this can be as low as 0.9. According to the aforementioned design code the so called material safety factor is a product of a number of other factors in the range

(0.95 -1.3). Finally the consequence of failure factor is normally taken to be 1.1 [13]. Utilizing equation (19), explicit knowledge of Classical Lamination Theory (CLT) from composite materials mechanics that composites and the maximum strain failure criterion, a maximum tip deflexion of 10% of the blade length can be determined. For the blade modeled in this work therefore the design criterion was such that the blade could not deflect more that 420 mm under the test load of 7600 N.

During static testing the applied test load is greater than the design load to account for in influences from temperature, humidity, production variations and other environmental aspects during the life of the blade. The test load, $F_T$, is determined as:

$$F_T=\gamma_{su}.\gamma_f.F_k \qquad (20)$$

where, $\Upsilon$su is the blade-to-blade variation factor considered to be 1 (worse case testing), $\Upsilon_f$ is the load factor and $F_k$ is the characteristic load (i.e. the in-situ load). The characteristic load for this blade is ~5050N, rendering the aforementioned test load of ~7600N [12].

## Model verification (benchmarking)

The benchmarking case was selected to show the confidence in the analytical model developed by comparing the results between the benchmarking cases in Whitty et.al. [1], the ANSYS-FE model data, this comparison is shown in Figure 4. Here, the benchmarking case used the original symmetrical aerofoil, under a point load of 8500 N (Figure 4). The work described in this paper slightly underestimates

---

[1]obtained by setting $\omega$ to zero.
[2]Sometimes erroneously referred to as the Dirac-delta function.

| Parameter | Nomenclature | Value | Units |
|---|---|---|---|
| Wind Speed | $v_n$ | 68 | m/s |
| Lift Coefficient | $C_L$ | 1 | - |
| Rotational speed | $\omega$ | 26.16 | rad/s |
| Drag Speed | $v_\theta$ | 0.5 | m/s |

**Table 2:** Modeling parameters.

the deflexion when compared to the original published work. This is due to previous work allows the chord lengths and aerofoil maximum height to vary independently, however the current work varies these parameters simultaneously in an attempt to maintain the aerodynamic properties of the aerofoil.

The Whitty et.al. aerofoil [1] varies the chord length and the height[3] of the blade linearly using two separate functions which are independent of each other and therefore the deflexion is greater as the blade is modeled to be less stiff; a comparison of the maximum deflexion values is shown in Table 3.

The fact that the FE method over estimates the stiffness of any continuum system coupled with the shell element formulation employed in the model explains the lower predictions from the ANSYS software. The new analytical method predicts 13.6% greater deflexion than the FE model, whereas the method employed in the reference [1] suggests a value 22.7% greater than the FE model. These modeling methods suggest a displacement of $567 \pm 89$ mm[4].

Using the new analytical model the maximum strain on the blade occurs not at the root of the blade, as originally expected, but it occurs at ~2250mm along the blade giving strain between about ~0.36%-0.41% depending upon the method employed. When compared to the results obtained in [1] (Figure 5) then it is clear that although the strain is expected to occur at that position, there is also an increase in the maximum strain prediction. Prima-facia this is unexpected as the maximum displacement is now less due to the increased overall stiffness (Table 3). However, this may be attributed to blade not being in a fully strained condition as reported in the reference [1]. That is, up to half the span of the blade the newly developed analytical model is most probably stiffer than that reported previously and ispo-facto is more optimally strained. Accordingly CLT dictates that de-lamination should not occur until much larger strains are realized.

When comparing the strain fields, it is important to note, for all results discussed in the proceeding section, that ANSYS-FE model maximum strain values at the root are too high due to the Saint Venant edge effect and therefore a more realistic maximum strain must be interpolated.

## Result

The following results were all obtained using the parameters displayed in Table 2 and Table 4, both the developed analytical and ANSYS-FE models used these parameters.

### Displacement field results

Figure 6 shows the application of equations (13) through (16); the calculated deflexion of the adjusted aerofoil under the various loading cases (Table 1). As expected, the significant reduction in the aerofoils second moment of areas due to the asymmetry of the new design, greater deflexions are predicted by all the modeling procedures.

The maximum deflexion is observed on application of DDL, where the total force is distributed such that the moment is at its greatest. The lowest deflexion is realized using the SDL, where the centroid of the load distribution is closer to the rotor axis when compared to all others. Table 5 illustrates the maximum deflexion under each of these load distributions.

As the loading cases change as does the centroid of the loading distribution this affects the deflexion. The model predicts that the further along the centroid of the load distribution from the rotor axis, the greater the deflexion.

### Strain field results

Analytical model results for the strain fields are shown in Figure 7. As expected, the largest strains are observed on the application of a point load, ~0.72%. The strain due to the point load, unlike the other loading cases, shows that the maximum strain occurs 2250 mm from the root. This was expected when consulting literature on classical lamination [14] and previously developed tapered beam theory [1].

The maximum strain for the remainder of the loading cases all occur at the root (Table 6), however there is a trend, as the centroid of the load distribution moves further from the root of the blade the strain field becomes increasingly non-linear.

Tables 5 and 6 compare the strain fields which are generally increase with displacement fields, as expected. These results also hold when considering the failure over time, from [5] it is known that wind turbine blades tend to fail due to maximum strain which occur at the root. The only loading case contradicting this phenomenon at present is the point load but as demonstrated previously the point load is not a realistic in-situ loading condition.

### Stiffening web

A single 5mm thick stiffening web, in the centre of the aerofoil, along the full length of the blade was added (Figure 8) in line with the literature [1,2] and current industrial practices. The analytical and ANSYS_FE model comparisons including the stiffening rib are shown in Figure 9. Application of a point load renders the maximum deflexion predicted is ~630 mm and 720mm respectively for the analytical and FE model calculations respectively.

The points in Figure 9 show the deflexion predictions from the ANSYS-FE UDL case and the PL case. Following this solution, in line with previous published work [1], the second moment of area of the stiffening web was increased using the analytical model such that the DDL deflexion field was within the design criteria of less than 10% deflexion, therefore reducing proportionally the delfexion the other loading cases considered (Figure 10).

It was found the analytical model increases the second moment of areas from 12.5 million mm[4] to 15 million mm[4] when the stiffening rib is included (Figure 11).

## Conclusions

A mathematical method for solving the deflexion and strain fields of composite wind turbine blades has been reviewed. The model makes

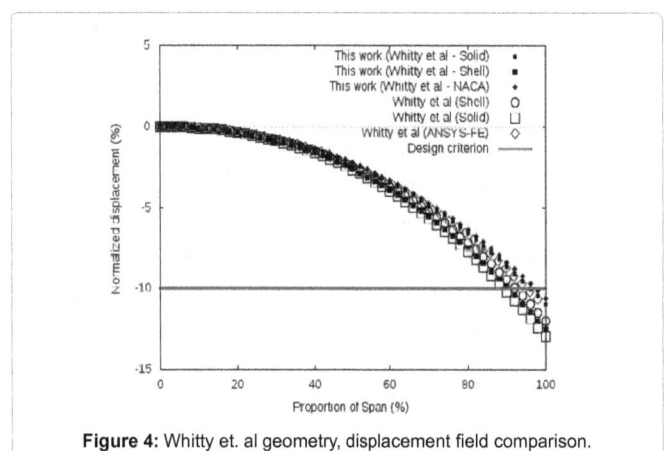

**Figure 4:** Whitty et. al geometry, displacement field comparison.

| Model | Deflexion |
|---|---|
| Analytical (this work) | 525 mm |
| Whitty et.al. | 567 mm |
| ANSYS | 462 mm |

**Table 3:** Comparison of maximum deflexion under point load.

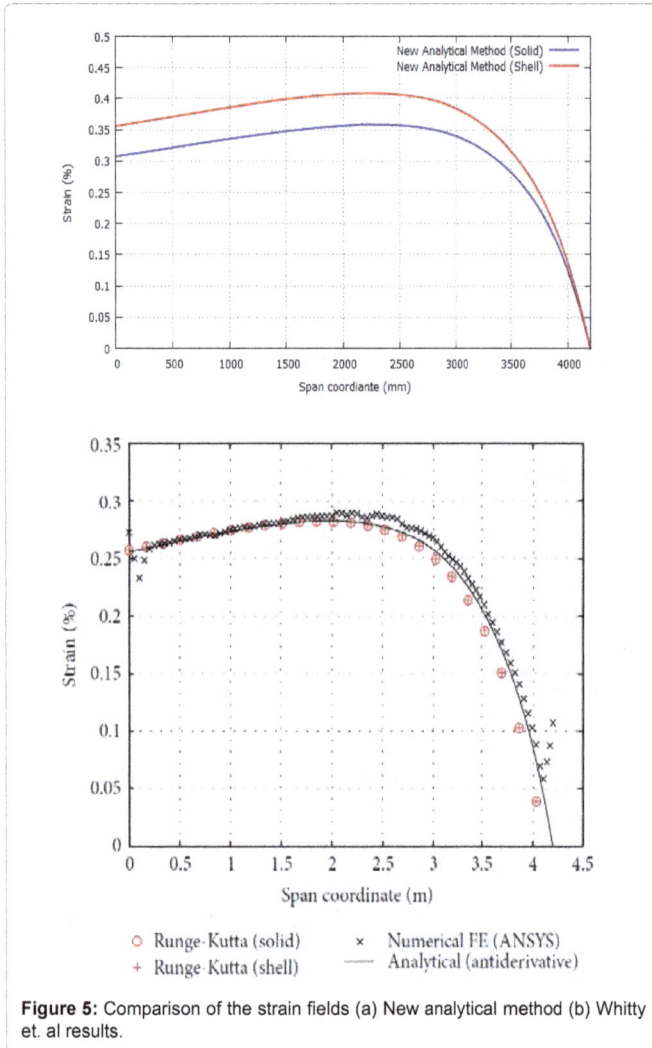

**Figure 5:** Comparison of the strain fields (a) New analytical method (b) Whitty et. al results.

| Parameter | Nomenclature | Value | Units |
|---|---|---|---|
| Blade length | L | 4200 | mm |
| Chord length at the root | $b_r$ | 600 | mm |
| Chord length at the tip | $b_t$ | 200 | mm |
| Total Force | W | 7600 | N |
| Shell thickness | t | 5 | mm |

**Table 4:** Model parameters.

use of a generalized moment functional (equation 6) which determines the moment at any point on a blade or beam utilizing of a in-situ load cases, e.g. equation 11). The models shown use freely available, open source software (e.g. via Maxima website: http://maxima.sourceforge. net/). Work is now progressing within the Computational Mechanics Research group at the University of Central Lancashire to allow minor adjustments to the code together with the use of external aerofoil

[3]the height being the distance for the neutral axis to the outer most fibre of the aerofoil.

[4]calculated at the 95% confidence level.

coordinate data (from such websites as: http://airfoiltools.com). This development will enable the deflexion and strain fields of any type of aerofoil constructed from quasi-isotropic materials. This being indiscriminate, whether said aerofoil is for a wind turbine, commercial aircraft or indeed other such industrial aerodynamic component. The salient conclusions gained from the work depicted in this paper are as follows:

The asymmetric nature of the blade sections designs reduces the second moment of area significantly from between 20.5 - 24 million mm4 for the symmetrical aerofoil to between 9.5 - 12 million mm[4], the asymmetry results in a reduction of between around 51% - 55%.

Due to the reduction of the second moment of area, the stiffness of the blades is reduced, rendering a higher predicted deflexion. The

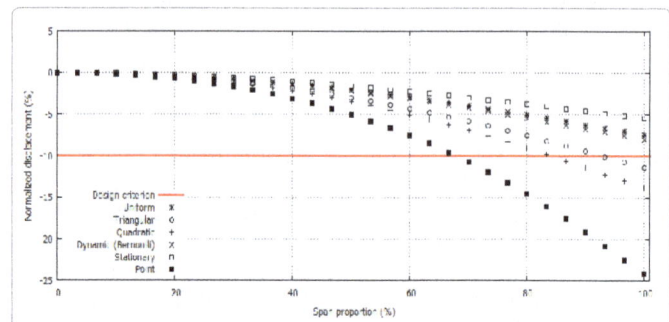

**Figure 6:** Displacement fields of the NACA function Aerofoil depending on the various Loading cases.

| Loading Case | Maximum deflexion (mm) | % Deflexion |
|---|---|---|
| BSDL | 231 | 5.5 |
| UDL | 315 | 7.5 |
| BDDL | 336 | 8 |
| RTDL | 462 | 11 |
| QDL | 588 | 14 |
| DDL | 1008 | 24 |

**Table 5:** Maximum deflexions due to loading cases.

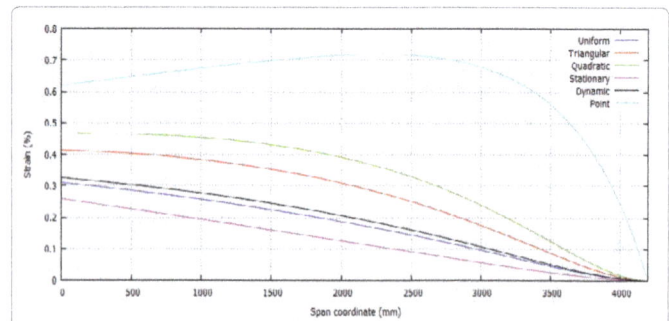

**Figure 7:** Strain fields of the NACA function Aerofoil depending on the various Loading cases.

| Loading Case | Maximum Strain, % |
|---|---|
| BSDL | ~0.25 |
| UDL | ~0.31 |
| BDDL | ~0.33 |
| RTDL | ~0.41 |
| QDL | ~0.47 |
| DDL | ~0.72 |

**Table 6:** Maximum strain values for corresponding loading cases.

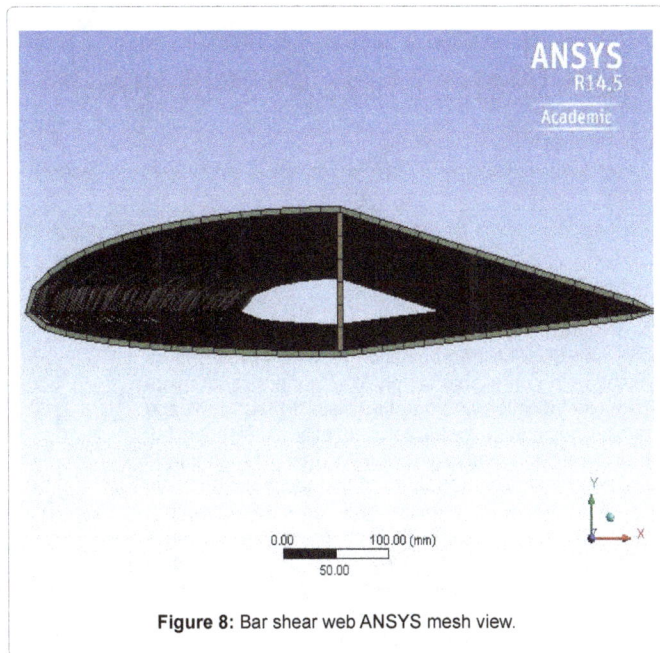

Figure 8: Bar shear web ANSYS mesh view.

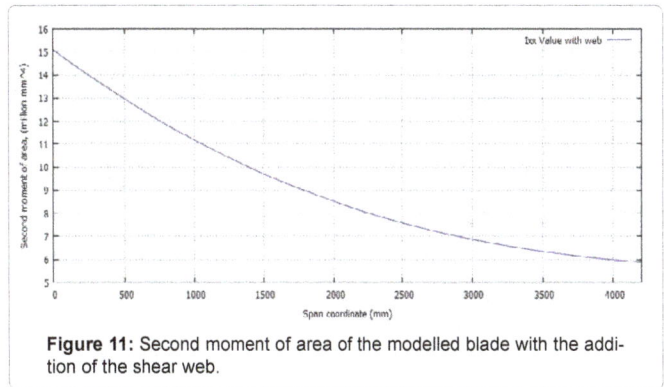

Figure 11: Second moment of area of the modelled blade with the addition of the shear web.

| Loading Case | Maximum deflexion, (mm) | Maximum Strain, (%) |
|---|---|---|
| BSDL | 231 | ~0.25 |
| UDL | 315 | ~0.31 |
| BDDL | 336 | ~0.33 |
| RTDL | 462 | ~0.41 |
| QDL | 588 | ~0.47 |
| DDL | 1008 | ~0.72 |

Table 7: Maximum deflexion and strain results for the asymmetric blade, under individual loading cases.

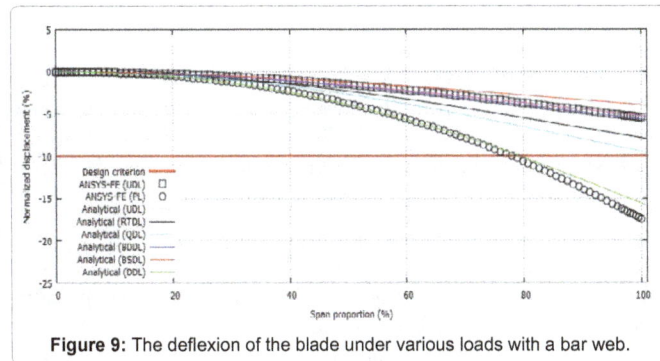

Figure 9: The deflexion of the blade under various loads with a bar web.

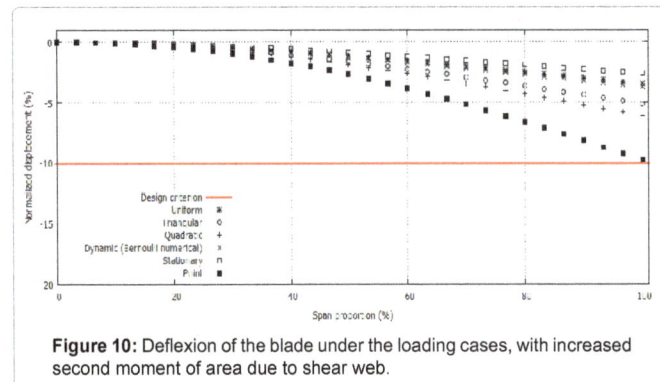

Figure 10: Deflexion of the blade under the loading cases, with increased second moment of area due to shear web.

maximum deflexion of the asymmetrical blade and their corresponding maximum strain values obtained from the analytical model are seen in Table 7.

In general for this and the previous work [1,2] the solid formulation predicts greater than the shell counterparts, indicating that the shell formulations are stiffer, as expected.

The most realistic loading condition for a dynamic blade is that produced by the BDDL, equation (16). Whereas the most realistic loading condition for a stationary blade is that produced by the BSDL equation seen in equation (16). It is most important to note however, the lift force, equation (8), is unique to the blade. That is the lift force is unique to the optimal rip speed ratio, blade dimensions and wind speed.

### References

1. Whitty J, Haydock T, Johnson B, Howe J (2014) On the deflexion of anisotropic structural composite aerodynamic components. J Wind energy, 2014:13.

2. Habali SM, Saleh IA (1999) Local design, testing and manufacturing of small mixed airfoil wind turbine blades of glass ®ber reinforced plastics part i - design of the blade and root. Energ Convers Managem 41:249-280.

3. Manwell JF, McGowan JG, Rogers AL (2009) Wind Energy Explained. Wiley, USA.

4. Ragheb M, Ragheb AM (2011) Wind turbines theory - the betz equation and optimal rotor tip speed ratio. Technical report, University of Illinois at Urbana-Champaign.

5. Burton T, Sharpe D, Jenkins N, Bossanyi E (2001) Wind Energy Hand Book. (2nd edn.) Wiley, USA.

6. Tong W (2010) Wind Power Generation and Wind Turbine Design. WIT Press, Southampton, UK.

7. http://airfoiltools.com/.

8. Milne-Thomson LM (1973) Theoretical aerodynamics. (4th edn.) Dover Publishing, USA.

9. Hansen MOL (2008) Aerodynamic of Wind Turbines. Earthscan.

10. Gere JM, Goodno BJ (2009) Mechanics of Materials. Cengage Learning.

11. Bardsley AM (2014) On the structural analysis of fibreglass asymmetric aerodynamic components with in-situ loading conditions. Master's thesis, University of Centreal Lancashire.

12. DNV-DS-J102 (2010) Design and manufacture of wind turbine blades, offshore and onshore wind turbines.

13. DS/IEC/TS-61400-23 (2001) Wind turbine generator systems - part 23 - full-scale structural testing of rotor blades.

14. Gibson RS (2011) Principles of Composite Material Mechanics. (3rd edn.) CRC Press, Cambridge Mass USA.

# Direct Current and Alternative Current Based Solar Micro-Grid in Rural Energy Infrastructure

**Vivek Kumar Singh\*, Lakshman Ravi Teja and Jitendra Tiwari**

*University of Coimbra, MIT Portugal Program, Coimbra, Portugal*

## Abstract

In India kerosene uses as primary source of lighting- 43% of rural and 7.1% of urban households, even most populated African countries (Uganda, Ethiopia, and Kenya) more than 60% of the population relies on kerosene as the primary lighting fuel. Though they have access to grid electrification the unreliable supply leaves them in complete darkness and hence compelling them to think of alternatives. Poor electrification and unreliable supply lead to usage of kerosene based lighting devices such as kerosene lamps, which not only provides lowly quality illumination but also results in hazardous smoke. Off-grid electrification can provide an alternative solution for many low demand users at lower cost than grid extension and market growth of rural energy service. Costs of off-grid technologies have decreased significantly over the last few years hence making it affordable. Morden micro grid has come up with a way to make lighting system accessible to most of the rural households. Solar based renewable energy technologies extensively development in the last decade, the Solar photo voltaic cell converts light energy into direct current using the photovoltaic effect. Battery devices storing the extra power generated at day time and used during nights. Inverters and power Store Electronic Systems are used to convert direct current power generated by solar photo voltaic systems to alternative current, and utilization of direct current /alternative current in renewable energy power system at higher capital cost. But the internal rate of interest makes it superior to other system. This paper a latest researcher's review of assessment of Direct current (DC) and Alternative current (AC) based Solar Micro-grid in rural community. The paper is therefore structured as follows: overview of micro-grid AC and DC system with case studies. Sensibility analysis in Compression with decentralized Diesel Genset (DG) based on Net present value (NPV) Internal rate of interest (IRR) for DG set 92% (not included as environmental economics), the IRR for solar micro-grids 50% and solar micro-grids with Government subsidy 80%. Opportunities and challenges involved in the implementation of solar mini grid in rural energy infrastructure.

**Keywords:** Decentralized rural energy; Micro grid; Rural energy infrastructure; Solar photovoltaic

## Introduction

Renewable energy sources deliver 16% of the total world energy demand [1]. For 10% of all energy from biomass, [2] and 3.4% from hydroelectricity and 3% accounted for another new renewables (small hydro, modern biomass, wind, solar, geothermal and biofuels) [3]. Renewable energy sources that meet the requirements of domestic energy have the potential to provide energy services with zero or almost zero emissions of both air pollutants and greenhouse gases [4]. Harvesting the renewable energy in decentralized manner is one of the options to meet the rural and small scale energy needs in a reliable, affordable and environmentally sustainable way [5,6].

Photovoltaic (PV) technology is one of the first of several renewable energy technologies, [7] which has been adopted worldwide as well as in India to meet the basic energy needs of rural areas that are not connected to the network. Even so, the growth rate of 30-50% of the global annual average PV is mainly driven by the markets of industrialized countries such as Spain, Germany, USA, Italy, South Korea and Japan [8].

The use of solar photo voltaic system to electrify rural areas has both social and economic benefits [9]. The power generated using this system can be used to operate motor and pumps used for irrigation, storing consumable agricultural products, and many usages [10]. As for the social benefits are concerned electricity allows school going children to study at night, women do some entrepreneurial activity and so on; it contributes to better health allows switching from inferior biomass fuels to clean electricity thereby enhancing indoor air quality [11]. It also gives rural people, lots of opportunities to get access telecommunications and mass media [12]. It also helps to reduce carbon footprint and in order

as to minimize fossil fuel consumption. In India for many decades rural electrification has been an important policy agenda for both the central as well as the state governments. In November 2009, India ministry of new renewable energy approved the Jahwahrlal Nehru National Solar mission (JNNSM) the goal of deploying 20 gigawatts (GW) of grid - connected solar power and 2 GW of off grid solar by 2022 [13].

In India, in 2011, an estimated one in three households reported kerosene as their primary source of lighting-43% of rural and 7.1% of urban households). In the lowest four socioeconomic declines of India, 60% of households use kerosene for lighting [14]. In several of the most populated African countries, including Uganda, Ethiopia, and Kenya, more than 60% of the population relies on kerosene as the primary lighting fuel [15]. Though they have access to grid electrification the unreliable supply leaves them in complete darkness and hence compelling them to think of alternatives. Poor electrification and unreliable supply lead to usage of kerosene based lighting devices such as kerosene lamps, which not only provides poor quality illumination but also results in

**\*Corresponding author:** Vivek Kumar Singh, University of Coimbra, MIT Portugal Program, Coimbra, Portugal, E-mail: viv.jsingh@gmail.com

hazardous smoke [15]. Off-grid electrification can provide an alternative solution for many low demand users at lower cost than grid extension and market growth of rural energy service. Costs of off-grid technologies have decreased significantly over the last few years hence making it affordable for some users. But many still cannot afford it. Taking this fact into account the Modern micro grid has come up with a way to make lighting system accessible to most of the rural households.

This paper a latest researchers' review of assessment of Direct current (DC) and alternative current (DC) based Solar Micro-grid in rural community. The paper is therefore structured as follows: overview of micro-grid AC and DC system with case studies.Sentivibity analysis in Compression with decentralized Diesel Genset (DG). Opportunities and challenges involved in the implementation of solar mini grid in rural energy infrastructure

## Overview of Solar Micro Grid System

A Microgrid is any small or local electric power system that is independent of the bulk electric power network. For example, it can be a combined heat and power system based on a natural gas combustion engine (which cogenerates electricity and hot water or steam from water used to cool the natural gas turbine), or diesel generators, renewable energy, or fuel cells. A Microgrid can be used to serve the electricity needs of data centers, colleges, hospitals, factories, military bases, or entire communities (i.e., "village power") [16].

## Solar photovoltaic system

The amount of sunlight striking the earth's atmosphere continuously is 1.75x105 TW. Considering a 60% transmittance through the atmospheric cloud cover, 1.05 x 105 reach the earth's surface continuously. If the irradiance on only 1% of the earth's surface could be converted into the electric energy with 10% efficiency, it would provide a resource base of 105TW, whereas the total global energy needs for 2050 are projected to be about 25-30 TW. The present state of solar energy technologies is such as single solar cell efficiencies have reached over 20%, Photovoltaic (PV) is a technology that converts light, directly into electricity [17]. To explain the Solar Photovoltaic panel in Simple terms, Photovoltaics is the direct conversion of light into electricity at the atomic level. Semiconductor materials exhibit a property known as the photoelectric effect that causes them to absorb photons of sunlight and higher state of energy release electrons, crating Direct Current (DC) electricity. The efficiency of system is about 4-40% against to semiconductor used in construction of PV arrays. The PV arrays made up of mono crystalline converts 15% of solar power reaching its surface into electricity, while multi crystalline converts only 12% of solar power reaching its surface into electricity this percentage further drop down to 6% and 4% in case of amorphous silicon cells (also called as thin film PV cells), cadmium telluride and copper indium PV cells. The efficiency of the cell increases with increase in cost of the PV arrays [17]. Recent development in PV technology has led to a development of high efficient PV arrays know as multi-junction PV, which operates at an efficiency of 40% [17]. The efficiency of the PV arrays can be improved by 30–50% by using 2 axis solar tracking system in sunny days and by 50% by using horizontal axis orientation instead of 2 axis solar tracking system during cloudy days [18]. The efficiency of the material can be further improved my using mirrors and lens to concentrate solar rays into PV arrays [19].

## Battery

The solar light is available throughout the day the excess energy generated must be stored to provide power during night [20]. Batteries are devices which stores electrical energy in the form of chemical energy and convert that energy into electricity, the batteries used in PV system are charged and discharged, often; hence they are specially designed to meet the stronger requirements than regular batteries. It also helps to produce constant output from PV system where the input is often fluctuating [17]. The specially design Solar tubular acid led batteries 3-5 years life span depends on connected load to system, charging cycle and operating temperature. The climate can be extremely hot, dry and dusty, which can affect the productivity of the panel and create additional wear and tear to the equipment. Hot climate reduces the life increasing the performance on the contrary low temperate increases life but reduces the performance. In order to optimize the performance and the life of the battery entrepreneurs were trained not to utilize the battery to its fullest capacity in order to avoid over-draining.

## Inverters

A power inverter, is an electronic device or circuitry that convert direct current (DC) to alternating current (AC), to meet the connected AC power demand in load side. The input voltage, output voltage and frequency, and overall power handling depend on the design of the specific device or circuitry. The inverter does not produce any power; the power is provided by the Solar PV System.

## MPPT Maximum Power Point Tracking (MPPT)

Recently the researcher developed a component to improve the charging cycle of batteries, MPPT the influence of various parameters on the performance of a photovoltaic system. MPPT system that produces a non-linear output efficiency between solar irradiation, temperature and total resistance (load) to obtain maximum power for any given environment.

## Power Distribution Network (PDN)

The Power distribution network design is essential for proper service on demand side. A power distribution network (PDN) consists of poles, conductors, insulators, wiring/cabling; service lines, internal wiring and appliances to individual households like compact fluorescent lamp, television, fan, radio, etc. [8] 40-50% cost of PDN plays a huge role while determining the project planning a micro grid PV system.

## Micro Grid System Technology

### Alternative current based Solar Micro-grid

AC Micro grid system consists of a solar panel, charge controller MPPT, inverter, batteries and wiring/cabling: service line. AC micro grid system design based on the connected load and appliance. AC Micro grid system technology point out as huge losses in a transmission line. While during mapping of system layout for avoiding the maximum losses by using copper wiring costly solution or small power distribution network). An AC mini-grid is expandable and modular with standardized system components. A major advantage of such a system is that it can easily grow to meet increased consumption demands simply by adding more producers to the electricity. A potential disadvantage to AC mini-grids as slightly lower system efficiency. This is a result of the more frequent conditioning of the electricity because all energy stored in the batteries must be rectified during charging and inverted during discharging [21-23] (Figure 1).

### Direct current based solar micro-grid

DC Micro grid system consists of a solar panel, charge controller MPPT, batteries and wiring/cabling: service line. DC micro grid system operation best suitable performs in a rural area with only DC appliance

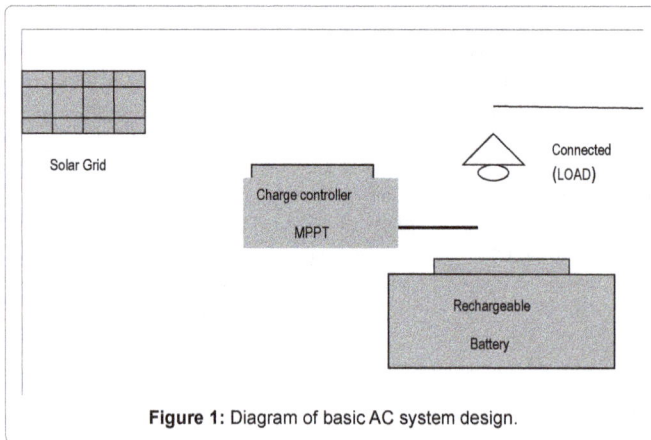

**Figure 1:** Diagram of basic AC system design.

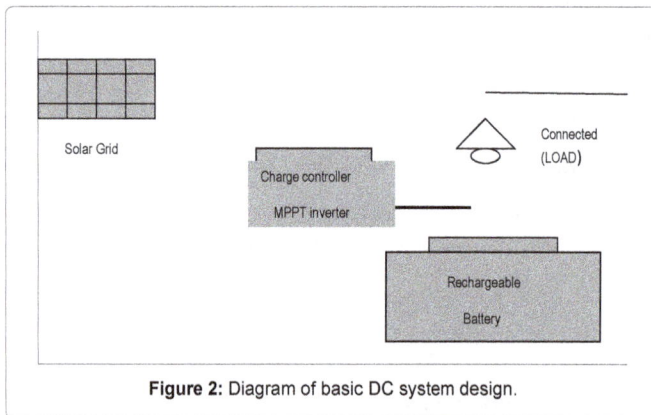

**Figure 2:** Diagram of basic DC system design.

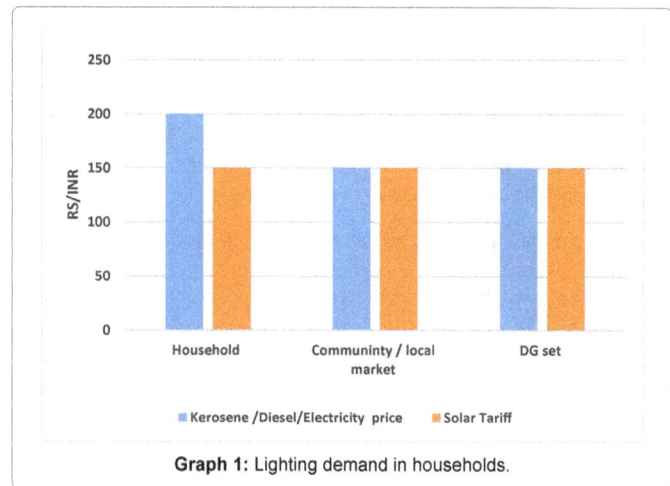

**Graph 1:** Lighting demand in households.

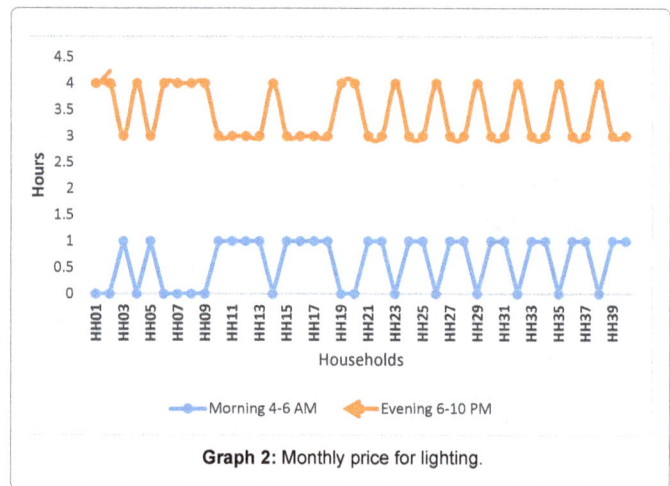

**Graph 2:** Monthly price for lighting.

as mobile, lamp, radio etc. Batteries variable play huge role in DC micro grid system better power supply on 24 voltage with copper wiring/cable in distribution line. DC based Solar Micro-grid comparatively easy to install and can be set up by local qualified technicians with some training (Figure 2).

## Lighting Demand and Price in Rural Household/Community

A survey study by TERI 2009, household need 5 hours complete lighting in the rural area morning 4-6 AM and 6-10 PM (Graph 1) according to winter and summer season, this study also compile for the 3- 5 liter kerosene according to household uses the application of kerosene to firing the firewood in the cooking section (UNDP 2003). A comparative study lighting cost of kerosene including the electricity price in (Graph 2). System specification of comparing three lighting system in Table 1 as household needs two numbers of lamp one for cooing and other one of study or rest work and it also replace with solar LED 1W, 2W respectively. And 15-18 CFL lighting sources of diesel replace with 3W solar LED.

## Compression with Decentralized DG SET (Sensitivity Analysis)

Unreliable supply of grid electricity in rural areas that leading to promotion of decentralized Diesel Genset in small rural market. Decentralized DG SET entrepreneur growth increasing to sell electricity to local level. The diesel generator operators in other rural markets have expressed their interest to install DC based solar micro-grids as they believe that the negligible operational cost balances the high investment cost (Table 2).

A real case study case we investigate has the following characteristics. A system of this size would meet the electricity needs of a community 40-100 households/shops for connected sufficient energy to 4 hours peak load. For each case, we estimate the initial and ongoing capital and O & M costs of the equipment. Modules, distribution wires are excluded from this analysis. Solar micro grid System cost break up for 40 households (Table 3) and Diesel genet system cost break up (Table 4 and Table 5).

Using Excel's NPV function, Sensitivity analysis considers the net present value (NPV) of the different options based on the differences in costs and efficiencies discussed above, assumed discount rate (10%) and life span of solar batteries 5 years and solar photovoltaic 20 years (Table 6).

## Solar Micro Grid and Rural Infrastructure

In rural India, technologies for off-grid rural electrification in combination with proper financial balance promises environmentally friendly access to electricity at a lower cost than conventional technologies. This ensures rural users social benefits and improving living standards. At the moment, kerosene and candles are used for lighting in rural areas, while dry cell batteries are used for radio. Both are expensive (1 liter of kerosene cost at least Rs 35 and lasts for one week or about 20 hours) and giving poor quality lighting.

Most of energy users (households and public uses) have an access to grid connections. Though they have access to grid electrification

| Sources of Energy | Lamp Number | Watts | | lumens (Distance 150 Centimetres at 90°) |
|---|---|---|---|---|
| Kerosene Household's | 2 | | | 25-35 |
| Kerosene / Diesel/ Electricity -Community | 1 | 15 -18 W CFL | | 320 |
| Solar- household's | 2 | 1 W LED | 2 W LED | 150-240 |
| Solar -Community | 1 | 3 W LED | | 350 |

**Table 1:** Specification of lighting system.

| Case | Working Assumption | |
|---|---|---|
| Diesel Genset | Annual Inflation | 10% |
| | Working days per month | 28 |
| | Tariff per day (Year 1) | 5.0 |
| | NPV Discount Rate | 10% |
| | Diesel Cost per litre (INR) | 45 |
| Solar Micro-grid | Annual Inflation | 10% |
| | Working days per month | 28 |
| | Tariff per day (Year 1) | 5.0 |
| | NPV Discount Rate | 10% |

**Table 2:** Base-case assumptions.

### System Cost Breakup

| Item Description | Multiplier | Per unit cost | Total |
|---|---|---|---|
| Diesel GENSET, 7.5 KVA | 1 | 35,000 | 35,000 |
| Wiring, Cable, etc. | 1 | 10,000 | 10,000 |
| Total Capital Cost INR | | | 45,000 |

**Table 3:** System cost break up with Diesel genset.

### System Cost Breakup

| Item Description | Multiplier | Per unit cost | Total |
|---|---|---|---|
| Solar module 50 Wp | 4 | 5,000 | 20,000 |
| Solar Battery 12V-100Ah | 2 | 7,500 | 15,000 |
| Charge Controller Unit 24V, 10 A | 1 | 5,000 | 5,000 |
| LED Luminary 3 W | 40 | 800 | 32,000 |
| Cable, Panel Mounting frame, Switches, Misc. | 1 | 15,000 | 15,000 |
| 5 Year AMC charges for power system/ distribution network | 1 | 10,000 | 10,000 |
| Installation Cost | 1 | 3,000 | 3,000 |
| Total Capital Cost INR | | | 100,000 |

**Table 4:** System cost break up with solar mini grid.

| Capital Cost Breakup | | Capital Cost Breakup | |
|---|---|---|---|
| MNRE Subsidy (Rs. /Watt) | 150 | MNRE Subsidy (Rs. /Watt) | 0 |
| System Capacity (Wp) | 200 | System Capacity (Wp) | 200 |
| MNRE subsidy | 30,000 | MNRE subsidy | - |
| Self-investment | 20,000 | Self-investment | 20,000 |
| Bank Loan | 50,000 | Bank Loan | 80,000 |
| ANNUAL Rate of interest | 12.0% | ANNUAL Rate of interest | 12.0% |
| Payment period (years) | 5 | Payment period (years) | 5 |
| Number of installments | 60 | Number of installments | 60 |
| Monthly EMI | 1112 | Monthly EMI | 1780 |
| Total loan repayment amount INR | 66,734 | Total loan repayment amount INR | 106,774 |

**Table 5:** Solar micro grid as considering with subsidy and without subsidy in capital cost break for both options.

| Parameters | Diesel genset | Solar micro grid | Solar micro grid with subsidy |
|---|---|---|---|
| Net Cash Flow INR | 45000 | 20000 | 20000 |
| Net present value INR | 407,151 | 172,334 | 197718 |
| IRR | 92% | 50% | 80% |

**Table 6:** Net present value of different options.

capacity to rural households for 4 hours daily from 6pm to 10pm when they need the light most. Costing around Rs 150 per month for each household. The project also focused on the commercial users who also need lighting solution the most from 6pm to 10 pm for their business. They charge Rs 6 per day irrespective of the shop being opened or closed.

## Solar Micro Grid Challenge

### Solar panel maintenance

Solar panels require constant cleaning to maximize their utility and life cycle. Productive capacity drops when dust or film accumulates on their surface. Concentrated areas, sustained over a period of time, can lead to failure of an entire segment of a panel. Every panel has a rating factor which indicates its expected readings under healthy conditions. A constant read of the expected value is a strong sign that the panel is functioning well, whereas a sudden dip or a consistently low reading during daytime is a sign of a potential problem. The underlying causes for a dip in the power output of a panel can be one of several: an unclean panel, a faulty battery, poor climate conditions, improper panel orientation.

### Charge controller maintenance

Charge controllers vary from system to system, but they all regulate

the unreliable supply leaves them in complete darkness and hence compelling them to think of alternatives.

Off-grid electrification can provide an alternative solution for many low demand users at lower cost than grid extension and market growth of rural energy service. Costs of off-grid technologies have decreased significantly over the last few years hence making it affordable for some users. But many still cannot afford it. Taking this fact into account the team has come up with a way to make lighting system accessible to most of the rural households.

In North Uttar Pradesh villages The Energy and Resources Institute facilitate solar micro -grid electrification light solution of 1W and 2W

current from the panel to the battery until it is fully charged. At that point they either cut power completely or transmit the minimum energy needed to keep the battery topped off.

## Quality of off-grid lighting products

This remains one of the key challenges. Rapid technological advancements and innovations, as well as a multitude of products available at the market, make quality control a difficult task. The consequences of bad quality products are a reputation loss for a whole product range-endangering market development. Located in sparsely populated and remote areas, it has limited accessibility due to bad roads and other factors. The geographical distance between houses in sparsely populated areas increases the cost of maintenance as the connecting wires are more and might get damaged due to storms and trees.

The long-term goal is a rapid scale-up of access to clean, reliable and affordable modern off-grid lighting services. Rural electrification has a number of barriers that are addressed by the training and development of entrepreneurs.

## Training and Development of entrepreneurs

The success of the project lies in the development of entrepreneurs in maintaining the grid and in-house installations. Intensive training program was also developed for the entrepreneurs for product maintenance to the long-term sustainability of the solar market.

## Conclusion

Energy demand, costs, and user satisfaction are particularly interwoven in the case of mini grids, making technology assessment difficult. Mini-grids are sized according to estimated consumption and then can be extended or revaluated. This problem can be solved at the planning phase by over sizing systems or restriction of consumption per user. Opportunities and challenges were a to find an entrepreneur to invest money; the lengthy procurement process due to a lack of suppliers in the local market; the lengthy legal processes involving in getting a micro-finance and loan guarantee agreement signed with the local bank; and limited awareness of improved lighting products amongst rural residents. Even after developing good rapport with the bank the entrepreneurs were not convinced into taking a loan. Lack of knowledge about the market demand made them skeptical.

## References

1. World energy outlook 2012 global energy trends.

2. International energy agency energy for cooking in developing Countries 2005.

3. REN21 (2011) Renewables 2011: Global Status Report.

4. Panwar NL, Kaushik SC, Kothari S (2011) Kothari Role of renewable energy sources in environmental protection: a review. Renew Sust Energ Rev 15: 1513-1524.

5. Ravindranath NH, Hall DO (1995) Biomass, energy, and environment: A developing country perspective from India. Oxford University Press.

6. Reddy AKN, Subramanian DK (1979) The design of rural energy centers. Proceedings of the Indian Academy of Sciences 2: 395-416.

7. Chu Y, Meisen P (2011) Review and Comparison of Different Solar Energy Technologies. Global Energy Network Institute (GENI).

8. Chaurey A, Kandpal TC (2010) Assessment and evaluation of SPV based decentralized rural electrification: an overview. Renew Sust Energ Rev 14: 2266-2278.

9. UNDP and World Health Organization (WHO) (2009) The energy access situation: a review focusing on the least developing countries and sub-Saharan Africa.

10. Andreas K (2006) Regional disparities in electrification of India do geographic factors matter? Centre for Energy Policy and Economics, Swiss Federal Institute of Technology. CEPE Working Paper No. 51.

11. Bhattacharyya SC (2006) Energy access problem of the poor in India: is rural electrification a remedy. Energy Policy 34: 3387-3397.

12. Government of India, Ministry of Power, Annual Report 2009-10.

13. Government of India, census 2011.

14. http://www.indiaenvironmentportal.org.in/files/file/Energy%20Sources%20of%20Indian%20Households.pdf

15. Lam NL, Smith KR, Gauthier A, Bates MN (2012) Kerosene: a review of household uses and their hazards in low- and middle-income countries. J Toxicol Environ Health B Crit Rev 15: 396-432.

16. Campbell RJ (2012) Weather-Related Power Outages and Electric System Resiliency.

17. Kalogirou SA (2009) Solar energy engineering: processes and systems.

18. Manoj Kumar MV, Banerjee R (2010) Analysis of isolated power systems for village electrification. Ener Sustain Dev 14: 213-222.

19. Zahedi A (2011) Review of modelling details in relation to low-concentration solar concentrating photovoltaic. Renew Sust Energ Rev 15:1609-1014.

20. Raman P, Murali J, Sakthivadivel D, Vigneswaran VS (2012) Opportunities and challenges in setting up solar photo voltaic based micro grids for electrification in rural areas of India. Renew Sust Energ Rev 16: 3320-3325.

21. http://www.homepower.com/view/?file=HP109_pg8_TOC

22. https://www.esmap.org/sites/esmap.org/files/FR26303India.pdf

23. TERI (2009) Sustainable Development through - Research, Customization Demonstration of Technology in Jagdishpur Block, district Sultanpur, Uttar Pradesh.

# Relative Humidity Effect on the Extracted Wind Power for Electricity Production in Nassiriyah City

**Abdul-Kareem Mahdi Salih[1]* and Abdullh Saiwan Majli[2]**

[1]*Department of Physics, College of Science, Thi-Qar University, Iraq*
[2]*Department of Electronics, College of Engineering, Thi-Qar University, Iraq*

## Abstract

The relative humidity effect on the wind power was extracted as a renewable energy for electricity production in Nassiriyah city - south of Iraq investigated for three years (2011-2013) by theoretical calculations. The study showed that the effect of relative humidity on the annual average of extracted wind power at the minimum altitude which is feasible for electricity production in this city (32 meter for α = 0.4 and 44 meter for α = 0.3) is limited, but it increase with the altitude to be noticeable at high altitude. The percentage loss on the annual average of power for moist air (due to relative humidity effect) uneffective and vary between (0.847% and 1.106%) at altitudes (15 m and 71 m) respectively.

**Keywords:** Physics; Renewable energy; Wind energy

## Introduction

The temperature and humidity are important climatic variables that effect on the wind power [1,2]. The previous paper [3] studied the temperature effect on the extracted wind power as a renewable energy in Nassiriyah city (in Iraq) which is located at 31 (east) with 46 (north) intersection lines. This paper is focusing on the study to simulate the relative humidity effect on the extracted wind power in this district. The results are necessary to complement obtained results in reference [3] because of near the district location from Euphrates river, marshes and Shutt Al-Arabe, which make sometimes humid climate features in Nassiriyah city. No data of relative humidity reported at different altitudes in the district just that measured at 10 meter elevation [4]. Software program prepared in this study to compute the relative humidity and investigate its effect on wind power as a function of altitude.

## Theory

According to the ideal gas law, a cubic meter of air has a certain number of molecules, and each of those molecules has a certain weight. Whenever water vapour molecules are added to the air, they displaces some other molecules in the volume of air. Nitrogen molecule is the most abundant in the air molecules, and it has an atomic weight approach to 14, then the molecules have a weight (atomic mass) approach to 28. For oxygen, the $N_2$ atomic weight is 16, so an $O_2$ molecule has a weight of 32. Hydrogen has an atomic mass of 1, so the molecule of $H_2O$ has a weight of (1+1+16=18). Hence the water molecule is lighter in weight than either the nitrogen molecule or oxygen molecule. Therefore, the volume of air that contains some water molecules will be weight less than the same volume of air without water molecules. This lead to decrease density which means that the moist air has less density than dry air at the same temperature [1,5]. Water is extremely found pervasive in the air as vapour. Humidity is the quantity of water vapour present in air, and there are many varied ways of expressing it such as an absolute humidity, specific humidity, and relative humidity. Relative humidity is the amount of water vapour in the air relative to the maximum amount possible, thus it is at all temperatures and pressures defined as the ratio of the water vapour pressure to the saturation water vapour pressure, and it is defined mathematically by the following expression [6-8].

$$RH = \frac{e_a}{e_s(T)} \times 100\% \tag{1}$$

Where RH is the relative humidity, $e_a$ is the actual water vapour pressure, it is contributes to the total atmospheric pressure, defined by the Antoine equation [9].

$$\log_{10} e_a = A - \left(\frac{B}{C+T}\right) \tag{2}$$

Where $A = 8.07131$, $B = 1730.63$, $C = 233.426$, $e_a$ is in mmHg (1mm Hg=133.322 Pascal), $T$ is the temperature in degrees Celsius (C˚). $e_s(T)$ is the saturation vapour pressure (Pascal) at the same temperature $(T)$, When air enclose above an evaporating water surface, an equilibrium is reached between the water molecules escaping and returning to the water reservoir. At this moment, the air is saturated since it cannot store any extra water molecules. Many algorithms for determining the saturation vapour pressure [10,11]. Herman Wobus developed Albeit formula for determining saturation vapour pressure [1,12] as the following expression.

$$e_s(T) = \frac{e_{s0}}{p^8} \tag{3}$$

Where,

$$e_{s0} = 6.1078$$

$$P = (C_0 + T*(C_1 + T*(C_2 + T*(C_3 + T*(C_4 + T*(C_5 + T*(C_6 + T*(C_7 + T*(C_8 + T*(C_9)))))))))),$$

$$C_0 = 0.99999683 \quad C_1 = -0.90826951E-02$$

$$C_2 = 0.78736169E-04 \quad C_3 = -0.61117958E-06$$

$$C_4 = 0.43884187E-08 \quad C_5 = -0.29883885E-10$$

*\*Corresponding author:* Abdul-Kareem Mahdi Salih, Department of Physics, College of Science, Thi-Qar University, Iraq
E-mail: karimmahdisalih@yahoo.co.ukoo.fr

$$C_6 = 0.21874425E - 12 \quad C_7 = -0.17892321E - 14$$

$$C_8 = 0.11112018E - 16 \quad C_9 = -30994571E - 19$$

For predict the temperature as a function of altitude (z), the following expression used [13].

$$T(z) = T_0 - R_a(z - z_0) \qquad (4)$$

Where $T_0$ is the temperature at the lower altitude $(Z_0)$ (10 m in the study which is represent the tower elevation of Nassiriyah meteorology station), $R_a$ is the temperature laps rate (0.0065 $C^0m^{-1}$).

From the following equation (Poisson equation) [13].

$$\frac{T_{(z)}}{T_0} = (\frac{P_{(z)}}{P_0})^{\frac{R}{C_P}} \qquad (5)$$

Where $P_0$ is the pressure at the lower altitude, then the pressure of dry air at any altitude (z) can be calculated by the following expression [13].

$$P(z) = P_0(\frac{T_{(z)}}{T_0})^{C_P/R} \qquad (6)$$

Where $R$ the universal gas constant (8.31432), $C_p$ is the constant - pressure specific heat of air, the amount of $\frac{C_p}{R} = 3.49$. The density of mixture of dry air molecules and water vapour molecules may be written in term of total pressure, temperature and actual water vapour pressure by the following expressions [12]:

$$\rho = (\frac{P}{R_d * T})(1 - \frac{0.378 * e_a}{P}) \qquad (7)$$

Where $\rho$ is the moist density ( $(Kg/m^3)$, $e_{a} = RH^{'} e_s(T)$, $P = P_d + e_a$ is the total pressure , $P_d$ is the pressure of dry air (partial pressure), $R_d$ is the gas constant for dry air ($J/Kg.K$), $T$ is the temperature in (K).

The density of dry air as a function of altitude ($\rho$ (z)) calculated by the following equation [14].

$$\rho(z) = \rho_0 e^{-(\frac{0.297 * Z}{3048})} \qquad (8)$$

Where ($\rho_0$) is the air density (1.255 Kg/m³), (z) is the altitude. The wind speed calculated by the flowing equation [13,14].

$$v = v_0(\frac{Z}{Z_0})^{\alpha} \qquad (9)$$

Where ($z_0$), (z) are represents the lower altitude ((10m) in this study) and other under study altitudes respectively, $v$ and $v_0$ are the wind speed at altitude (z) and ($z_0$) respectively, $\alpha$ is the ground surface friction coefficient. The net power of a practical wind turbine can be described by the following equation [15].

$$P_W = \frac{C_P \rho A v^3}{2} \qquad (10)$$

Where $A$ the swept area of turbine blades in $m^2$ is, $v$ is the wind speed in $m/Sec$. $C_p$ is the power coefficient, it's maximum theoretical value is a brooch to 0.59, but in practical designs the maximum value below 0.5 [13-16].

## Calculations

The relative humidity, temperature and wind speed data of Nassiriyah city where measured at (10 m) altitude only (tower elevation of Nassiriyah meteorology station) for three years' time interval (2011-2013). These data where feed to Q-Basic program prepared in this study for simulation of air density, power, power density, pressure, temperature, relative humidity, saturation vapor pressure, actual vapor pressure and percentage losses of power because of relative humidity in moist air as a function of different altitudes (10-71 m) to estimate the effect of relative humidity on the extracted wind power at the feasible altitudes for electricity production in Nassiriyah city (32 meter for $\alpha$ = 0.4 and 44 meter for $\alpha$ = 0.3) [17]. Other data used in computations listed as: a blade radius (r) =10m, air density $\rho$ = 1.225 Kg/m³), power coefficient ($C_p$) = 0.5 [13-16], from reference [15] obtained the following data : ground surface friction coefficient ($\alpha$) = 0.3, shape factor (k) = 2 m/sec, scalar factor (C) = 7 m/sec.

The computer subroutine programs execute the average daily computations for temperature, pressure, and the wind speed by using equations (4, 6, 9) respectively. The results has been utilized in equations (2, 3, 1), then it is possible to compute the density of moist and dry air by using equations (7, 8) respectively. The result of equation (7, 8) feed to compute the power by equation (10).

## Results and Discussion

For computation checking; (Figure 1 (A, B, C)) shows the profile of relative humidity daily average for air according of measured and calculated data at (10 m) altitude for the time interval study (2011-2013) respectively. The figures appeared good consistence behaviour, the relative humidity daily average values decreases at the number of days (200 ± 20) which are characterized by high temperature degree. This behaviour is agree with mathematical relations and physical ideas. (Figure 2A-2C) confirms the inverse relation between the temperature and the relative humidity, and they shows the time variation of the daily average of each relative humidity and temperature at 35 m altitude for the years (2011-2013) respectively, the figures are in good identity with the published literature. While the (Figure 3A-3C) shows the relation between the annual average of the relative humidity and the temperature as a function of altitude, and they displayed the increasing of relative humidity with the increasing of altitude, on the other hand they displayed the decreasing of annual average of temperature with the increasing of altitude, that is related to the decreasing of the air pressure with increment of altitude.

(Figure 4A-4C) shows the comparison of annual average values of dry and moist air density as a function of altitude for the years (2011-2012). The figures clarified that the moist air density was less than the dry air density that is related to the moist air mass decrement comparing with the dry air, but each of their (moist and dry air density) increasing with altitude increment. For explain; there are two important factors (temperature and relative humidity) affecting on air density, therefore we have two cases of behaviour, the first one is resulting from the effect of temperature degree only (dry air density behaviour) thus it seems the density increasing with the increment of altitude due to decrease of temperature. The second one is resulting from the relative humidity and the temperature degree effects (moist air density behaviour) thus it seems the density increasing with the increment of altitude. The study explains that the temperature degree is the dominator factor from the relative humidity (the effect of relative humidity less than the temperature effect).

Figure 5 shows the percentage of loss occurring in the annual

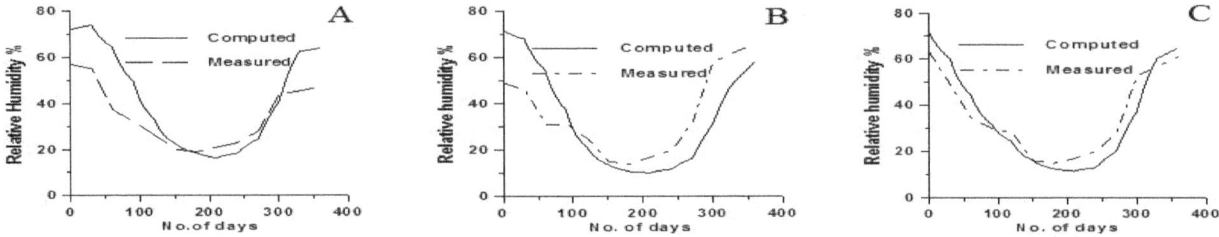

**Figure 1:** The profile of computed and the measured values of daily average of relative humidity at 10 m altitude for the years 2011-2013 respectively.

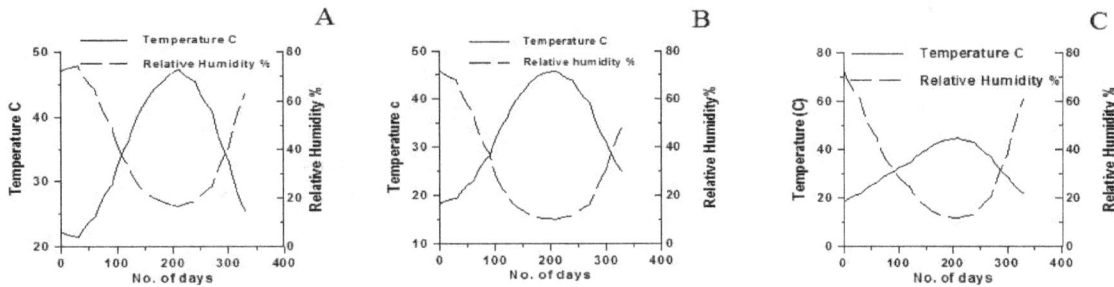

**Figure 2:** The behavior of the daily average for temperature and relative humidity at 35 m altitude for the years 2011-2013 respectively.

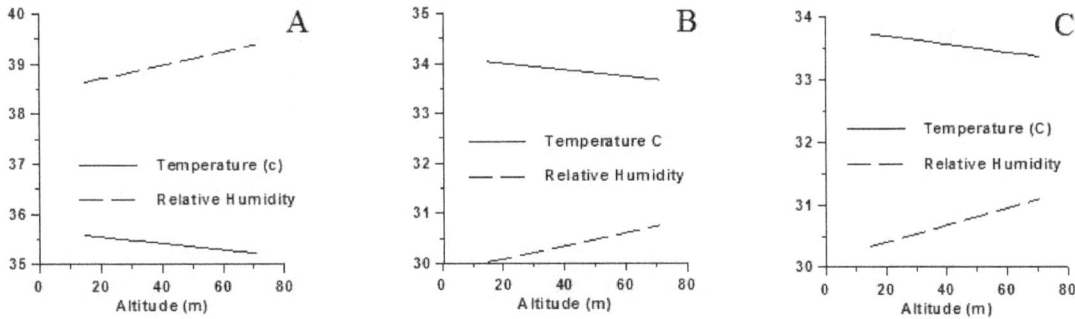

**Figure 3:** The profile of the annual average of temperature and relative   humidity as a function of altitude for the years 2011-2013 respectively.

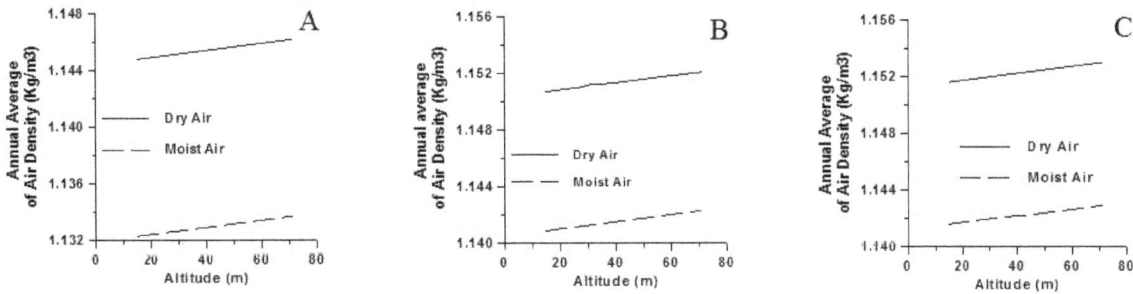

**Figure 4:** The profile of the annual  average of the density for dry and moist air as a function of altitude for the years 2011-2013 respectively.

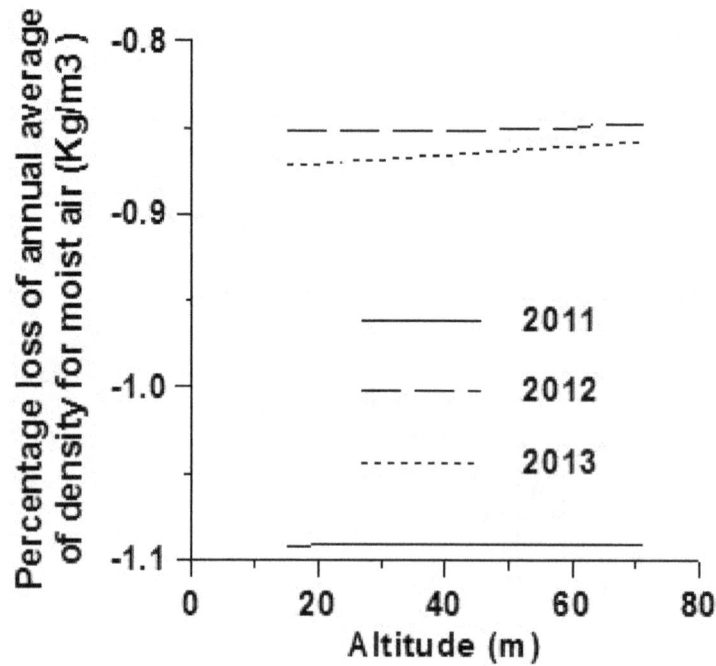

**Figure 5:** The percentage of loss of the annual average of density for moist air as a function of altitude for the years 2011-2013 respectively.

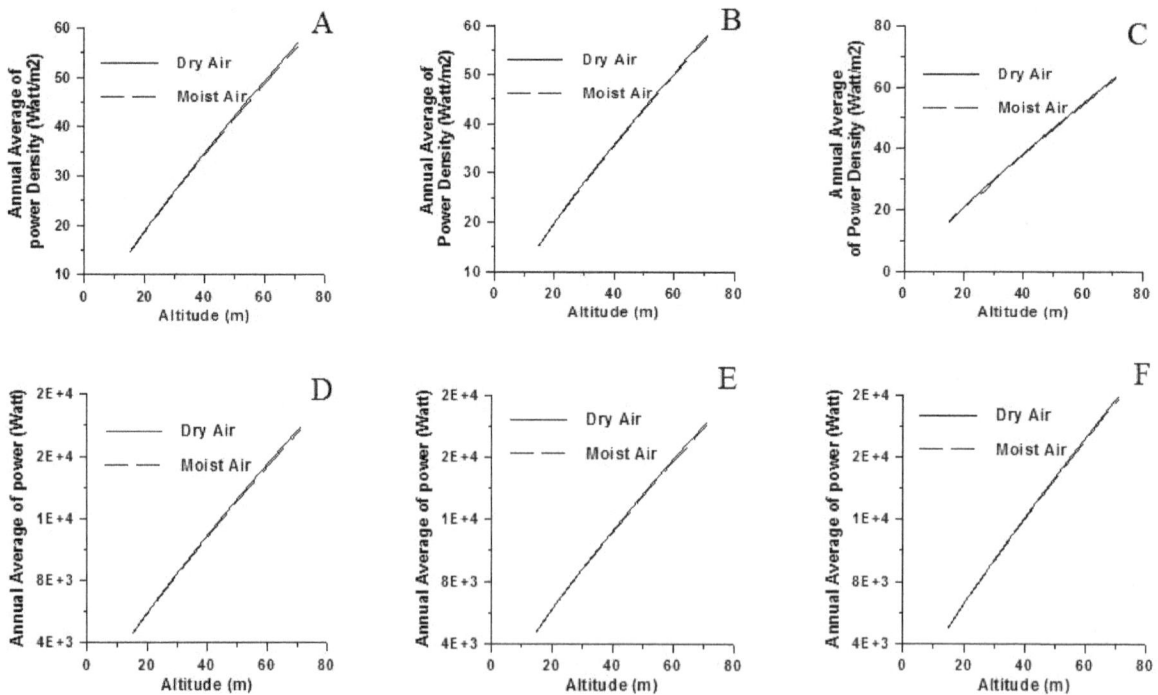

**Figure 6: A, B, C.** The profile of the annual average of power density for the dry and moist air as a function of altitude for the years 2011-2013 respectively. **D, E, F.** The profile of the annual average of power for the dry and moist air as a function of altitude for the years 2011-2013 respectively.

average of air density due to relative humidity as a function of altitude for the study time interval (2011-2013) respectively. It is appear that the max percentage of loss at the low altitude and vice versa (1.092%, 0.852%, 0.868% at (15 m) altitude for years (2011-2013) respectively (note: the negative sign refer to losses), while these values

became 1.09%, 0.848%, and 0.858% at (71 m) altitude respectively. The study explains that related to mass decrement effect on air density due to humidity comparing with the volume decrement effect due to temperature decrease with the altitude increasing. In general, cane concludes that the effect of relative humidity on air density in Nassiriyah

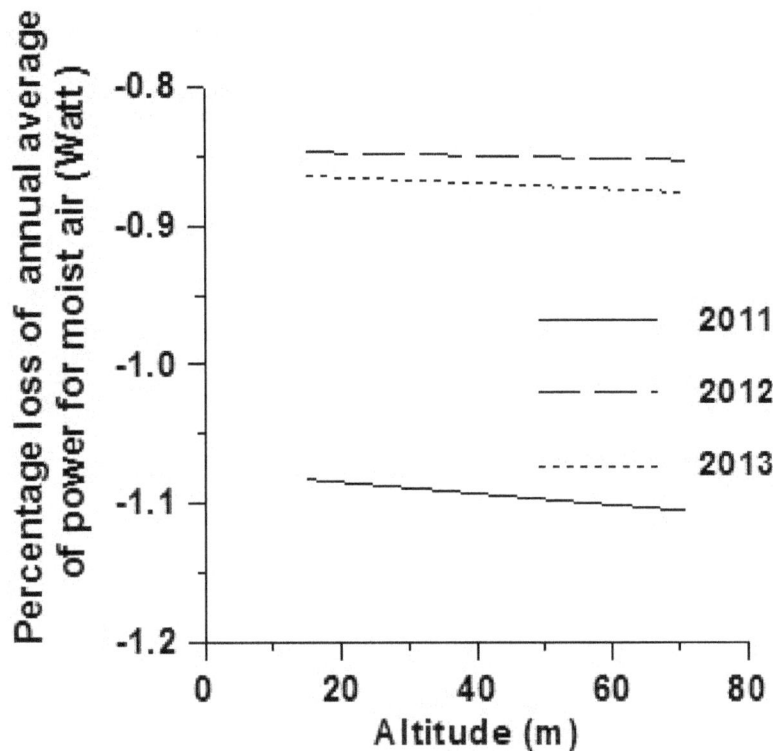

**Figure 7:** The percentage of loss of annual average of power for moist air as function of altitude.

city is a little.

(Figure 6A-6C) shows the variation of the annual average of power density as a function of altitude for the study interval time (2011-2013) respectively. It appears that the annual average of the power density of moist air less than dry air. (Figure 6D-6F) are enhancing (Figure 6A-6C) and shows the variation of the annual average of wind power for dry and moist air as a function of altitude for the years (2011-2013) respectively. It appears that the annual average of the power of moist air less than dry air and the variance increase with the altitude to be somewhat noticeable at high altitude, the computations results appear variance between the dry and the moist air at (15 m) altitude 51 watt, 41 watt, and 44 watt for the years (2011-2013) respectively, while this variance to be approach to 194 watt, 155 watt, and 174 watt for the same years respectively at 71 m altitude.

Figure 7 shows the percentage of loss occurring on the annual average of power due to relative humidity as a function of altitude for the study time interval (2011-2013) respectively. It is appear that the minimum percentage of loss at the low altitude and vice versa (1.083%, 0.847%, 0.864% at (15 m) altitude for years (2011-2013) respectively, (note: the negative sign refer to losses) while these values became 1.106%, 0.853%, and 0.876% at (71 m) altitude respectively. In general cane concludes that the effect of relative humidity on the annual average of wind power which is extracted in Nassiriyah city is a little, and the percentage of loss on the annual average of power for moist air (due to humidity effect) is increase with the increment of altitude comparing with the dry air.

## Conclusion

The loss in the extracted power from wind as a renewable energy for electricity production due to relative humidity in Nassiriyah city - south of Iraq it is un effective and vary between (0.847% and 1.106%) at altitudes (15 m , 71 m ) respectively.

### References

1. Modelling of the Variation of Air Density with Altitude through Pressure, Humidity and Temperature.

2. Nolan P, Lynch P, McGrath R, Semmler T, Wang S (2011) Simulating climatechange and its effects on the wind energy resorce of Irelend. Wind Energy 15: 593-608.

3. Abdul-Kareem MS (2015) Theoretical study of temperature effect on the wind power extracted as a renewable energy for electricity production in Nassiriyah district. J C Bio Phy Section D 5: 2.

4. Meteorological station of Nassiriya city, Thi- Qar, Iraq.

5. Energy Management Handbook- Moist air (Psychrometry). 6th edition, Lulea University of Technology.

6. The Rotronic Humidity Handbook.

7. Lawrwnce MG (2005) The Relationship between Relative Humidity and the Dewpoint Temperature in Moist Air: A Simple Conversion and Applications. Bull Amer Meteor Soc 86: 225-233.

8. Oyj V (2013) Humidity conversion formulas.

9. Jaime Wisniak (2001) Historical development of the vapor pressure equation from Dalton to Antoine. J Phase Equilibria 22: 622-630.

10. Murphy DM, Koop T (2005) Review of the vapour pressures of ice and supercooled water for atmospheric applications. Quart J Royal Met Soc 131: 1539-1565.

11. Guide to Meteorological Instruments and Methods of Observation.

12. Shelguiost R (2015) An Introduction to air density and density altitude Calculations.

13. Johnson GL (2006) Wind energy system, Electronic Edition.

14. Patel MR (1999) Wind and Solar Power System. U.S. Merchant Marine Academy, New York, CRC Press LLC.

15. Betsch P, Siebert R, Sanger R (2012) Natural coordinates in the optimal control of multibody system. J Comput Nonlinear Dynamic 7: 317-326.

16. Erich H (2000) Wind turbines -Fundamentals, Technologies, Application, Economics. Energy Tech.

17. Abdul-Kareem MS (2015) Simulation of wind power dynamic for electricity production in Nassiriyah district. IOSR J Appl Physics 7.

# 6

# Life Cycle Exergy Analysis of Solar Energy Systems

**Mei Gong[1] and Göran Wall[2]***

[1]*School of Business and Engineering, Halmstad University, PO Box 823, SE-30118 Halmstad, Sweden*
[2]*Oxbo gard, SE-43892 Härryda, Sweden*

## Abstract

Exergy concepts and exergy based methods are applied to energy systems to evaluate their level of sustainability. Life Cycle Exergy Analysis (LCEA) is a method that combines LCA with exergy, and it is applied to solar energy systems. It offers an excellent visualization of the exergy flows involved over the complete life cycle of a product or service. The energy and exergy used in production, operation and destruction must be paid back during life time in order to be sustainable. The exergy of the material that is being engaged by the system will turn up as a product and available for recycling in the destruction stage. LCEA shows that solar thermal plants have much longer exergy payback time than energy payback time, 15.4 and 3.5 years respectively. Energy based analysis may lead to false assumptions in the evaluation of the sustainability of renewable energy systems. This concludes that LCEA is an effective tool for the design and evaluation of solar energy systems in order to be more sustainable.

**Keywords:** Exergy analysis; Sustainable development; Life cycle analysis

## Introduction

With a dependence on finite natural resources, increasing energy demands and increasing environmental problems, the use of renewable resources becomes even more important. This is one part of a sustainable development. The ultimate source of most of our renewable energy supplies is the sun. The world's primary energy use is easily exceeded by the solar energy by a factor of about 10000 [1]. The renewable energies except geothermal and tidal energy depend on solar radiation. Thus, it is essential for future energy systems to rely on energy from the sun and presently, the use of direct solar energy is rapidly increasing. Solar energy systems can be classified into direct-solar systems and indirect-solar systems [2]. Here solar energy systems refer to direct-solar systems.

Exergy is a useful concept in the work towards sustainable development [3]. Exergy accounting of the use of energy and material resources provides important knowledge on how effective and balanced a society is regarding conserving nature's capital. This knowledge can identify areas in which technical and other improvements should be undertaken, and indicate the priorities, which should be assigned to conservation measures, efficiency improvements and optimizations. Hepbasli [4] offers a careful review of exergy analysis applied to renewable energy resources for a sustainable future. Thus, the exergy concept and exergy tools are essential to the creation of a new engineering toolbox or paradigm towards sustainable development.

In the literature, a number of studies applies energy and exergy analyses to the whole or part of solar energy systems, e.g. solar heater [5,6], solar power plant [7], solar photovoltaic (PV) system [8] combined solar photovoltaic and thermal (PV/T) plants [9,10]. Exergy of solar radiation is often regarded as input into the energy systems of these studies, and the overall energy efficiency is about 25% for solar power plant, 14-15% for PV/T system and less than 12% for PV system.

Life cycle analysis/assessment (LCA) is a tool used to evaluate the total environmental impact and total energy resource use of a product or service during its complete lifetime or from cradle to grave. It covers three steps – construction, operation and destruction. In previous studies LCA has been applied to solar energy system, e.g. PV system [11-14], solar thermal power system [15], solar heating system [16] and PV/T system [17,18]. However, none of these studies made use of exergy analysis. Some of the papers [11,13,18] estimated material recycling with present technology and future developed technology as well as using recycled material that will reduce the need for energy in the construction stage.

Two different methods combining LCA and exergy analyses have been proposed, e.g. Exergetic Life Cycle Analysis (ELCA) [19] or Life Cycle Exergy Analysis (LCEA) [20]. Cumulative Exergy Consumption (CExC) was introduced by Szargut et al. [21] to calculate the sum of all exergy input in all steps of a production process. Ayres with co-authors [22] stated the advantage of using exergy in the context of LCA, and concluded that exergy is appropriate for general statistical use, both as a measure of resource stocks and flows and as a measure of waste emissions and potential for causing environmental harm, which was also indicated by Wall in 1977 [3]. However, no detailed comparison has been made with existing methods, like the LCA. ELCA [19,23] introduced by Cornelissen is based on the framework of LCA with exergy applied to the inventory analysis and the impact assessment. Finnveden and Östlund [24] used exergy consumption as an indicator in LCA. Several metal ores and other natural resources were analyzed with system boundaries compatible with LCA. Since the cumulative exergy consumption index is just the sum of the chemical exergy contents of all original input flows, Valero [25] has introduced "exergetic cost for replacement of material" into the analysis. Lombardi (2001) performed an ELCA and a classical environmental LCA for a carbon dioxide low emission power cycle in which exergy was considered to be an indicator of resource depletion.

Life cycle exergy analysis (LCEA) includes sustainability aspects [20,26]. LCEA uses the same framework as LCA, but makes an important distinction between renewable and non-renewable resources. In LCEA, renewable resources as solar energy are excluded in the cost

*Corresponding author: Göran Wall, Oxbo gard, SE-43892 Härryda, Sweden
E-mail: gw@exergy.se

of calculation due to being free of charge, and/or otherwise wasted. LCEA has also been applied to industrial processes [27,28] and to wind power systems [29].

Thus, LCEA is a powerful tool in the design of sustainable systems, especially in the design of renewable energy systems. The application of LCEA to different solar energy systems offers an excellent visualization of the exergy flows involved during the life time of the system. The analyzed plants are net producers of exergy, since the exergy consumed can be paid back during their life time, however, in a varying degree. The exergy of material that is part of the system in various components during its operation will turn up as a product in the destruction phase and is depicted in the LCEA diagram. The recycling of this material will considerably reduce the payback time for future energy systems.

In this paper we present the first application of LCEA to solar systems together with a proposal of how to evaluate the recycled material in the LCEA method. The aim is to: (1) increase the awareness of and encourage the use of LCEA, (2) show the advantages of exergy instead of energy in systems analysis, (3) apply LCEA to solar energy systems, and (4) introduce a new way to evaluate recycled material in LCEA.

## Method

### Concept of Exergy

The exergy concept originates from works of Carnot [30], Gibbs [31], Rant [32] and Tribus [33] and the history is well documented [34]. Exergy of a system is [3,35]

$$E = U + P_0V - T_0S - \sum_i \mu_{i0}n_i \qquad (1)$$

where U, V, S, and $n_i$ denote extensive parameters of the system (energy, volume, entropy, and the number of moles of different chemical materials i) and $P_0$, $T_0$, and $\mu_{i0}$ are intensive parameters of the environment (pressure, temperature, and chemical potential). Analogously, the exergy of a flow can be written as:

$$E = H - T_0S - \sum_i \mu_{i0}n_i \qquad (2)$$

where H is the enthalpy.

The exergy of material substances can be calculated by [3]

$$E = \sum_i (\mu_i - \mu_0)n_i \qquad (3)$$

where $\mu_i$ is the generalized chemical potential of substance i in its present state.

The exergy of solar radiation is related to the exergy power per unit area of black body radiation e, which is [36]

$$e = u\left[1 + \frac{1}{3}\left(\frac{T_0}{T}\right)^4 - \frac{4}{3}\frac{T_0}{T}\right] \qquad (4)$$

Where u is energy power emission per unit area which can be calculated according to Stefan-Boltzmann law, T is taken to equal the solar radiation temperature 6000K.

All real processes involve the conversion and consumption of exergy, thus high efficiency is of utmost importance. This implies that the exergy use is well managed and that effective tools are applied. Presently, an excellent online web tool for calculating exergy of chemical substance is also available [36,37].

Energy is always in balance, however, for real processes exergy is never in balance due to irreversibilities, i.e. exergy destruction that is related to the entropy production by

$$E_{in}^{tot} - E_{out}^{tot} = T_0\Delta S^{tot} = \sum_j (E_{in} - E_{out})_j \rangle 0 \qquad (5)$$

where $\Delta S^{tot}$ is the total entropy increased,

$E_{in}^{tot}$ is the total input energy

$E_{out}^{tot}$ is the total output energy and $(E_{in} - E_{out})_j$ is the exergy destruction in process j.

The exergy loss, i.e. exergy destruction and exergy waste, indicates possible process improvements. In general "tackle the biggest loss first" approach is not always appropriate since every part of the system depends on each other, so that an improvement in one part may cause increased losses in other parts. As such, the total losses in the modified process may in fact be equal or even larger, than in the original process configuration. Also, the use of renewable and non-renewable resources, as well as recycled resources must be considered. Therefore, the problem needs a more complete and careful approach.

### Exergy factor and the reference state

Exergy factor is defined as the ratio of exergy to energy, and is sometimes referred to as quality factor, exergy coefficient and exergy quality factor.

The exergy factor for electricity and solar radiation is 1 and 0.93 respectively according to Eq. 4 with the temperature of the sun and the earth 6000K and 300K respectively, more detailed calculation can be found in [38].

When the heat capacity is independent of temperature and temperature decrease from T to $T_0$, the exergy factor for a heat flow can be calculated by [38]:

$$\frac{E}{Q} = \left|1 - \frac{T_0}{T - T_0}\ln\frac{T}{T_0}\right| \qquad (6)$$

The exergy factor depends on the environment or reference state. The exergy reference state is carefully analyses by Gaudreau [39], and Dincer and Rosen [40] offer a summary of models of reference-environment state. The reference-substance model [40] is applied in this study, and the local environment temperature is used as reference temperature.

### Exergy analysis

In engineering, Sankey diagrams are often used to describe the energy or exergy flows through a process. The energy/exergy efficiency is defined as the ratio of output energy/exergy to input energy/exergy of the systems.

Figure 1 shows a medium-temperature solar thermal power plant with solar collector, heat exchanger, turbine, condenser, regenerator and pump, its main components and roughly the main energy and exergy flows of the plant. This diagram shows where the main energy and exergy losses occur in the process, and also whether exergy is destroyed from irreversibility or whether it is emitted as waste, often waste heat, to the environment. In the energy flow diagram energy is always conserved, the waste heat carries the largest amount of energy into the environment, far more than is extracted as work in the turbine. However, in the exergy flow diagram the temperature of the waste heat is close to ambient so the exergy becomes much less than the energy.

In the solar collectors the energy efficiency is assumed to be about 55%. This depends on type of collector, average temperature difference between absorber and environment, and saturation temperature in the boiler [41]. The exergy efficiency of the concentrated medium-

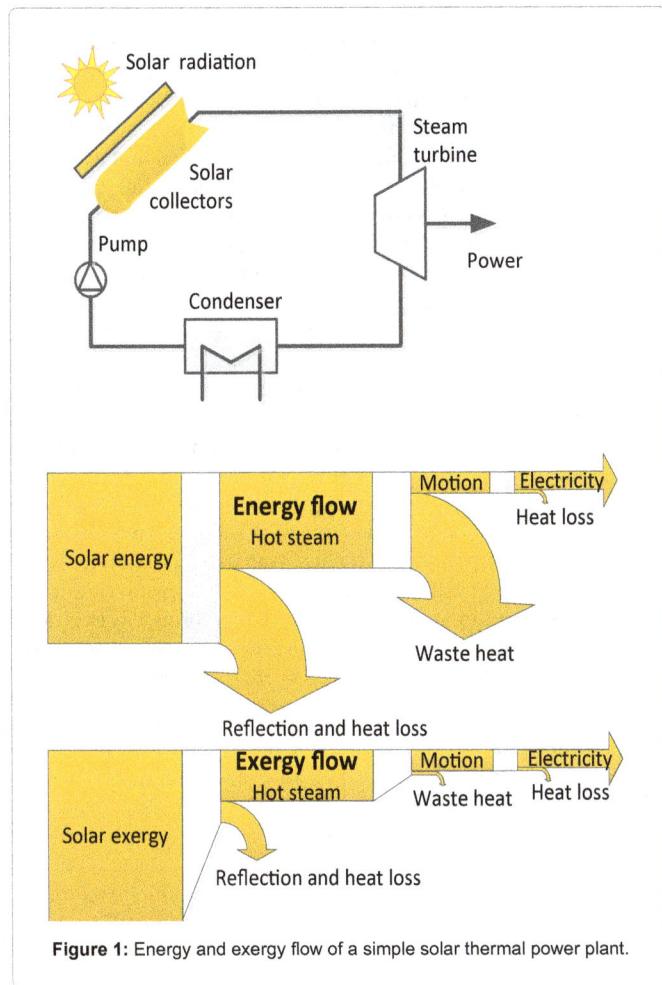

**Figure 1:** Energy and exergy flow of a simple solar thermal power plant.

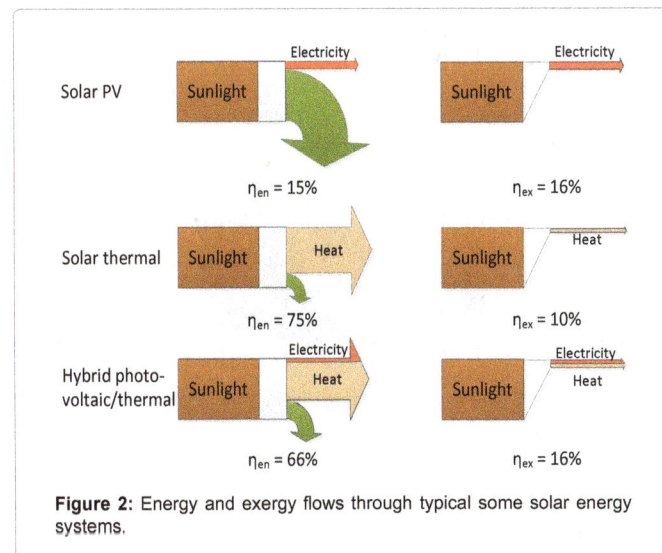

**Figure 2:** Energy and exergy flows through typical some solar energy systems.

temperature solar collectors is much lower or about 30% due to the low saturation temperature needed in order to drive a steam turbine. The exergy efficiency of solar collector, PV and hybrid solar collector are about 4%, 11% and 13% respectively [42].

Figure 2 illustrates the energy and exergy flows of solar PV cells,

solar thermal and hybrid PV/thermal as an example. The produced heat is used for hot water and/or space heating. In the solar PV systems, the energy and exergy efficiencies are almost the same or 15% and 16% since for solar radiation exergy is 93% of the energy [38] and for the outflow of electricity both energy and exergy is identical. In a PV cell solar radiation is directly transferred to electricity by means of photons of light exciting electrons into a higher energy state to act as carriers of an electric current. The low energy efficiency of a PV cell is partly due to physical limitations in the photo-electric conversion, and the energy losses are mainly due to this that instead becomes heat radiation to the environment. A solar thermal converter has an energy efficiency of about 75%, however the exergy efficiency is very low or 10% because the temperature of the heat is close to ambient and thus of low exergy. In the case of hybrid PV/thermal systems, the energy and exergy efficiency is about 66% and 16% respectively.

In Figure 1, a solar thermal power plant, the exergy efficiency is about the same as for a PV plant (Figure 2). This can be better understood from the exergy diagrams. The main exergy loss in the thermal power plant occurs in the conversion of solar radiation into high temperature heat for the turbine. The total exergy efficiency depends on the quality of heat, i.e. the temperature and pressure of the heat. The exergy efficiency would be higher in a concentrating power plant (CSP) due to higher temperatures and pressures.

### Energy/Exergy payback time

Energy/exergy pay-back time means time to recover primary energy/exergy use throughout its life cycle by the energy/exergy of the product, which is calculated as ratio of total energy/exergy input to the energy/exergy of the annual production.

Stepanov [43] compiles some of the different methodologies proposed for analyzing solar energy systems. The models developed by Valero and Lozano [44] for obtaining the chemical exergy of fossil fuels are applied here. However it must be pointed out that more complex calculation procedures do not necessarily mean more reliable results. Both, the experimental error associated to the determination of the heating values and the error associated to the correlations are in an interval close to ±2% [45]. Additionally, the chemical exergy of fuels is approximated to the higher heating value (HHV). The HHV assumes that water is in liquid state after combustion.

In this paper all primary thermal energy inputs are converted into primary electrical energy, with an assumed efficiency of 35%, i.e. 1 MJ (1/3.6 kWh), primary thermal energy becomes 0.097 kWh primary electrical energy. Thus, primary energy and primary exergy value becomes the same. Since the efficiency from primary thermal to primary electric energy varies from country to country local conditions are recommended.

## Life Cycle Exergy Analysis (LCEA)

### Life Cycle Analysis (LCA)

To estimate the total exergy used in a process, it is necessary to take all the different inflows of exergy to the process into account. This type of budgeting is often termed Exergy Analysis [3,35], Exergy Process Analysis, see Figure 3, or Cumulative Exergy Consumption [46], and focuses on a particular process or sequence of processes for making a specific final commodity or service. It evaluates the total exergy use by summing the contributions from all the individual inputs, in a more or less detailed description of the production chain.

Life Cycle Analysis or Assessment (LCA) is common to analyze

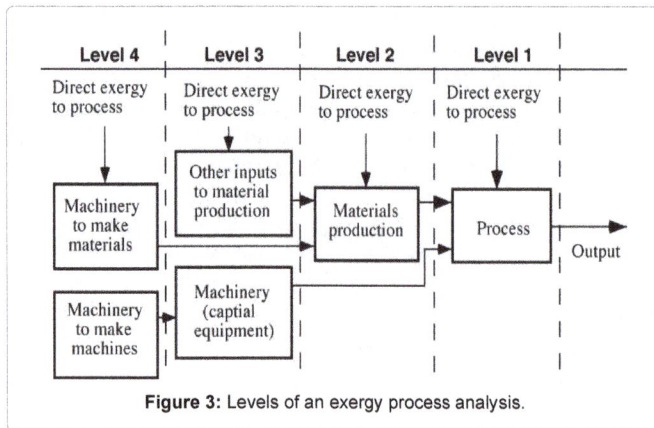

**Figure 3:** Levels of an exergy process analysis.

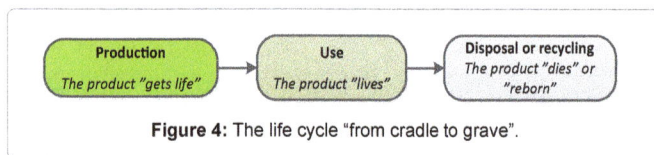

**Figure 4:** The life cycle "from cradle to grave".

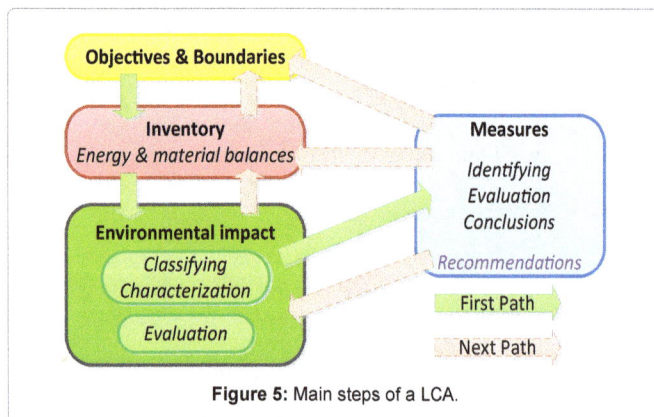

**Figure 5:** Main steps of a LCA.

environmental impacts associated with three "life processes": production, use and disposal or recycling of products or product systems, or as it is sometimes named "from cradle to grave", see Figure 4.

For every "life process" the total inflow and outflow of energy and material is calculated, thus, LCA is similar to Exergy Analysis. In general Exergy Analysis and LCA have been developed separately even though they are very similar. In LCA the environmental burdens are associated with a product, process, or activity by identifying and quantifying energy and materials used, and wastes released to the environment. This inventory of energy and material balances is then put into a framework in order to assess the impact on the environment, Figure 5. Four parts in the LCA can be distinguished: (1) objectives and boundaries, (2) inventory, (3) environmental impact, and (4) measures. These four main parts of an LCA are indicated by boxes, and the procedure is shown by arrows. Green arrows show the initial path and red dashed arrows indicate suitable next paths, in order to further improve the analysis.

## Life Cycle Exergy Analysis (LCEA)

The multidimensional approach of LCA causes large problems when it comes to comparing different substances, and general agreements are crucial. This problem is avoided if exergy is used as a common quantity, which is done in a Life Cycle Exergy Analysis (LCEA) [26].

In this method we distinguish between renewable and non-renewable resources. The total exergy use over time is also considered. These kinds of analyses are of importance in order to develop sustainable supply systems of exergy in society. The exergy flow through a supply system over time, such as a power plant, usually consists of three separate stages (Figure 6). At first, during the construction stage ($0 \le t \le t_{start}$) exergy is used to build a plant and put it into operation. The exergy is spent, of which some is accumulated or stored in materials, e.g. in metals, as well as exergy used for transportation etcetera. Secondly the system need to be maintained during time of operation ($t_{start} \le t \le t_{stop}$), and finally the cleaning up and disposal stage during destruction stage ($t_{stop} \le t \le t_{life}$). Eventually, some material, i.e. stored exergy, can be recycled. These time periods are analogous to the three steps of the life cycle of a product in an LCA. The exergy input used for construction, maintenance and destruction are called indirect exergy Eindirect and it is assumed that this originates from non-renewable resources. By using recycled material in the production stage, the indirect exergy may be considerably reduced. If exergy is recovered by recycling in the destruction stage, this is accounted for as an additional product of the system, $E_{rec}$. When a power plant is put into operation, it starts to deliver a product, e.g. electricity with exergy power $E_{pr}$, by converting the direct exergy power input into demanded energy forms, e.g. electricity. In Figure 6, the direct exergy is a non-renewable resource, e.g. fossil fuel and in Figure 7 the direct exergy is a renewable resource, e.g. solar radiation.

In the first case as shown in Figure 6, the system is not sustainable, since the system use exergy originating from a non-sustainable resource and it will never reach a situation where the total exergy input will be paid back, simply because the situation is powered by a depletion of resources, i.e. $E_{pr}+E_{rec}<E_{in}+E_{indirect}$. In the second case, as shown in Figure 7, at time t= $t_{payback}$ the produced exergy that originates from a natural flow has compensated for the indirect exergy input, i.e.

$$\int_{t_{start}}^{t_{payback}} E_{pr}t(dt) + E_{rec} = \int_{0}^{t_{life}} E_{indirect}(t)dt = E_{indirect}$$

Since the exergy input originates from a renewable resource, we

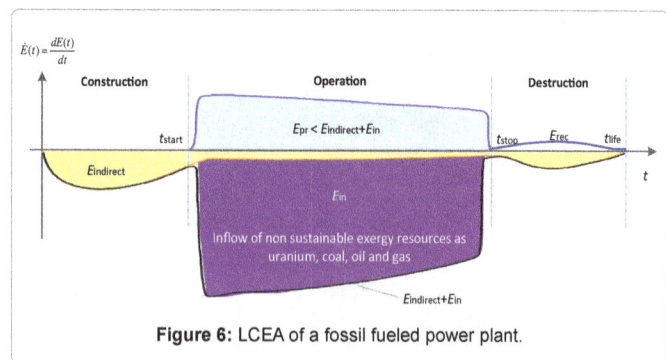

**Figure 6:** LCEA of a fossil fueled power plant.

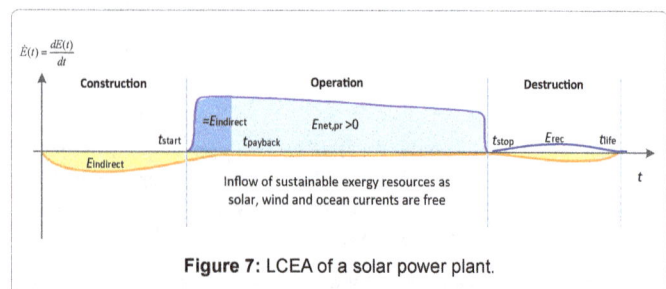

**Figure 7:** LCEA of a solar power plant.

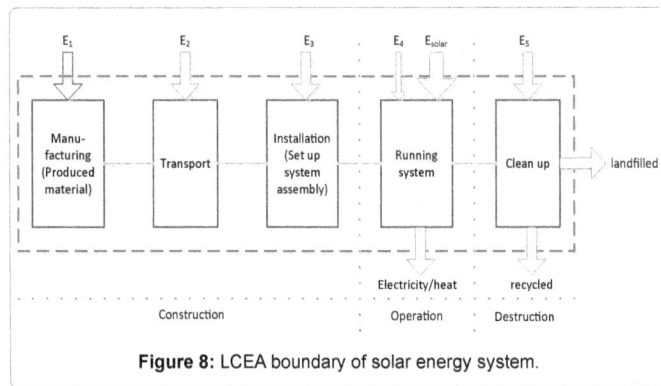

**Figure 8:** LCEA boundary of solar energy system.

do not account for it. By regarding renewable resources as free, then after $t = t_{payback}$ there will be a net exergy output from the plant, which will continue until it is closed down, at $t = t_{stop}$. Then, exergy has to be used to clean up and restore the environment, which accounts for the last part of the indirect exergy input Eindirect. Exergy in recycled materials Erec now turns up as an additional product of the system. By considering the total life cycle of the plant the net produced exergy becomes $E_{net,pr} = E_{pr} - E_{indirect} + E_{rec}$. These areas representing exergies are indicated in Figure 7. Assume that, at time t=0, the building of a solar PV/power plant starts and at time t = tstart the plant is completed and put into operation. At that time, a large amount of exergy has been used in the construction of the plant, which is indicated by the area of Eindirect between t = 0 and t = $t_{start}$. Then the plant starts to produce electricity, which is indicated in Figure 7 by the upper curve Epr. At t = $t_{payback}$ the exergy used for construction, maintenance and destruction has been paid back. The payback time will be further reduced if exergy is recycled from the destruction. For solar PV/power plants this time is only some months. Then the system has a net output of exergy until it is closed down, which for a solar energy system usually last for 20-25 years. Thus, LCEA diagrams could be used to show if a power supply system is sustainable.

LCEA is very important in the design of sustainable systems, especially in the design of renewable energy systems. Take a solar panel, made of mainly aluminum and glass that is used for the production of hot water for household use, i.e. about 50°C. Then, it is not obvious that the exergy being spent in the production of this unit ever will be paid back during its use, i.e., it might be a misuse of resources rather than a sustainable resource use. The production of silicon, aluminum and glass require a lot of exergy as electricity and high temperature heat or several hundred degrees Celsius, whereas the solar panel delivers small amounts of exergy as low temperature heat. LCEA must therefore be carried out as a natural part of the design of sustainable systems in order to avoid this kind of misuse. Another case to investigate is the production of biofuels in order to replace fossil fuels in the transport sector. This may not necessarily be sustainable since the production process uses a large amount of fossil fuels, directly for machinery or indirectly as fertilizers, irrigation and pesticides. This would be well described by a LCEA.

In order to be sustainable, energy supply systems must be based on a use of renewable resources in such a way that the input of non-renewable resources will be paid back during its life time, i.e. $E_{pr} > E_{in} + E_{indirect} - E_{rec}$. In order to be truly sustainable, the used deposits must also be completely restored or, even better, not used at all. Thus, by using LCEA and distinguishing between renewable and non-renewable resources we have an operational method to estimate the sustainability of energy systems.

LCEA diagrams are of particular importance in the planning of large scale renewable energy systems of multiple plants. Initially, this system will consume most of its supply within its own constructions phase. However, sometime after completion it will deliver at full capacity. Thus, the energy supply over time is heavily affected by internal system dynamics.

## LCEA of solar systems

### LCEA of a solar energy system

Figure 8 indicates a LCEA of a solar energy system where the red dashed box indicates the system boundary and blue dotted lines indicate the three steps of a life cycle. Construction includes manufacturing, transport and installation in order to set up the system.

The indirect exergy $E_1$ can be exergy of used electricity, fuels and material from natural resources. The produced materials for solar systems contain the PV module, metal, and electrical equipment. For large systems there are also electrical substations, fence and land. Fabrication of PV modules includes silicon production, PV cell manufacturing and supporting structures. Electrical equipment has inverters, transformers, cables and low and medium voltage switchboards, charge regulations (control panel) and bank of batteries (only for standard-alone-system). In the case of solar thermal system the produced materials are solar collector, storage tank, pipes and so on. The electricity used for manufacturing material during indirect use can be from both non-renewable and renewable energy. In this study electricity is, for practical reasons, assumed to be only from non-renewable resources.

The indirect exergy $E_2$ is exergy used for transportation of material from the manufactures to the installation site, and $E_3$ is exergy consumption during installation.

During operation phase the indirect exergy $E_4$ can be material used for maintenance. The direct exergy from solar is not accounted for during LCEA analyses since it is renewable energy and would other vice most probably be wasted. The product is electricity and/or heat. Part of the electricity production may be used for the control system.

The destruction phase restores the process to its original state. The indirect exergy $E_5$ is the exergy consumption for cleaning up; land filled and recycled material, such as aluminum and the PV module. Land filled and recycling depends on technology applied.

The total indirect exergy is $E_1+E_2+E_3+E_4+E_5$ and recycled exergy is an additional product of the system from its destruction.

### LCEA applied to PV solar energy systems

There are numerous publications on LCA of different solar energy systems [11-18]. The energy use for construction, operation and destruction are often presented. However, some do not consider recycling at the destruction phase. The use of recycled material will often reduce the amount of energy needed from using fresh resources. Two studies use recycled material in the construction [11,15].

Solar exergy is often used to produce heat and/or electricity. The amount of solar exergy captured and converted by solar collectors and/or solar PV cells depend on the location, type of solar PV cell and collector, and working conditions. For a solar thermal plant, the produced heat often has low temperature which means low exergy values. For a solar thermal power plant, e.g. concentrate solar power (CSP) plant, the heat usually have high temperature, e.g. 500-1000°C for solar power tower, as steam with high exergy to drive a turbine and

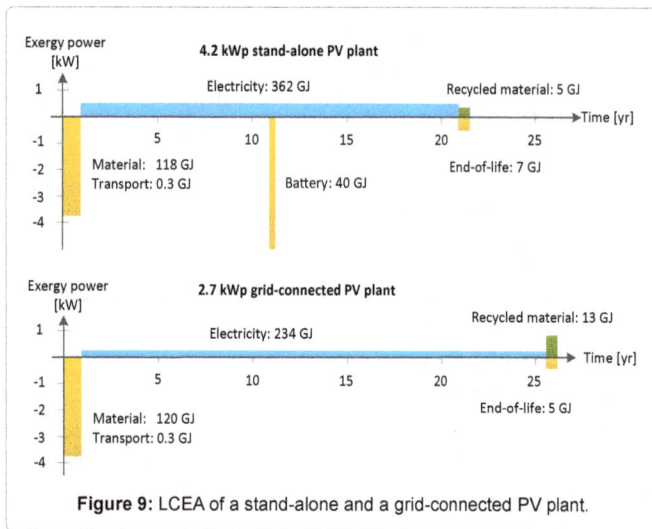

**Figure 9:** LCEA of a stand-alone and a grid-connected PV plant.

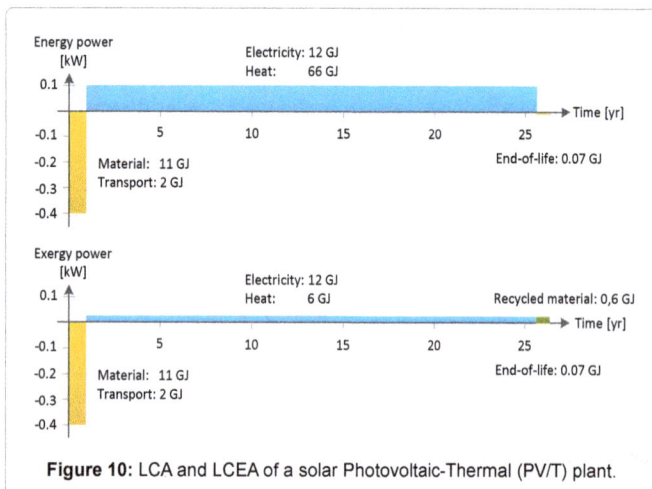

**Figure 10:** LCA and LCEA of a solar Photovoltaic-Thermal (PV/T) plant.

generator to produce electricity. The efficiency of PV cells is usually less than 20%, and the rest is mostly converted into waste heat to the environment. It is also very sensible to the working temperature, e.g., the efficiency of crystalline silicon (c-Si) cells typically decreases 0.4% when temperature rise 1 degree [47]. A PV/solar thermal (PV/T) system directly converts solar exergy to electricity and thermal exergy. In such systems the PV cells also work as thermal collectors in which the cell temperature can be controlled in order to prevent the decrease of efficiency.

Garcia-Valverde et al. [11] analyzed a 4.2 kWp stand–alone photovoltaic system in Spain. The system consists of 40 mono-crystalline modules (24V, 106Wp) mounted on a building rooftop. In the construction phase the exergy requirements for production is divided into two parts: manufacturing with and without recycling. Recycled materials use less exergy in its production. It is assumed that 35% of aluminum, 50% of the lead-acid batteries, 90% of the steel, and 43% of the copper came from recycled materials in the original. The highest exergy requirements in the production relates to PV modules and lead-acid batteries. In the operation phase the only exergy needed is to replace the lead-acid batteries after 10 years. The electricity for regulators and inverters come from the system itself. In the destruction phase exergy is required in recycling and transport to the recycling plant, and then PV modules will become landfill.

Kannan et al. [12] studied a 2.7 kWp grid-connected mono-crystalline solar PV system in Singapore. The system consists of 36 mono-crystalline modules (12V, 75Wp) mounted on a building rooftop. Most primary exergy was used in the production of PV modules which is about 81.4% of the life cycle primary exergy use. In the operation phase, no external exergy is used since produced electricity is transferred to the electric grid; however, this implies special conditions and consequences. In the destruction phase, it is assumed that the solar PV module would be landfill and 90% of aluminum frames are recycled with 90% recovery rate.

LCEAs of these two PV plants are depicted in Figure 9. By use of some recycled material during construction phase, the indirect exergy is about 118 GJ [11] for the stand-alone plant and if we assume no use of recycled material this would instead become about 130 GJ. For the grid-connected plant that does not make use of recycled material in the construction about 120 GJ is used. In addition about 40 GJ are used to replace batteries after 10 years in the stand-alone plant. The exergy of the metal will show up as a product as we have indicated in the diagrams, 5 GJ for the stand-alone and 13 GJ for the grid-connected plant. This material and exergy is available for recycling. In the grid-connected plant the authors assumed that 90% of the metal is able to recycle [12]. Probably, in the future more material will be recycled.

The exergy pay-back time becomes about 7 years for the stand-alone plant and about 9 years for the grid-connected plant. With improved PV module production technology and increased use of recycled material the pay-back time could be further reduced. These results relate to a thermoelectric conversion efficiency of 35% as indicated in Section 2.4 above. However, if the thermoelectric conversion efficiency varies from 30 to 40% the energy and exergy payback time will vary from 6 to 11 years and 8 to about 11 years, respectively.

### Compare between LCA and LCEA with a PV/T system

A photovoltaic/thermal (PV/T) system in Hong Kong is investigated by Chow and Jie [18]. PV/T is a combination of photovoltaic and solar thermal system that produces both electricity and heat. In the construction phase, energy is used for the collector panels and PV module. For the solar thermal system, a water tank is needed. The yearly production thermal energy and electricity energy is 2650 MJ and 473 MJ respectively. The year average temperature in Hong Kong is 23°C, the temperature from solar collector is 85°C. Thus, the yearly thermal exergy production becomes only 244 MJ due to the relatively low temperature of the heat. Figure 10 shows an energy based LCA, the upper energy diagram, and an LCEA, the lower diagram, of this plant. In both cases the input of solar energy is excluded. Clearly, the LCA and LCEA show a large difference during operation phrase. The output of electricity and heat amounts to 78 GJ in the energy case but only 18 GJ in the exergy case, due to the low exergy value of the heat. In addition, the exergy of the metal of the equipment used turn up as a product in the destruction phase, i.e. about 0.6 GJ is available for recycling. The total input of non-renewable resources amounts to 12.8 GJ in both cases. The total energy output becomes more than 6 times the input. However, this is misleading since the value of heat does not reflect its true physical value. Instead, from the LCEA the output and input is more or less the same. If materials used at construction stage came from non-renewable mostly, this implies an inefficient energy usage. Such system can hardly be regarded sustainable. From a pure resource conservation perspective, it may be better to use the input of non-renewable resources directly for other purposes with less conversion energy loss instead. Considering that energy is also used for transport and destruction or end-of-life, the energy and exergy payback time is 3.5 and 15.4 years respectively.

Thus, the exergy payback time is more than four times longer than the energy payback time since the production consist of both electricity and thermal energy. Thus, the LCEA offers a better tool than LCA in the evaluation of energy systems for a sustainable future.

## Conclusion

In a solar thermal plant the energy payback time is much shorter than the exergy payback time since the exergy of the output of heat is much lower than its energy value. This may lead to false assumptions in the evaluation of the sustainability of renewable energy systems. From Figures 6 and 7 we see the advantage of LCEA when applied to systems based on non-renewable and/or renewable energy resources.

Solar energy systems producing electricity have less exergy back time than system for heat production. The exergy payback time of PV/T system is much longer than the energy payback time or about 4 times longer. This indicates that LCEA gives a completely different view of these systems that is of essential importance in scientific evaluations.

Among the three solar thermal power plants the PV/T plant has the shortest energy payback time or 3.5 years. However, by applying exergy the 4.2 kWp stand-alone PV plant has the shortest payback time or 7 years, since the product in this case has a higher exergy value, i.e., more of electricity. This system is also the larger of the two pure PV systems.

LCEA is shown to be advantages in the study of solar based energy systems and is recommended as a suitable tool for the design and evaluation of renewable energy systems.

## Nomenclature

E: Exergy, J

e: Exergy power per unit area, W m$^{-2}$

E*: Exergy power, W

H: Enthalpy, J

n: The number of moles of substance, mol

P: Pressure, Pa

Q: Heat or thermal energy, J

S: Entropy, J K-1

t: Time, s

T: Temperature, K

u: Energy power per unit area, W m-2

U: Internal energy, J

V: Volume, m3

$\mu$: Chemical potential, J mol-1

$\eta$: Efficiency

### Superscripts

$^{tot}$ :Total system, i.e. the system and the environment

Subscripts

$_0$ : Reference or state time

$_{en}$ : Energy

$_{ex}$ : Exergy

$_i$ : Unit for different chemical materials

$_{in}$: Input

$_{indirect}$: Indirect input

$_j$ : Unit for different process

$_{life}$: When a process or product exist

$_{net,pr}$ : Net of product

$_{out}$ : Output

$_{payback}$ : All input is paid back

$_{pr}$: Product

$_{rec}$: Recycled material

$_{start}$: Operation starts

$_{stop}$ : Operation stops

$_{tot}$ :Total

$_{waste}$ : Not used products

## References

1. Laughton C (2010) Solar domestic water heating. London, England: Earthscan.

2. Torío H, Schmidt D (2010) Framework for analysis of solar energy systems in the built environment from an exergy perspective. Renew Energ 35: 2689-2697.

3. Wall G (1977) Exergy — a Useful Concept within Resource Accounting. Chalmers University of Technology: Institute of Theoretical Physics.

4. Hepbasli A (2008) A key review on exergetic analysis and assessment of renewable energy resources for a sustainable future. Renew Sust Energ Rev 12: 593-661.

5. Oztop HF, Bayrak F, Hepbasli A (2013) Energetic and exergetic aspects of solar air heating (solar collector) systems. Renew Sust Energ Rev 21: 59-83.

6. Alta D, Bilgili E, Ertekin C, Yaldiz O (2010) Experimental investigation of three different solar air heaters: Energy and exergy analyses. Appl Energ 87: 2953-2973.

7. Xu C, Wang Z, Li X, Sun F (2011) Energy and exergy analysis of solar power tower plants. Appl Therm Eng 31: 3904-3913.

8. Akyuz E, Coskun C, Oktay Z, Dincer I (2012) A novel approach for estimation of photovoltaic exergy efficiency. Energ 44:1059-1066.

9. Joshi AS, Tiwari A (2007) Energy and exergy efficiencies of a hybrid photovoltaic–thermal (PV/T) air collector. Renew Energ 32: 2223-2241.

10. Sarhaddi F, Farahat S, Ajam H, Behzadmehr A (2010) Exergetic performance assessment of a solar photovoltaic thermal (PV/T) air collector. Energ Build 42: 2184-2199.

11. García-Valverde R, Miguel C, Martínez-Béjar R, Urbina A (2009) Life cycle assessment study of a 4.2kWp stand-alone photovoltaic system. Solar Energ 83:1434-1445.

12. Kannan R, Leong KC, Osman R, Ho HK, Tso CP (2006) Life cycle assessment study of solar PV systems: An example of a 2.7kWp distributed solar PV system in Singapore. Solar Energ 80: 555-563.

13. Desideri U, Proietti S, Zepparelli F, Sdringola P, Bini S (2012) Life Cycle Assessment of a ground-mounted 1778kWp photovoltaic plant and comparison with traditional energy production systems. Appl Energ 97: 930-943.

14. Stoppato Av (2008) Life cycle assessment of photovoltaic electricity generation. Energy 33: 224-232.

15. Cavallaro F, Ciraolo L (2006) A life cycle assessment (LCA) of a paraboloidal-dish solar thermal power generation system. 2006 First International Symposium on Environment Identities and Mediterranean Area 2:127-132.

16. Kalogirou S (2009) Thermal performance, economic and environmental life cycle analysis of thermosiphon solar water heaters. Solar Energ 83: 39-48.

17. Cellura M, Grippaldi V, Brano VL, Longo S, Mistretta M (2011) Life cycle assessment of a solar PV/T concentrator system. Life cycle managment conference (LCM), Berlin.

18. Chow TT, Ji J (2012) Environmental Life-Cycle Analysis of Hybrid Solar Photovoltaic/Thermal Systems for Use in Hong Kong. Int J Photoenerg:1-9.

19. Cornelissen RL (1997) Thermodynamics and Sustainable Development – The use of exergy analysis and the reduction of irreversibility. Enschede, The Netherlands: University of Twente.

20. Gong M, Wall G (1997) On exergetics, economics and optimization of technical processes to meet environmental conditions. Conference On exergetics, economics and optimization of technical processes to meet environmental conditions, Beijing, China.

21. Szargut J, Morris DR, Steward FR (1988) Exergy analysis of thermal, chemical, and metallurgical processes. Hemisphere publishing corporation, USA.

22. Ayres RU, Ayres LW, Martinás K (1998) Exergy, waste accounting, and life-cycle analysis. Energ 23: 355-363.

23. Cornelissen RL, Hirs GG (2002) The value of the exergetic life cycle assessment besides the LCA. Energ Convers Manag 43:1417-1424.

24. Finnveden G, Östlund P (1997) Exergies of natural resources in life-cycle assessment and other applications. Energ 22: 923-931.

25. Valero A (1986) Thermoeconomics as a conceptual basis for energy-ecological analysis. Conference Thermoeconomics as a conceptual basis for energy-ecological analysis, Porto Venere, Italy: 415-444.

26. Gong M, Wall G (2001) On exergy and sustainable development—Part 2: Indicators and methods. Exergy 1: 217-233.

27. Gong M (2004) Using exergy and optimization models to improve industrial energy systems towards sustainability. Linköping University, Linköping, Sweden.

28. Gong M (2005) Exergy analysis of a pulp and paper mill. Int J Energ Res 29: 79-93.

29. Wall G (2011) Life Cycle Exergy Analysis of Renewable Energy Systems. The Open Renew J 4:72-77.

30. Carnot S (1824) Réflections sur la puisance motrice du feu et sur les machines propres a développer cette puissance. R.Fox, Bachelier, Paris, French.

31. Gibbs JW (1873) A Method of Geometrical Representation of the Thermodynamic Properties of Substances by Means of Surfaces. Trans Conn Acad 2: 382-404.

32. Rant Z (1956) ein neues Wort für 'technische Arbeitsfähigkeit (Exergy, a New Word for Technical Available Work). Forschungen im Ingenieurwesen 22: 36-37.

33. Tribus M (1961) Thermostatics and Thermodynamics. Van Nostrand, New York, USA.

34. Sciubba E, Wall G (2007) A brief commented history of exergy from the beginnings to 2004. Int J Thermodynamics. 10:1-26.

35. Wall G (1964) Exergy — a Useful Concept. Chalmers University of Technology, Göteborg, Sweden.

36. Petela R (1964) Exergy of heat radiation. J Heat Trans 86:187-192.

37. The Exergoecological Portal.

38. Wall G, Gong M (2001) On exergy and sustainable development—Part 1: Conditions and concepts. Exerg 1:128-145.

39. Gaudreau K (2009) Exergy Analysis and Resource Accounting . University of Waterloo, Waterloo, Canada.

40. Dincer I, Rosen MA (2013) Exergy, Energy, Environment and Sustainable Development. (2nd edn) Elsevier , Oxford, UK.

41. You Y, Hu EJ (2002) A medium-temperature solar thermal power system and its efficiency optimisation. Appl Therm Eng 22:357-364.

42. Saitoh H, Hamada Y, Kubota H, Nakamura M, Ochifuji K, et al. (2003) Field experiments and analyses on a hybrid solar collector. Appl Therm Eng 23: 2089-2105.

43. Stepanov VS (1995) Chemical energies and exergies of fuels. Energ 20: 235-242.

44. Valero A, Lozano MA (1994) Curso de Termoeconom´ıa. Universidad de Zaragoza.

45. Valero A, Valero A (2011) The actual exergy of fossil fuel reserves. Conference The actual exergy of fossil fuel reserves, Novi Sad, Serbia: 931-938.

46. Szargut J, Morris D (1987) Cumulative exergy consumption. Int J Energ Res 11: 245-261.

47. Fraisse G, Ménézo C, Johannes K (2007) Energy performance of water hybrid PV/T collectors applied to combisystems of Direct Solar Floor type. Solar Energ 8:1426-1438.

# 7

# Fuzzy Logic Control of a SEPIC Converter for a Photovoltaic System

**Meryem Oudda\* and Abdeldjebar Hazzab**

*Department of Electric and Electronics Engineering, Tahri Mohamed Bechar University, Algeria*

### Abstract

In this work, a fuzzy logic controller is used to control the output voltage of a photovoltaic system with a DC-DC converter; type Single Ended Primary Inductor Converter (SEPIC). The system is designed for 210 W solar PV (SCHOTT 210) panel and to feed an average demand of 78 W (24V). This system includes solar panels, SEPIC converter and fuzzy logic controller. The SEPIC converter provides a constant DC bus voltage and its duty cycle controlled by the fuzzy logic controller which is needed to improve PV panel's utilization efficiency. A fuzzy logic controller (FLC) is also used to generate the PWM signal for the SEPIC converter.

**Keywords:** Photovoltaic system; DC-DC converter; Duty cycle

**Abbreviations**: PV: Photovoltaic; SEPIC: Single Ended Primary Inductor Converter; FLC: Fuzzy Logic Controller

## Introduction

The non-renewable source of energy is depleting rapidly and the demand for power is increasing day by day. To overcome this problem, generation of electric power from renewable source of energy should be made effective and efficient [1].

The energy source which the society can depend on is renewable energy since it is clean, pollution free, and endless. Photovoltaic (PV) system is one of power generations that utilize renewable energy [2]. To reduce consumption of conventional energy, then the PV system must be connected to grid, either directly or through back-up battery bank. However, the PV system has low efficiency because of the power generated from PV systems depends on the irradiation and temperature variation [2].

For the control of the PV systems, there are various types of DC-DC converters such as, Buck converter, Boost converter and Buck-Boost converter. The output of buck converter is less than the input voltage whereas the boost converter output is greater than the input voltage. The polarity of buck-boost converter is inversed of input signal. Yusivar et al. have been proposed Buck-Converter Photovoltaic Simulator [3]. Whereas Single Ended Primary Inductor Converter (SEPIC) is a special type of DC-DC converter which maintains a constant output voltage even under varying input conditions and load conditions [4].

From the literature survey it can be understood that SEPIC is widely used converter topology in renewable source based energy generation. Venkatanarayanan et al. are presented photovoltaic energy system with SEPIC converter [5]. The SEPIC converter is proposed also by El Khateb et al. [6]. Another recent search [7] a SEPIC converter for a standalone PV system is chosen. SEPIC converter also overcomes the drawback of buck-boost converter. The performance of SEPIC converter can further be improved by using a suitable control scheme [2].

The control of this SEPIC converter is a much discussed and invested very subject. Indeed, this converter is nonlinear in nature and different approaches have been used to control it. Conventional control modes such as voltage mode control and current mode control require a good knowledge of the converter and therefore a fairly accurate model [1]. Proportional-integral control for SEPIC converter is presented by Venkatanarayanan et al. [8], Current mode control and PI controller have been proposed by Reddy et al. [9].

These controllers are easy to implement and simple to design, but their performance generally depends on the operating point so that too large disturbance, wide load variation ranges or supply voltage variations can make the choice of the parameters very difficult for different operating conditions.

However, a very different approach is offered by the fuzzy logic control (FLC), which does not require precise mathematical model or complex calculation [10]. The fuzzy control technique is based primarily on human understanding of the process control and on qualitative rules. The objective of this research is to develop a fuzzy voltage regulator for a SEPIC converter.

This paper is organized as follows. In section 2, the photovoltaic array model is presented. Section 3 presents the SEPIC converter, while the design of the fuzzy logic controller for the SEPIC converter has been done in section 4. Simulation results are shown in section 5. Finally conclusion is given in section 6.

## Photovoltaic Array Model

Photovoltaic is the field of technology and research related to the devices which directly convert sunlight into electricity using semiconductors that exhibit the photovoltaic effect.

The photovoltaic panel is composed of many cells, placed in series $N_s$ or in shunt $N_{sh}$. Where it can be modelled by current source connected in parallel with diode according with shunt and series resistor noted by $R_{sh}$ and $R_s$ as illustrated in Figure 1 [11].

The output current is given by the following equations:

$$I = I_{ph} - I_D \qquad (1)$$

---

**\*Corresponding author:** Meryem Oudda, Department of Electric and Electronics Engineering, Tahri Mohamed Bechar University, Bechar, 08000, Algeria, E-mail: ouddameryem@gmail.com

$$I = I_{ph} - I_0 \left[ \exp\left( \frac{q(V+R_sI)}{AK_BT} \right) - 1 \right] - \frac{V+R_sI}{R_{sh}} \qquad (2)$$

$I_{ph}$: Photo-current; A: Ideality factor; $K_B$: Boltzmann's constant; T: Cell temperature; $I_D$: Diode current; $R_{sh}$ Series resistance; $I_0$: Saturation current; q: Electronic charge; $R_{sh}$: Shunt resistance; V: Cell voltage; I: Cell current

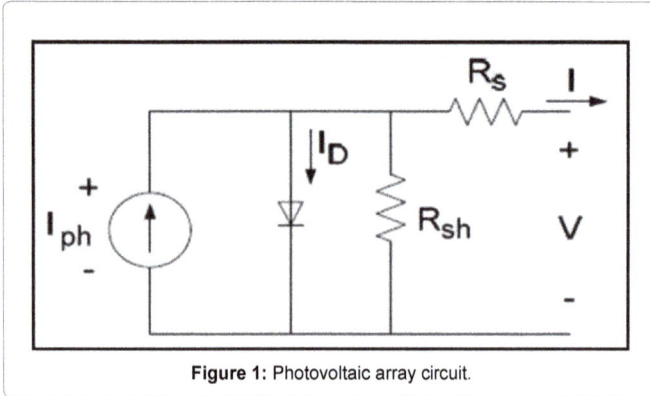

**Figure 1:** Photovoltaic array circuit.

Typically, the shunt resistance ($R_{sh}$) is very large and the series resistance ($R_s$) is very small [12]. Therefore, it is common to neglect these resistances in order to simplify the solar cell model. The resultant ideal voltage-current characteristic of a photovoltaic cell is given by equation 3.

$$I = I_{ph} - I_0 \left[ e^{\left( \frac{qV}{KT} \right)} - 1 \right] \qquad (3)$$

Figures 2 and 3 show the behavior of a photovoltaic module simulation used in this study, in accordance to solar radiation variation and at a constant temperature.

The typical output power characteristics of a PV array under various degrees of irradiation is illustrated by Figure 2. It can be observed in Figure 3 that there is a particular optimal voltage for each irradiation level that corresponds to maximum output power. Therefore by adjusting the output current (or voltage) of the PV array, maximum power from the array can be drawn.

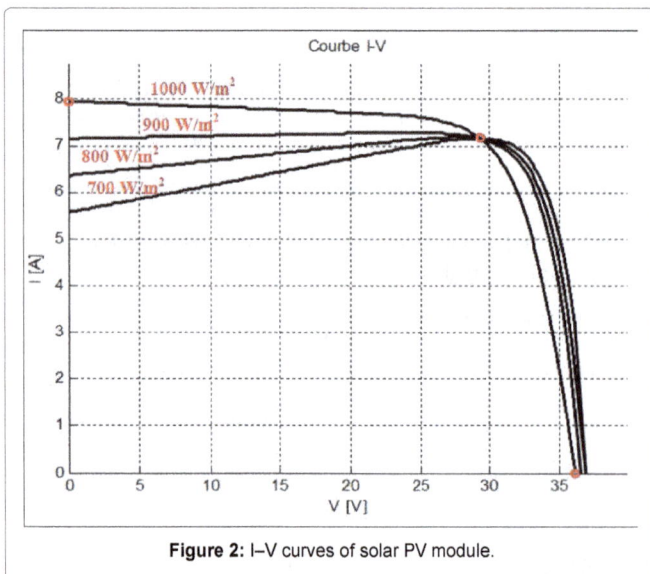

**Figure 2:** I–V curves of solar PV module.

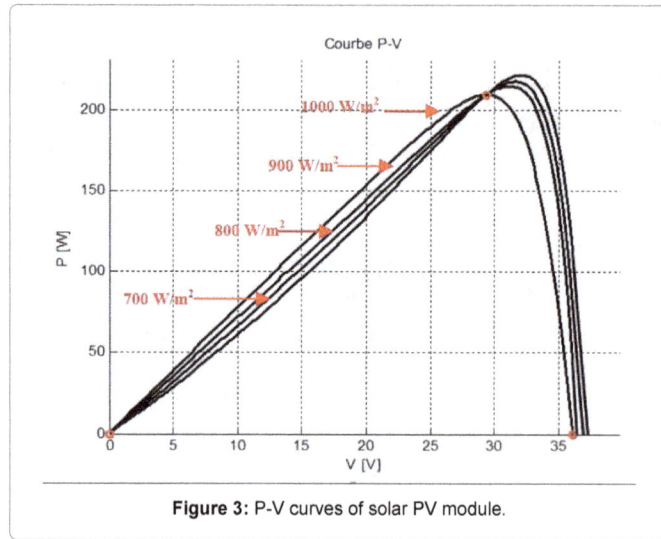

**Figure 3:** P-V curves of solar PV module.

As we can see in the curve of the Figure 2, the current increase is highly affected by the solar radiation.

## DC-DC/SEPIC Converter

A DC–DC converter with simpler structure and higher efficiency has been an active research topic in the power electronics [13]. The proposed converter is based on DC to DC converter to maintain the constant output voltage [1].

Single Ended Primary Inductor Converter (SEPIC) converter consists of a switch S with duty cycle α, a diode, two inductors $L_1$ and $L_2$, two capacitors $C_1$ and $C_2$ and a load Resistor. The circuit diagram of a SEPIC converter is shown in Figure 4.

A SEPIC is a type of DC-DC converter allowing the electrical potential (voltage) at its output to be less than, greater than, or equal to that at its input; the output of the SEPIC is controlled by the duty cycle of the control transistor. SEPIC is effectively a boost converter followed by a buck-boost converter, consequently it is like to a conventional buck-boost converter, other than has advantages of having non-inverted output (the output has the same voltage polarity as the input), passing through a series capacitor to couple energy from the input to the output (and thus can respond more gracefully to a short-circuit output), and being able of factual shutdown: after the switch "S" is turned off, its output drops to 0V, following a rather hefty transient abandon of charge [14].

Figure 4 shows a simple circuit diagram of a SEPIC converter can both step up and step down the input voltage, while maintaining the same polarity and the same ground reference for the input and output.

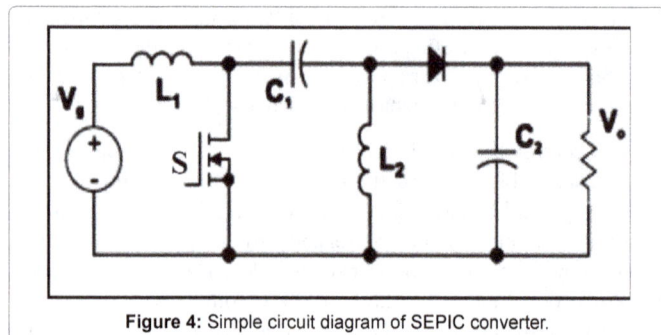

**Figure 4:** Simple circuit diagram of SEPIC converter.

Figure 5 shows the circuit when the power switch is turned on and off (respectively in Figure 5A and Figure 5B). When the switch is turned on; the first inductor; $L_1$ is charged from the input voltage source during this time. The second inductor $L_2$ takes energy from the first capacitor $C_1$, and the output capacitor $C_2$ is left to provide the load current.

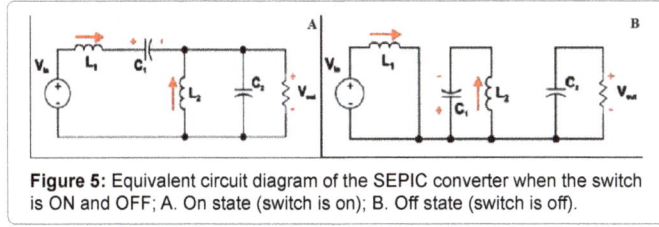

**Figure 5:** Equivalent circuit diagram of the SEPIC converter when the switch is ON and OFF; A. On state (switch is on); B. Off state (switch is off).

When the switch is turned on, the input inductor is charged from the source, and the second inductor is charged from the first capacitor. No energy is supplied to the load capacitor during this time. Inductor current and capacitor voltage polarities are marked in Figure 4.

When the power switch is turned off, the energy stored in inductor $L_1$ is transferred to $C_1$. The energy stored in $L_2$ is transferred to $C_2$ through the diode and supplying the energy to load [9], as shown in Figure 5B. The second inductor $L_2$ is also connected to the load during this time.

The output capacitor sees a pulse of current during the off time, making it inherently noisier than a buck converter.

The amount that the SEPIC converters increase or decrease the voltage depends primarily on the Duty Cycle and the parasitic elements in the circuit.

The output of an ideal SEPIC converter is:

$$V_{OUT} = \frac{D}{1-D} V_{in} \qquad (4)$$

A SEPIC converter is to process the electricity from the PV system. This converter either increases or decreases the PV system voltage at the load. The proposed SEPIC converter operates in buck mode.

## The Fuzzy Logic Controller for the SEPIC Converter

In fuzzy logic controller (FLC) design, one should identify the main control variables and determine the sets that describe the values of each linguistic variable. The input variables of the FLC are the output voltage error ($eV_{out}$) and the change of this error ($\Delta eV_{out}$) of the SEPIC converter. The output of the FLC is the duty cycle of ($\alpha$) of the PWM signal, which regulates the output voltage. The triangular membership functions are used for the FLC for easier computation. A five-term fuzzy set, i.e., negative big (NB), negative small (NS), zero (Z), Positive small (PS), and positive big (PB), is defined to describe each linguistic variable. The fuzzy rules of the proposed SEPIC DC-DC converter can be represented in a symmetric form (Table 1). For the output variable ($\alpha$), five-term fuzzy is defined to give sharpness to the regulation: negative big (NB), negative small (NS), zero (Z), Positive small (PS) and positive big (PB).

The variables of the FLC for the SEPIC converter are as follow:

- The first input is the error in the output voltage ($eV_{out}$):

$$eV_{OUT}(k) = V_{Ref} - V_{out}(k) \qquad (5)$$

- The second input is the variation (the change) in error ($\Delta eV_{out}$):

$$\Delta eV_{OUT}(k) = eV_{out}(k) - eV_{out}(k-1) \qquad (6)$$

- The single output variable ($\alpha$) is duty cycle.

Where:

- $V_{Ref}$: is the reference output voltage.

- $V_{out}(k)$ is the measured output voltage in the $K^{th}$ sample.

The Block diagram of the control SEPIC converter with the Fuzzy Logic controller is presented in Figure 6, where $\alpha(k)$ is sent to the PWM generator. PWM generator generates the necessary switching signal for the switch in the converter.

$K_{eVout}$ is the control gain of input $eV_{out}$.

$K_{\Delta eVout}$ is the control gain of input $\Delta eVout(k)$.

$K\alpha$ is the control gain of output. $\alpha$.

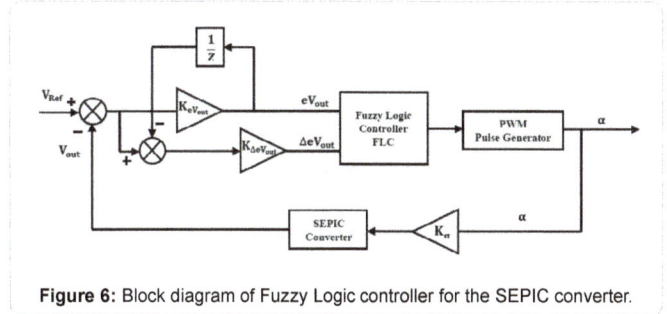

**Figure 6:** Block diagram of Fuzzy Logic controller for the SEPIC converter.

The membership functions $\mu eV_{out}$ and $\mu\Delta eV_{out}$ for $eV_{out}$ and $\Delta eV_{out}$ respectively are represented in Figure 7A. The membership function for the output variable $\alpha$ is represented in Figure 7B. All the functions are defined on a normalized interval [−1, 1].

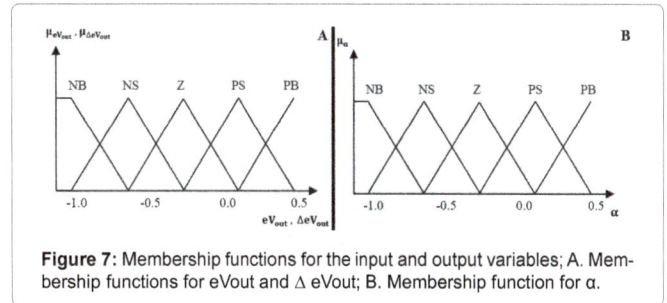

**Figure 7:** Membership functions for the input and output variables; A. Membership functions for eVout and Δ eVout; B. Membership function for α.

The rules of the fuzzy logic controller are shown in Table 1.

| $\Delta eV_{out}$ \ $eV_{out}$ | NB | NS | Z | PS | PB |
|---|---|---|---|---|---|
| NB | NB | NB | NB | NS | Z |
| NS | NB | NB | NS | Z | PB |
| Z | NB | NS | Z | PS | PB |
| PS | NS | Z | PS | PB | PB |
| PB | Z | PS | PB | PB | PB |

**Table 1:** Fuzzy control rules.

Fuzzy control rules are derived from the analysis of the converter behavior:

When the output of the converter is far from the set point (Positive Big or Negative Big), the duty cycle should be close to zero or one so as to bring to the set point quickly.

When the output of the converter is approaching to the set point (Negative Small or Positive Small), a small change of the duty cycle is necessary.

When the output of the converter is approaching very close to the set point, duty cycle must be kept constant in order to prevent the overshoot.

## Simulation Results

The photovoltaic array that we use in this paper is SCHOTT Solar (SCHOTT POLYTM 210); the characteristics of this array are given in Table 2.

| Maximum Power | $P_{max} = 210$ W |
|---|---|
| Open circuit voltage | $V_{oc} = 36.1$ V |
| Short circuit current | $I_{sc} = 7.95$ A |
| Voltage at maximum power | $V_{max} = 29.3$ V |
| Current at maximum power | $I_{max} = 7.16$ A |

**Table 2:** Electrical data apply to standard test conditions (STC): (T = 25°C, G = 1000 W/m²).

Figures 8 and 9 show the simulation result of the PV model.

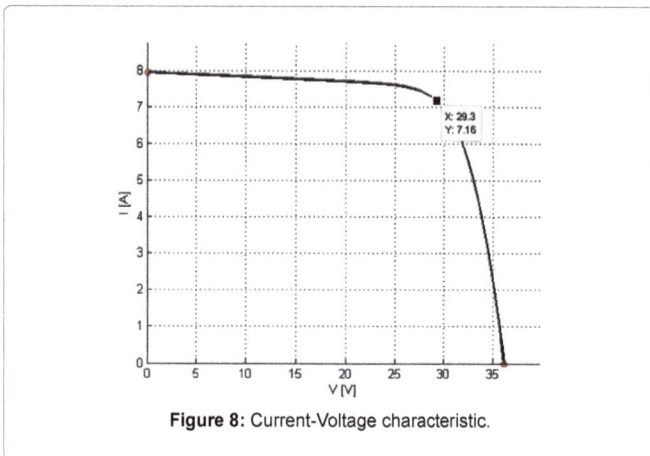

**Figure 8:** Current-Voltage characteristic.

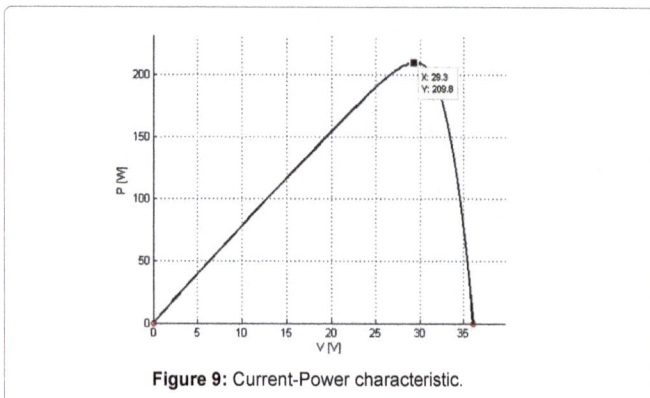

**Figure 9:** Current-Power characteristic.

The parameters of the SEPIC converter used in this study are given in Table 3.

| Switching frequency | 20 KHz |
|---|---|
| Input Voltage ($V_{in}$) | $V_{in} = 29.3$ V |
| Output Voltage ($V_{out}$) | $V_{out} = 24$ V |
| Load resistance | $R_{load} = 7.38$ Ω |
| Inductance L1 | $L1 = 460$ µH |
| Inductance L2 | $L2 = 460$ µH |
| Capacitor C1 | $C1 = 8.4$ µF |
| Capacitor C2 | $C2 = 0.0163$ F |

**Table 3:** The simulation parameters of SEPIC converter.

In the full model; the SEPIC converter is connected to the PV panel, and the duty cycle of this is controlled using the Fuzzy Logic Controller. The results are provided under standard test conditions; G = 1000 W/m² and T = 25°C.

Figure 10 shows the simulation results obtained from the model; $V_{out} = 24$V; $I_{out} = 3.25$A and Power = 78 W.

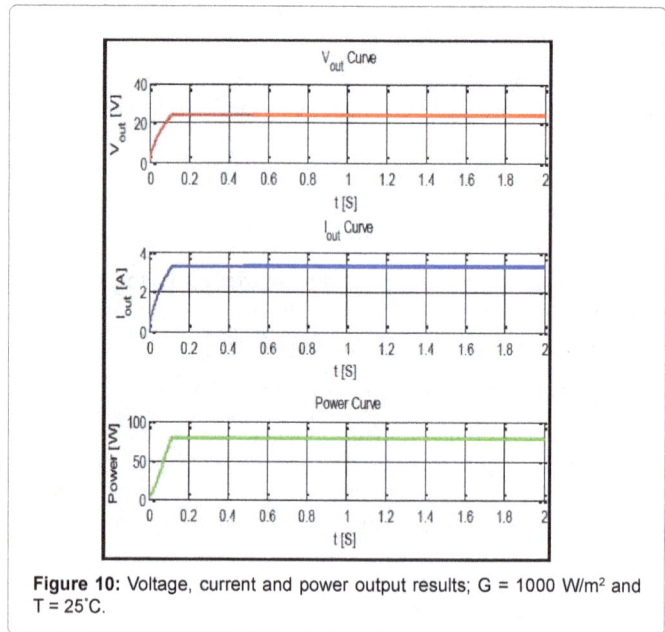

**Figure 10:** Voltage, current and power output results; G = 1000 W/m² and T = 25°C.

According this (Figure 10); we notice that since the time 0.1S the model gave a stable voltage (24 V), and thus a current and a power so stable.

To prove the efficiency of the integration of SEPIC converter controlled by the fuzzy logic controller for the photovoltaic system, we have studying the influence of the temperature and the solar irradiation separately.

### Influence of temperature

For stable irradiation (G = 1000 W/m²), and for a variable temperature, we obtain the results shown in the figures below.

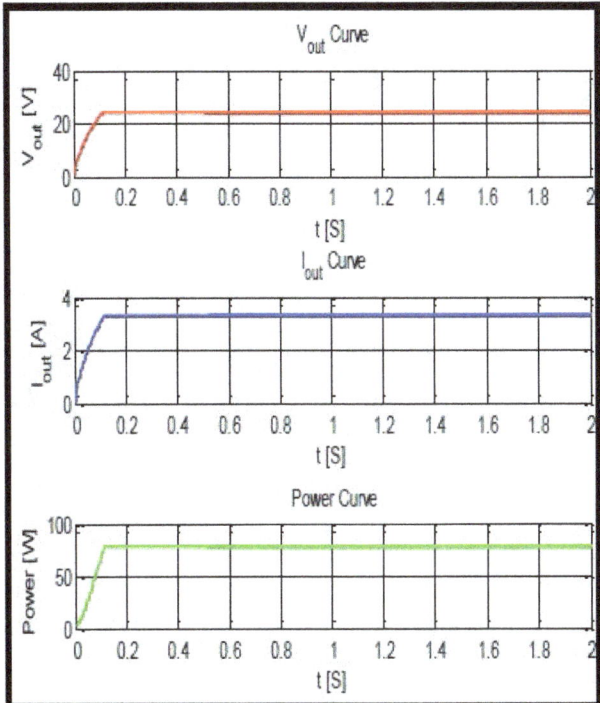

**Figure 11:** Voltage, current and power output results; G = 1000 W/m² and T = 35˚C.

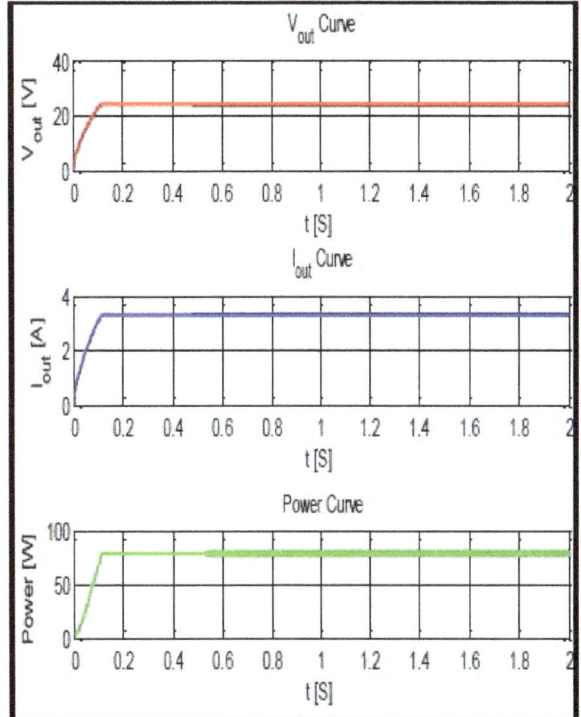

**Figure 12:** Voltage, current and power output results; G = 1000 W/m² and T = 45˚C.

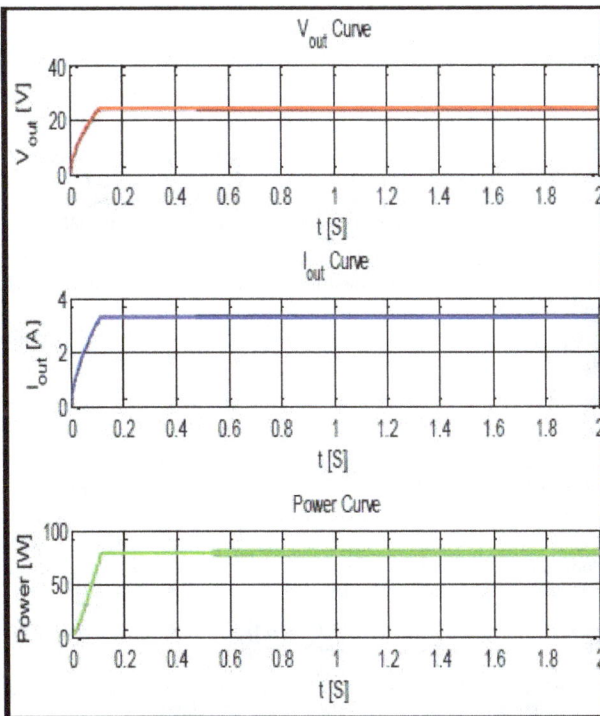

**Figure 13:** Voltage, current and power output results; G = 1000 W/m² and T = 55˚C.

From this three figures above (Figures 11-13); it is clear that for a variable temperature value, we obtain the desired voltage (24 V), and it is so stable since 0.1S.

## Influence of irradiation

For stable temperature (T = 25˚C), and for a variable solar irradiation, we obtain the results shown in the figures below.

Figures 14-16 shows the simulation results obtained from the model (T = 25˚C); for G = 900W/m², G = 800W/m² and G = 700W/m² respectively.

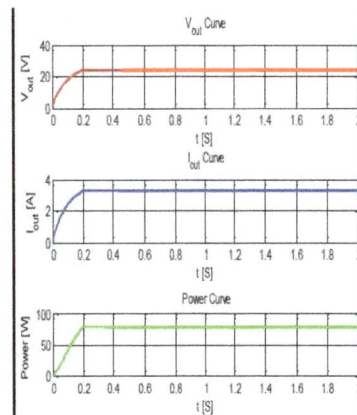

**Figure 14:** Voltage, current and power output results; T = 25˚C and G = 900 W/m².

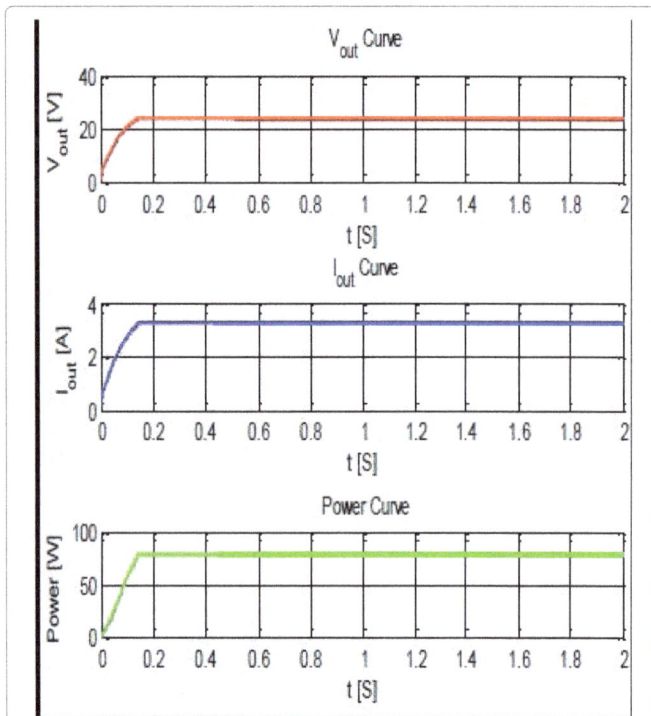

**Figure 15:** Voltage, current and power output results; T = 25°C and G = 800 W/m².

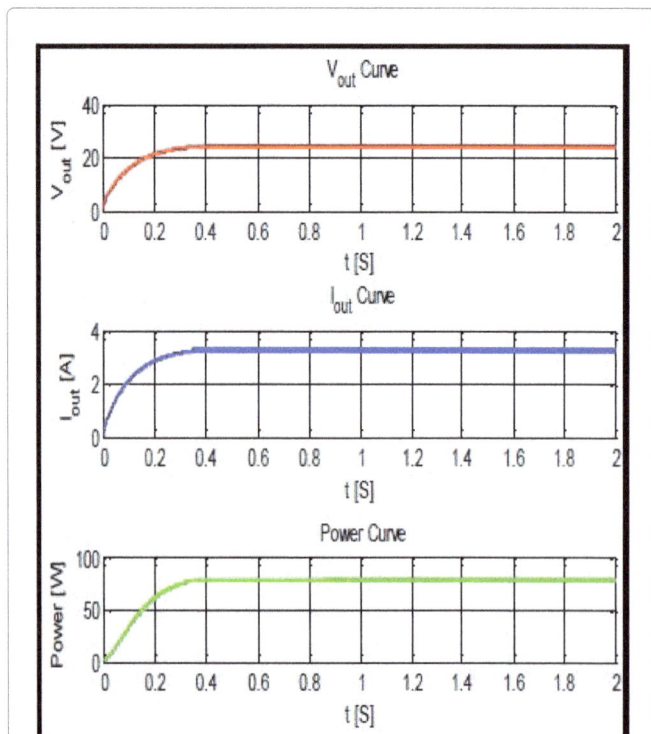

**Figure 16:** Voltage, current and power output results; T = 25°C and G = 700 W/m².

From the figures 14, 15 and 16; it is notable that the output voltage is 24V, even if there are a variation in the solar irradiance.

The following table (Table 4) summarizes the results given from the study.

The temperature and solar irradiation variations can be addressed by the FLC controller, and this is observed from the corresponding table, and generally the steady state is not exceeding 0.4s.

| G and T variation | | $V_{out}$ (V) | $I_{out}$ (A) | P (W) |
|---|---|---|---|---|
| 25°C | 900 W/m² | 24.03 | 3.256 | 78.23 |
| | 800 W/m² | 24.04 | 3.257 | 78.29 |
| | 700 W/m² | 24.03 | 3.257 | 78.27 |
| 1000W/m² | 35°C | 24.04 | 3.257 | 78.29 |
| | 45°C | 24.03 | 3.257 | 78.28 |
| | 55°C | 24.03 | 3.257 | 78.28 |

**Table 4:** Temperature and solar irradiation influence results.

## Conclusion

The SEPIC performs the voltage conversion from positive source voltage to positive load voltage. Due to the time variations and switching nature of the power converters, their dynamic behavior becomes highly non-linear. In this paper; a stand-alone solar-PV energy generation system with a SEPIC; DC-DC converter controlled by a Fuzzy Logic Controller has been designed and the efficiency of the system has been presented under variation in temperature and solar irradiation. This study has successfully demonstrated the design, analysis and suitability of fuzzy logic controller for SEPIC converter.

### References

1. Mahendran G, Kandaswamy KV (2013) Ant Colony Optimized Tuned DC-DC Converter ", Int J Comp Appl 2013: 16-21.

2. Dunia J, Mwinyiwiwa BM (2013) Performance Comparison between ĆUK and SEPIC Converters for Maximum Power Point Tracking Using Incremental Conductance Technique in Solar Power Applications. Int J Electrical, Computer, Energetic, Electronic and Communication Engineering 7: 1638-1643.

3. Yusivar F, Farabi MY, Suryadiningrat R, Ananduta WW, Syaifudin Y (2011) Buck-Converter Photovoltaic Simulator. Int J Power Electronics Drive Sys 1: 156-167.

4. Srivastav S, Singh S (2011) An Introduction To Sepic Converters. Int Referred Res J 2: 14-15.

5. Venkatanarayanan S, Saravanan M (2014) Fuzzy logic based PV energy system with SEPIC converter. J Theoretical Appl Inform Technol 59: 89-95.

6. El Khateb AH, Rahim NA, Selvaraj J (2012) Fuzzy logic control approach of a maximum power point employing SEPIC converter for standalone photovoltaic system.

7. Thambi G, Prem Kumar S, Murali Krishna Y, Aruna M (2015) Fuzzy-Logic-Controller-Based SEPIC Converter for MPPT in Standalone PV systems. Int Res J Engineering Technol 2: 492-497.

8. Venkatanarayanan S, Saravanan M (2014) Proportional-integral Control for SEPIC Converter. Res J Appl Sci Engineer and Technol 8: 623-629.

9. Krishna Reddy TGL, Ramamurthyraju P, Revanth KV (2013) Analysis of SEPIC for PV-Applications using PI Controller and Current Mode Control. Int J Sci Res Dev 1: 1821-1824.

10. Kung YS, Liaw CM (1994) A Fuzzy Controller improving a Linear Model Following Controller for Motor Drives. IEEE Transactions on Fuzzy Systems 2: 194-202.

11. Bouzeria H, Fetha C, Bahi T, Abadlia I, Layate Z, et al. (2015) Fuzzy Logic

Space Vector Direct Torque Control of PMSM for Photovoltaic Water Pumping System. Energy Procedia 74: 760-771.

12. Chen YM, Liu YC, Hung SC, Cheng CS (2007) Multi-Input Inverter for Grid-Connected Hybrid PV/Wind Power System. IEEE Transactions on Power Electronics 22: 1070-1077.

13. Ganesh S, Janani J, Angel GB (2014) A Maximum Power Point Tracker for PV Panels Using SEPIC Converter. Int J Electri Comp Energetic, Electron Commun Engineer 8: 356-361.

14. Ramkumar A, Shini Florence SV (2015) Analysis of Single Phase AC-DC SEPIC Converter using Closed Loop Techniques. Int J Adv Res Electri Electron and Instrumen Engineer 4: 756-766.

# 8

# Potential of Wind Energy Development for Water Pumping in Ngaoundere

**Ruben M. Mouangue**[1]*, **Myrin Y. Kazet**[1,2], **Daniel Lissouck**[3], **Alexis Kuitche**[2] and **J.M. Ndjaka**[4]

[1]*Department of Energetic Engineering, UIT, University of Ngaoundere, Cameroon*
[2]*Departments of GEEA, PAI, ENSAI, University of Ngaoundere, Cameroon*
[3]*Department of Renewable Energy, HTTTC, Kumba, Cameroon*
[4]*Department of Physics, Faculty of Sciences, University of Yaounde 1, Cameroon*

## Abstract

The modelling of wind energy conversion systems is of great importance if one intends to develop water pumping applications for a sustainable development. This paper presents a technical assessment based on the measured wind data in which we investigate the possibility of coupling piston pump, roto-dynamic pump and electric pump with wind rotors for water pumping applications. Weibull distribution is used to model the monthly mean wind speed for a location in the town of Ngaoundere. It is found that there is a good agreement between the predicted values of the mean wind speed and those obtained from data suggesting that the Weibull distribution can be used to provide accurate estimation of the mean wind speed. The mean electric power and energy are computed based on the power curves of Vestas V25 and V100. Taking into account the wind regime characteristics of our site, we provide the amount of water which can be expected from each type of wind pumps. The monthly amount of water has minimum and maximum average values of 422 m³ and 674 m³ for the piston pump, 1275 m³ and 1982 m³ for the roto-dynamic pump and 31334 m³ and 100042 m³ for the electric pump. From the results, it is clear that electric pump offer better performances than piston and roto-dynamic pumps.

**Keywords:** Weibull distribution; Wind power; Water pumping; Ngaoundere

## Abbreviations:

Cp: Power coefficient; g: Acceleration due to gravity, $m.s^{-2}$; H: Head of water, m; Q: Water flow rate, $m^3.h^{-1}$; η: Efficiency of transmission; f (V): Probability density function of the Weibull distribution; F (V): Cumulative distribution function; Γ ( ): Gamma function; QVE: Instantaneous discharge of the electric pump, $m^3.s^{-1}$; Vm: Mean wind speed, $m.s^{-1}$; G: Gear ratio; λd: Design tip ratio; DT: Rotor diameter, m; PR: Rated power, W; Npd: Pump speed at design point, rps; UN: United Nations; ρa: Air density, $kg.m^{-3}$; ρw: Water density, $kg.m^{-3}$; k: Weibull shape parameter; C: Weibull scale parameter, $m.s^{-1}$; V: Instantaneous wind speed, $m.s^{-1}$; QVP: Instantaneous discharge of the piston pump, $m^3.s^{-1}$; PH: Hydraulic power, W; PV: Developed power, W; QVR: Instantaneous discharge of the roto-dynamic pump, $m^3.s^{-1}$; T: Time, s; VI: Cut-in wind speed, $m.s^{-1}$; VO: Cut-out wind speed, $m.s^{-1}$; VR: Rated velocity, $m.s^{-1}$; Vd: Design wind velocity, $m.s^{-1}$; A: Swept surface, $m^2$

## Introduction

In sub-Saharan Africa, operation and control of water in both rural and urban areas have become strategic issues for population growth, diversification of economic activities and the current degradation of the environment. Sensitive and complex subject, water should be open to all without any distinction but there is however still in 2015, a severe water crisis in Africa [1]. This is the continent where access to quality water is more limited, according to the 4th UN report on water [1]. Nevertheless, Africa has seventeen major rivers and hundreds of lakes, as well as significant groundwater. Despite all the efforts consent by the governments of African countries, it is clear that much remains to be done in terms of investment.

One of the solutions proposed to solve this problem could be to invest in the acquisition of infrastructures for pumping and water distribution. But first, we should carry out technical studies to provide effective assistance to investors as to the decisions concerning the planning and implementation of projects to supply drinking water.

Wind energy, which since 2009 is the second global source of renewable energy [2], remains unexploited in Cameroon. The wind pumping could be an interesting solution for people in rural areas because it uses a free energy source and does not contributes to increased greenhouse gas emissions [3].

In this paper, we investigate the possibility of coupling piston pump, roto-dynamic pump and electric pump with wind rotors for water pumping applications.

Although the performance of piston and wind-driven roto-dynamic pumps were estimated in some studies as the one of Mathew and Pandey in 2003 [4], the distribution of wind speed in a regime was modelled by using the Rayleigh distribution which is a particular case of the Weibull distribution and has not yet been investigated for the site of Ngaoundere. However, the Weibull distribution has been found satisfactory for the modelling of the observable wind speed frequency in general [5-7] and for the case of Ngaoundere site in particular [8]. Approximation which consists of assuming that the variation between two nodes of the power-wind speed curve is linear is also considered here for the computation of pumped volume flow rate of an electric pump. Taking into account the wind regime characteristics of our site, we provide the amount of water which can be expected from each type of wind pumps and we compare them.

*Corresponding author: Ruben M. Mouangue, Department of Energetic Engineering, UIT, University of Ngaoundere, P. O Box 455, Ngaoundere, Cameroon, E-mail: r_mouangue@yahoo.fr

## Material and Methods

### Weibull distribution

The probability density function (PDF) of the Weibull distribution is used to characterize the frequency distribution of wind speeds over time and is expressed mathematically as [5,6]:

$$f(V) = \frac{k}{C}\left(\frac{V}{C}\right)^{k-1} exp\left[-\left(\frac{V}{C}\right)^{k}\right] \tag{1}$$

This distribution is characterized by two parameters: a dimensionless shape parameter $k$ and a scale parameter $C$ (m/s).

The cumulative distribution function (CDF) is express in equation 2 as [6,9]:

$$F(V) = 1 - exp\left[-\left(\frac{V}{C}\right)^{k}\right] \tag{2}$$

### Mean wind speed

One of the most important wind characteristic is the mean wind speed. The probability density function and Weibull parameters are related to the mean speed by equation 3 [5,9].

$$V_m = C\tilde{A}\left(1+\frac{1}{k}\right) \tag{3}$$

Where

$$\tilde{A}(n) = \int_0^{\infty} x^{n-1}e^{-x}dx \tag{4}$$

$\Gamma$ ( ) is the gamma function.

### Extrapolation of weibull parameters

Most modern wind turbines have hub heights considerably higher than measurement heights of meteorological towers. Hence, the measured wind characteristics at the lower measurement height must be extrapolated to the hub height of the turbine. For this task we choose the use of the Justus and Mikhail law [6,9,10]. The values of Weibull's parameters ($k_h$ and $C_h$) are then evaluated at any desired height ($Z_h$) by the following equations (5) and (6):

$$k_h = k_a \left[1 - ALn\left(\frac{Z_h}{Z_a}\right)\right]^{-1} \tag{5}$$

$$C_h = C_a \left(\frac{Z_h}{Z_a}\right)^{n} \tag{6}$$

The exponent $n$ is express as:

$$n = B - ALnC_a \tag{7}$$

Where constants $A = 0.0881$ and $B = 0.37$.

$k_a$ and $C_a$ are, respectively, the shape parameter and the scale parameter at the anemometer height $Z_a = 10$ m.

### Wind pump discharges

**Piston pump:** Piston pumps are types of pump with reciprocating motion. The overall performance coefficient of a wind rotor coupled to a reciprocating pump can be modelled as [11]:

$$C_p\eta = 4C_{pd}\eta(T,P)\left[1 - K_O\left(\frac{V_I}{V}\right)^2\right]K_O\left(\frac{V_I}{V}\right)^2 \tag{8}$$

where $C_p\eta$ is the overall (wind to water) efficiency of the system, $C_{pd}$ is the power coefficient of the rotor at the design point, $\eta$ (T,P) is the combined transmission and pump efficiency and $K_O$ is a constant taking care of the starting behaviour of the rotor pump combination.

The power developed by the system in pumping water $P_V$ is given by equation (9).

$$P_V = \frac{1}{2}C_p\eta\rho_a AV^3 \tag{9}$$

From Equations (8) and (9), $P_V$ can be expressed as:

$$P_H = \rho_w gQ_{VP}H \tag{10}$$

The hydraulic power $P_H$, needed by the pump for delivering a discharge of $Q_{VP}$ against a head H is estimated by the equation below:

$$P_H = \rho_w gQ_{VP}H \tag{11}$$

By equalizing $P_H$ and $P_V$, the instantaneous discharge $Q_{VP}$ of the system at any velocity $V$ can be deduce.

$$Q_{VP} = 2C_{pd}\eta(T,P)\frac{\rho_a}{\rho_w}\frac{\pi D_T^2}{4gH}V^3\left[1 - K_O\left(\frac{V_I}{V}\right)^2\right]K_O\left(\frac{V_I}{V}\right)^2 \tag{12}$$

The wind regime characteristics are integrated with the model in order to provide the pumped volume flow rate expected over a time period.

$$Q_P = T\int_{V_I}^{V_O} Q_{VP}(V)f(V)dV \tag{13}$$

$F(V)$ is the probability density function of the Weibull distribution. $V_I$ and $V_O$ are respectively cut-in wind speed and cut-out wind speed.

The pumped volume flow rate expected from a wind driven piston pump, installed at a given site, over a period T is obtained by computing equation (14).

$$Q_P = 2C_{pd}\eta(T,P)\frac{\rho_a}{\rho_w}\frac{\pi D_T^2}{4gH}T\int_{V_I}^{V_O}V^3\left[1 - K_O\left(\frac{V_I}{V}\right)^2\right]K_O\left(\frac{V_I}{V}\right)^2\left(\frac{k}{C}\right)\left(\frac{V}{C}\right)^{k-1}*exp\left[-\left(\frac{V}{C}\right)^{k}\right]dV \tag{14}$$

**Roto-dynamic pump:** According to the paper of Mathew and Pandey [4], the discharge of an ideal roto-dynamic pump at any speed $V$ can be estimated by using the equation (15) below:

$$Q_{VR} = \frac{1}{8}C_{Pd}\eta_{Pd}D_T\frac{\rho_a}{\rho_w}\frac{V_d^3}{gH}\frac{G\lambda_d}{N_{Pd}}V \tag{15}$$

Where $C_{pd}$ is the design power coefficient, $\eta_{pd}$ is efficiency, $D_T$ is the diameter of the rotor, $G$ is the gear ratio, $\lambda_d$ is the design tip ratio, $N_{Pd}$ is the pump speed at design point and $V_d$ is the design wind velocity.

If we integrate the wind regime characteristics with the model, the output providing the pumped volume flow rate expected over a time period from a wind-powered water pumping system at a given site can be expressed by equation (16):

$$Q_R = T\int_{V_I}^{V_O} Q_{VR}(V)f(V)dV \tag{16}$$

From equation (15) and (16), the integrated system output can be expressed as:

$$Q_R = \frac{1}{8}kC_{Pd}\eta_{Pd}D_T\frac{\rho_a}{\rho_w}\frac{V_d^3}{gH}\frac{G\lambda_d}{N_{Pd}}T\int_{V_I}^{V_O}\left(\frac{V}{C}\right)^{k}exp\left[-\left(\frac{V}{C}\right)^{k}\right]dV \tag{17}$$

The integrated output of the roto-dynamic pump is then estimated by the computation of equation (17).

**Electric pump:** To estimate the discharge of an electric pump, the power developed by the wind turbine generator $P(V)$ may be equated with the corresponding hydraulic power $P_H$ demand.

$$P_H = \frac{\rho_w gQ_{VE}H}{\eta} = P(V) \tag{18}$$

Hence, discharge of the pump at any speed $V$ can be express as follow:

$$Q_{VE} = \frac{\eta}{\rho_w gH}P(V) \tag{19}$$

Where $P(V)$ is the electric power generated at a given speed $V$.

Values of $P(V)$ are provided by the manufacturer of the wind turbine generator through the power curve.

In this work, we have made the choice of two power curves: Vestas V82 [12] with a hub height of 40 m (Figure 1) and Vestas V100 [13] with a hub height of 80 m (Figure 2) because of their relatively low cut-in wind speed (Table 1).

It is possible to carry out an approximation which consists of assuming that the variation between two nodes of the power-wind speed curve is linear [14]. Then, given two points $i$ and $i + 1$ of the power curve, power as a function of speed can be written as equation (20).

$$P(V) = \frac{P_{i+1} - P_i}{V_{i+1} - V_i}(V - V_i) + P_i \tag{20}$$

By using this approximation, power curves of both wind turbines is plotted as shown in Figures 1 and 2.

Then, the amount of water expected over a time period is obtained by integrating the wind regime characteristics into the electric pump discharge equation.

$$Q_E = T \int_{V_I}^{V_O} Q_{VE} f(V) dV \tag{21}$$

By substituting equation (19) in (21) we have:

$$Q_E = \frac{\eta T}{\rho_w g H} \int_{V_I}^{V_O} P(V) f(V) dV \tag{22}$$

Substituting equation (1) and (20) in (22), the pumped volume flow rate of an electric pump is then estimated by solving numerically equation (23) bellow:

$$Q_E = \frac{\eta T}{\rho_w g H} \left\{ \int_{V_I}^{V_I} \left[ \frac{P_{i+1} - P_i}{V_{i+1} - V_i}(V - V_i) + P_i \right] \left( \frac{k}{C} \right) \left( \frac{V}{C} \right)^{k-1} exp \left[ -\left( \frac{V}{C} \right)^k \right] dV + P_R \int_{V_R}^{V_O} \left( \frac{k}{C} \right) \left( \frac{V}{C} \right)^{k-1} exp \left[ -\left( \frac{V}{C} \right)^k \right] dV \right\} \tag{23}$$

In the present study, the computation of equations was made through Fortran codes that we wrote using for this task Eclipse Juno which is an open source integrated development environment [15]. Specifications of the three systems considered for the analysis are given in (Table 1) [4].

## Results and Discussion

Results of Ngaoundere wind resource analysis using one year wind data as well as the vertical wind shear analysis and the water discharge at various wind turbine hub heights are given in this section.

### Monthly wind speed variation

Monthly mean wind speed variation is found in the range 1.292-1.794 m/s as shown in Figure 3. The lowest wind speed of 1.292 m/s occurs in October and the highest wind speed 1.794 m/s in February. The maximum monthly mean wind speeds occur during February, March and May with 1.794 m/s, 1.786 m/s and 1.788 m/s, respectively. Wind speeds decrease to 1.292 m/s in October and slightly gain strength as 1.470 m/s, 1.523 m/s, and 1.576 m/s during November, December and January, respectively.

In Figure 3, the predicted values of the mean wind speed are plotted side by side to those obtained from data. As we can see, there is a good agreement between them. It suggests that the Weibull distribution can be used to provide accurate estimation of the mean wind speed.

### Observed wind speed histogram

The observed wind speed histogram is well approximated by the Weibull probability density function. This is shown by Figure 4. The

Pearson's correlation coefficient computed is 0.9 suggesting that the wind regime of a site can be well described by the Weibull distribution.

## Wind regime at different heights

After the Weibull parameters have been extrapolated at 40 m and 80 m heights, probability density functions using only global values for each of them are plotted side by side in Figure 5.

Cumulative distribution functions for the whole year of the site are also plotted at different altitudes. This is shown in Figure 6.

The trend regarding the evolution of the PDF curve function of altitude (Figure 5) is similar to that obtained by Safari and Gasore [16]. This trend reflects the fact that the probability of significant wind increases with altitude. The trend is the same for CDF curves shown in Figure 6.

## Wind speed shear analysis

The vertical wind profile is calculated at hub heights of wind turbines at 40 m and 80 m from one year data. The monthly mean wind speed variations at different hub heights are shown in Figure 7.

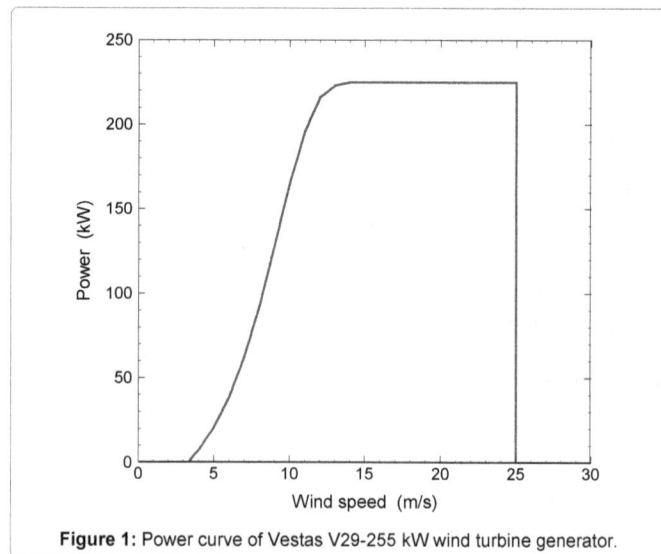

**Figure 1:** Power curve of Vestas V29-255 kW wind turbine generator.

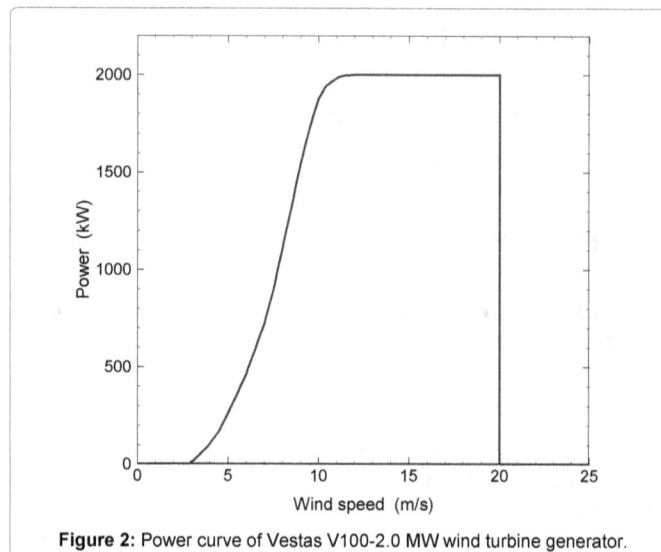

**Figure 2:** Power curve of Vestas V100-2.0 MW wind turbine generator.

| Pump specifications | Piston | Roto-dynamic | Electric | |
|---|---|---|---|---|
| | | | V29 | V100 |
| Coefficient K | 0.25 | - | | - |
| Cut-in wind speed (m/s) | 2.5 | 2.5 | 3 | 3.5 |
| Cut-out wind speed (m/s) | 10 | 10 | 20 | 25 |
| Design power coefficient | 0.3 | 0.38 | | |
| Design tip speed ratio | - | 2 | | - |
| Design wind velocity | - | 6 | | - |
| Efficiency (pump + transmission) | 0.95 | 0.558 | | |
| Gear ratio | - | 19.8 | | - |
| Rated power | | | 255 kW | 2000 kW |
| Rated wind speed | | | 14 m/s | 12 m/s |
| Pump speed at design point (rps) | - | 40 | | - |

**Table 1:** Specifications of the three wind-driven pumps considered for the performance computation.

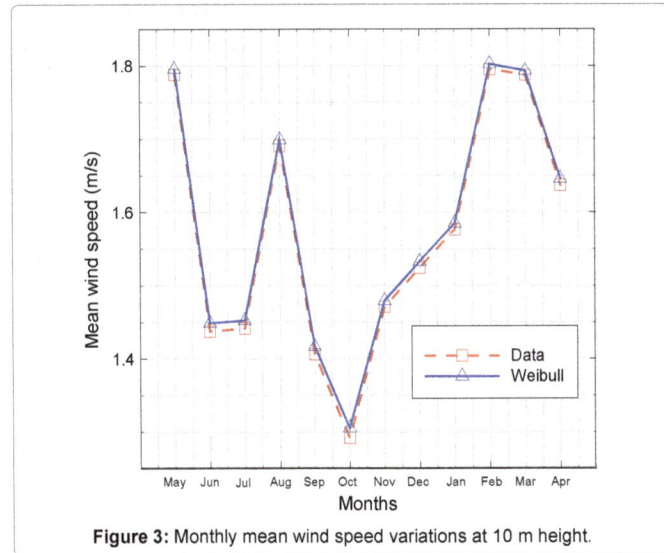

**Figure 3:** Monthly mean wind speed variations at 10 m height.

The monthly reference mean wind speeds during the period May 2011 to April 2012 vary from 1.292 m/s to 1.794 m/s which is increased from 2.762 m/s to 3.772 m/s at the 80 m hub height in October and February.

These wind speeds are found to be suitable for micro wind turbines with lower cut-in speeds. The one year extrapolated wind speeds at this location at 40 m height are not found suitable for large turbines for electricity power generation as they require at least annual mean wind speeds of 3-7 m/s. Figure 8 shows that at some hours of the day, the wind speed is higher than the cut-in wind speed. The wind turbine for this period will be able to produce electricity to supply the pumps and to recharge batteries.

## Monthly power and Energy output

The monthly and annual electric power developed $Pu$ and energy $E$ is calculated and the results are shown in Table 2.

At 40 m height, the maximum value of monthly mean electric power is computed as 14.04 kW in March and the minimum value is computed as 4.40 kW in October. However, the monthly energy has maximum average value of 10.453 MWh in March while the minimum value is obtained as 3.274 MWh in October.

At 80 m height, the maximum and the minimum values of the monthly mean electric power are respectively 287.94 kW in February and 123.70 kW in October while the monthly energy has maximum

and minimum average values of 213.471 MWh in March and 92.036 MWh in October, respectively.

It can be also observed from Table 2 that the power generated by the wind turbine increases with altitude. This result confirms the trend observed above that the probability of significant wind increases with altitude. Thus, the wind resource becomes increasingly important when the altitude increases.

## Monthly and annual water output

Results obtained from the computation of volume flow rate equations are shown in Table 3. It presents the monthly and annual volume flow rate and amount of water for the three types of pump considered at the same height.

For the piston pump, the maximum value of the volume flow rate is computed as 0.917 $m^3h^{-1}$ in February and the minimum value is computed as 0.567 $m^3h^{-1}$ in October. However, the monthly amount of water has maximum average value of 674 $m^3$ in March while the minimum value is obtained as 422 $m^3$ in October.

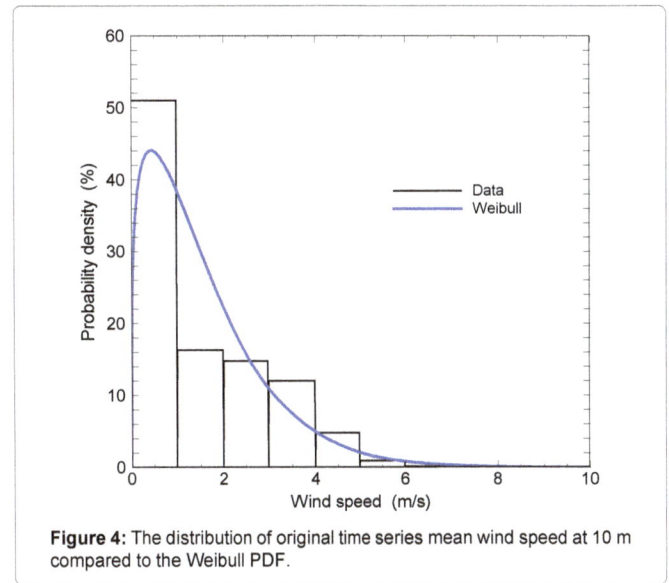

**Figure 4:** The distribution of original time series mean wind speed at 10 m compared to the Weibull PDF.

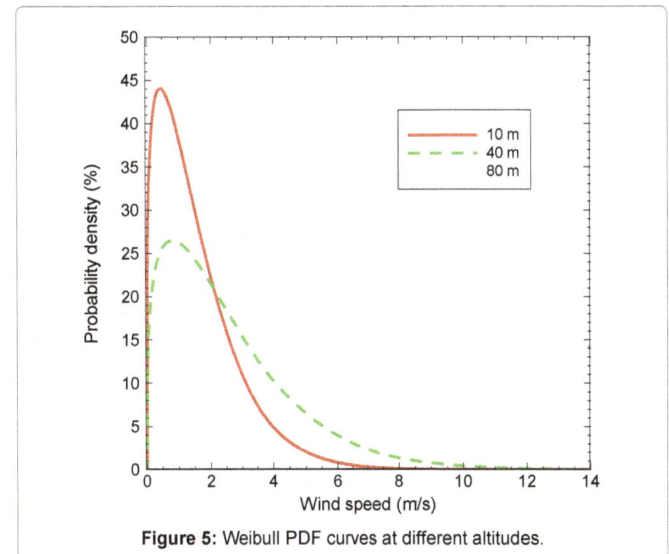

**Figure 5:** Weibull PDF curves at different altitudes.

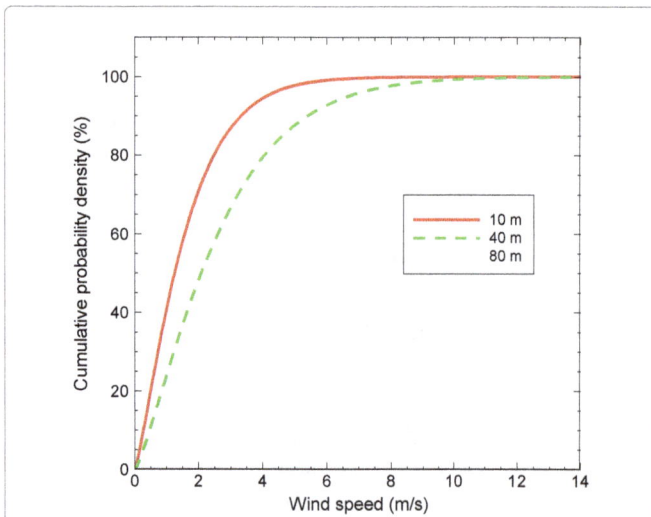

**Figure 6:** Weibull cumulative distribution function curves at different altitudes.

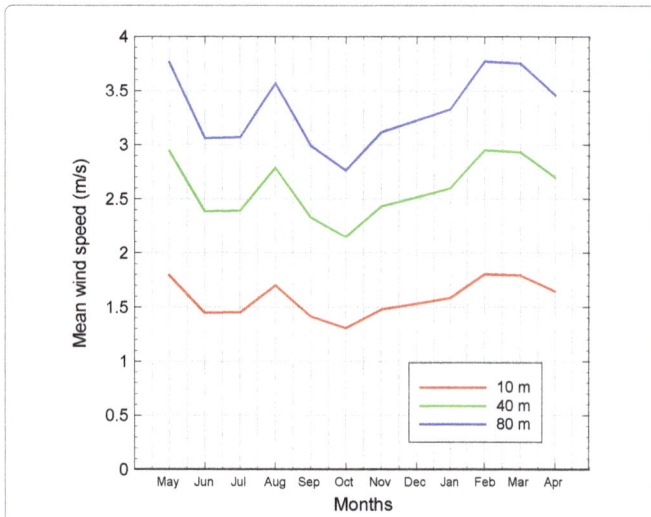

**Figure 7:** Vertical wind profile at different heights.

Then, the roto-dynamic pump has 2.700 m³h⁻¹ and 1.713 m³h⁻¹ as maximum and minimum values of the volume flow rate, respectively. On the other hand, it has 1982 m³ and 1275 m³ as maximum and minimum values of the amount of water, respectively for the same months as above.

For the electric pump finally, the maximum and the minimum values of the volume flow rate are respectively 134.465 m³h⁻¹ in March and 42.116 m³h⁻¹ in October while the monthly amount of water has maximum and minimum average values of 100042 m³ in March and 31334 m³ in October, respectively.

As a result of (Table 3), monthly pumped volume flow rate and water amount at 40 m height of piston, roto-dynamic and electric pumps are plotted in Figures 9 and 10. It can be seen that the rate of discharge patterns are similar to each other. By looking at Figure 8 it is clear that at the same height, the monthly discharge rate of the electric pump is upper than the one of roto-dynamic pump which is itself upper than the piston one. One can also note that an important gap exists between the volume flow rate of the electric pump and the two other types of pump. On the other hand, it is less important between

volume flow rates of piston and roto-dynamic pumps. In the months of May, Aug, Feb, Mar and Apr monthly mean wind speeds are upper than 2.7 m/s and therefore more wind energy can be captured by wind turbines and pumped water flows rates are consequent.

Monthly water amount histogram is found in the ranges 422 - 674 m³ for the piston pump, 1275 - 1982 m³ for the roto-dynamic pump and 31334 - 100042 m³ for the electric pump as shown in Figure 9. The lowest amount occurs in October and the highest amount in March.

At the same height, results show that electric pump offer better performances than piston and roto-dynamic pumps. Roto-dynamic and piston pumps are therefore recommend for low head of water.

In practice, piston and roto-dynamic pumps are located near water points. If the wind resource is not enough at the water point, there will be no good outputs. This disadvantage could be overcome by the use of an electric pump. The turbine being able to be installed far from the site of the pump, it could be therefore installed at the most important resource place. Furthermore, if storage devices like batteries are combined to the system, water could be pumped at any time of the day independently of the wind availability.

## Conclusion

In this study the potential of the development of water pumping using wind energy in Ngaoundere is carried out. Energy produced by the rotor in wind regimes following the Weibull distribution and the amount of water output from three types of pump are estimated. The most important outcomes of the study can be summarized as follows:

The monthly mean wind speeds are recorded as 1.292 and 1.794 m/s in October and February for the minimum and maximum values respectively at 10 m height; the monthly mean electric power and energy at 40 m height are computed as 4.40 kW, 3.274 MWh in October and 14.04 kW, 10.453 MWh in March for the minimum and maximum values. At 80 m height, these values are respectively 123.70 kW, 92.036 MWh in October, 287.94 kW in February and 213.471 MWh in March. These wind speeds, power and energy are found to be suitable for micro wind turbines with lower cut in speeds.

There is a good agreement between the predicted values of the mean wind speed and those obtained from data suggesting that the

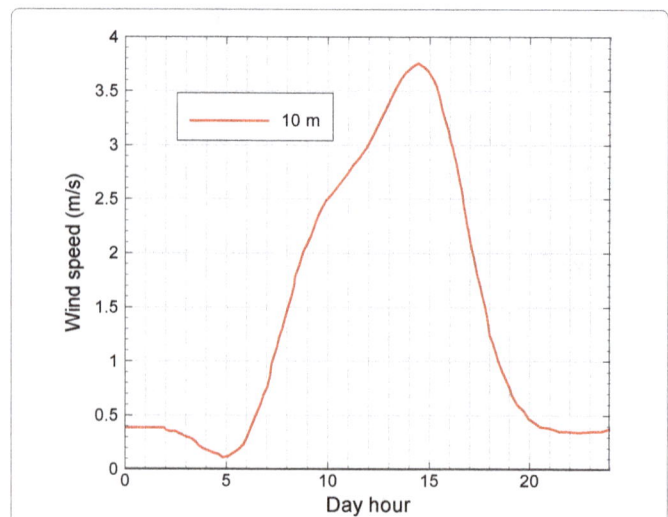

**Figure 8:** Vertical wind profile for a day at 10 m height.

| Months | Wind Turbine Generators | | | |
|--------|------------|------------|------------|------------|
| | Vestas V29 - 255 kW | | Vestas V100 - 2 MW | |
| | 40 m | | 80 m | |
| | Pu | E | Pu | E |
| | (kW) | (MWh) | (kW) | (MWh) |
| May-11 | 11.93 | 8.883 | 270.42 | 201.197 |
| Jun-11 | 6.46 | 4.654 | 166.03 | 119.546 |
| Jul-11 | 6.11 | 4.55 | 161.78 | 120.368 |
| Aug-11 | 9.91 | 7.378 | 236.28 | 175.796 |
| Sep-11 | 5.92 | 4.266 | 155.69 | 112.102 |
| Oct-11 | 4.4 | 3.274 | 123.7 | 92.036 |
| Nov-11 | 8.05 | 5.795 | 188.73 | 135.89 |
| Dec-11 | 8.69 | 6.467 | 202.25 | 150.474 |
| Jan-12 | 9.95 | 7.408 | 221.46 | 164.769 |
| Feb-12 | 13.93 | 9.698 | 287.94 | 200.41 |
| Mar-12 | 14.04 | 10.453 | 286.92 | 213.471 |
| Apr-12 | 9.24 | 6.652 | 221.86 | 159.744 |
| | | | | |
| Global | | | | |
| 1 year | 8.97 | 78.773 | 210.59 | 1849.83 |

Table 2: Monthly and annual electric power and energy at different altitudes.

| Months | Piston pump | | Roto-dynamic pump | | Electric pump | |
|--------|-------------|----------|-------------------|----------|---------------|----------|
| | QP (m³h⁻¹) | Amount of water (m³) | QR (m³h⁻¹) | Amount of water (m³) | QE (m³h⁻¹) | Amount of water (m³) |
| May-11 | 0.941 | 700 | 2.781 | 2069 | 114.269 | 85016 |
| Jun-11 | 0.683 | 492 | 2.046 | 1473 | 61.863 | 44542 |
| Jul-11 | 0.684 | 509 | 2.054 | 1528 | 58.535 | 43550 |
| Aug-11 | 0.875 | 651 | 2.596 | 1932 | 94.913 | 70615 |
| Sep-11 | 0.657 | 473 | 1.973 | 1421 | 56.707 | 40829 |
| Oct-11 | 0.567 | 422 | 1.713 | 1275 | 42.116 | 31334 |
| Nov-11 | 0.705 | 507 | 2.098 | 1510 | 77.037 | 55467 |
| Dec-11 | 0.743 | 553 | 2.21 | 1644 | 83.197 | 61899 |
| Jan-12 | 0.776 | 577 | 2.3 | 1711 | 95.297 | 70901 |
| Feb-12 | 0.917 | 638 | 2.7 | 1879 | 133.362 | 92820 |
| Mar-12 | 0.905 | 674 | 2.664 | 1982 | 134.465 | 100042 |
| Apr-12 | 0.833 | 600 | 2.477 | 1784 | 88.43 | 63670 |
| Global | | | | | | |
| 1 year | 0.779 | 6839 | 2.314 | 20324 | 85.827 | 753905 |

Table 3: Monthly and annual pumped volume flow rate and water amount for each type of pump at 40 m height.

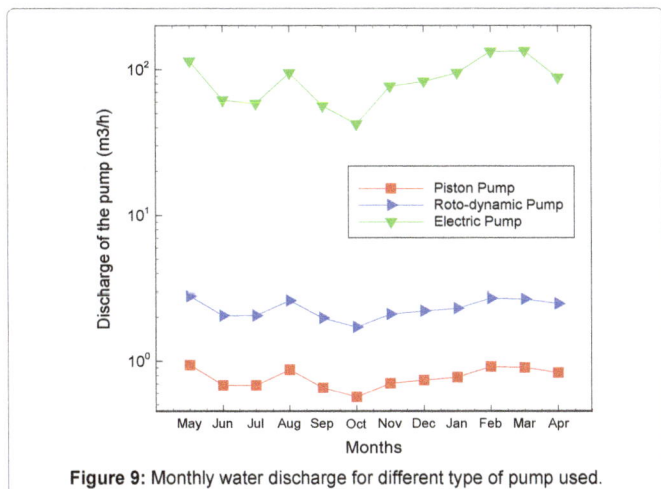

Figure 9: Monthly water discharge for different type of pump used.

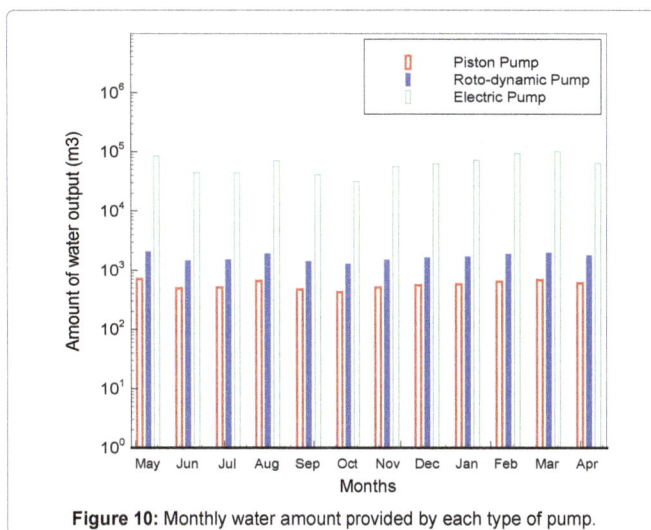

Figure 10: Monthly water amount provided by each type of pump.

Weibull distribution can be used to provide accurate estimation of the mean wind speed.

The monthly amount of water has minimum and maximum average values of 422 m³ and 674 m³ for the piston pump, 1275 m³ and 1982 m³ for the roto-dynamic pump and 31334 m³ and 100042 m³ for the electric pump.

At the same height, results show that electric pump offer better performances than piston and roto-dynamic pumps.

Underground water can be pumped using wind power and this could be an initial solution for people living in remote areas.

### Acknowledgments

Authors would like to thank the ASECNA weather service of Ngaoundere for providing meteorological data used in this work. Thanks for the reviewers for their constructive comments.

### References

1. The United Nations World Water Development Report 2014.

2. Observ ER (2013) Electricity production in the world. Fifteenth inventory edition.

3. Li M, Li X (2005) MEP-type distribution function: a better alternative to Weibull function for wind speed distributions. Renew Energy 30: 1221-1240.

4. Mathew S, Pandey KP (2003) Modelling the integrated output of wind-driven roto-dynamic pumps. Renew Energy 28: 1143-1155.

5. Bataineh KM, Dalalah D (2013) Assessment of wind energy potential for selected areas in Jordan. Renew Energy 59: 75-81.

6. Boudia SM, Benmansour A, Ghellai N, Benmedjahed M, Tabet Hellal MA (2013) Temporal assessment of wind energy resource at four locations in Algerian Sahara. Energy Conversion and Management 76: 654-664.

7. Omer AM (2008) On the wind energy resources of Sudan. Renew Sust Energ Rev 12: 2117-2139.

8. Mouangue RM, Kazet MY, Kuitche A, Ndjaka JM (2014) Influence of the Determination Methods of K and C Parameters on the Ability of Weibull Distribution to Suitably Estimate Wind Potential and Electric Energy. Int J Renew Energy Develop 3: 145-154.

9. Ohunakin OS, Akinnawonu OO (2012) Assessment of wind energy potential and the economics of wind power generation in Jos, Plateau State, Nigeria. Energy Sustain Develop 16: 78-83.

10. Justus CG, Mikhail A (1976) Height variation of wind speed and wind distributions statistics. Geophys Res Lett 3: 261-264.

11. Mathew S, Pandey KP (2000) Modelling the integrated output of mechanical wind pumps. J Solar Energy Engineering 122: 203-206.

12. Vestas V29 documentation.

13. Vestas V100 documentation.

14. Carta JA, Ramírez P, Velázquez S (2008) Influence of the level of fit of a density probability function to wind-speed data on the WECS mean power output estimation. Energy Convers Management 49: 2647-2655.

15. Eclipse for Parallel Application Developers.

16. Safari B, Gasore JA (2010) Statistical investigation of wind characteristics and wind energy potential based on the Weibull and Rayleigh models in Rwanda. Renewable Energy 35: 2874-2880.

# 9

# Using Angstrom-Prescott (A-P) Method for Estimating Monthly Global Solar Radiation in Kashan

**Ali Razmjoo[1]\*, S. Mohammadreza Heibati[2], Mohammad Ghadimi[3], Mojtaba Qolipour[4] and Javad Rezaei Nasab[5]**

[1]Department of Energy Systems Engineering, Islamic Azad University, South Tehran Branch, Tehran, Iran
[2]Department of Mechanical Engineering, Islamic Azad University, Pardis Branch, Iran
[3]Department of Mechanical Engineering, Islamic Azad University, Roudehen Branch, Iran
[4]Department of Industrial Engineering, Yazd University, Iran
[5]Islamic Azad University, Bushehr Branch, Iran

### Abstract

Today's world is witness to growing use of renewable energy as solutions for reduction of air pollution and provision of a healthier environment. Of all available types of renewable energy, solar energy is perhaps the one that can make the best contribution to this cause. In this article, first the data pertaining to solar radiation in all cities of Kashan region were gathered from Iran's national meteorological organization. Then, Angstrom-Prescott (A-P) method and MATLAB software were used to calculate the monthly radiation, maximum monthly radiation, constant coefficients, and solar radiation on a flat surface in each specific site. All cities in Kashan region showed a mean annual solar radiation of 8.32 hours a day, and constant coefficients were found to be 0.30 and 0.49. In conclusion, results show that Kashan region has a high solar radiation and thus solar energy generation potential.

**Keywords:** Solar energy; Estimate; Sun radiation; Mat lab

**Nomenclature:** H0: Monthly average daily extraterrestrial radiation on horizontal surface (MJ/m²); H: Monthly average daily global radiation on horizontal surface (MJ/m²); GSC: Solar constant; So: Monthly average of the maximum possible daily; s: Monthly average of the solar radiation; a: Regression coefficients; b: Regression coefficients; L: Longitude of the location

**Greek letters:** δ: Solar declination; φ: Latitude of the location; ωS: Sunset hour angle

## Introduction

Adequate supply of energy always has been one of the important factors of economic growth and development [1]. On one hand, today's world is faced with rapidly increasing population and energy consumption, which gives rise to more extensive usage of fossil fuels. On the other hand, the global warming phenomenon triggered by the excessive use of these fuels has now become a pressing global issue. Hence, researchers have introduced and promoted the use of sustainable energy resources as a solution for tackling these major problems [2,3]. At present, most countries have developed or adopted national plans for decreasing the reliance on fossil energy and promoting the development of renewable energy and this is reflected in the multitude of renewable energy projects implemented in the last two decades in different points of the world [4]. Renewable energy resources are clean and widely available and provide a number of environmental and economic benefits which distinguishes them from conventional energy resources. Recent studies have shown a steady growth in use of renewable energy around the world [5,6].

There are different ways to estimate solar radiation and forecasting it, Angstrom- Prescott, klein, Doorenbos and Pruitt, Allen etc. presented a method to evaluate solar radiation, recently many of researchers demonstrated variety of methods for estimation solar radiation such as a novel and efficient 2-D model approach, using of intelligent ANN modeling for prediction of solar energy which is a precise method than before and using of MLP neural networks, Wavelet network, Adaptive Neuro-Fuzzy Inference System for prediction solar radiation, application of using hybrid renewable energy. To evaluate solar radiation in an area different parameters such as average global Solar radiation, global horizontal Irradiance, diffuse horizontal Irradiance ,daily average of global Solar radiation, daily average of GSR during month, maximum and minimum daily average GSR during month can be used [7-13]. Today, investigation of renewable energy potentials is part of planning policies of many countries [14-16]. Figure 1 shows Global Power Plant Market Shares in % and MW/a, 2004-2013 years. In this figure, the consumption of fuel gas that produce by power plants in 2004 year is 45% while with growth of the renewable energy this amount reduced and reached to 16% in 2013 year [17].

Today, the world is witness to a steady and continuous rise in the use of solar energy [18]. In the past decade the issues related to solar energy, including the estimation of solar and hybrid energy generation potentials have been the subjects of many researches [19-22]. Germany is one of the leading countries in regard with development and deployment of renewable energy technologies and the competition triggered by Germany as well as other developed countries has led to a significant drop in prices of solar energy generation equipment. In 1980 solar photovoltaic modules had an average price of 30 US \$/Wpeak but by 2013 this price was reduced by 97% to 0.9 US \$/Wpeak. In Germany, cost reductions were achieved primarily due to the Feed-in Law. Figure 2 shows the price trend of solar photovoltaic module [23].

Today, the use of solar energy has become a global trend and as a result an increasing number of companies are becoming engaged in production of solar panels [23]. Solar energy is perhaps the best-known types of renewable sources and one of the most important resources

**\*Corresponding author:** Ali Razmjoo, Department of Energy Systems Engineering, Faculty of Engineering, Islamic Azad University-South Tehran Branch, Tehran, Iran, E-mail: razmjoo.eng@gmail.com

$\omega$

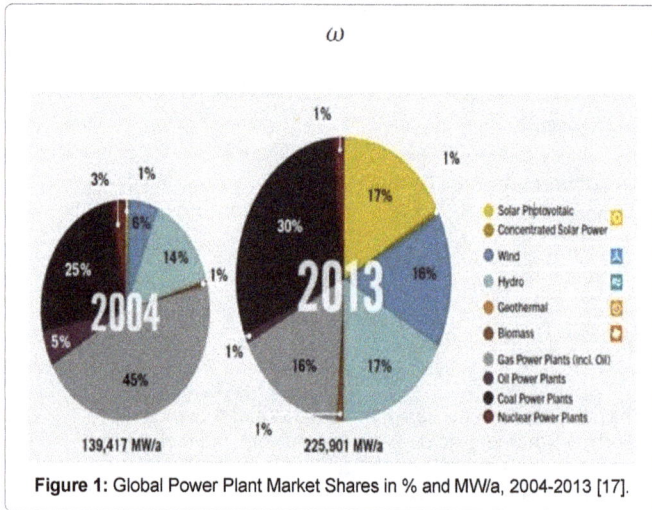

**Figure 1:** Global Power Plant Market Shares in % and MW/a, 2004-2013 [17].

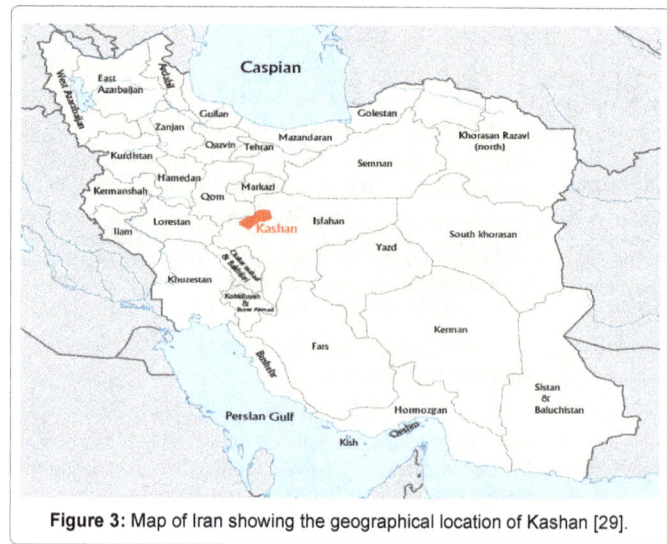

**Figure 2:** Development of Solar Photovoltaic Module Prices [17].

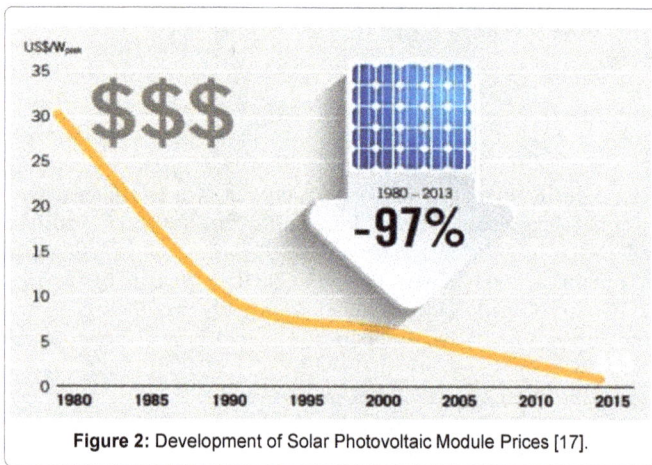

**Figure 3:** Map of Iran showing the geographical location of Kashan [29].

$$\frac{H}{H0} = a + b\left(\frac{s}{S0}\right) \tag{1}$$

Where H0 Monthly average daily extraterrestrial radiation on horizontal surface (MJ/m^2 day), H Monthly average daily global radiation on horizontal surface (MJ/m^2), n is the mean daily number of hours of bright sunshine per month and N is the mean value of day length per month in a specific region, we also use the following equation to calculate the coefficients 'a' and 'b' [32,33]:

Parameters a and b are the regression coefficients.

$$a = -0.110 + 0.235\cos\varphi + 0.323\left(\frac{s}{S0}\right) \tag{2}$$

$$b = 1.449 - 0.553\cos\varphi - 0.694\left(\frac{s}{S0}\right) \tag{3}$$

H0 is expressed as: [34]

$$H0 = \frac{24*3600*GSC}{\pi}[1+0.033\cos\frac{360n}{365}](\cos(\varphi) \tag{4}$$

$$\cos(\delta)\sin(\omega S) + \frac{2\pi\omega S}{360}\sin(\varphi)\sin(\delta))$$

where, H0 is the Monthly average daily extraterrestrial radiation on horizontal surface (MJ/m^2day), GSC is the solar constant and is equal to 1367 Wm^2, δ is the solar declination, ωS is the mean hour angle at sunrise in a given month, solar declination (δ) and mean hour angle at sunrise, and ωS can be calculated by equations (22) and (24) [35]. The following formula is used to obtain δ .The solar declination (δ) and the mean sunrise hour angle (ωS) can be calculated by the following equations (5) and (6). In order to obtain the amount of solar declination (δ), there is an expression: [36]

$$\delta = 23.45\sin\left(360\frac{284+n}{365}\right) \tag{5}$$

Also mean sunrise hour angle can be obtained by:

$$\omega S = \cos^{-1}(-\tan\varphi\tan\delta) \tag{6}$$

The monthly average of the maximum possible daily can be

for production of economic, clean and sustainable energy [24]. In Iran, photovoltaic cells (PV) have proved to be a high performance technology with great potential as an element of large-scale energy generation schemes [25]. Sabziparvar et al. have studied and estimated the global solar radiation in central arid deserts of Iran [26].

## Area Studied

Kashan is a city in the Isfahan province (Iran), with population 335875 and 9647 km² area respectively. The mean of kashan back to a tribe with name kasho that after years it changed and converted to kashan. This city is located at 51.27° N and 33.59° E and at an altitude of 982.3 meters above sea level. Kashan has a temperature 28° and approximately cities such as Yazd with 26°, Isfahan 27° and arak 25° has similar climate as kashan. The region where Kashan is located has a warm and relatively dry climate, but due to a lot of variety ancient places and historical attractions has high visitor during year, actually Kashan is widely known as a touristic city and is famous for its ancient and beautiful houses [27,28]. Figure 3 shows the location of the studied area on the map of Iran. As can be seen, Kashan is located in the central parts of Iran and it's obvious with red color [29].

## Solar Analysis

Solar radiation estimate is an important factor for analysis and assessment of solar radiation potential of a given region. There is an equation called Angstrom- Prescott (1940) which can be used for this purpose [30,31]:

obtained [37]:

$$No = \frac{2}{15} cos^{-1} (-tan\ \varphi\ tan\ \delta) \qquad (7)$$

Table 1 presents the constant coefficients, solar radiation on a flat surface in a specific region, mean daily solar radiation and maximum daily radiation.

Table 2 presents the regression coefficients, solar radiation on a flat surface, Monthly average daily extraterrestrial radiation on horizontal surface (MJ/m²day), Monthly average daily global radiation on horizontal surface (MJ/m² in Kashan city.

Figure 4 illustrates the mean solar radiation per month. As this figure shows, maximum and minimum radiation in February with 6.06 and January with 10.97 was obtained respectively.

Figure 5 depicts the monthly average daily global radiation on horizontal surface and the monthly average daily extraterrestrial radiation on horizontal surface. According to this figure, maximum (44.28) and minimum (16.69) monthly average daily global radiation on horizontal surface have been observed in January and May. Maximum and minimum monthly average daily extraterrestrial radiations on horizontal surface are 24.12 and 9.67 and have been observed respectively in August and December.

Figure 6 shows the monthly solar radiation in Kashan. In this figure, minimum (178.71) and maximum (340.23) radiations can be observed in February and August.

## Conclusion

Global warming and environment problems are crucial factors that need to be considered in clean energy utilization decisions. The present paper was a statistical analysis of solar radiation for the city of Kashan by the use of Angstrom-Prescott (A-P) method. First, the data pertaining to solar radiation in all cities within Kashan region were gathered from Iran's national meteorological organization. Next, the Angstrom-Prescott (A-P) method, which is a known method of solar radiation analysis, was used along with MATLAB software to calculate the amount of radiation per month, the maximum amount of possible radiation per month, constant coefficients of equation, Monthly average daily extraterrestrial radiation on horizontal surface (MJ/ m^2day), Monthly average daily global radiation on horizontal surface (MJ/m^2), for the specified areas. After analysis, the average amount of annual solar radiation for the cities of Kashan was calculated to 8.32 hours a day. Moreover, the constant coefficients obtained for

| Time | a | b | Ho (KJ/m²-day) | H (KJ/m²-day) | H/Ho | s | S0 | $\frac{s}{S0}$ |
|------|------|------|------|------|------|------|------|------|
| Jan | 0.22 | 0.68 | 44.28 | 17.13 | 0.38 | 6.25 | 14.23 | 0.43 |
| Feb | 0.27 | 0.58 | 20.54 | 10.12 | 0.49 | 6.06 | 10.28 | 0.58 |
| Mar | 0.26 | 0.61 | 41.23 | 19.36 | 0.46 | 7.47 | 13.72 | 0.54 |
| Apr | 0.28 | 0.56 | 35.96 | 18.42 | 0.51 | 7.83 | 12.81 | 0.61 |
| May | 0.37 | 0.35 | 16.69 | 10.93 | 0.65 | 8.92 | 9.79 | 0.91 |
| June | 0.36 | 0.38 | 31.21 | 20.09 | 0.64 | 10.58 | 12.14 | 0.87 |
| July | 0.41 | 0.27 | 18.54 | 12.98 | 0.70 | 10.45 | 10.14 | 1.03 |
| Aug | 0.34 | 0.43 | 39.17 | 24.12 | 0.61 | 10.97 | 13.56 | 0.80 |
| Sep | 0.33 | 0.46 | 37.07 | 22.25 | 0.6 | 9.90 | 13.02 | 0.76 |
| Oct | 0.36 | 0.38 | 17.52 | 11.27 | 0.64 | 8.58 | 9.84 | 0.87 |
| Nov | 0.27 | 0.58 | 32.13 | 15.84 | 0.49 | 6.99 | 12.01 | 0.58 |
| Dec | 0.22 | 0.69 | 25.32 | 9.67 | 0.38 | 5.95 | 10.98 | 0.42 |
| Year | 0.30 | 0.49 | 29.97 | 16.01 | 0.54 | 8.32 | 11.87 | 0.70 |

**Table 2:** Monthly solar radiation, regression coefficients, mean monthly solar radiation, Monthly average daily extraterrestrial radiation on horizontal surface (MJ/m²day), Monthly average daily global radiation on horizontal surface (MJ/m²).

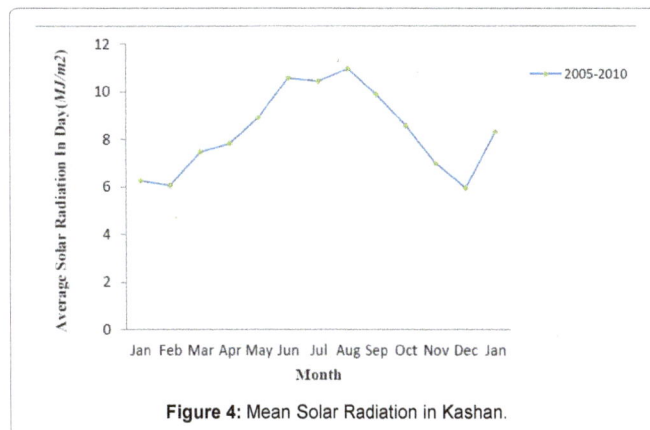

**Figure 4:** Mean Solar Radiation in Kashan.

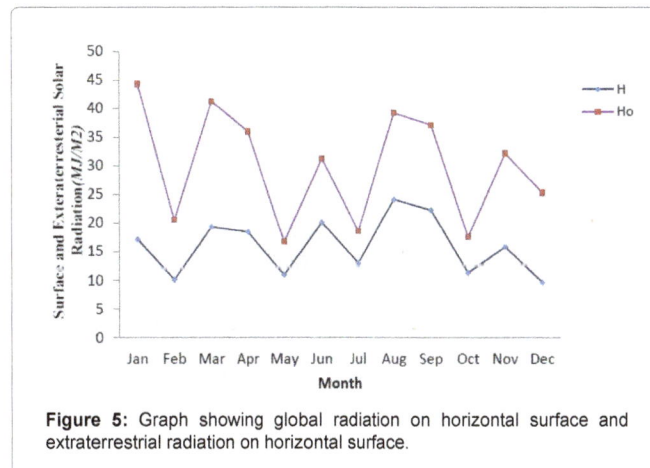

**Figure 5:** Graph showing global radiation on horizontal surface and extraterrestrial radiation on horizontal surface.

| Month | ωS | δ | Mean solar radiation per month |
|-------|------|------|------|
| Jan | 106.74 | 23.44 | 187.53 |
| Feb | 77.15 | -18.50 | 182.01 |
| Mar | 102.93 | 18.62 | 224.16 |
| Apr | 96.08 | 9.06 | 242.76 |
| May | 73.44 | -23.21 | 276.75 |
| Jun | 91.04 | 1.58 | 328.16 |
| Jul | 76.05 | -19.94 | 324.06 |
| Aug | 101.7 | 16.98 | 340.23 |
| Sep | 97.65 | 11.33 | 307.11 |
| Oct | 73.84 | -22.72 | 257.63 |
| Nov | 90.11 | 0.17 | 209.96 |
| Dec | 82.36 | -11.31 | 178.71 |

**Table 1:** Mean monthly global solar radiation and its input parameters in Kashan.

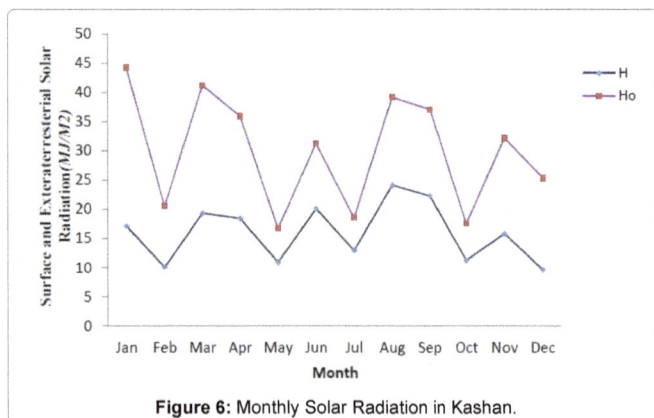

**Figure 6:** Monthly Solar Radiation in Kashan.

Kashan were 0.30 and 0.49. Also, Monthly average of the maximum possible daily calculated to be 14.23 and 9.79 (KJ/m^2-day). Finally, the technical and economic analysis demonstrated that investment on solar energy of this area is justified. The results indicate that Kashan region has a high solar radiation and thus solar energy generation potential.

## References

1. Ekren O, Ekren BY (2011) Size Optimization of a Solar-wind Hybrid Energy System Using Two Simulation Based Optimization Techniques.

2. Fazelpour F, Soltani N, Rosen MA (2014) Wind resource assessment and wind power potential for the city of Ardabil. Iran Int J Energy Environ Engineering 6: 431-438.

3. Ali M, Ahmad S, Morteza G, Yagob D, Mojtaba M, et al. (2013) Evaluation of wind energy potential as a power generation source for electricity production in Binalood, Iran. Renew Energy 52: 222-229.

4. Yazdanpanah MA (2014) Modeling and sizing optimization of hybrid photovoltaic/wind power generation system. J Ind Eng Int 10: 49.

5. Nevzat O, Sedat E (2011) Analysis of wind climate and wind energy potential of regions in Turkey. Energy 36: 148-156.

6. Bilal BO, Ndongo M, Kebe CMF, Sambou V, Ndiaye PA (2013) Feasibility study of wind energy potential for electricity generation in the northwestern coast of Senegal. Energy Procedia 36: 1119-1129.

7. Dumas A, Andrisani A, Bonnici M, Madonia M, Trancossi M (2014) A new correlation between solar energy radiation and some atmospheric parameters. Atmospheric and Oceanic Physics.

8. Hocaoglu FO, Gerek ON, Kurban M (2009) A Novel 2-D Model Approach for the Prediction of Hourly Solar Radiation. Computational and Ambient Intelligence 5: 749-756.

9. Hocaoglu FO, Gerek ON, Kurban M (2008) Hourly solar radiation forecasting using optimal coefficient 2-D linear filters and feed-forward neural networks. Solar Energy 82: 714-726.

10. Besharat F, Dehghan AA, Faghihm AR (2013) Empirical models for estimating global solar radiation: A review and case study. Renew Sustain Energy Rev 21: 798-821.

11. Olatomiwa L, Mekhilef S, Shamshirband S, Petković D (2015) Adaptive neuro-fuzzy approach for solar radiation prediction in Nigeria. Renew Sustain Energy Rev 51: 1784-1791.

12. Kumar N, Sharma SP, Sinha UK, Nayak YK (2016) Prediction of Solar Energy Based on Intelligent ANN Modeling. Int J Renew Energy Res 6: 1.

13. Ahmed EA, Adam ME (2013) Estimate of Global Solar Radiation by Using Artificial Neural Network in Qena, Upper Egypt. J Clean Energy Technol 1: 2.

14. Zhoua Y, Wenxiang W, Liu G (2011) Assessment of Onshore Wind Energy Resource and Wind-Generated Electricity Potential in Jiangsu, China. Energy Procedia 5: 418-422.

15. Archer CL, Caldeira K (2009) Global Assessment of High-Altitude Wind Power. Energies 2: 307-319.

16. Sadeghi M, Gholizadeh B (2012) Economic analysis of using of wind energy, Case study: Baladeh city, North of Iran. Int J Agriculture Crop Sci 4: 666-673.

17. REN21 (2004) Renewable energy policy network for 21st century.

18. Srinivas T, Reddy BV (2014) Hybrid solar-biomass power plant without energy storage. Case Studies Thermal Engineering 2: 75-81.

19. Enda F, Kevin MD, Fionnuala M, Ger D (2011) Feasibility Analysis of Photovoltaic Solar Power for Small Communities in Ireland. The Open Renewable Energy Journal 4: 78-92.

20. Ahmad MJ, Tiwari GN (2009) Optimization of Tilt Angle for Solar Collector to Receive Maximum Radiation. The Open Renew Energy J 2: 19-24.

21. Ganguli S, Singh J (2010) Estimating the Solar Photovoltaic generation potential and possible plant capacity in Patiala. Int J Appl Engineering Res 1: 253.

22. Muralikrishna M, Lakshminarayana V (2008) Hybrid (solar and wind) energy systems for rural electrification. ARPN J Engineering Appl Sci 3: 50-58.

23. Perez R, Seals R, Ineichen P, Stewart R, Menicucci D (1987) A new simplified version of the perez diffuse irradiance model for tilted surfaces. Sol Ener 39: 221-231.

24. Ng KM, Adam NM, Inayatullah O, Zainal M, Kadir AA (2014) Assessment of solar radiation on diversely oriented surfaces and optimum tilts for solar absorbers in Malaysian tropical latitude. Int J Energy Environ Eng 5: 75.

25. Sharma P, Harinarayana T (2013) Solar energy generation potential along national highways. Int J Energy Environ Engineering 4: 16.

26. Sabziparvar AA (2007) Simple formula for estimating global solar radiation in central arid Deserts of Iran. Renew Energy 33: 1002-1010.

27. Wikipedia (2009) Kashan.

28. Iran Census 2006 (2006) Islamic Republic of Iran.

29. Resistance economy, action "Leader".

30. Maraj A, Londo A, Firat C, Karapici R (2014) Solar Radiation Models for the City of Tirana, Albania. Int J Renew Energy Res 4: 2.

31. Sarsah EA, Uba FA (2013) Monthly-Specific Daily Global Solar Radiation Estimates Based On Sunshine Hours In Wa, Ghana. Int J Scientific Technol Res 2: 246-254.

32. Gana NN, Akpootu DO (2013) Angstrom Type Empirical Correlation for Estimating Global Solar Radiation in North-Eastern Nigeria. Int J Engineering Sci 2: 58-78.

33. Rajput AK, Tewari RK, Sharma A (2012) Utility Base Estimated Solar Radiation at Destination Pune, Maharashtra, India. Int J Pure Appl Sci Technol 13: 19-26.

34. Garba AA, Amusat RO, Ngadda YH (2016) Estimation of global solar radiation using sunshine-based model in Maiduguri, north east, Nigeria. Appl Res J 2: 19-26.

35. Saffaripour MH, Mehrabian MA, Bazargan H (2013) Predicting solar radiation fluxes for solar energy system applications. Int J Environ Sci Technol 10: 761-768.

36. Toğrul IT (2009) Estimation of solar radiation from angstroms coefficient by using geographical and meteorological data in Bishkek, Kyrgyzstan. J Thermal Sci Technol 29: 99-108.

37. Li H, Lian Y, Wang X, Ma W, Zhao L (2011) Solar constant values for estimating solar radiation. Energy 1-5.

# Numerical Investigations of Coupling a Vacuum Membrane Desalination System with a Salt Gradient Solar Pond

**Samira Ben Abdallah\*, Nader Frikha and Slimane Gabsi**

*Samira Ben Abdallah, Research Unit of Environment, Catalysis and Analysis Processes, University of Gabes, National Engineering school of Gabès, Street Omar Ibn ElKhattab, 6029 Gabès, Tunisia*

## Abstract

This work proposes new configurations for the desalination of salt water using systems based on coupling of membrane distillation with solar energy. This study is a comparison between two coupling configurations of the vacuum membrane distillation (VMD) hollow fiber module with salinity gradient solar pond (SGSP). The first configuration is a module membrane in series with SGSP and the second one is a hollow fiber module immersed in the SGSP. Two models describing the heat and mass transfer in the hollow fiber module and in the SGSP will be developed. The coupling of the two models allows the determination of the instantaneous variation of temperature and salinity in the SGSP and the permeate flow variation. A comparison of each module production was carried out. The mathematical model shows that the immersed module production presents more than one and a half times that of the separated module, their production reached 75 kg.day$^{-1}$ per m² of the membrane in the third year. Thus, immersing the module in the solar pond improves the performance of the hollow fiber module.

**Keywords:** Desalination; Hollow fiber module; Modeling; Solar pond

## Abbreviations

### Nomenclature

AM: Sun's path; alt: altitude; $C_c$: Conversion factor of the isotropic fraction; $C_p$: The specific heat capacity of the water; $D_i$: Diffuse radiation on a flat surface; D: Coefficient of salt diffusion; $D_h$: Horizontal diffuse radiation; F: Fraction of isotropic diffusion; $G_h$: Horizontal global radiation; h: Height of the sun; $I_0$: Incidence solar radiation; $P_s$: Vapor pressure; $P_{vacuum}$: Vacuum pressure; S: Salinity; $Q_m$: Mass flow; Q: Heat flow; R: The ideal gas constant; T: Temperature; $X_{NaCl}$: Salt molar fraction; $I_n$: Normal incidence solar radiation

### Greek letters

$\alpha_d$: Distance correction coefficient; $\alpha_{th}$: Heat expansion coefficient; $\alpha_{eau/NaCl}$: Water activity coefficient; $\beta$: Disturbing factor; $\beta_m$: Expansion coefficient; $\mathcal{E}$: Membrane porosity; $\Phi$: Inclination angle; $\lambda$: Heat conductivity; $\mu$: Dynamic viscosity; $\mu_{ex}$: Extension coefficient; $\theta_{inci}$: Angle of incidence; $\rho$: The salt water density; $\tau$: Membrane tortuosity

### Indices

acc: accumulated; amb: ambient; abs: absorbed; f: feed; o: out; sw: Sea water

## Introduction

The VMD process is based on the evaporation of solvents through hydrophobic porous membranes promoted by applying vacuum or low pressure on the permeate side [1]. Vacuum is applied in the permeate side of the membrane module by means of a vacuum pump. The applied vacuum pressure is lower than the saturation pressure of volatile molecules to be separated from the feed solution. Permeate condensation takes place outside the module, inside a condenser [2]. This configuration presents a high permeate flow compared to other membrane distillation (MD) configurations, in addition to advantages of low thermal conduction loss [3-5].

MD was considered as a promising technology and has appeared as a process more attractive than any other popular separation process

due to its lower operating temperatures and hydrostatic pressures [6]. Additionally, MD is capable of integrating with various renewable energy sources such as solar energy, geothermal energy and waste heat source [7,8]. The utilization of the renewable energy is worth further research in order to bring the technology closer to the process intensification. So far, the solar energy has been frequently studied in MD by many researchers [9-12]. However, MD requires high heat energy [13]. Being capable of directly using solar thermal energy, the solar membrane distillation desalination system has developed as a promising green technology to reduce the water resource problem [14]. Saffarini et al. have shown that heater systems costs account for over 70% of the total cost for all systems, suggesting the desirability of using alternative sources of thermal energy, such as solar energy [15].

Based on different methods used to impose a vapor pressure difference on either membrane side to drive the permeate flow, there are essentially four types of membrane distillation configurations: direct contact membrane distillation (DCMD), sweeping gas membrane distillation (SGMD), air gap membrane distillation (AGMD) and vacuum membrane distillation (VMD). The majority of research studies concern the coupling of solar collectors with the other configurations of the MD such as DCMD [14,16,17] and AGMD [18,19]. In desalination, the AGMD, the DCMD and the VMD configurations have been successfully applied, providing fresh water as permeate [12,18,20]. Wang et al. were among the first researchers to couple VMD with solar energy [18]. Their study shed the light on a designed and tested solar-heated hollow fiber based VMD system. The largest permeate flux obtained is 32.19 L.h$^{-1}$ per m² of membrane with an 8 m² solar energy

---

**\*Corresponding author:** Samira Ben Abdallah, Research Unit of Environment, Catalysis and Analysis Processes, University of Gabes, National Engineering school of Gabès, Street Omar Ibn ElKhattab, 6029 Gabès, Tunisia
E-mail: abdallahsamira@yahoo.com

collector. The existing configurations are essentially coupled with flat plate collectors and the module was separated to the solar collector. But for the other DM configurations the only coupling with a solar pond is carried out by Nakao et al. [12], which study the utilization of DCMD coupled to a salt gradient solar pond. So the coupling of a flat membrane module separated from the pond was realized. This study shows that the module production reached 2 l/m²/hr at the end of June, 2014. Therefore, Mericq et al. [21] studied the possibility of submerging the plate DMV membrane in the SGSP. The use of SGSP does not only seem to be the most interesting solution but also allows a high permeate flux to be reached with membrane fed by waters from lower convective zone (LCZ) of the SGSP. The study of different coupling configurations has confirmed the benefits of the membrane module immersion in the solar collector [18,22-24]. This possibility reduces heat loss and has a compact installation, which leads to improved productivity.

This research work accounts for the possibility of coupling the VMD membrane module with an SGSP. It presents a follow-up of the hollow fiber module production along three years. It also provides a comparison of the water production of two coupling possibilities, the first of which is a module fed with SGSP (Figure 1A) and the second one is a hollow fiber module immersed in the SGSP (Figure 1B). The latter is developed in order to minimize the heat loss and maximize water production. The approach presented in this paper was used to choose the most efficient coupling possibility.

## Modeling

The desalination system is composed of an SGSP and a hollow fiber module. Vacuum membrane distillation is a complicated physical process in which both heat and mass transfers are involved. Indeed, the coupling of the heat and mass transfer equations in the module and the SGSP leads to the establishment of a model describing the functioning of each configuration. The model was developed to calculate the effect of the solar energy on the permeate flux. The variations of the temperature and product distillate during the day were determined.

### Hollow fiber module

The membrane module is a hollow fiber membrane which can have a large effective area compared with other membrane module types. The hollow fiber module configuration is external-internal. The feed solution flows from outside the hollow fibers and the permeate is collected inside the hollow.

MD is a thermally driven process based on the principle of vapor/liquid equilibrium and coupled heat and mass transfer [18]. The heat transfer simultaneously occurs with mass transfer whose process influences the rate and coefficients of the heat transfer process, giving birth to a complex heat transfer model [23]. The transfer through the membrane is caused by a partial pressure difference on either side of the membrane. The vapor water molecule was transported through

the membrane pores from the higher pressure to lower pressure side. Generally in the VMD the vapor water molecules transfer through the membrane pores is given by the mechanism of Knudsen diffusion where the mean free path of the molecules is very large relative to the average pore [9,25-27]. Indeed in VMD, the mean free path of water can reach relatively high values. Then molecule-pore wall collisions are dominant in membranes with small pores. In addition Knudsen diffusion dominates in VMD if the vacuum is sufficiently pushed. In our case we used small pore size membrane (less than 0.45 microns) and a very high vacuum does not exceed the 4000 Pa. So the diffusion of the vapor through the membrane pores according to a Knudsen mechanism.

The water vapor flux through the internal interface membrane-water (kg.s⁻¹.m⁻²) is described by the following equation:

$$J_v = K_m \Delta P \tag{1}$$

The driving force of the pressure difference ΔP can be expressed as follows:

$$\Delta P = P_i - P_{vacuum} \tag{2}$$

Partial pressure $P_i$ was written as a function of the activity coefficient and the interfacial concentration.

$$P_i = \alpha_{water/NaCl}(1 - X_{NaCl})P_s \tag{3}$$

$\alpha_{water/Nacl}$ is the water activity coefficient. This coefficient depends on the water concentration in the treated solution [25].

$$\alpha_{water/NaCl} = 1 - 0.5X_{NaCl} - 10X_{NaCl}^2 \tag{4}$$

and $P_s$ is the interfacial vapor pressure of pure water and can be evaluated using the Antoine equation:

$$= \exp(23.238 - \frac{3841}{45}) \tag{5}$$

where $T_i$ is the corresponding interfacial temperature in Kelvin.

The coefficient of the membrane permeability or the Knudsen permeability $K_m$ can be related to the membrane structural properties such and the membrane interface temperature ($T_i$) [9,25,26].

$$K_m = \frac{2\,\varepsilon\,r_p}{3\,\tau\,e_m}\frac{1}{R\,T_i}\sqrt{\frac{8\,R\,T_i}{\pi}} \tag{6}$$

So

$$J_V = K_m\left[\alpha_{eau/NaCl}(1 - X_{NaCl})\exp(23.238 - \frac{3841}{T_i - 45}) - P_{vacuum}\right] \tag{7}$$

The mass transfer inside the module is coupled with a heat transfer through the membrane. The establishment of a rigorous model describing the heat and mass transfer inside the hollow fibre module is very complex. Based on some assumptions and the heat transfer equation for a flowing liquid in a cylindrical conduct, a model

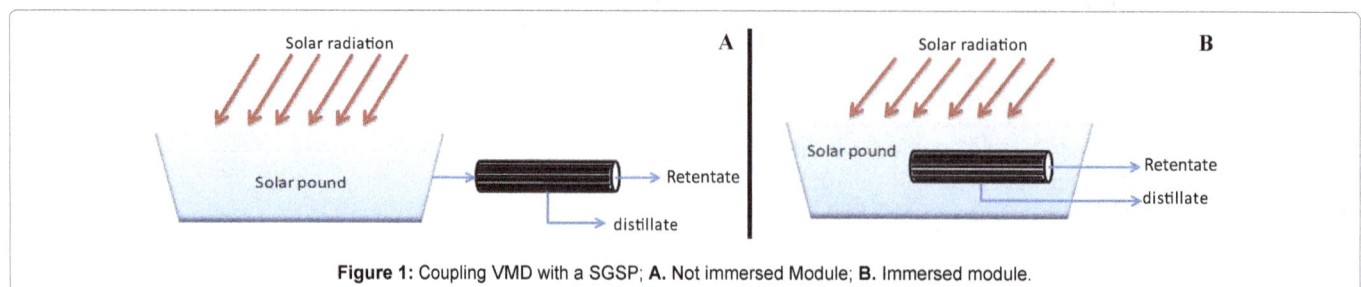

**Figure 1:** Coupling VMD with a SGSP; **A.** Not immersed Module; **B.** Immersed module.

describing the heat and mass transfer in the hollow fiber module was developed [22,24].

To solve the set of model equations we have used of a program using the Matlab calculation software which allows the equations to be presented in the obtained model. The resolution is fully developed by the Ruge Kutta method using the predefined function in Matlab ode 23. This function is executed to solve non-stiff differential equations, low order method.

The obtained model allows to determine the temperature profile inside the module, the module output temperature as well as the flow of distillate produced as a function of different parameters.

## Solar radiation model

The study of solar radiation is the starting point of any solar energy investment. A model describing the different irradiations depending on climatic parameters was developed. This calculus is based on the EUFRAT model which, in turn, is based on the synthesis of various research works, especially those Brichambant, Kasten and Hay [28]. The atmosphere does not transmit the entire solar radiation to the ground:

- Direct radiation is the one that passes through the atmosphere without modification. This radiation at normal incidence is determined from the following equation:

$$I_n = I_0 \; \alpha_d \exp\left[\frac{-\,AM\;\beta}{0.9\;AM + 9.4}\right] \tag{8}$$

with AM the sun's path that is given by the following expression:

$$AM = \frac{(1-0.1\,alt)}{\sinh} \tag{9}$$

- The global radiation on a horizontal plane which is the sum of direct and diffuse radiation is described by the following equation:

$$G_h = \alpha_d \;(1270\text{-}56\;\beta)\;\sin(h)^{\frac{\beta+36}{33}} \tag{10}$$

- The diffuse radiation is the part of the solar radiation diffused by the solid or liquid particles suspended in the atmosphere. It has no preferred direction.

The diffuse radiation is determined from the global radiation:

$$D_h = G_h - I_n\sin(h) \tag{11}$$

The diffuse radiation on a flat surface inclined at an angle ($\Phi$) relative to the horizontal (the tilt angle) and oriented to a direction at an angle $\theta_{inci}$ with the South (incidence angle) is determined using a conversion factor of the isotropic fraction:

$$C_c = \frac{1+\cos\varphi}{2} \tag{12}$$

So

If $\dfrac{\theta_{inci}}{\sinh} > C_c$ we have $D_i = D_h [F\; C_c + \dfrac{(1-F)\,\theta_{inci}}{\sinh}]$ (13)

If $\dfrac{\theta_{inci}}{\sinh} \leq C_c$ we have $D_i = D_h C_c$ (14)

with F as the fraction of isotropic diffusion: $F = 1 - \dfrac{I_n}{I_0\,\alpha_d}$ (15)

Then, the global radiation on an inclined plane is deduced as the sum of two terms:

$$G_i = I_n \cos\theta_{inci} + 0.2\;D_i\;G_h\;\frac{(1-\cos\phi)}{2} \tag{16}$$

The resolution of these equations was carried out using a program that is developed on the Matlab calculation software. The simulation of this program allows us to determine the different types of radiation (direct, diffuse and global) for a given day.

## SGSP model

Practically the solar pond consists of three distinct zones as shown in Figure 2. The first zone, located at the top of the pond, contains the low density saltwater mixture. This zone is called the upper convective zone (UCZ) which is the absorption and transmission region. The second zone which contains a variation of salinity increasing with depth is the gradient zone or non-convective zone (NCZ). This zone acts as an insulator to prevent heat from escaping to the UCZ, maintaining higher temperatures at lower zones. The bottom zone is the heat storage zone or lower convective zone (LCZ) with uniform salinity.

Solar ponds produce relatively low grade thermal energy (less than 100°C) and are generally considered suitable for thermal distillation processes. The solar pond is subjected to solar radiation and heat exchange only through its upper surface. It is assumed that the pond is a priori 'artificially stabilized' so that the convection currents can be considered non-existing and remains as such during the entire period of time under consideration [29].

The energy balance on the volume element of the pond ($\Delta V = A_{pond}\Delta x$) is given by the following equation:

$$Q_{acc}(t,x) = Q_f(t,x) - Q_o(t,x) + Q_{abs}(t,x) \tag{17}$$

This equation reflects that the heat flow accumulated in the volume element is equal to the sum of the inlet flow and absorbed flow minus the out flow.

The energy flow accumulated in the pond volume element (saltwater) is given by the equation:

$$Q_{acc}(t,x) = \Delta x \frac{\partial(\rho_{sw}\,Cp_{sw}T_{sw}(t,x))}{\partial t} \tag{18}$$

The difference between the feed and output energy flow is expressed by the following equation:

$$Q_f(t,x) - Q_o(t,x) = \lambda_{sw}\,A_{pond}\frac{\partial^2 T_{sw}(t,x)}{\Delta^2 x}\Delta x \tag{19}$$

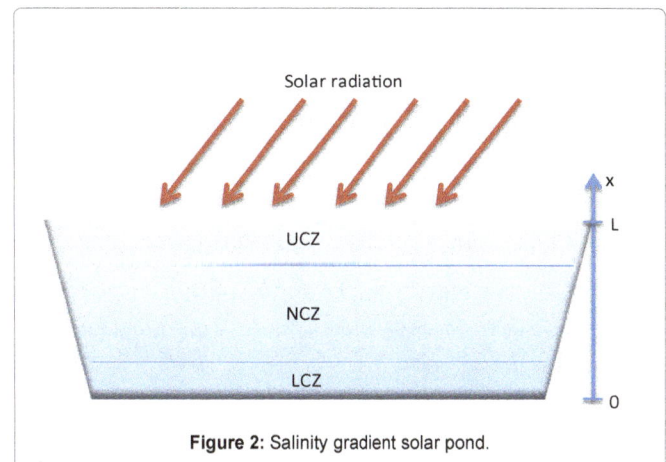

**Figure 2:** Salinity gradient solar pond.

The absorbed energy flow is the rate of energy production in a salt water layer, which results from the solar radiation absorption.

$$Q_{abs}(t,x) = \frac{dI}{dx} A_{pond} \Delta x \tag{20}$$

I is the solar flux received by the higher interface of the pond. The term $\frac{dI}{dx}$ represents, in fact, the rate of energy generation per unit volume in a water layer, which results from the solar radiation absorption. In the present study, we have assumed that the amount of solar radiation, mostly in short wave radiation ranges, reaching the depth x within the pond, suffers an exponential decay as follows [30]:

$$I = 0,6 \, G_i \, e^{-\mu_{ex}(L-x)} \tag{21}$$

Where $G_i$ is the solar radiation incident on the water free surface, which is assumed to be normal to that surface, $\mu_{ex}$, called the extinction coefficient, represents the transparency of the saline solution and L the solar pond depth.

The absorbed flow expression is then as follows [31]:

$$Q_{abs}(t,x) = 0,6\mu_{ex}G_i \, e^{-\mu_{ex}(L-x)} A_{pond} \Delta x \tag{22}$$

The heat transfer process within the solar pond is governed by the following general equation:

$$\frac{\partial(\rho_{sw} Cp_{sw} T_{sw}(t,x))}{\partial t} = \lambda_{sw} \frac{\partial^2 T_{sw}(t,x)}{\partial^2 x} + 0,6\mu_{ex}G_i \, e^{-\mu_{ex}(L-x)} \tag{23}$$

In addition, the mass balance in the volume element of the pond is expressed by the following equation:

$$Q_{m,acc}(t,x) = Q_{m,x+dx}(t,x) - Q_{m,x}(t,x) \tag{24}$$

The mass flow in the pond is described by Fick's law:

$$Q_{m,x} = -D \, A_{pond} \frac{\partial(\rho_{sw}S)}{\partial x} \tag{25}$$

Then, the difference between the inlet and exit mass flow is expressed by the following equation:

$$Q_{m,acc} = D \, A_{pond} \frac{\partial^2(\rho_{sw}S)}{\partial^2 x} \Delta x \tag{26}$$

The accumulated salt flux in the volume element of the pond is also expressed by the following equation:

$$Q_{m,acc} = A_{pond} \Delta x \frac{\partial(\rho_{sw}S)}{\partial t} \tag{27}$$

So, the mass transfer process within the solar pond is governed by the following general differential equation:

$$\frac{\partial(\rho_{sw}S)}{\partial t} = D \frac{\partial^2(\rho_{sw}S)}{\partial^2 x} \tag{28}$$

All fluid properties vary as a function of both temperature and salinity and are evaluated using known formulae.

With the equation of state of the saline solution given as follows [31]:

$$\rho_{sw} = \rho_{ref}(1 - \alpha_{th}(T - T_{ref}) + \beta_m(S - S_{ref})) \tag{29}$$

$\alpha_{th}$ and $\beta_m$ are thermal and salt expansion coefficients published in the literature. The subscripts 'ref' refers to the reference temperature of 25°C.

Finally, the process of mass and heat transfer in solar pond is described in the following system of equations:

$$\begin{cases} \dfrac{\partial(\rho_{sw} Cp_{sw} T_{sw}(t,x))}{\partial t} = \lambda_{sw} \dfrac{\partial^2 T_{sw}(t,x)}{\Delta^2 x} + 0,6\mu_{ex}G_i \, e^{-\mu_{ex}(L-x)} \\ \dfrac{\partial(\rho_{sw}S)}{\partial t} = D \dfrac{\partial^2(\rho_{sw}S)}{\partial^2 x} \end{cases} \tag{30}$$

We developed a calculation program by using the Matlab software computation which solves the set of differential equations. We used the ode 45 function based on the explicit Runge-Kutta method. This function is generally used for the lower order systems. The resolution allows to determine the variations of the different temperatures and the daily distillate flow.

## Results and Discussion

### Solar radiation

The solar radiation model allows us to determine the instantaneous variation of different solar radiation for any day. Figure 3 represents the global solar radiation variation for the four typical days. Figure 3 shows that, for any day, the global solar flux follows the same shape. It increases from sunrise to reach a maximum at noon. It reaches a maximum of 980 W/m² at noon for the June 21, which is the highest while that of December is the lowest. Sunshine depends on the day and month, it is higher in summer.

After the theoretical modeling and simulation, experimental validation step is a milestone for the model performance evaluation and its subsequent exploitation. The Sunshine experimental values used for the model validation are taken using a station equipped with a sunshine Lambrecht installed in the region of Gabes (33.8933° of latitude and 10.1029° of longitude).

To validate the model results, a comparison between the simulated values and experimental data measured by our research team is made. Figure 4 shows a superposition of the two curves for the June 21. The present irregularity in the daily progress of the experimental curve is due to the cloudy crossing that could interrupt the running of experiments. The simulated values are a bit underestimated at the beginning and at the end of day.

The calculated average deviation between the two curves does not exceed 20%, while the average spread of solar flux is 38 W/m². We can consider that the model correctly describes the evolution of the solar flux along the day. The results obtained in this section will be used for the simulations of the pond temperature variation.

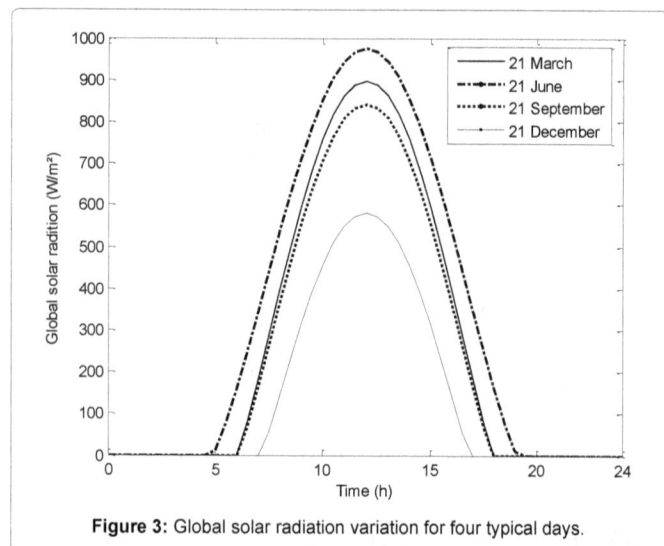

**Figure 3:** Global solar radiation variation for four typical days.

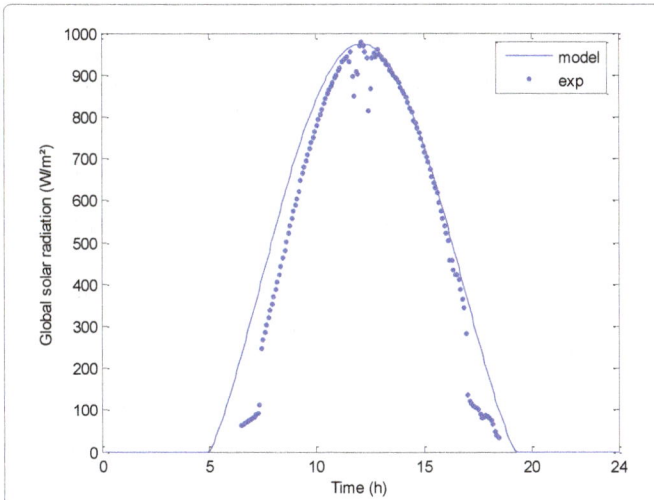

**Figure 4:** Experimental and modeling global solar radiation variation for the 21 June.

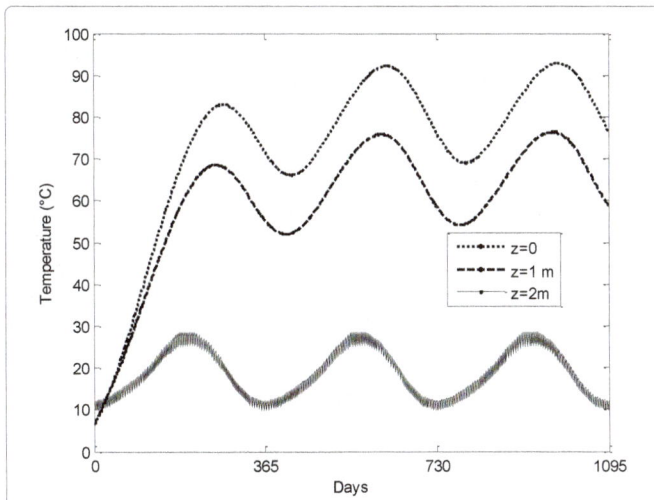

**Figure 5:** instantaneous temperature variation for three years.

## Solar pond

Solving the pond model equations allows determining the temperature instantaneous variation in the SGSP. The SGSP depth is of 2m. We propose to present the simulation results for three years of pond operation to interpret the pond behavior along time and study the stability and the effect of the energy accumulation. Figure 5 presents the instantaneous temperature variation for different depths in bottom of the pond (z=0m), in the middle (z=1m) and in the top (z=2m). The SGSP operation started on 1st, January of the first year where the temperature in the SGSP was the ambient temperature. The temperature reaches its maximum during the summer when the solar radiation is maximal. The temperature at the end of the year is greater than that of the beginning, which is explained by the energy accumulation in the pond. Figure 5 shows that the upper zone temperature is near ambient temperature since the losses are higher in this zone. The temperature fluctuations at the interface that are due to variations in ambient temperature gradually disappeared, approaching the bottom of the pond. The highest temperature is the pond bottom; it is in the order of 95°C for the third year. For the same date, the temperature increases from one year to another due to the energy

stored in the pond. Consequently, the maximum temperature increases from 84°C for the first year to 92°C for the second year to 94°C for the third. So, this elevation decreases from one year to another until the temperature stabilizes at the end of the third year. Figure 6 presents the temperature according to the depth for four typical days for three successive years. It should be noted that these profiles were taken at a specific time, say at 12: 00h of each day.

Figure 6 shows that, when approaching the SGSP bottom, the temperature increases with the increase in salinity. Thus, the largest quantity of energy is absorbed by the concentrated salt solution at the bottom of the SGSP. Similarly, for the bottom, the temperature varies slowly so that the temperature becomes almost constant, thereby forming the LCZ zone. On the other hand, the temperature varies differently for four typical days, in which it varies slowly from one year to another in the cold months (September and December), while the variation is important for hot months (March and June). The solar heating effect within the pond is more important than for the cold season. Besides, for the cold season, in particular for the winter, heat losses towards the ambient temperature are more important due to cold air.

For the top portion (UCZ), the temperature is close to the ambient temperature. The cooling effect of water can be clearly observed in addition to the effect of the energy accumulated within the pond for March and June. This deviation disappeared for September and December, thus forming the UCZ zone.

For the intermediate portion (NCZ), the salinity increases with depth. This zone behaves as transparent heat insulation, and crossed by the solar radiation which is absorbed and trapped by the very salty water at the bottom. For this zone, the temperature profile remains essentially linear, then it slowly varies and the solar heating effect depends on the month. Despite the low ambient temperature and solar radiation for September, the temperature is higher than in June reaching 92°C. The solar heating effect due to the heat loss (low ambient temperature and

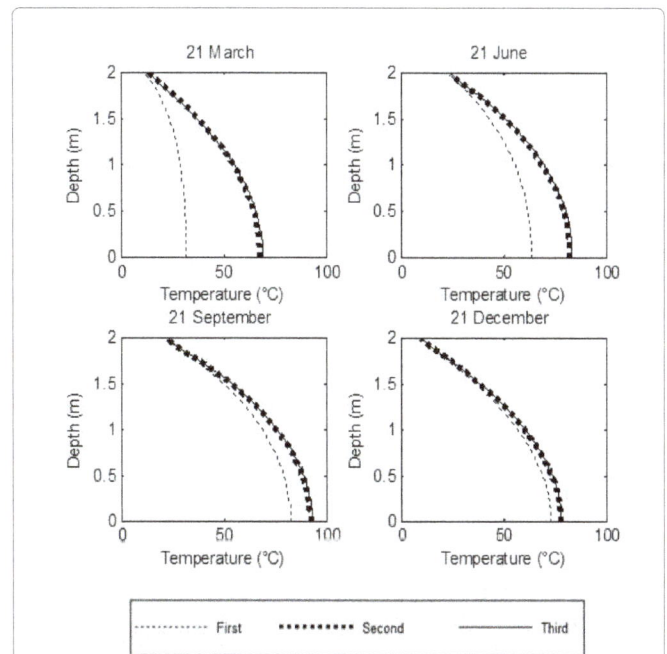

**Figure 6:** Temperature variation according to the depth for four typical days for three successive years.

solar radiation) is about 5°C from September to December. Figure 6 shows that the temperature increase is important for March and June from the first to second year, whereas it does not exceed 5°C for December. Secondly, the temperature stabilizes in December of the third year.

Figure 7 displays salinity according to the depth for four typical days, showing the effects of temperature and heat diffusivity on the salinity variation. It is observed, at first, that the instantaneous salinity variation remains very low and seems almost imperceptible along the year. Thus, the profiles for the upper and central zone were nearly identical, except for the surface and the bottom zone where only a slight change can be noticed. So, the salinity varied slowly during the year reaching stability, thus the salinity profiles remained in the S form then it became linear.

## Module production

The fibres module characteristics were selected to support high temperatures at the membrane wall and have a good permeability. Table 1 shows the characteristics and operating conditions of hollow fibre module.

It is to be noted that the most important temperature is in LCZ part, exceeding 80°C which is the allowable temperature membrane. On the other hand, this zone is characterized by a high salinity, which presents a risk of clogging. For these two reasons, we came to the conclusion that the module feed or the module immersion in this zone is not recommended, hence we chose to couple the module with the middle part (NCZ). Therefore, two configurations were studied:

a hollow fiber module fed from the NCZ water and the hollow fiber module immersed in the NCZ zone.

Following the temperature, the permeate flow varied slightly along the day (Figure 8) such that it varies less than one kilogram along day. This permeate flow variation is related to the low variation of the temperature which does not exceed 1 to 2°C along the day. This low temperature variation is due to the energy accumulation in the pond which can be explained by the great inertia of the pond and the fact that the pond is stabilized after two operation years.

Table 2 shows that the highest daily production of the first year is that of June. However, the lowest production was that of March 21 because the temperature level was still low. Although the March solar radiation is higher than that of September and December, we can see that the production remains the lowest such that the March production for the third year is lower than that of December of the first year. On the other hand, the daily production of September is the greatest for the three year, reaching 346 kg per m² of membrane. This difference does not only depend on the temperature and the heat lost in the top of SGSP and the received solar flux, but also on the energy stored in the SGSP bottom. In fact, the March production was relative to previous months, which corresponds to low energy storage in winter months while the September production is relative to summer months. Consequently, the daily production increases from one year to another, especially for June and September, in which it multiplies of two times and half from the first year to the third.

To compare the two configurations of the fiber module, the daily production for the same operating conditions was calculated. Figure 9 reveals the daily production variation along three years. Module production began only at 80 days when the desired temperature level was reached. Indeed, the temperature did not exceed 30°C which is the evaporating temperature for 4000 Pa vacuum pressure. Figure 9 shows

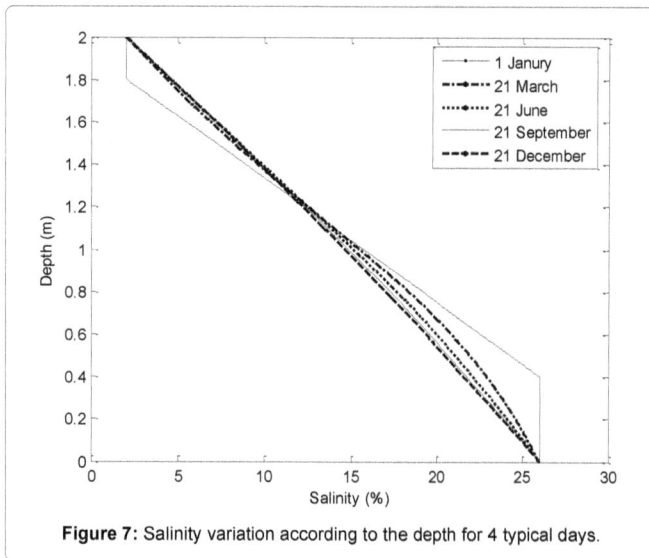

**Figure 7:** Salinity variation according to the depth for 4 typical days.

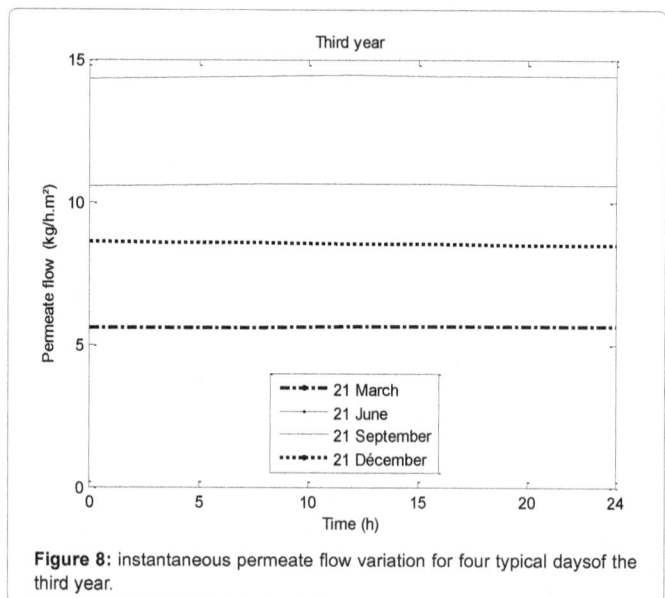

**Figure 8:** instantaneous permeate flow variation for four typical daysof the third year.

| Nature | PVDF |
|---|---|
| Internal fibre diameter (mm) | 2.6 |
| Membrane thickness (mm) | 0.4 |
| Length (m) | 1 |
| Fibres number | 18 |
| Diameter module (cm) | 2 |
| Maximum temperature (°C) | 80 |
| Permeability (ms$^{-1}$) ($T_i$ in K) | $Km = 7.8\ 10^{-6} * T_i^{-0.5}$ |
| Vacuum pressure (Pa) | 4000 |
| Water flow rate (m/s) | 0.5 |

**Table 1:** Characteristics of hollow fibre module.

| Day | 21-Mar | 21-Jun | 21-Sep | 21-Dec |
|---|---|---|---|---|
| First year | 3,5 | 100 | 242 | 163 |
| Second year | 125 | 243 | 339 | 202 |
| Third year | 135 | 255 | 346 | 205 |

**Table 2:** Daily separated module production for four typical days of three successive years (Kg/m²).

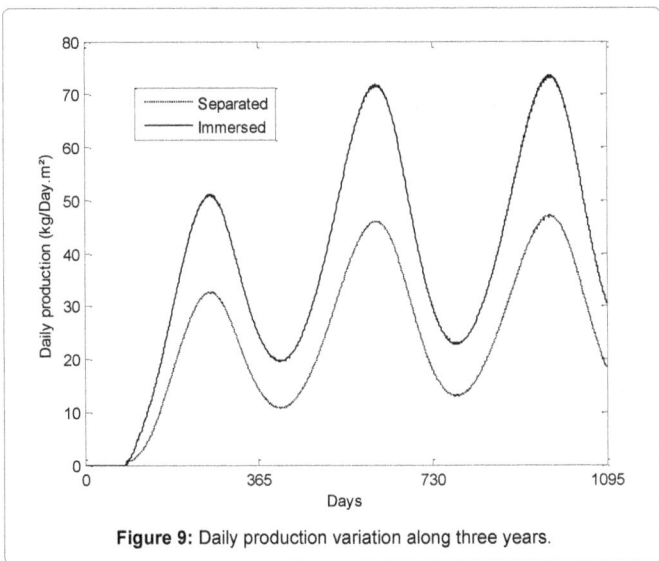

**Figure 9:** Daily production variation along three years.

that the daily production of the immersed module was greater than that of the separated module. The immersed module production presents more than one and a half times that of the separated module which exceeded the 75 kg/day per m² of membrane. This production presents an average productivity of 3.2 kg/h per m² of membrane which was close to that found by J.P. Mericq for a plan module immersed in SGSP (3.7 kg/h per m² of membrane) [21].

The third yearly immersed module production is 16.8 m³ representing more than one and half times the separated module production. Thus, immersing the module in the solar pond improves the performance of the hollow fiber module.

## Conclusion

The present work theoretically investigates the possibility of coupling the DMV module with a SGSP. In order to study the contribution of solar energy effect and choose the most appropriate module configuration, a model describing the transfer in the module and the SGSP was developed. These models allow us to determine the instantaneous variation of hollow fiber module. The obtained results have confirmed that the daily production of the immersed module is higher than that of the separated module. It is to be noted that for this study, we have neglected the circulation conduit heat losses in the separated module, which is not the case for the immersed one, because the module immersion in the SGSP reduces the heat losses. Therefore, a more detailed study of the entire desalination system should be under taken in order to show the benefits of membrane module immersion in the SGSP. However, immersion is facing several technical limitations. Indeed, given that it must necessarily be put in contact with the module, it is necessary to use noble materials for the construction of the module. This problem does not arise in the case of the configuration where the membrane module is separated from the SGSP.

## References

1. Khayet MM, Godino MP (2004) Heat and mass transfer in vacuum membrane distillation. Int J Heat Mass Transfer 47: 865-875.

2. Cabassud CC, Lahoussine-Turcaud V (1998) A new process to remove halogenated VOCs for drinking water production: vacuum membrane distillation. Desalination 117.

3. Darwish AN, Hilal N (2012) Mambrane distillation. Desalination 287: 2-8.

4. Jin Z, Yang DL, Jian XG (2008) Hydrophobic modification of poly (phthalazinone ether sulfone ketone) hollow fiber membrane for vacuum membrane distillation. J Membrane Sci 310: 20-27.

5. Cabassud C, Wirth D (2002) Water desalination using membrane distillation: comparison between inside/out and outside/in permeation. Desalination 147: 139-145.

6. Zuo G, Guan G, Wang R (2014) Numerical modeling and optimization of vacuum membrane distillation module for low-cost water production. Desalination 339: 1-9.

7. Zrelli A, Chaouchi B, Gabsi S (2011) Simulation of vacuum membrane distillation coupled with solar energy: Optimization of the geometric configuration of a helically coiled fiber. Desalin Water Treat 36: 41-49.

8. Mericq JP, Laborie S, Cabassud C (2010) Vacuum membrane distillation of seawater reverse osmosis brines. Water Res 44: 5260-5273.

9. Abu-Zeid M, Zhang Y, Dong H, Hou L (2015) A comprehensive review of vacuum membrane distillation technique: Desalination 356: 1-14.

10. Khayet M (2011) Membranes and theatrical modeling of membrane distillation: a review. Advances Colloid Interface Sci 164: 56-88.

11. Sarbatly R, Chel-Ken C (2013) Evaluation of geothermal energy in desalination by vacuum membrane distillation. Appl Energy 112: 737-746.

12. Nakoa K, Kawtar R, Date A, Aliakbar A (2015) An experimental review on coupling of solar pond with membrane distillation. Solar Energy 119: 319-331.

13. Susanto H (2011) Towards practical implementations of membrane distillation. Chemical engineering and processing: Process intensification 50: 139-150.

14. Khayet M (2012) Solar desalination by membrane distillation: Dispersion in energy consumption analysis and water production costs (a review). Desalination 308: 89-101.

15. Saffarini RB, Summer EK, Arafat HA, Lienhard VJH (2012) Economic evaluation of stand-alone solar powered membrane distillation systems. Desalination 299: 55-62.

16. Banat F, Jwaied N, Rommel M, Wieghaus M (2007) Performance evaluation of the "large SMADES" autonomous desalination solar-driven membrane distillation plant in Aqaba, Jordan. Desalination 217: 17-28.

17. Shirazi MA, Kargari A (2015) Water Desalination: Solar-Assisted Membrane Distillation. Encyclopedia of Energy Engineering and Technology 2: 2095-2109.

18. Wang X, Zhang L, Yang H, Chen H (2009) Feasibility research of potable water production via solar-heated hollow fiber membrane distillation system. Desalination 247: 403-411.

19. Galvez JB, Garcia-Rodriguez L, Martin-Mateos I (2009) Seawater desalination by an innovative solar-powered membrane distillation system. Desalination 246: 567-576.

20. Criscuoli A, Carnevale MC (2015) Desalination by vacuum membrane distillation: The role of cleaning on the permeate conductivity. Desalination 365: 213-219.

21. Mericq JP, Laborie S, Cabassud C (2011) Evaluation of systems coupling vacuum membrane distillation and solar energy for sea water desalination. Chemical Eng J 166: 596-606.

22. Abdallah SB, Frikha N, Gabsi S (2013) Simulation of solar vacuum membrane distillation unit. Desalination 324: 87-92.

23. Zhang Y, Peng Y, Ji S, Chen P (2015) Review of thermal efficiency and heat recycling in membrane distillation processes. Desalination 367: 223-239.

24. Abdallah SB, Frikha N, Gabsi S (2013) Study of the performances of different configurations of seawater desalination with a solar membrane distillation. Desalination Water Treat 52: 1-10.

25. Lawson KW, Lloyd DR (1997) Membrane distillation. J Membrane Sci 124: 1-25.

26. Sun L, Wang L, Wang Z, Wang S (2015) Characteristics analysis of cross flow vacuum membrane distillation process. J Membran Sci 488: 30-39.

27. Dao TD, Mericq JP, Laborie S, Cabassud C (2013) A new method for permeability measurement of hydrophobic membranes in Vacuum Membrane Distillation process. Water Res 47: 2096-2104.

28. Climatic data handbook for Europe.

29. Mansour RB, Nguyen CT, Galanis N (2004) Numerical study of transient heat and mass transfer and stability in a salt-gradient solar pond. Int J Ther Sci 43: 779-790.

30. Alsaadi AS, Francis L, Amy GL, Ghaffour N (2014) Experimental and theoretical analyses of temperature polarization effect in vacuum membrane distillation. J Membran Sci 471: 138-148.

31. Mansour RB, Nguyen CT, Galanis N (2006) Transient heat and mass transfer and long-term stability of a salt-gradient solar pond. Mechanics Res Commun 33: 233-249.

# 11

# Enhanced Biosolids Drying with a Solar Thermal Application

**Domènec Jolis[1] and Natalie Sierra[1*]**

[1]San Francisco Public Utilities Commission 750 Phelps Street, San Francisco, California, USA

## Abstract

Covered, green-house type biosolids solar drying facilities provide a low-energy option with operational simplicity and reduced cost. However, their large footprint consequence of low water evaporation rates make them unattractive for large wastewater treatment plants in urban areas with limited available space. This project investigated whether recent advances made in solar thermal technology conferred sufficient benefit in water evaporation rates that solar drying of wastewater biosolids may be feasible. A demonstration solar drying chamber was constructed with warm air from a solar thermal panel being routed to the chamber to aid in evaporation. Experiments were conducted with water alone to measure water evaporation rates in a range of weather conditions and to develop a regression model for evaporation. Experiments were also conducted with digested, dewatered biosolids to measure evaporation rates when drying biosolids. Total solids concentration in biosolids samples reached 42.3% after 102 hours in the dryer. Data showed that evaporation rates strongly depend on the temperature inside the dryer chamber but also on biosolids mixing. Measured evaporation rates were more than twice those previously reported in the literature for solar dryers and imply that with an experimental setup optimized for mixing, humidity control and energy recovery, still higher rates could be achieved. If confirmed in larger scale demonstration projects, the results from this study would allow for compact solar dryers to be located in urban settings.

**Keywords:** Solar drying; Solar thermal; Beneficial reuse; Biosolids; Compact treatment technologies; Evaporation rates

## Introduction

Solar drying has long been used as a method employed in the reduction of water and pathogen content from biosolids. Traditionally, this took the form of open air drying beds, but as odor and air emissions have become more pertinent issues for many facilities, covered solar drying beds and greenhouse-type installations have replaced some of these installations. In addition, as biosolids transportation costs have risen, utilities have sought ways to reduce the volumes of biosolids hauled to final disposal/reuse sites. While conventional thermal drying is a proven solution to volume reduction, solar drying provides a low-energy alternative with operational simplicity.

Solar drying technology makes use of renewable solar energy to dry biosolids in greenhouse type installations. In this type of installation, biosolids are loaded into a greenhouse manually or via a conveyer, and dried in a batch or continuous process. The greenhouse serves to capture and contain heat generated by solar radiation. In addition to enhancing the available heat generated by solar energy, the greenhouse helps to contain odors that might be generated by the drying biosolids. To enhance drying, the newer generation of solar dryers employs automated mixing and control of the climate within the greenhouse [1]. There are three commercial examples, distinguished by mixing systems and number of commercial scale installations. One company uses an "electric mole," or small robot, to mix the biosolids, a second uses a system of conveyer belts, and a third uses a proprietary mixing machine. The system using the electric mole is able to process solids with solids concentrations as low as 3%, while the other two methods require a total solids concentration (TS) of 20% or greater [2]. Overall, these systems have been used in smaller plants, ranging from those serving 1000 population equivalents (PE) to those serving 300,000 PE [1].

Evaporation factor is a function of outdoor solar radiation, outdoor air temperature, and ventilation flux [3]. In sizing the area required for solar drying in any geographical location, the solar radiation is a key factor. The corresponding temperature and relative humidity inside

the greenhouse are also pertinent factors. As solar radiation can vary throughout the year, the evaporation rate will vary leading either to a variation in the moisture content of the dried cake or in the time required to achieve target percent solids. The ventilation system set-up in any biosolids solar drying unit should allow free exchange of air between the interior and exterior of the greenhouse gas unit, in order to ensure that the air absorbing moisture from the biosolids does not reach a point of saturation so that the drying process continues to be driven by humidity. Figure 1 shows an image of a typical ventilation system in a solar drying unit.

While different models of evaporation rate have been explored in the literature, generally speaking, evaporation rate can be correlated to solar radiation, ventilation rate, air temperature, and relative humidity [2,4]. Performance, as defined by evaporation rate, thus varies widely in the literature, depending on the location and climate of the experiment, and maximum evaporation rates reported range from 1 kg/$m^2$-day to 8 kg/$m^2$-day [2,5-8].

Seginer and Bux [4] have put forth several models for evaporation rates from solar dryers. In general, they describe the evaporation rate as a function of weather, "state of the sludge" (e.g. dry solids content, sludge temperature), and control within the greenhouse (e.g. ventilation rate, mixing rate) [9-11]. Seginer and Bux used the vapor balance method in their initial modeling efforts, which consists of measuring the humidity ratio, w, of the ventilating air at the inlet and outlet of the unit; multiplying the difference, $w_o - w_{in}$, by the density of air and the

*Corresponding author: Natalie Sierra, RMC Water & Environment, 222 Sutter Street, San Francisco, California, USA, E-mail: djolis@sfwater.org

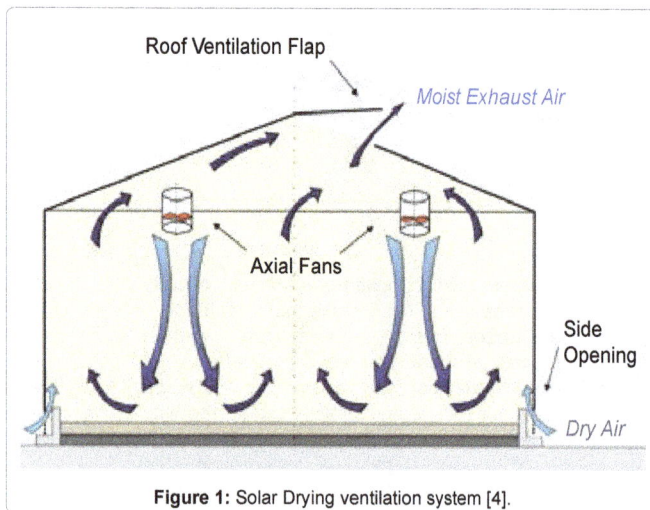

**Figure 1:** Solar Drying ventilation system [4].

discharge of the ventilation fans. Based on experimental data, they also proposed a linear equation, for evaporation rate as follows:

$$E = 0.000461R_o + 0.001010Q_v + 0.00744T_o - 0.220\sigma + 0.000114Q_m$$

Where:

$E$ = evaporation rate (mm/h)

$R_o$ = outdoor solar radiation (W/m²)

$Q_v$ = ventilation rate (m³/m²-h)

$T_o$ = air temperature (°C)

$\sigma$ = dry solids content of the sludge (kg solids/kg sludge)

$Q_m$ = air mixing (m³/m²-h) [4]

The first three variables demonstrated a strong effect on the evaporation rate, while the last two demonstrated a smaller effect, based on the researchers' available data.

Solar drying installations have advanced in the last decade, largely by introducing automation to mixing and ventilation within a controlled greenhouse setting. The environmental controls offered by these greenhouse systems has enabled researchers to better predict evaporation rates, given measurements of solar radiation, temperature, ventilation rate, dry solids content of the sludge, and mixing rate. While individual results for evaporation rate vary, researchers reported rates between 1-9.6 kg/m²-day. The ability to model evaporative behavior within the greenhouse setting further enables the optimization of such an installation depending on the site specific characteristics and needs of the agency employing this drying technology.

## Objectives

This project investigated whether recent advances made in solar thermal technology conferred sufficient benefit in water evaporation rates that solar drying of wastewater biosolids may be feasible in densely populated urban areas. This work was done as proof of concept for a potential urban solar drying facility.

## Materials and Methods

### Chamber construction

A chamber (Figure 2) was constructed to allow a supported solar panel to heat the interior, thus drying the biosolids. This chamber

was constructed of wood and measured 122 cm×46 cm×61 cm. The chamber design reflected the air-volume to sludge-area ratio as well as the air cross-flow to sludge-area featured in the low temperature (60°C) tunnel sludge drying technology developed by Aquology STC, Castellon, Spain. Aquology STC has tunnel sludge drying installations in France, Ireland and Spain with capacities of up to 500 metric tons per day. However, for simplicity, ease of construction and cost, no energy recovery features were incorporated in the experimental chamber beyond passive insulation, and mixing of the sludge inside the chamber was done manually. A four foot solar panel designed and sized to provide heat to a small bedroom, was mounted on the chamber, and with warm air generated by the panel traveling via a short length of four-inch ductwork to the chamber. The solar panel selected was designed to absorb 95% of the available solar energy and produce up to 100W of heat energy per linear foot. The panel was outfitted with a 12 volt DC fan attached to the intake vent; the fan is intended to turn on automatically when the inside temperature reaches 38°C. The chamber was insulated on the inside with commercially available household insulation. Temperature inside and outside of the chamber was measured using a data logger capable of taking and recording periodic temperature measurements. Relative humidity inside the chamber was also measured through the data logger. The unit was installed and tested outdoors in San Francisco.

### Experiments to evaluate water evaporation rates

The initial phase of work sought to optimize the maximum achievable temperature inside the chamber. Factors that were tested include vent fan speed, insulation, and angle of the solar panel. The second phase of work sought to measure potential evaporation rates across representative outdoor temperatures and sun exposure. Six plastic cups were filled with approximately 50 mL of water each morning and their weights were recorded along with the current weather condition and time. Each cup measured 7.5 cm tall and had a diameter of 6.5 cm, providing an identical surface area (33.16 cm²) to ensure uniform testing conditions. Three of the cups were placed inside the drying chamber of the solar dryer at three different locations. Cups 2 and 3 were offset on either side of the chamber's center where the sludge holding pans can be seen in Figure 2, while Cup 1 was placed between Cup 2 and the wall towards the outside edge of the dryer chamber. The two remaining cups were set outside of the drying chamber with Cup 4 placed in the shade and Cup 5 receiving direct sunlight.

### Experiment to evaluate biosolids drying efficiency

Once adjustments to the chamber had been made and approximate

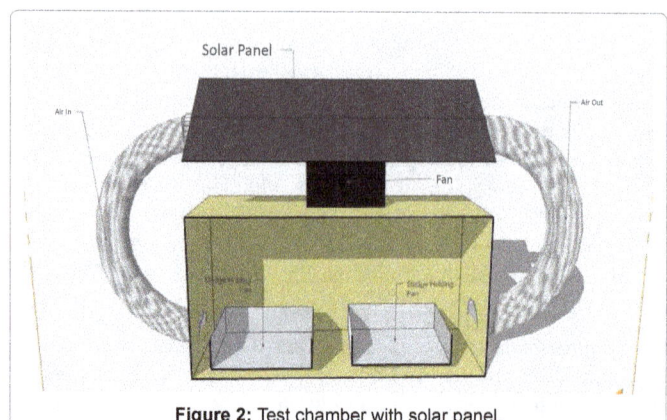

**Figure 2:** Test chamber with solar panel.

evaporation rates established, the efficacy of solar thermal drying on biosolids was tested. Three aluminum tins (33cmx23cmx5cm) were filled to a depth of two and one half centimeters with biosolids collected from the belt presses at one of San Francisco's wastewater treatment plants. The treatment plant incorporates a pure-oxygen activated sludge system and anaerobic digesters operated at mesophilic temperature and more than 20 days of hydraulic retention time. After digestion, the total solids (TS) and volatile solids (VS) of the stabilized sludge range between 2.3 and 2.7%, and 60 and 65%, respectively. Since the temperatures inside the drying chamber were not sufficient to significantly volatize the organic fraction of the biosolids, the VS content of the dried cake remained unchanged at 60-65%. Two of the tins were placed inside the solar dryer and the third was left as a control in the partial sun. Biosolids were mixed twice daily in one of the tins inside the solar dryer while the other tin inside the solar dryer and the outside control tin were left unmixed. The tins were left in the solar dryer 24 hours per day for four days.

### Experimental conditions

As relative humidity, temperature, and solar radiation can all affect the efficacy of drying, establishing evaporation rates under different weather conditions was critical to the understanding of the applicability of a larger solar thermal installation. Table 1 summarizes the conditions for the three experiments conducted.

### Data collection and statistical analyses

During Experiments 1 and 2, the five cups were removed every afternoon, weighed, and the evaporation losses were calculated.

Temperature data were recorded each day in three different places: inside the drying chamber, outside the chamber in the ambient air (with sensor placed on top of the drying chamber), and inside the conduit that delivers the heated air from solar heater to the drying chamber. Temperatures were recorded using a data logger with data recorded every ten minutes. The rate of evaporation was measured in kg/m²-day. The total weight of water evaporated each day was divided by the total number of hours the dryer was run per day to give an average rate of evaporation per surface area. Paired T-tests of the observed evaporation rates were used to determine whether results for Cup 1 were different from Cup 2 or Cup 3 results, given the uneven exposure to air flow inside the dryer chamber.

During Experiment 3, biosolids samples were collected each afternoon at approximately 3 PM to be analyzed for %TS. As biosolids dry from top to bottom and the non-mixed tins dry unevenly, care was taken to sample at a depth of 1.2cm to achieve an average %TS for the sample. Temperature data were recorded inside the drying chamber every ten minutes using a data logger. Ambient temperature data was recorded every 15 minutes.

### Result

An increase in temperature, often doubling, was observed in the chamber during daylight hours. Figure 3 shows this difference for a typical experimental day. Since the chamber's insulation was not optimized, diurnal variation, correlated with peak daytime temperatures and sun exposure, is evident. Figure 4 shows the temperature and average evaporation rates for Cups 2 and 3 inside the chamber, along with the average evaporation rates for cups placed outside the dryer in the shade and direct sun. The data presented in Figure 4 does not represent the evaporation rates recorded for Cup #1. Lower evaporation rates were observed in Cup #1 and are most likely

attributed to its corner position in the drying chamber, putting it out of reach of the air currents generated by the fan. In fact, paired T-tests of the observed evaporation rates revealed that Cup 1 results were statistically different from Cup 2 or Cup 3 results at the 95% confidence level (T=0.0005<$T_{crit}$=2.3) and will not be further considered.

The results of Experiment 3 are detailed in Table 2. At the end of five days of testing, the mixed dryer sample was 42.3% solids (a 25 % increase), the unmixed dryer sample was 34.7% solids (a 17.4% increase), and the unmixed outside sample was 32.9% solids (15.6% increase).

### Discussion

Initial work performed indicated that the interior of the chamber could get up to 38°C during daylight hours, with significant heat losses overnight. Evaporation rates (daily averages) for greenhouse applications range from 1 kg/m²-day (Bux and Bauman 2003) to 2.2 kg/m²-day [6]. Success of the pilot unit was therefore measured against these industry figures, while also acknowledging that the chamber was not constructed as an ideal solar dryer would be. For example, a more demonstration scale model would be better insulated to protect against

| | Experiment #1 | Experiment #2 | Experiment #3 |
|---|---|---|---|
| Media tested | Water | Water | Biosolids mixed and unmixed |
| Ambient temperature range, °C | 11.5-19.1 | 12.7-26.8 | 5.7-17.4 |
| Maximum Solar Irradiation, W/m² | 855 | 771 | 505 |
| Maximum wind speed, m/s | 5.2 | 5.4 | 3.3 |
| Weather | Partly cloudy | Sunny | Cloudy |

**Table 1:** Summary of experimental conditions.

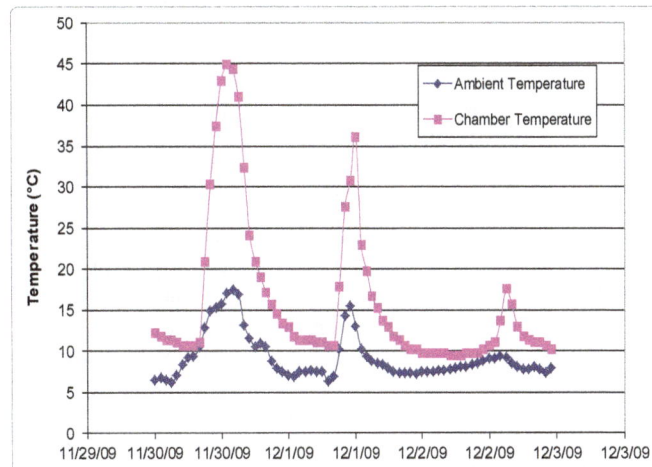

**Figure 3:** Typical Difference Between Dryer and Ambient Temperature during Experiments.

| Date | Elapsed Time (hour) | Mixed Dryer (% TS) | Unmixed Dryer (% TS) | Unmixed Outside (%TS) |
|---|---|---|---|---|
| 30-Nov-09 | 0 | 17.31 | 17.31 | 17.31 |
| 1-Dec-09 | 30 | 21.29 | 20.09 | 18.60 |
| 2-Dec-09 | 54.25 | 27.92 | 25.16 | 23.75 |
| 3-Dec-09 | 67.75 | 37.54 | 32.86 | 29.18 |
| 4-Dec-09 | 102 | 42.32 | 34.72 | 32.92 |

**Table 2:** Increase in %TS Over Experiment.

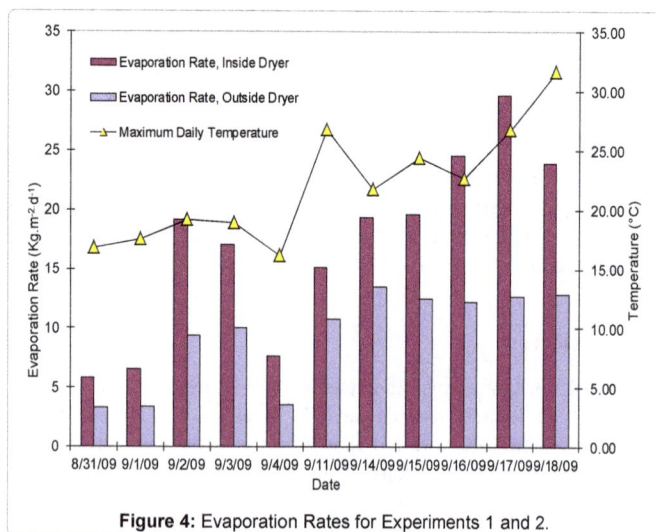

**Figure 4:** Evaporation Rates for Experiments 1 and 2.

the heat losses experienced during this experiment. Figure 3, however, indicates the increase in temperature experienced inside the chamber as compared to ambient temperature, often nearly doubling during peak sunlight hours. If the insulation of the chamber had been better optimized, this difference would likely have been greater.

Evaporation rate data in Experiment 1 is somewhat flawed due to the strong winds present at the time, which likely aided the evaporation rates of the cups set outside the chamber. The evaporation rates observed for Experiment 1 ranged from a low of 5.9 kg/m²-day to a high of 19.2 kg/m²-day. On cooler, cloudy days (8/31, 9/1, 9/4) these evaporation rates ranged from 78% to 116% improvement over the cups placed outside (which were not shielded from wind effects). On sunny days, the difference between the chamber samples and the outside samples is similar, ranging between 70% and 105% greater than the cups placed outside, indicating that the dryer's superior performance was maintained over a range of weather conditions. The evaporation rates measured during Experiment 2 ranged from 15.1 kg/m²-day to 29.7 kg/m²-day, a considerable difference from the values seen under the cooler conditions of Experiment 1 and much higher than those reported in the literature for biosolids.

As expected, a strong correlation exists between evaporation rate and the temperature inside the chamber for Experiments 1 and 2, which explains almost 80% of the variation in the data; the remaining variation is likely dependent on variations in relative humidity (not controlled in our experiments) and fan operation (i.e, air circulation). The regression equation (Figure 5) confirms the strong suspicion that evaporation rates for samples outside of the chamber were grossly overestimated, most likely due to evaporative cooling. Thus, for Experiment 2, the Cup #4 results would have required air temperatures inside the chamber between 24 and 27°C, but the recorded air temperatures outside the chamber ranged between 10 and 17°C.

Data in Table 2 can be used to calculate the evaporation rates for the duration of the experiments. These were 5.3 kg/m²-d for the mixed sample and 4.6 kg/m²-d for the unmixed one, clearly indicating the importance of mixing. Although the biosolids drying experiment could not be repeated under warmer conditions, it can be reasonably assumed that the evaporation rates in the biosolids samples could be nearly double when conditions similar to those of Experiment 2 were present.

However the limitations of the experiment as conducted, the evaporation rate for the mixed sample is 121% higher than current state of the art for solar biosolids dryers [6].

For Experiment #3, the higher rates of TS% increase inside the solar dryer are primarily due to the heating of the air inside the drying chamber as well as the flow of air over the biosolids produced by the solar dryer's fan. When the air inside the heating unit in the solar panel reaches a trigger (set at 75% for this study), a fan is switched on which circulates the heated air into the drying chamber and recycles the air into the heating unit. Insulation was included to contain the heated air inside the drying chamber and minimize losses to the ambient air.

Combining temperature profiles such as that shown in Figure 4 with the regression equation developed in Figure 5, the evaporation rates for different temperature and solar irradiation conditions can be calculated. Using this same approach for the five days, from November 30th to December 4th, when the biosolids drying experiment was conducted yields an average evaporation rate of 6.2 kg/m²-d which compares favorably with the observed value for the mixed sample (17% difference) and validates the approach. Better mixing of the biosolids would have likely improved the rate of evaporation during the experiment, since the difference in rates between the mixed (twice per day) and unmixed samples was 14%, and would have brought it even closer to the calculated value.

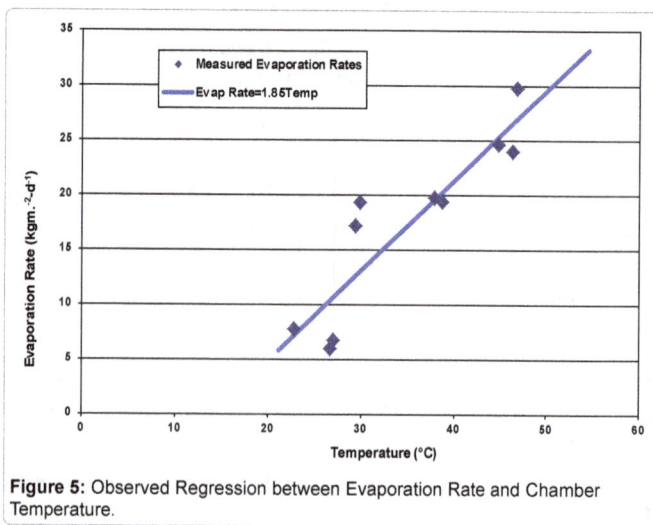

**Figure 5:** Observed Regression between Evaporation Rate and Chamber Temperature.

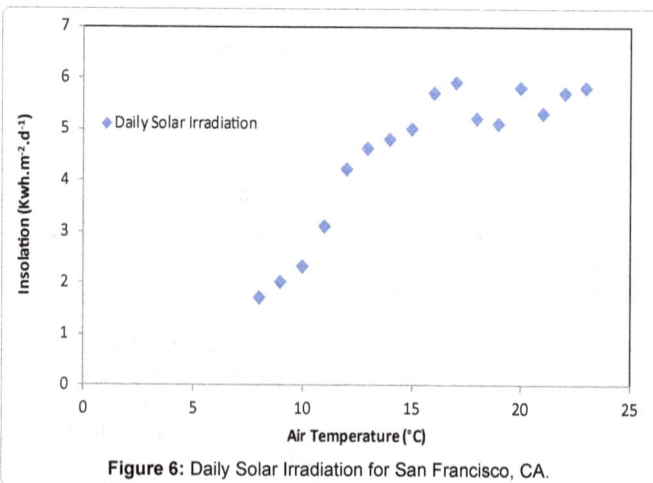

**Figure 6:** Daily Solar Irradiation for San Francisco, CA.

The solar dryer consistently provided higher evaporation rates inside the drying chamber over a wide range of weather conditions, but at times the increase was small while at other times the benefit was substantial. The weather along the coast in San Francisco can be extremely varied with the coldest temperatures seen in the fog-filled summer and rainy winter and the warmer temperatures seen in the sunnier spring and fall. Due to the varied weather conditions in San Francisco, the feasibility of installing a large scale solar biosolids dryer hinges in part on the unit's drying effectiveness in relation to the outside weather and temperature.

The solar dryer provides higher increases above ambient air temperature on days when the ambient temperature is higher due to increased solar irradiation. On November 30th, when the maximum ambient air temperature reached 18°C, the difference between ambient and chamber temperatures reached 82%. On December 2nd, the coldest day with a maximum ambient temperature of 9°C, the temperature difference only reached 31%. This relationship exists because high ambient air temperatures are typically correlated with high levels of solar energy reaching the earth's surface from the sun, so the solar heater runs for greater periods of time. Also, the fan is operated on a switch that turns on when the air inside the fan unit reaches approximately 24°C. So with increased solar energy the solar heater delivers more warm air and the fan is turned on for longer periods of time, creating more air flow at higher temperatures. These two changes create the maximum temperature differences between the chamber air and the ambient air and thus maximize the increased biosolids TS percent rates. A weakness of this experiment is that the relative humidity within the chamber could not be controlled, as would be true in a full-size installation. Greater evaporation rates would be observed if this factor could be optimized.

In addition, higher evaporation rates would be expected for more typical San Francisco weather conditions. In fact, the average temperature (9°C) and solar irradiation (2.0 Kwh/m²-d) during the period of November 30th to December 4th were at the low end of annual values for San Francisco, CA (Figure 6). If instead, the median conditions for San Francisco of 14.6°C and 4.8 Kwh/m²-d were used to estimate the temperature inside the chamber throughout the day and then the equation in Figure 5 is employed to calculate the evaporation rate for those temperatures, the predicted result is 17.8 kg/m²-d or 236% higher than the value for the period of the biosolids drying experiment. Certainly, these estimates will need to be confirmed with further experimental results.

## Conclusions

The solar dryer consistently provided elevated evaporation rates over a wide range of weather conditions. The evaporation rate measured during Experiment 3 (i.e., biosolids drying with mixing)

was more than twice the current state of the art for solar dryers. The solar dryer increases biosolids TS% rates by altering two main factors inside the solar chamber: temperature and air flow. The solar dryer maximizes its heating efficiency on hot, sunny days when more solar energy reaches its panel and both the heating device and fan run for longer periods of time. The influence of local weather, particularly fog and cloud cover, often diminished the efficacy of the panel, as would be expected. Due to the varied weather conditions in San Francisco, the feasibility of installing a large scale solar biosolids dryer hinges in part on the unit's drying effectiveness in relation to the outside weather and temperature. Further studies to establish conservative drying rates (e.g. in the absence of idealized insulation and humidity conditions) and preliminarily size such a unit for a large wastewater treatment plant (500,000 to 1,000,000 PE) are needed.

### Acknowledgements

The authors would like to acknowledge David Glucs and Zachary Wheeler for their assistance with field work and data recording.

### References

1. Bux, M, Baumann R (2003) Performance, Energy Consumption and Energetic Efficiency Analysis of 25 Solar Sludge Dryers. Proceedings of the Water Environ Federat 10: 522-524.

2. Horn S, Barr F, McLellan J, Bux M (2005) Accelerated Air-Drying of Sewage Sludge using a Climate-Controlled Solar Drying Hall. Brisbane Water-University of Hohenheim, Stuttgart, Germany.

3. Mehrdadi N, Joshi SG, Nasrabadi T, Hoveidi, H (2007) Application of Solar Energy for Drying of Sludge from Pharmaceutical industrial Waste Water and Probable Reuse. Int J Environ Res 1: 42-48.

4. Seginer I, Bux M (2005) Prediction of Evaporation Rate in a Solar Dryer for Sewage Sludge. Agri Eng Int 7.

5. Bux M, Baumann R, Philipp W, Conrad T, Muhlbauer W (2001) Class A By Solar Drying Experiences in Europe. Proceedings of the Water Environment Federation, WEFTEC 2001.

6. Mangat S, McTaggart M, Marx J, Baker S, Luboschik U (2009) Introduction of Solar Drying Technology to Trinidad and Tobago. Proceedings of the Water Environment Federation, WEFTEC 2009.

7. Mathiodoukis VL, Kapagiannidis AG, Athanasoulia E, Diamnatis VI, Melidis P, et al. (2009) Extended Dewatering of Sewage Sludge in Solar Drying Plants. Desalin 248: 733-739.

8. Nathan S, Clarke B (1992) SolarMix' – Innovation in Drying Technology. Proceedings of the 8th Annual Drying Symposium, Montreal.

9. Seginer I, Ioslovich I, Bux M (2007) Optimal Control of Solar Sludge Dryers. Dry Technol 25: 401-415.

10. 10.Shanahan EF, Roiko A, Tindale NW, Thomas MP, Walpole R, et al. (2010) Evaluation of Pathogen Removal in a Solar Sludge Drying Facility Using Microbial Indicators. Int J Environ Res 7: 565-582.

11. Sinton LW, Hall CH, Lynch PA, Davies-Colley RJ (2002) Sunlight inactivation of fecal indicator bacteria and bacteriophages from waste stabilization pond effluent in fresh and saline waters. Appl. Environ. Microbiol 68: 1122-1131.

# 12

# Understanding Various Technologies with Perspective of Energy and Environmental Audit

**Harwinder Singh[1], Aftab Anjum[2]\*, Mohit Gupta[3], Aadish Jain[4] and Amrik Singh[5]**

*Department of Mechanical Engineering, Delhi Technological University, India*

### Abstract

The present paper is based on understanding the energy audit in a holistic approach. India is a great developing country, economy and energy plays a vital role as fuel for development. As the demand is increasing rapidly, production of energy on the same slope is somewhat difficult. With exponential rise in demands, energy deficit is also rising with each passing day. Hence, judicial use of energy and energy audit becomes an important. The energy audit helps in improving the overall efficiency of the system thereby lowering down energy consumption and the costs in terms of economics. It also helps us determining the energy inflows, outflows and inefficient components. The work has been emphasized on environmental audit along with its importance on protecting the environment. This paper explains various technological aspects, needs and positives related to energy and environmental audit thereby giving an overview of it.

**Keywords:** Energy audit; Environmental audit; Water conservation; Electrical devices; Mechanical equipment

## Introduction

The cost of the energy is going up with the depletion in the current energy resources. The industrial sector is a major contributor towards this trend. So, an overall need arises to reduce down the energy costs which can be considerably achieved by enhancing the working efficiency of the various processes, equipment and operations involved in the industry. The energy efficiency can be further improved by switching over to the renewable sources of energy, creating awareness and a dedicated monitoring of the systems. The Government of India has passed various legislations and amendments to shift the focus towards energy efficiency. Bureau of Energy Efficiency (BEE) was made for defining certain energy standards, providing guidelines, accrediting various organizations for energy audits, carrying out the projects related to energy efficiency and carrying out the research activities related to the energy efficiency equipment [1].

## Energy and Environmental Audit

### Energy audit

Energy audit enables us to work upon the various inefficiencies in the existing system and analyzing the components thereby ensuring the energy conservation. It tells about the utilization of energy in a given system.

### Environmental audit

Environmental audit surge in pollution levels in our country India has ensued in higher environmental risks. Hence the need for sustainable development is emphasized in every quarter be it policy formations, project clearances, future planning or management of already existing infrastructure and industry. While energy audit of a setups concentrates mainly on the energy efficiency of the system. Environmental audit will take care of preserving environment, natural resources such as water, air, land etc. Environmental audit cater to the issues such as Material balance, non-hazardous waste generation and disposal, hazardous waste generation and disposal, characteristics of waste for recycle or reuse along with short term and long term impact on local environmental conditions including air and water quality [1].

### Main aim

Main aim of energy and environmental audit is to minimize the energy requirement and waste production with least environmental effect and without affecting the output during any process like residential, industrial, governmental and commercial or any other process.

Positives and need for energy and environmental audit are:

- Energy cost reduction
- Preventive maintenance and
- Quality control programs
- Waste prevention and reduction
- Compliance with regulatory requirements
- Option for adoption energy conservation technologies
- Profit maximization
- Optimization of available resources

## Energy and environmental audit can be classified into the following two types

- Preliminary audit
- Detailed audit

**Preliminary audit:** Preliminary audit is a relatively simple exercise:

- Make energy consumption chart for every process in the organization
- Identify the scope for improvement in effectiveness (and the easiest areas for attention)

---

**\*Corresponding author:** Aftab Anjum, Department of Mechanical Engineering, Delhi Technological University, Bawana Road, Delhi-110042, India
E-mail: aftabanjum915@gmail.com

- Look out for immediate (low cost) improvements

- Spot areas for more detailed and process specific study

**Detailed audit:** Detailed energy auditing is carried out in three phases: Phase I, II and III.

- **Phase I - pre audit phase:** Establishment is identified for audit. Basic information is collected regarding location, surrounding, climatic conditions, inputs of establishment in form of energy and raw materials, outputs, waste generation, management and disposal.

- **Phase II - audit phase:** Identification type of energy consumed in various departments include major processes, preparation of material balance sheets (includes raw materials, products, by products etc.), data regarding energy cost and tariff and latest regional energy policy. Process and material flow charts include sources of energy supply (e.g. electricity from the grid or self-generation). Potential for alternative energy sources such as solar energy, geothermal etc., process improvements and modifications and the induction of co-generation systems (combined heat and power generation).

- **Phase III - post audit phase:** After conducting audit energy manager/energy auditor should report and file observations to the top management about findings and possible recommendations if any, for effective communication and implementation.

The methods like present value criterion, average rate of return criterion and payback period criterion etc. are used for the economic analysis of the investments.

## Electrical Energy Devices

Electrical energy devices collectively consume large amount of electrical power and also incurring large amount of distributed losses. In the current scenario when the energy deficit is large energy efficiency and audit are the need of the hour to plug these distributed losses one by one. Energy audit of these systems becomes essential so as to identify the inefficiencies in the system and can help in reducing this energy deficit.

## Motors

An electric motor converts electrical energy into mechanical energy. They are classified as DC (Direct Current) motors and AC (Alternating Current) motors. Further AC motors are classified as AC induction and AC synchronous motors. They are mainly used to provide power to the various equipment. While selecting an electric motor for a process, aspects like reliability, inventory, availability, breakdown torque and operating conditions should be considered.

The efficiency of an electric motor can be written mathematically as:

$$Efficiency(\%) = \frac{Watts(input) - Watts(output)}{Watts(input)}$$

**Losses:** Losses can be described as the energy fees charged by the motor in order to convert electrical energy into the mechanical energy. They can be further classified into four categories

- Power Losses includes stator and rotor losses which can account for even more than half of the total losses occurring in the motor.

- Friction losses

- Magnetic core losses

- Stray load losses

Conductivity of the material, operating conditions, overall weight of the machine and lubrication are the major factors influencing the efficiency of motors. By giving considerable attention to above losses can be minimized.

**Conservation of energy:** The amount of energy consumed can be saved by operational improvements which include - balanced supply, improved controls, maintenance and retrofit improvements which look for the replacement of old, oversized and inefficient motors with the new and efficient ones.

Induction motors operate at lagging power factor because of the magnetizing current that is drawn by the motor, the incorporation of capacitors helps in enhancing the power factor thereby reducing energy charges, kVA demand and voltage drop and increased overall efficiency. Use of flat belt drives, variable speed drives and energy-saving controllers also contributes towards the saving of energy.

## Illumination

A lumen can be defined as a measurement of light output from a tube or a bulb whereas illumination is the distribution of light on a horizontal surface which is measured in lux or foot-candles. Efficacy can be defined as the ratio of the light output to the electric power the lamp consumes. Efficacy is measured in Lumens per watt (LPW). It should be noted that efficacy should not be confused with the term efficiency which is unit-less quantity.

**Quality of illumination:** Quality is basically the visibility factor during work conditions and denotes how much people feel visually comfortable. Glare, which is the extra brightness from a direct light source, should be removed for having good quality lighting. The ability of the light source to render the same colors as sunlight does is termed as CRI (Color rendering index).

**Types of illumination source:** The four basic types of lighting lamps are as follows

- **Incandescent:** Most common types of incandescent lights are tungsten halogen, reflector lamps and standard incandescent.

- Fluorescent

- Compact Fluorescent

- **High intensity discharge:** Common types are mercury vapour, high pressure sodium and metal halide.

**Increasing efficiency:** Energy efficiency can be improved by replacing the lamps or entire fixtures thereby lowering the wattage, using day lighting, improving light controls, preserving illumination and ensuring simple maintenance. The devices used for regulating the lights are called lighting control like snap switches, timers, photocells, occupancy sensors and dimmers etc.

## Transformer

Transformers are used to step up or step down the AC (Alternating Current) voltage. Transformers operate on the Faraday's law of induction. Operating losses in the transformers are namely no-load loss and load loss. No-load losses can be further categorized as eddy current losses, hysteresis losses, $I^2R$ losses and di-electric losses whereas load losses are basically the coil losses.

The net losses for a transformer can be described mathematically as:

$$L_P = L_N + (P/P_R)^2 \times L$$

Where,

$L_P$ = Total power loss in kW

$L_N$ = No-load loss in kW

L = Load loss in kW

P = Actual load on transformer in kVA

$P_R$ = Rated power of transformer in kVA

Identifying the load is an important factor in determining system load curve which will also provide us with the information of peak loads and base load periods. Proper monitoring of the load curves can help to manage the peak loads.

The demands can be controlled in a centralized manner, in a timed bound manner and can also be controlled manually. The prime objective of the transformer audit is to ensure that quality of power to different load centers is high and overall efficiency is better.

## Power quality

Power quality is the severity of variations in the power supplied to the customers and deviations from the standard frequency of 50 Hz. If the power received has a pure sinusoidal waveform, it is considered as the clean power. Nowadays a term 'dirty power' is used to describe the contaminations in the clean power.

A harmonic is basically a part of the waveform having a frequency which is an integral multiple of fundamental power line frequency. Harmonic problems should be corrected or prevented from an industrial facility.

Harmonics can be reduced by the following methods:

- Power system design
- Isolation transformer
- Line reactors
- Harmonic trap filters
- K-factor transformers
- Applying tuned filters

## Energy Management

Energy Management is the program aimed at reducing the organization's electricity bills and negative environmental impacts. The important hardware components of any energy management information system are:

- Field Transducers
- Power monitor modules
- Programmable logic controllers
- Communications network
- Centralized personal computer

## Mechanical Energy Devices

### Boilers

Boilers are employed in the industries to act as a heat source for various process applications. Boilers are basically classified as fire tube and water-tube boilers.

**Evaluation of boiler efficiency:** Efficiency of the boilers can be evaluated by either the direct methods or the indirect methods.

The formula used for the direct method is

Boiler Efficiency = (Heat output/ Heat input) × 100 = (Steam flow rate × (Steam enthalpy- feed water enthalpy) × 100)/ (Fuel firing rate × Gross calorific value).

Indirect methods aims at calculating the various heat losses, namely dry flue gas loss ($L_{fg}$), heat loss due to the moisture in the fuel ($L_{mf}$), heat loss due to moisture from burning hydrogen in the fuel ($L_{hf}$), heat loss due to the moisture in air ($L_{ma}$), heat loss due to Carbon monoxide (CO) in flue gas ($L_{co}$), heat loss due to un-burnt carbon in bottom ash ($L_{ubb}$), heat loss due to un-burnt carbon in fly ash ($L_{ubf}$), heat loss due to sensible heat in fly ash ($L_{sf}$) and loss due to surface radiation and convection ($L_{rc}$). All these losses are calculated in terms of percentage, added and then subtracted from 100 to obtain the thermal efficiency of the boiler.

Thermal efficiency of boiler (%) = 100-(Total losses in %)

**Energy conservation measures for a boiler:**

- **Excess air flow rate:** Stoichiometric flow rates are possible only for theory calculations and certain amount of excess air flow rates are required for the complete combustion of fuel. This additional amount of air is known as excess air flow which is also to be controlled as large amount of excess air can negatively impact the working of boiler by reducing the average temperature of the fire ball, reduces radiative heat transfer, increased stack losses by reducing the flue gas temperature and increase in the flue gas flow rate will also contribute to losses.

- **Burner and nozzle size:** The size of burner and nozzle must be adequately designed for optimal performance and cost effective maintenance. We should replace the nozzles with the better designed advanced burners where ever possible for advantage.

- **Control of temperatures of working fluid as well mass flow rate of heat carrier:** There are circumstances when a single source of heat is used for different process and there temperature may be different. Hence its adequate design may maximize efficiency of the system.

- **Proper insulation:** This is the first and easiest way to minimize losses. Where ever required proper insulation must be given to minimize heat loss to the environment as heat losses mainly occur at higher temperatures and this result in direct loss in availability of the system.

**Waste heat recovery:** Waste heat recovery from the flue gases can be done by:

- **Feed water preheating using economizer:** The water entering the boiler is pre-heated using the thermal energy of the flue gases in the equipment known as economizer before the flue gases are exhausted.

- **Combustion air-preheat:** The combustion chamber is provided with the air which is pre-heated in the air preheater utilizing the waste heat from the flue gases.

- **Improvement of condensate recovery:** Feeding the boiler with the steam condensate will also reduce the load of sensible heating for boiler.

- **Optimization of blow-down:** Though blow- down is necessary for the efficient working of boiler but frequency of blow-down is also an important factor.

## Compressed air network

Compressed air is used in almost all types of industrial applications and accounts for a major share of electricity used in some of plants. Compressors are mainly classified into dynamic or centrifugal compressors and positive displacement types which are further divided into categories on the basis of principle of functioning. The distribution of compressed air is equally important as its generation. Normally the compressors are located at a central place in the plant and the compressed air travels a long distance through pipes to the point of usage. This results in large pressure drops due to long pipe lines and large number of pipe fittings. An efficient compressed air distribution system must include the proper pipe sizing, surface finish of the duct, minimum number of sudden bends and fittings for minimum pressure drops and leakages [2].

**Compressor selection:** Selection of compressor must be done considering the main factors [1] such as number of units required, cost of operation, installation cost and quality of air to be delivered by the compressor.

**Monitoring performance and testing**: Monitoring performance and testing of the working of compressor must be done on the factors stated [1]:

- Pump-up or capacity test

- Specific power consumption

- Compressor efficiency or isentropic efficiency

- Quality of air intake

- Compressor cooling

- Compressor operating pressure

- Capacity control and power consumption

## Steam distribution systems

Steam has always been the most preferred working fluid for numerous plants for heating and drives, but the ever rising cost of fossil based fuels has made it necessary to adopt measures to reduce energy losses in steam distribution network. Some of the major considerations for designing a steam distribution network are named below [3,4]:

- **Pipe layout in plant:** Layout network must be so designed to minimize the steam travel from point of steam generation to utilization. Generally underground piping is not preferred.

- Pipe sizing

- Steam quality

- Moisture separation

- Insulation

All sections of hot pipes, valve bodies, unions, flanges and mechanical traps, such as floats, buckets and bodies of disc traps should be insulated.

## Refrigeration and air-conditioning

The process of taking heat from a source of lower temperature and delivering it at a relatively high temperature level with the help of an external agent is known as Air conditioning and same concept can be used to either keep the desired space at a temperature lower or higher than surroundings depending on the requirement.

**Main mechanisms:** Mainly two mechanisms namely vapour compression system and vapour absorption system are employed for the job.

**Vapour compression system:** The heat absorbed by the evaporation of a liquid refrigerant in the evaporator at a controlled lower pressure after that increase the pressure of low-pressure vapour coming from the evaporator, with the use of compressor, heat removed from the high-pressure vapour in the condenser so that vapour can liquefy or condenses after this reduces the pressure of the high-pressure liquid to a level needed in the evaporator by using the throttling device [5].

**Vapour absorption system:** This refrigeration system is a heat-operated system. In this system two pressure levels like evaporating and condensing pressure levels are to be created. Compressor in this system is replaced by absorber and generator. A solution is known as absorbent is circulated between the absorber and generator with the help of a solution pump. The absorbent in the absorber draws the refrigerant vapour formed in the evaporator, thus maintaining a low pressure in the evaporator. In generator, the absorbent is heated, therefore releasing the refrigerant vapour at a high pressure and to be condensed in condenser. Thus the suction function of the compressor is performed by the absorbent in the absorber and the generator performs the function of compression and discharge [6].

An air-conditioning and refrigeration plant is efficient when all the system components like compressor and the condenser, the evaporator and the condenser cooling are working in matched conditions. This means that under peak operating conditions they must perform to their optimum output.

Energy conservation opportunities [1]:

**Changes on system operating parameters:**

- Increasing the chilled water temperature set point

- Installation of variable speed drives

- At air-handling units in fan motors

- At cooling tower fan motors

**Changes on system operation:**

- Enthalpy control/dry bulb economizer

- Dry bulb economizer

- Enthalpy control

- Reduced minimum outdoor air

- Exhaust air control

**Changes on system design:**

- Retrofit of central fans for variable air volume usage

- Forward-curved fan systems

- Fans with vortex vanes

- Conversion of dual-duct constant volume to variable air volume

- Heat recovery system

- Use evaporative cooling for comfort cooling in dry areas

## Fans and blowers

Fans and blowers are widely used in industrial and commercial applications for ventilation and industrial process requirements. These are categorized into two different types based on the path of the airflow (centrifugal and axial) and blade shape (centrifugal and axial fan). Blowers are also classified into centrifugal and positive displacement type.

**Performance:** Performance is typically defined by plot of developed pressure and power required over a range of fan generated airflow. Performance of fan is also strictly depends upon the fan enclosure and duct design

**Fan efficiency:** Fan efficiency is the ratio of power imparted to the air streams to the power delivered by the motor.

In any fan system, the resistance to airflow (pressure) increases when the flow of air is increased. Bends and elbows in the inlet or outlet ducting can change the velocity of air, thereby changing the fan characteristics. Fans often serve a wide range of operating conditions. To accommodate demand changes, flow is controlled by four principle methods: vanes, pitch, outlet dampers and fan speed control [1].

## Cooling tower

Cooling tower is a waste heat rejection device which is employed to reject the heat into the atmosphere. It basically removes the heat with the help of cooling water. Hence the performance of the tower is directly related to the temperature of the cooling water.

**Categories:** Cooling towers can be further categorized according to the way in which air is fed to the tower as 'natural draft' and 'mechanical draft'.

### Mechanical draft:

- **Forced draft:** In this type of tower the fan is placed at the bottom from which air is forced into the tower. It is well suited for indoor applications.

- **Induced draft:** The movement of air in this arrangement is either counter flow or crossed flow.

- **Wet dry tower:** It employs an air-cooled heat exchanger, suited for high temperature applications.

### Natural draft:

- **Atmospheric tower:** The movement of air in this type of tower is in horizontal fashion.

- **Spray pond:** Two arrangements with fan which involves spraying of the warm water for evaporation and without fan which works on the ejector principle.

**Range, approach, cooling tower effectiveness and Performance of the cooling tower:** Range of a cooling tower can be mathematically expressed as,

Range (°C) = Temperature of the water out from the cooling tower - Temperature of the inlet water

Whereas approach is,

Approach (°C) = Temperature of the water out from the cooling tower- Design wet bulb temperature

Cooling tower effectiveness (%) = Range/ (Range + Approach)

Performance of the cooling tower depends upon the heat dissipation(Kcal/h), heat load, approach, wet bulb temperature, size of the tower and fill media (film fill, low-clog film fill and splash fill).

There are various instruments which are used during the energy audit of cooling towers.

- **Anemometer:** Used to measure the flow rate of air.

- **Power analyser:** It analyses the electrical parameters of the pump and fan.

- **Sling hygrometer:** Measures dry bulb and wet bulb temperatures of air.

- **Flow meter:** Measures water flow-rate.

- **Temperature indicator:** It uses a thermocouple and measures the water temperature.

## Alternate working fluids

Water and steam are the most common heat carriers. However there are problems such as high operating pressures along with leakage losses etc. associated with such systems.

Alternative fluids such as mineral oil, synthetic oil and various other nanofluids are also being used and systems are proved to be much more efficient in its functioning. Some of the advantages of such a system are:

- High operating Temperatures such as 300°C at atmospheric pressure

- Wide range of temperature for stable working

- No need for feed water treatment

- No heat loss related to condensation or Flash steam

- No risk of corrosion

- No need for anti-freeze agents

- Lower maintenance cost

- Quite in operations

- Easy to operate

- Uniform pressures for complete cycle

## Internal combustion engines

These engines although deliver High quality of power, but are relatively less efficient. Studies have shown that out of total energy supplied to the engine only 26-28% is available. Nearly 34% of energy is lost to exhaust, 30% in engine cooling and around 10% is dissipated to overcome frictional losses. There are several ways suggested to improve its efficiency by Fuel conservation by operating on design points, proper engine loading, proper lubrication which could reduce frictional losses. Waste heat recovery mechanisms such as turbo charger, hot water generation, vapour absorption refrigeration system are also advised for specific systems to maximize the efficiency of engine where ever possible. Further I.C. engines require regular servicing and timely renewal of lubricants and oil well for proper functioning.

## Water Audit and Conservation

Water audit and conservation with the exponential urban and

industrial development along with population, the available quantity of water and its quality has reduced drastically as compared to past decades. In a report by IWRS 2004 it was reported that there was demand of 22 bcm (billion cubic meters) of water in 1995-97 and according to National commission the estimated demand of water stand at 70 bcm by 2025 and 103 bcm by 2050. This data shows the urgency with which the issue of water conservation has to be dealt with:

## Classification

The water conservation measures can be broadly classified into technological advancements and habitual practices. The former is used in industrial and commercial users where new installations in water management can significantly affect the setup profitability in future.

**Technological advancements:** Technological advancements majorly include water recirculation, reuse and recycle. There are few applications such as low flow fixtures, water less urinals, planned landscape irrigation, etc. are used in many commercial projects.

**Habitual practices:** Changes in behavior of individuals and society as a whole can contribute to great amount of water conservation.

The measures for water conservation may also depend upon the factor of usage of the conservation system by social or industrial/commercial establishment. As a social establishment may use this as a policy towards environment and responsibility towards nature and future generations but industrial establishments may also have ulterior motives of profit maximization of reduction in input cost etc. [7,8].

## Conservation

There are several ways suggested for water conservations:

- Rain water harvesting
- Recycle, recirculate, reuse
- Low flow fixtures
- Planned landscape irrigation
- Alternative technologies

## Solar Energy

Many reviews have been conducted by engineers and researchers are very important and informative to understand solar technologies [9]. As India is energy deficient country and the demand of energy is enormous and the deficit is also very large. This energy deficit can't be fulfilled by any single energy source but to a great extent can be controlled with the help of solar energy technologies applied to various uses such as power generation [10] or energy source for various industrial applications. There is huge potential for solar thermal and solar photovoltaic (PV) technologies to be applied in the industries which can reduce the demand of energy on the grid.

Some solar thermal technologies are:

- Flat plate collector
- Evacuated tube collector
- Parabolic trough concentrator
- Parabolic dish concentrator
- Compound parabolic concentrator

Solar thermal and PV technologies can be used for power generation, air-conditioning, water heating or process heating, waste heat recovery and many other methods [11] as discussed by S. Mekhilef and R. Saidur, which directly or indirectly utilize solar heat as energy source and reduce the energy budget for the setup.

## Conclusion

Energy generation is a complicated issue than its expenditure and therefore it becomes very important to follow the judicious use of the available energy. Energy and environment conservation is not the responsibility of a single person or a group of persons. The society as a whole has to become aware and act aptly for sustainable development. Many government agencies are also working to spread awareness by means of advertisement campaigns such as Petroleum Conservation Research Association (PCRA), BEE and Green Energy Rating Agencies such as GRIHA etc. It has also been observed that with minor changes in Habitual practices by society major amount of energy waste can be plugged, this will also result in lower Energy bills for households as well as commercial establishments. Many electrical devices such as Motor, Transformer, illumination devices along with power quality issues have also been discussed.

### References

1. Abbi YP, Jain S (2006) Hand Book on Energy Audit and Environment Management. The Energy and Resources Institute (TERI) pp: 1-302.

2. Efficient Utilization of Steam - Energy Efficiency Office, U.K.

3. Bureau of Energy Efficiency.

4. Prakash A (2013) Improving the Performance of Vapour Compression Refrigeration System by using Sub- Cooling and Diffuser. IJEBEA 1: 88-90.

5. Kaushik S, Singh S (2014) Thermodynamic Analysis of Vapor Absorption Refrigeration System and Calculation of COP. IJRASET 2: 73-80.

6. Abbi YP, Jain S (2006) Hand Book on Energy Audit and Environment Management. The Energy and Resources Institute (TERI) pp: 1-302.

7. Feldbeuer SL (2005) Furnace Optimization; Meeting the Need to Reducing Cost. Abbott Furnace.

8. Indian Water Resources Society (2004) Efficiency of water resources.

9. Kesari JP, Gupta M, Jain A, Ojha AK (2015) Review of the Concentrated Solar Thermal Technologies: Challenges and Opportunities in India. IJRSI 2: 105-111.

10. Singh H, Singh P (2014) A Review About Solar Thermal Power Technology And Its Use In India.

11. Mekhilef S, Saidur R, Safari A (2011) A Review On Solar Energy Use In Industries. Renewable and Sustainable Energy Reviews.

# 13

# Radical Urban Development in the Egyptian Desert

**Abouelfadl S[1]\*, Ouda K[2], Atia A[3], AL-AMIR N[3], Ali M[3], Mahmoud S[3], Said H[3] and Ahmed A[3]**

[1]*Architectural department- College of Engineering, Assiut University, Asyut, Egypt*
[2]*Department of Geology- College of Science, Assiut University, Asyut, Egypt*
[3]*Architect, Asyut, Egypt*

### Abstract

Gardens' City is a new city in newly discovered area in the Egyptian western desert, which is rich to be developed. It lies in new Farafra Oasis. The site has different potential aspects for sustainable development; it has agricultural and industrial economic bases. The city center's area is designed to be about 5% of the city's area. The area of the industrial zone is about 22% of city area. This paper refers to the development of the city with a focus on the central and the industrial zones. The city center has the major managerial and commercial services. The industrial zone includes industrial areas as well as the major industrial education, training and managerial services. Renewable energy will be generated with different methods. This city will be the first step of development series opportunities in Egypt.

**Keywords:** Gardens' city; New Farafra; City centre; Industrial area; Egyptian desert

## Introduction

Egypt has the highest population in the Middle East, with about 85 million inhabitants as 2013 reports referred. Egyptian people are living along the Nile (notably Cairo and Alexandria), in the Delta and near the Suez Canal. Egypt's area is about 1 Million square kilometers, but the inhabited area is only 50 thousands kilometer. Nile Delta's area is about 37 thousands kilometer. It represents 74% of the inhabited area. The ratio of land used in agriculture is 3.74% of Egypt's area [1,2]. Several threats will face Egypt's densely populated coastal strip and Nile Delta by the probable dramatic increase of sea level due to global warming. These threats will badly affect Egypt's economy, agriculture and industry. The rise in sea level could turn millions of Egyptians into environmental refugees by the end of the century, as Nile Delta will turn to a wasted land if the sea level only rises 30 cm, which means that 70,000 agricultural jobs will be ended. Challenges have risen now, so we should be prepared for the expected global warming crisis and its impacts [3-5]. This paper discusses the development of Gardens' City in the Egyptian western desert with focus on its' central and industrial zones as a solution for development needs and to overcome the coming challenges. The methodology which is used, site analysis- to discover strength and weakness points- applying international planning ratios and achieving sustainable city planning.

## Newly Explored Areas in Egyptian Western Desert

There are different scenarios to help Egypt overcome these challenges. One of them is protecting the Nile Delta lowlands from the sea's incursion by building seawalls along the Delta's entire coastline to hold back the Mediterranean. Another one is to move the affected people to more suitable areas [4].

There is a new discovered area in the Egyptian western desert which can be developed. An Egyptian expedition found about 3.5 million Acres, which are rich to be developed in the Egyptian western desert [6-8]. Figure 1a shows the place of these newly discovered areas, while Figure 1b shows the new discovered oasis and plateaus. This area was always considered as a part of great sand sea.

## New Farafra Oasis

The New Farafra Oasis is a part of the newly explored area. It is the nearest Oasis to the existing urban areas. New Farafra Oasis- extends northeast between altitudes 27° 03' N and 26° 58' E to altitudes 27° 22'30 N and 27° 24' E, with a maximum NE length of 52 km and a maximum width of 20 km, attaining a total area of 932 km². The floor base of the

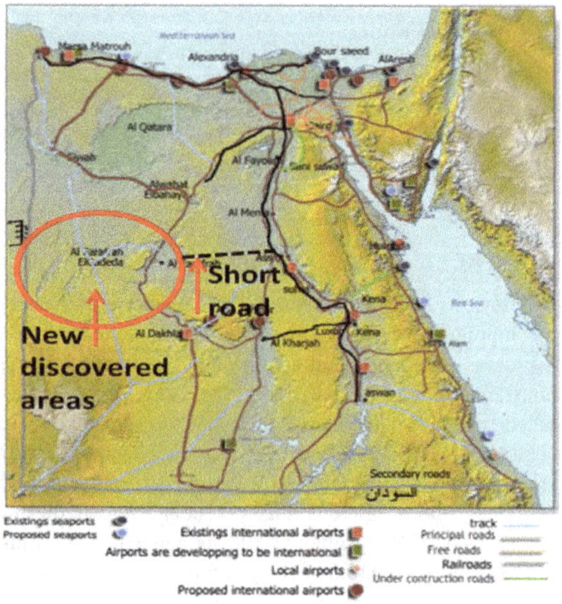

**Figure 1a:** Newly discovered areas western Egyptian desert.

**\*Corresponding author:** Abouelfadl S, Architectural department, College of Engineering, Assiut University, Asyut, Egypt E-mail: sabouelfadl@yahoo.com

Oasis is situated at a ground elevation varying from 60 m to 115 m and average 94 m above sea level [7,8].

The existing roads from the site to Nile valley are too long (500 km long) so, there are two suggested roads, 1st road which is about 180 km length connects the discovered site to Asyut city as shown in Figure 1a. The 2nd road will connect the discovered site with the existing urban areas around it, Cairo-New valley road from one side and to Baharia-Siwa road from the other side, (Figure 2) [9].

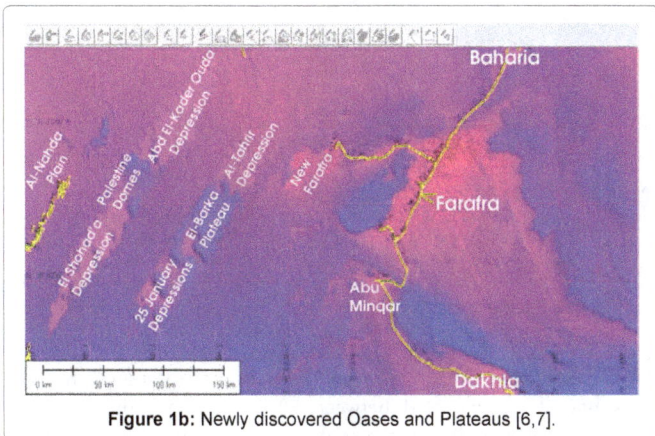

**Figure 1b:** Newly discovered Oases and Plateaus [6,7].

As a development green thinking: This site has a lot of potential aspects (Figures 3 and 4) [9-11]. These aspects are as follows:

- It lies in the Egyptian western desert which represents a high ratio of the Egyptian land that has not been well used yet.

- Flat land is available, which assures easy urbanization and development.

- It is away from sea flooding which may happen due to the global warming, which keeps it safe for the probable future climate migration in Delta and coastal areas in Egypt.

- It lies in Upper Egypt, a part which has been neglected for a long time.

- Water resources are available there; the Nubian sand stone there forms a big storage of water.

- There are natural springheads like Dalah springhead.

- The oasis area is 932 km$^2$ (222 thousand acres) which assures great areas of agriculture land.

- Solar and wind energy there are enough to generate clean renewable energy.

- Shale/clay soil is available there which allows agricultural and industrial development.

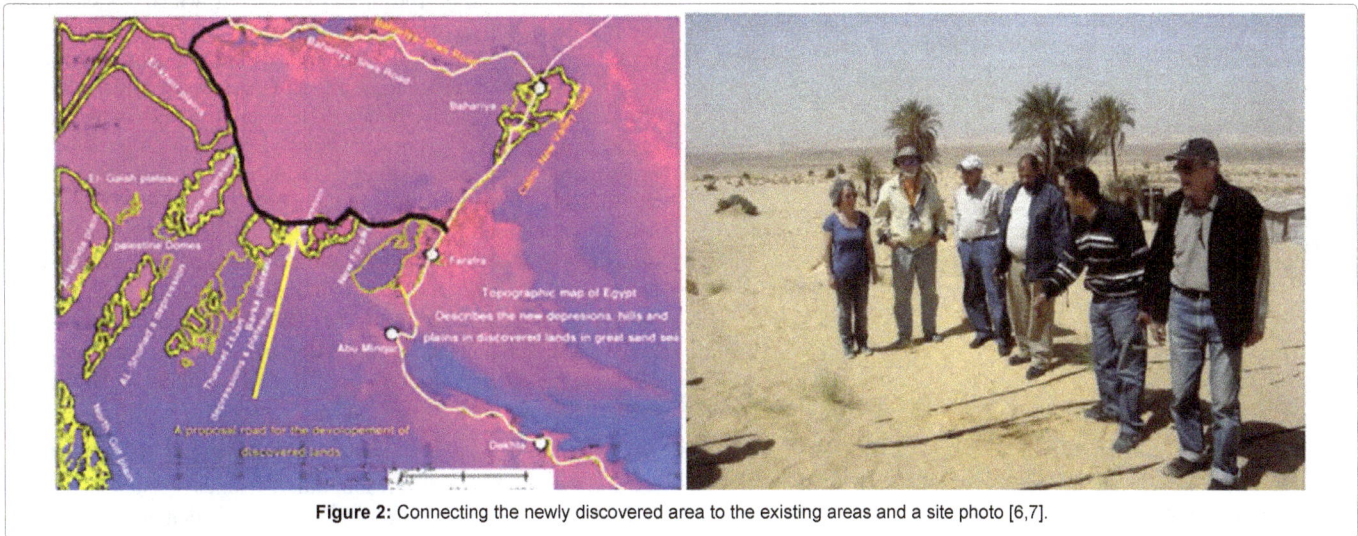

**Figure 2:** Connecting the newly discovered area to the existing areas and a site photo [6,7].

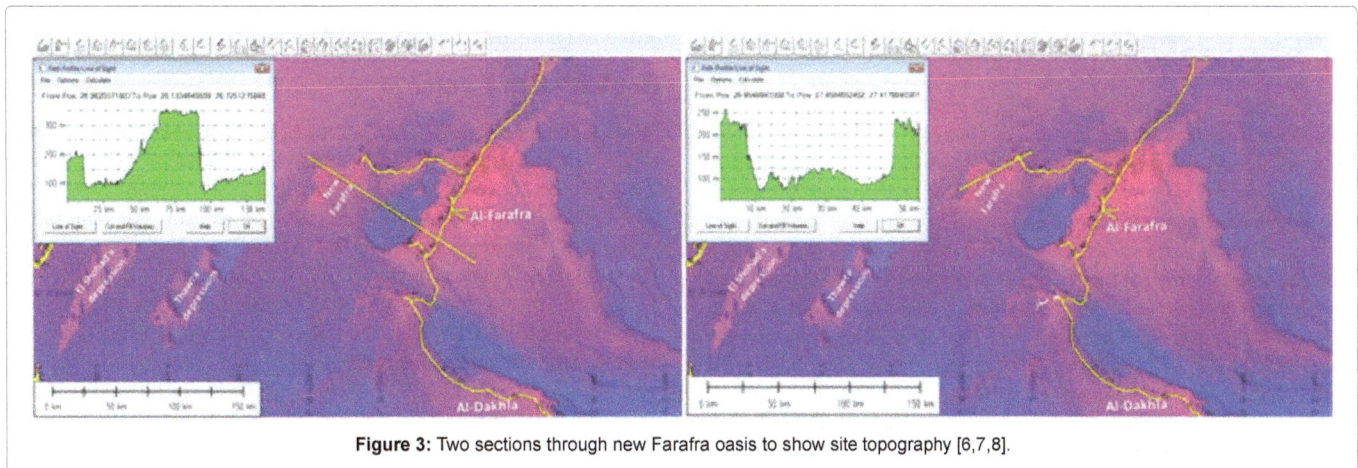

**Figure 3:** Two sections through new Farafra oasis to show site topography [6,7,8].

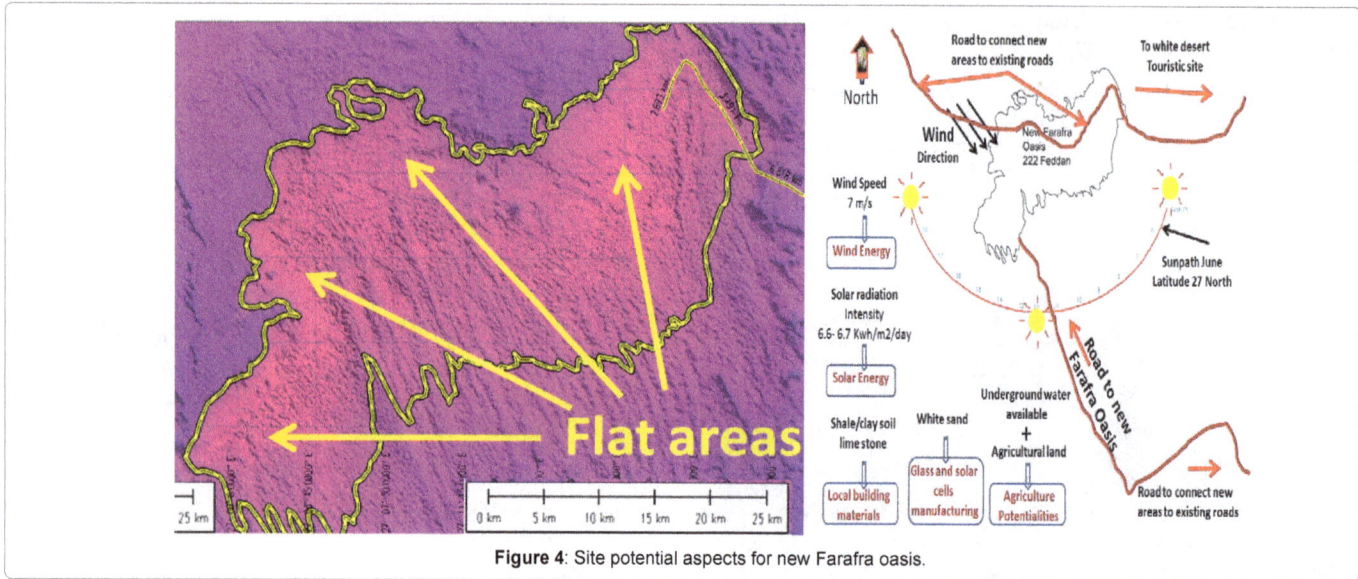

Figure 4: Site potential aspects for new Farafra oasis.

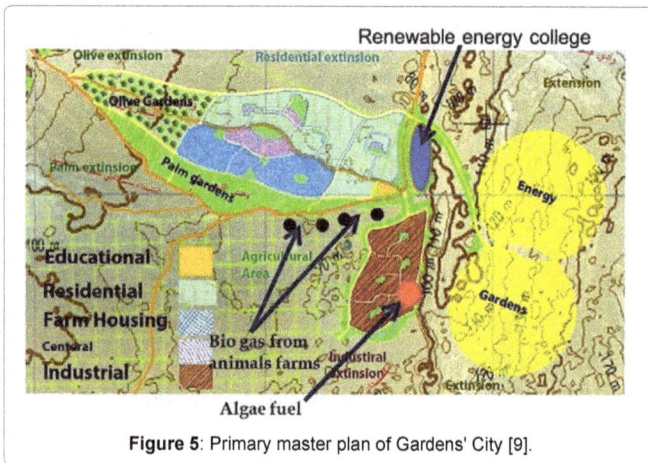

Figure 5: Primary master plan of Gardens' City [9].

- Shale/clay soil and lime stone are available, which facilitates local constructional material.

- The white desert, in the western desert, lies near the site, which allows geological tourism.

- White sand, which is suitable for glass and solar cell manufacturing, is found near the site.

## Gardens' City

Gardens' City is a new city, which is planned to be implemented in new Farafra Oasis [9,10]. The site of Gardens' city has been chosen to be located in the eastern northern part of the new Farafra Oasis. The plan benefits from some rough roads are already found in the site, one of these roads has been chosen to be a major axe of the city, and another one will work as a high way. The Egyptian desert, including the area around new Farafra Oasis, is famous for its good palm and olive, which is considered as an economical base for the oasis. A big palm garden is planned southern to the city, with an area enough for about 123 thousands palm trees. Another garden for olive is planned to the west of the city with an area enough for about 78 thousands olive trees. Agricultural area, about 15 thousand acres, will be planted with wheat rotated with other crops (southern to the city). As the city

site has elevations mostly between 70-90 m above sea level, housing areas have been allocated between 70-80 m above the sea level. The industrial zone lies in the southern east of the city. The energy production area will be 100 m above sea level, which lies eastern to the city with an extendable area up to 9000 acres. A primary master plan shows different zones in Gardens' city and future extension areas for different zones is in Figure 5. The city master plan, which has been also developed, adopts neighborhood community system with services inside each community. Local materials, like plants rest, will be used firstly to feed 750 thousands animals head in specific animal barns area, lies southern to the city, then the remains in addition to the animal residue will be used in producing biogas and fertilizers. Algae will be planted on a sanitary lake in the industrial area to produce biofuel. New Farafra Oasis will be developed in parallel with Gardens' City in the four developing flat spots (Figure 4). Gardens' City is planned for about 117 thousand inhabitants. Each developing spot in the oasis will have 250 thousand inhabitants, so the target will be a million inhabitants, so in the whole oasis [9].

Gardens' City will be sustainable in energy, agriculture, land use, rest and residue etc. It will have four different kinds of renewable energy and a yearly net profits about 63-90 Million Egyptian pound (LE). The whole new Farafra oasis will have around a million palm trees, 633 thousands olive tree and 60 thousand acres of wheat. The yearly net profits are estimated to be 394- 535 Million LE [10].

## Gardens' City Center

The city centre is 137 acres, which forms about 5% of city area without including the industrial area, which has its own central area. The city centre includes central managerial and commercial areas lies in 397 land pieces. These land pieces are divided into, 203 land pieces of about 400 m² area for commercial use and 194 land pieces of about 900 m² area for managerial use. The city centre has two places for car parking, they wide for about 1100 cars. In addition to, a big central park of about 10.3 acres, (Figure 6). shows land use, roads, and the big central city garden of the city central area.

## The Industrial Zone

Gardens' City area is 3250 acres including the industrial zone. The

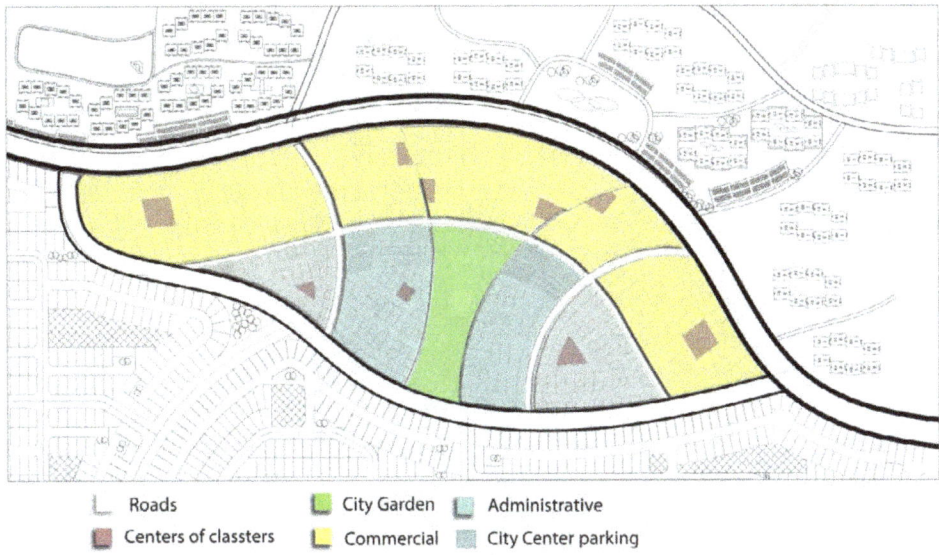

**Figure 6:** City centre.

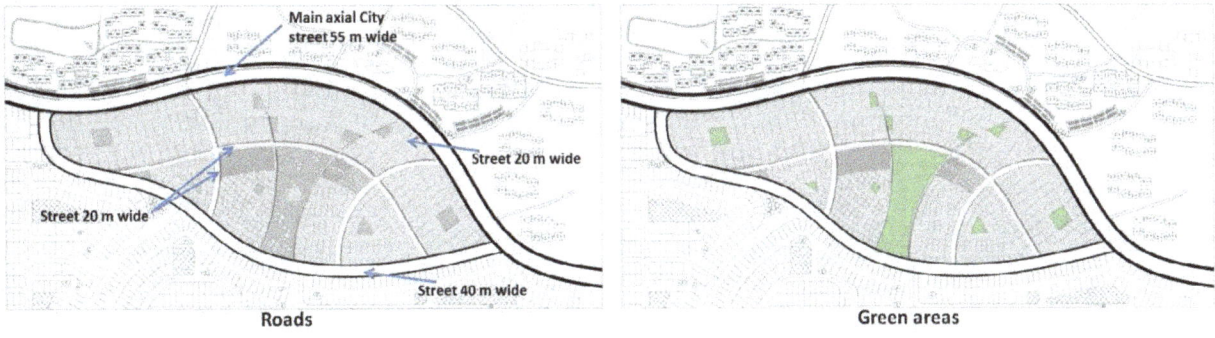

**Figure 7:** City centre land use, roads, sub centers and the city garden.

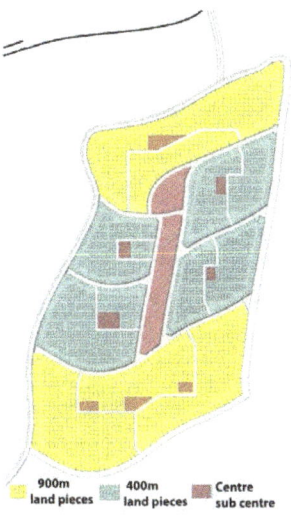

**Figure 8:** The industrial zone.

area of the industrial zone is 720 acres (22% of city area). It consists of 2401 land pieces. 1494 piece of about 400 m² each (262.3 acres), and 907 piece of about 900 m² each (334 acres). Total area of land pieces=596.7 acres. The area of streets is about 296 acres (41% of the industrial area). Each piece of land has a direct access on a street of at least 20 m wide. An industrial project might take any number of land pieces according to planning requirements of factory. The main streets and the central area of the industrial zone are designed to be 123.3 acres (17% of the industrial area). The industrial area centre is 38 acres (5.3% of industrial zone), (Figure 8). Tables 1-3 show services ratios and areas in industrial zone, Figure 9 shows the industrial zone's land use, roads and centre and sub centers. Services have been distributed in the central and sub central areas in the industrial zone. Figure 10 shows Gardens' City, its central and industrial zones and the neighborhood system used.

## Feasibility Study for Gardens' City

Gardens' City will be the first step in a series of development opportunities in Egypt. It will be sustainable in energy, agriculture, land use, rest and residue etc. It will have four different kinds of renewable energy. The estimated yearly net profit for the city would be 63-90 Million Egyptian pound (LE) and 394-535 Million LE yearly net profit for the whole new Farafra oasis from olive, palm and wheat only. Tables 4 and 5 shows that plants rest will be able to supply animal farms with the required animals' food [10].

| Land use | Land use ratios% | | |
|---|---|---|---|
|  | Minimum | maximum | Average |
| Industrial Land use | 60 | 70 | 65 |
| Services | 3 | 7 | 5 |
| Roads and car parking | 18 | 32 | 25 |
| Green and open areas | 3 | 7 | 5 |

**Table 1:** Land use ratios in industrial city [12,13].

| Services | Per unit | |
|---|---|---|
| Medium Industries | 15-25 craftsman/Acres | 14400 labour/720 Acres Divided to sex sector |
| Commercial | 0.35 m²/person in sector | 5040 m² for the sector |
|  | 0.56 m²/ person in neighbourhood | 1008 m² for a neighbourhood |
| Religious | 1.05/person in the sector | 15120 m² for the industrial sector |
|  | 1.4 m²/person in neighbourhood | 2520 m² for the neighbourhood |
|  | 1.2 m² For a cell in the neighbourhood | 600 m² for a cell |
| Recreational | For the sector | 30 Acres |
|  | For the neighbourhood | 5 Acres |
|  | For a cell | 1 Acres |
| Educational | A Kinder garden for 2000-3000 person with a service scope 400 m (an area of 0.15- 0.25 Acres) for 75-100 child. | 5-8 kinder garden for the industrial zone (0.75-2.0 Acres) - 375-800 child |
|  | Secondary school | 1 high technical school |
|  | Craftsmen 0.75-1.5 m²/person | Craftsmen village 10800-21600 m² |
|  | Vocational training centre | 1 Vocational training canter |
| Health | For the sector a Polyclinic for 20000- 50000 person | I Polyclinic with an area 0.6 Acres with service scope of 1000 m |
|  | For a neighbourhood a health canter for 5000- 10000 person with a service scope 250-300 m (an area 0.1 Acres) | 1 health canter |
| Secondary services | 1 Fire extinguisher Station | 1 Fire extinguisher Station, |
|  | 1 police station per 20000- 30000 person | 1 police station |
|  | 1 Ambulance station per 30000-40000 person (4-10 ambulance car) | 1 Ambulance station |
| Car parking | 15 place/1000 m² in mosque 5.5 place/100 commercial mall 100-200 place/1000 m² recreational |  |

**Table 2:** Industrial area (720 Acres–1 Acres= 4200 m²) [12,13].

| Services ratio/ industrial region | 5.3% (160750 m²=38 Acres) |
|---|---|
| services | % From region service area |
| Industrial services | 30-50 (40 average) (60480 m²) (general industrial management, marketing, Incubator projects, maintenance, ... |
| Social services | 20-40 (average 30) (45360 m²) |
| General services | 20- 40 (average 30) (45360 m²) (police station, civil defence, commercial, health, religious |

**Table 3:** Industrial services ration [12,13].

## Sustainable Community

Creating a sustainable community is not including the economic side only, but also other aspects of life: social justice, freedom, dignity, respecting rules and duties etc as shown in Figure 11 so, Gardens' City will be sustainable in different aspects as follows:

## Land owning

A part of the city strategy is to give youth about 3000 piece land pieces with economical price. Daily consumed crops like tomatoes, potatoes etc. will be planted there. This will help in attracting inhabitants, allow them produce their needs of vegetables and fruits, give them the chance to feel owner of the city, enhance their creativity and drive their

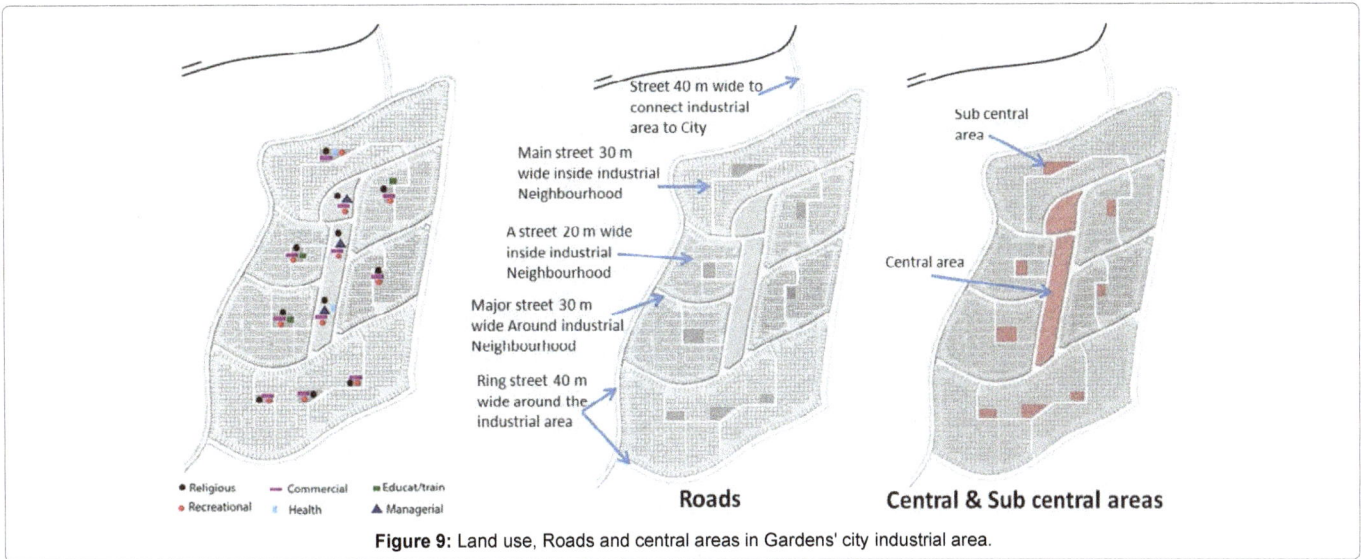

**Figure 9:** Land use, Roads and central areas in Gardens' city industrial area.

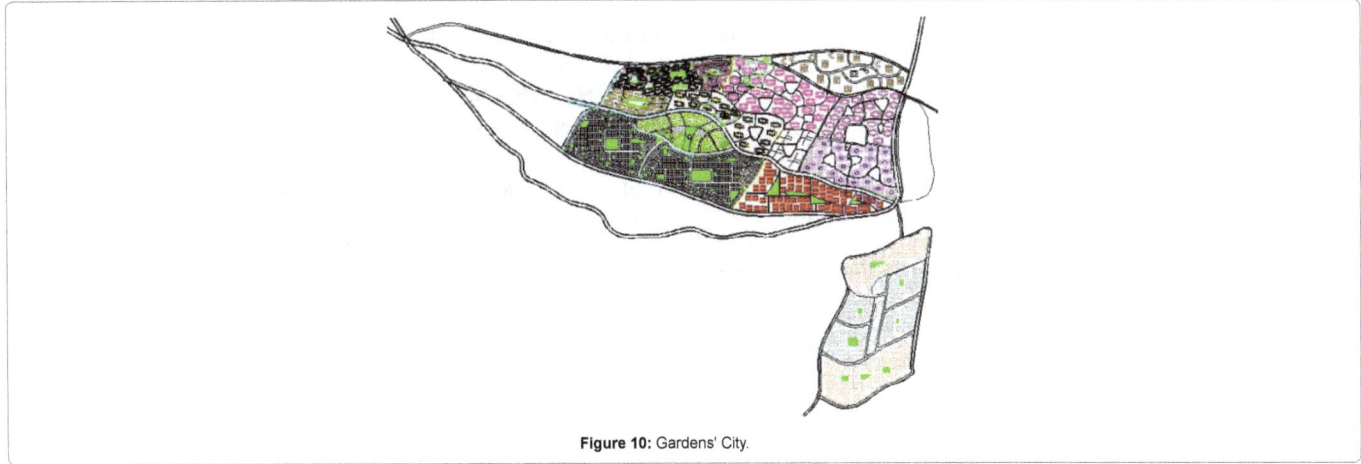

**Figure 10:** Gardens' City.

| Gardens City before extension | City production (Ton) | Agricultural rests (Ton/1000 Ton) | City Agricultural rests (Mio Ton) | City Agricultural rests after future extension (Mio Ton) | Oasis Agricultural rests (Mio Ton) |
|---|---|---|---|---|---|
| Wheat straw | 42000 | 60000-75000 | 2.520-3.150 | 2.520- 3.150 | 10.520-12.600 |
| Olive Leaves and branches | 2508 | 1000 | 25.080 | 50.16 | 200.64 |
| Olive geft | 2508 | 25000 | 62.700 | 125.4 | 501.6 |

**Table 4:** Plants rest in Gardens' city [14-16].

| Cows (head) | City needs - concentrated Feed (Thousand Ton/ day) | City need- concentrated Feed (Thousand Ton/ year) | City needs from dry fillers (Thousand Ton/ day) | City needs from dry fillers (Thousand Ton/ year) | Green Feed (Ton/ day) | Green Feed (Mio Ton/ year) |
|---|---|---|---|---|---|---|
| 375000 | 4.500-5.625 | 1.6-2.0 | 1.870 | 2.3-2.7 | 5625-5625 | 1.3-2.0 |

**Table 5:** Animal food needed in Gardens' City animals farms [17,18].

efforts forward. Legislation will prevent selling these land pieces before 30 years to prevent investing and price rising by selling them.

## Production and work

The produced crops will be collected by companies or organizations to be sold inside and outside the city. Homes will be used as residence, work and investment. From the profit, lands' prices will be paid.

People will find chances in planting palms, olives and other plants. Corporations will have opportunities in planting wheat and other national crops and other materials. Industrial expertise will help developing industry and energy in the oasis. Academics will develop theoretical and practical education for energy and industry. Industry will be initiated on crops and other materials. There will be more

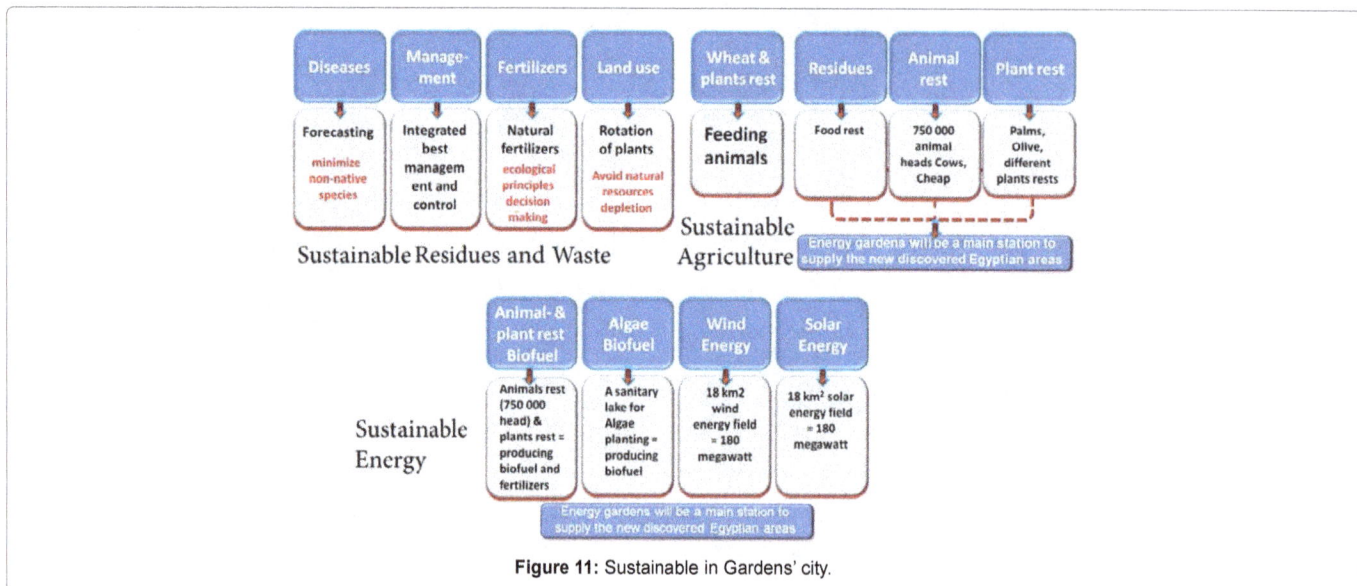

**Figure 11:** Sustainable in Gardens' city.

than a half million work opportunities for different people (educated, craftsmen, expertise, workers, etc.).

### Social reality

The city targets to attract different inhabitants. Both youth and expertise will find opportunities. Farmers and industrial worker will have chances. Investors and job seekers as well as, people who are newly coming back from abroad and those who are in danger due to global warming in Delta and coastal areas in Egypt will find places there. People will share societies and help serving them. All this will allow a stabilized, balanced society and assure dignity and social justice.

### Managing a sustainable community:

Decentralization mixed with centralization will be applied to assure a good city reward and Egyptian commercial and economical identity. Crops will be defined through councils from land owner themselves and expertise. Civil societies will solve problems and afford needs and technical help. City citizens will form councils in each city quarter. Expertise will help people to produce best quality products and market them. Prizes will be rewarding for citizens and the city. The city commercial chamber will be responsible for making commercial agreement in Egypt and abroad.

### Sustainability aspects: site, water, materials, infrastructure, energy

City site lies on flat land easy to be built. A beautiful touristic site near the oasis gives touristic opportunities. Water is available near ground surface. It will be used wisely. A new sanitary net design will be used to collect used water. Gray water will be reused to grow plants, palms olives, and in industry. Lime stone in the oasis with shale/clay soil form local building materials. Renewable energy will be produced to supply the newly discovered areas in the Egyptian western desert. Biofuel from plants and animals residues and from algae will be also produced [9,10].

### Steps on the Road

A conference has been held in September 2012 in the new valley governorate in Egypt announcing the start of new Farafra oasis development. Some investors suggested initiating the city infrastructure.

10 Millions Egyptian pounds have been assigned for experimental wells in the oasis. Another conference has been held at the same governorate on 27th November 2012 looking for financial support. A professor team from Assiut University is ready to develop the project. Professor from agricultural college will define agricultural priorities, best irrigation methods for different crops, local plants in the area to develop their species. Professors from civil engineering roads will design the road from Asyut to new Farafra and other roads. Professors from department of Mining and Metallurgical and civil material professors will study the soil to develop building material from local existing materials. Professors from sanitary specialty will study the city's sanitary system to make use of solid and liquid wastes and plant algae on sanitary water to produce Biofuel. Professor from survey department will develop maps to help plotting the city on land. Professors from mechanical and electrical departments will study making use of wind and solar power. From architecture department the planning team and from College of Science a geological team etc. Investing opportunities will be opened in the new Farafra oasis in different related fields.

### Conclusion

Gardens' City is a future city in the new Farafra Oasis in the Egyptian western desert. It represents a sustainable development to afford future growth needs and overcome the probable immigration from Delta and Nile regions due to climate change. Gardens' City is planned for 117 thousand inhabitants in a development spot system with a target of one million inhabitants in oasis. The city centre has 203 land pieces for commercial use, 194 land pieces for managerial use, a big central recreational garden and two big central parking areas. The industrial zone's area is 720 acres. It has 2401 land pieces for industrial use in six industrial neighborhoods with different services. It has a technical school, a craftsmen village and a vocational training centre. Its' central area has the main industrial managerial service. Gardens' City will have different sustainable options and the estimated yearly net profit for it would be 63-90 Million Egyptian pound (LE) and 394-535 Million LE yearly net profit for the whole new Farafra Oasis from olive, palm and wheat only. This city will be the first step that opens great development opportunities in Egypt.

## References

1. Worth K (2011) Egyptian Turmoil Bullish for Oil Stocks, But Not for Reasons Some Might think. Seeking Alpha.

2. http://en.wikipedia.org/wiki/Egypt-15 June 2011.

3. Abouelfadl S (2012) Global warming-causes, effects and solution's trials. J Engg Sci 40: 352-360.

4. Abouelfadl S, El-Lithy KH (2014) Impact Assessment of Global warming on Egypt. Int J Engg AdvTechnol 3: 352-360.

5. Ouda KH (2012) Atlas of risks of climate change on the Egyptian coasts and defensive policies. Geology of the Nile basin countries conference, Alexandria, Egypt: 95-103.

6. Ouda KH (2009) Fostering sustainable water management and agriculture in Egypt's New Valley.Presented conference research work, WaDImena Egypt.

7. Ouda KH, Sinosi M, Saber M, Gaber M, Hasan G (2011) Final report of new geological and underground water witness at the great sand sea in the western desert- Egypt, a push that leads to agricultural development and architectural societies at the western part of the western desert-Egypt.

8. Ouda KH (2012) Final report results of new geological and underground water witness at the great sand sea in the western desert–Egypt, a push that leads to agricultural development and architectural societies at the western part of the western desert-Egypt.

9. Abouelfadl S, Ouda KH, Atia A, Al-Amir N (2013) The Egyptian Gardens' City- A New City in Western Desert (EGCWD)-A primary master plan. International Conference on Energy and Sustainability, NED University of Engineering and Technology, Pakistan.

10. Abouelfadl S, Ouda Kh, Atia A, Al-Amir N, Ali M, et al. (2014) WIT Transactions on Ecology and the Environment. A new sustainable city in The Egyptian western desert-Gardens' City. Environmental Impact Conference, Ancona, Italy.

11. Ouda KH, Senosy M, Gad M, Hassan G, Saber M (2012) New Findings in Geology, Geomorphology, and Groundwater Potentiality of the Great Sand Sea, Western Desert, Egypt: An advantage which would lead to establishment of new societies in order to meet growing demands in Egypt. Geology of the Nile Basin Countries Conference GNBCC-2012: Geology and development challenges, Alexandria, Egypt.

12. Abokreen A (1997) Basics of the theories of urban planning. Alsafaa press Abuhelal-Elmenia, Egypt.

13. http://www.sidf.gov.sa/En/INDUSTRYINSAUDIARABIA/Pages/IndustrialDevelopmentinSaudiArabia.aspx

14. Harb M (2012) Agricultural wastes that can be used as animal feed in Jordan. Department of Animal Production. College of Agriculture - University of Jordan .

15. http://www.caaae-eg.com/new/index.php/2012-12-25-10-49-19/2010-09-18-17-00-51/2011-01-10-19-57-24/490-2011-12-13-10-12-34.html

16. http://www.mawhopon.net/ver_ar/news.php?news_id=593

17. http://website.paaf.gov.kw/paaf/ershad/ad46.jsp

18. http://aljsad.com/forum49/thread103743/

# 14

# Transmission Network Enhancement with Renewable Energy

**Oyedepo SO[1],\*, Agbetuyi AF[2] and Odunfa MK[1,3]**

[1]*Mechanical Engineering Department, Covenant University, Ota, Nigeria*
[2]*Electrical Engineering Department, Covenant University, Ota, Nigeria*
[3]*Mechanical Engineering Department, University of Ibadan, Nigeria*

## Abstract

Wind and solar energy play an important role in the de − carbonization of electricity generation. However, high shares of these Variable Renewable Energies (VREs) challenge the power system considerably due to their temporal fluctuations and geographical dispersion. This paper systematically reviews and analyzes transmission grid extensions as an integration measure for VREs. Effects of grid extensions for fundamental properties of the power system as a function of the penetration and mix of wind coupled with solar energy were revealed in the study. The paper also provides an overview of the system implication of wind and solar PV energy and investigates a way to partly overcome transmission grid extensions.

**Keywords:** Renewable energy; Transmission network; Integration; Electric power; Grid

## Introduction

Access to modern energy is considered one of the foremost factors contributing to the socio-economic and technological development of every nation. Energy use is a prerequisite for physical and socio-economic development in both rural and urban communities. There is a need to promote and guarantee energy security, availability and reliability to preserve any existing level of development and further new developmental strides for human comfort [1].

However, at the same time, energy production can contribute to local environment degradation, such as air pollution and global environmental problems, principally climate change. As a result of this, most developing nations of the world are looking towards renewable energy sources as a sustainable option [2]. Sustainable development has been at the center of recent policies and development plans of many developing countries. This is a pattern of development that delivers basic environmental, social and economic services without threatening the viability of natural, built and social systems upon which these services depend.

Today, the primary energy source all over the world is non-renewable fossil fuels that have been and will continue to be a major cause of pollution and climate change [3]. Moreover, fossil fuels-based conventional grid extension in developing countries from centralized power systems in urban centers to rural areas is usually capital intensive and in most cases not economically realistic. From a global perspective, more than a quarter of the human population experiences an energy crisis, especially those living in the rural areas of developing countries.

Because of these problems and dwindling supply of primary energy source such as petroleum, natural gas etc finding sustainable alternatives is becoming increasingly urgent.

With increasing energy demand and growing concerns for environmental impacts, renewable energy sources are receiving increased attention for their inherent low pollutant and greenhouse gas emissions. However, the intermittent and uncontrollable nature of renewable energy sources introduces new technical challenges for integration into electric power systems, especially as the market share of renewable energy becomes large [4].

The objectives of harnessing renewable energy (RE) as primary energy source are to focus on provision of sustainable energy to the economically subjugated fraction of the society, combat energy shortage, encourage the development of rural infrastructure and provide clean energy from the perspective of the Kyoto directive towards global decarbonization. This concept of RE has become a fast growing idea in the global power sector. The popularity of RE development can be directly allied to the growing trend of environmental concern and the rapidly depleting reserves of conventional energy resources due to the aggressive utilization. These emergent concerns call for a viable alternative solution to the contemporary environmental challenges and the energy crisis scenario through sustainable means [3]. Perhaps the greatest challenge in realizing a sustainable future is to develop technology for integration and control of renewable energy sources in smart grid distributed generation. The smart power grid distributed energy system provides the platform for the use of renewable sources as adequate emergency power for major metropolitan load centers and the ability to break up the interconnected power systems into the cluster smaller regions.

Among the available literatures on the potential of integrating renewable energy on grid include:

The concept based on an appropriate combination of solar, wind and biomass systems used by Jain [5] to prove that integrated renewable energy system (IRES) is reliable and viable concept from energy production and utilization point of view. Further, the small-scale decentralized IRES concepts were discussed in the paper by Ramakumar [6]. The study considered solar PV (SPV), solar thermal, wind, biomass and falling water as renewable resources. Another study by Ramakumar [7] showed that the concept of energization through resource-need matching has been found to be preferable as compared to straightforward rural electrification. Segurado et al. [8] used the hydrogen renewable energy source (H2RES) model to analyse different

---

**\*Corresponding author:** Oyedepo SO, Mechanical Engineering Department, Covenant University, Ota, Nigeria
E-mail: Sunday. oyedepo@covenantuniversity.edu.ng

scenarios with the objective of increasing the penetration of renewable energy in the electric energy system of the São Vicente Island in Cape Verde. An integrated approach was used to analyse the electricity as well as the water supply systems. The results showed that it is possible to have more than 30% of the yearly penetration of renewable energy sources in the electricity supply system, and to desalinate more than 50% of the water supplied to the population by the use of wind power. Bernal-Agustin and Dufo-Lopez [9] performed economic analysis on the Grid Connected (GC) solar PV system connected to the Spanish grid. Using Net Present Value (NPV) and Payback Period (PP) parameters, the profitability of the system was studied. The system was evaluated for its economic as well as environmental benefits and the results clearly showed that the system is profitable enough to be invested in, but very long pay back periods were dissuading the investors. The Ministry of New and Renewable Energy (MNRE) annual report [10] discussed how attractive GC systems are in the context of Clean Development Mechanism (CDM). Life Cycle Analysis (LCA) was performed to assess environmental benefits of solar PV systems, the effects of application Kyoto protocol, and reduction in emissions.

Furthermore, studies on the introduction of flexibility to electric power system to allow the penetration of fluctuating renewable energy to be increased have been carried out. One technology which is ideally suited for increasing energy flexibility is energy storage. Benitez et al. [11] analysed the impacts of additional wind capacity on the Alberta electricity network in Canada and concluded that when Pumped Hydroelectric Energy Storage (PHES) is added in conjunction with wind power it can provide most of the peak-load requirements of the system and thus, peak-load gas generators are no longer required. Dursun and Alboyaci [12] carried out a detailed review of wind - PHES studies and outlined how this solution could be employed in Turkey, by utilizing the mountainous areas around the Black Sea and electrical infrastructure to other hydro facilities. Black and Strbac [13,14] examined the benefits of PHES on the British energy system for a wind penetration of 20%, which equates to an installed wind capacity of 26 GW. After paying particular attention to reserve requirements and systems costs, the authors concluded that the value of PHES is very dependent on the flexibility of the conventional generation also on the system. The results also indicated that energy storage could reduce system costs, wind curtailment, and the amount of energy required for conventional generation.

The literature reveals that no comprehensive review has been reported on transmission network enhancement with renewable energy. The prime objectives of this study therefore are: (i) to review availability of renewable energy for electricity generation (ii) to review grid interface technology with renewable energy (iii) to assess possible ways of strengthening integration of renewable energy into electrical power systems and (iv) to proffer remedial actions aimed at addressing problems related to transmission network with renewable energy sources.

## Renewable Energy Sources Availability

Energy demand in both developed and developing countries is growing rapidly. As global energy demand increases, RE provides one means among many of adding energy assets to the system alongside growth of other resources [15–17].

In 2008, it was estimated that RE accounted for 12.9% of the total 492 EJ of primary energy supply globally. Among which, Biomass contributed the largest share of 10.2%, while Hydropower represented 2.3%, whereas other RE sources accounted for 0.4%. From global electricity supply point of view, RE contributed approximately 19% (i.e 16% hydropower and 3% other RE). The global road transport fuel supply in 2008, biofuels contributed 2% and traditional biomass (17%). Considering the total demand for heat in 2008, modern biomass contributed 8%, solar thermal and geothermal energy together contributed 27% [18,19].

In 2009, renewable electricity generation already accounted for 62% (17 GW) of all newly constructed power-generating capacity in Europe. In particular, wind energy installation, increasingly larger in size, accounted for 38% (10.2 GW) of all renewable energy growth [20]. Recently, the world's fastest energy source is wind power [21]. The decentralized and locally available nature of wind energy makes it particularly attractive to grid electrification. Wind power is growing at the rate of 30 per cent annually, with a worldwide installed capacity of 198 gigawatts (GW) in 2010, and is widely used in Europe, Asia, and the United States. Wind power accounts for approximately 19 per cent of electricity use in Denmark, 9 per cent in Spain and Portugal, and 6 per cent in Germany and the Republic of Ireland. The United States is an important growth area and installed US wind power capacity reached 25,170 MW at the end of 2008.

Research has shown that the available solar energy resources are 3.8 YJ/year (1, 20,000 TW) [19]. Less than 0.02 per cent of available resources are sufficient to entirely replace fossil fuels and nuclear power as an energy source. Solar PV uses and applications have been justified and strongly recommended for grid electrification [22]. In 2007 grid-connected photovoltaic electricity was the fastest growing energy source, with installations of all photovoltaic increasing by 83 per cent in 2009 to bring the total installed capacity to 15 GW. Nearly half of the increase was in Germany, which is now the world's largest consumer of photovoltaic electricity (followed by Japan). Solar cell production increased by 50 per cent in 2007, to 3800 MW, and has been doubling every two years.

Power generation through the use of biomass offers a viable and long-term solution to grid electrification; however it is inefficient use, biomass resources presently supply only about 20% of what they could if converted by modern, more efficient, available technologies [17]. In recent years, interest in biomass as a modern energy source, especially for electricity generation has been growing worldwide. Electricity produced from biomass sources was estimated at 44 GW for 2005. Biomass electricity generation increased by over 100 per cent in Germany, Hungary, the Netherlands, Poland, and Spain. A further 220 GW was used for heating (in 2006), bringing the total energy consumed from biomass to around 264 GW. The use of biomass fires for cooking is excluded.

World production of bio-ethanol increased by 8 per cent in 2007 to reach 33 billion liters (8.72 billion US gallons), with most of the increase in the United States, bringing it level to the levels of consumption in Brazil. Biodiesel increased by 85 per cent to 3.9 billion liters (1.03 billion US gallons), making it the fastest growing renewable energy source in 2007. Over 50 per cent is produced in Germany.

The total shares of all renewable for electricity production make up for about 19%, a vast majority (83%) of it being from hydroelectric power [23]. Worldwide hydroelectricity installed capacity reached 816 GW in 2005, consisting of 750 GW of large plants, and 66 GW of small hydro installations. In 2005, China, Brazil and India added large hydro capacity totaling 10.9 GW. There was a much faster growth (8 per cent) small hydro, with 5 GW added, mostly in China where some 58 per cent of the world's small hydro plants are now located. China is the largest hydropower producer in the world, and continues to add capacity.

The growth rate of renewable energy has been keeping at a double-digit for the recent 5 years. EIA (Energy Information Administration) estimated that there would be a 3.1% annually increase in the share of electricity generated from renewable energy during the period from 2008 to 2035 all around the world [24]. It means 45% of global electricity will be generated from renewable energy by the year 2035 [9,15].

## Features and Structures of Electrical Power Systems

The main elements in an electric power network are generators, transformers, switchgear and power factor correction components interconnected by a web of transmission lines of various voltage and current ratings and a considerable number of electrical joints [25].

The electricity generation (or power station) equipment has the capacity to convert a primary energy flux into an electrical energy flux and (in some cases) to maintain a voltage waveform at its point of connection. End - use equipment has the capacity to convert an electrical energy flux into an end-use energy flux, in the process providing an end-use energy service. However, neither generation nor end-use equipment can operate in isolation, particularly as they are usually at different geographical locations [26-28].

Electricity transmission and distribution networks provide current paths so that electrical energy can flow between the generation and end-use equipment to complete the energy conversion chain. The electrical transmission and distribution system is an essential part of every developed nation and failures within the system have the potential to cause immediate and unexpected power loss over a large geographical area with wide ranging implications [29,30].

Electrical power systems are operating under heavily loaded conditions due to various economic, environmental and regulatory changes. So with the increased loading and exploitation of the power transmission system, the problem of voltage stability and voltage collapse has been reoccurrence situation and this has attracts more attention recently. Hence, maintaining voltage stability has become a growing concern for electric power utilities [31-33].

The stability and security of the supply of electricity are assured when consumers can rely on electricity of a defined quality at any time of the day. The consumption and production of electricity must be balanced to secure the stability of the grid [34]. Among the potential measures to compensate for more frequent imbalances between supply and demand is introduction RE into electricity industries. Higher RES penetration will result in a significantly reduced load factor for conventional generation, as the RES technologies will replace a growing section of the electricity supply curve [35,36]. Electric systems that can accept lower levels of overall reliability may be able to manage the integration of RE into electrical power systems at lower costs than systems that demand higher levels of reliability, creating a trade-off that must be evaluated on a case-by-case basis [37].

### Renewable energy integration into competitive electricity industries

Electricity generated using renewable energy resources will, in the most part, be delivered to the point of end-use via large scale transmission and distribution systems [38]. Consequently, the successful integration of renewable energy generation into large power systems has become fundamental to successfully addressing climate change and energy security concerns [20,39].

Renewable energy technologies are suitable for off-grid services; they can serve remote areas of the world without expensive and complicated grid infrastructure.

The ability to integrate electricity generated from renewable into grid supplies is governed by several factors, including [40]:

- The variation with time of power generated
- The extent of the variation (availability)
- The predictability of the variation
- The capacity of each generator
- The dispersal of individual generators
- The reliability of plants
- The experience of operators
- The technology for integration
- The regulations and customs for embedded generation

Despite these difficulties, researches have shown that electricity from renewable can be integrated into grid supplies without significant financial penalty [39-41]. Several mature RE technologies, including wind turbines, small and large hydropower generators, geothermal systems, bio-energy cogeneration plants, bio-methane production, first generation liquid bio-fuels, and solar water heaters, have already been successfully integrated into the energy systems of some leading countries. Further integration could be encouraged by both national and local government initiatives. Over the longer term, integration of other less mature, pre-commercial technologies, including advanced bio-fuels, solar fuels, solar coolers, fuel cells, ocean energy technologies, distributed power generation, and electric vehicles, requires continuing investments in R &D, infrastructure, capacity building and other supporting measures.

The outstanding example of ever-increasing integration of renewable energy generation into the grid is Jutland, western Denmark [42]. In the early 1980s, the limit for wind power exported to the grid was considered to be 20% of total supply. However by 2003, about 40% of annual electricity supply was from wind, and at times, significant areas were supplied totally by wind power. The reason for the change was the willing application of new technologies and practices.

Successful integration of high shares of RE with energy systems in recent years has been achieved in both Organization for Economic Co-operation and Development (OECD) and non-OECD countries, including Brazil, China, Denmark, Spain, New Zealand and Iceland [43–48].

Access to some Renewable Energies (RE) resources is abundant in many parts of the world [49]. The characteristics of many of these resources distinguish them from fossil fuels and nuclear systems and have an impact on their integration [50]. Some resources, such as solar, are widely distributed, whereas others, such as large hydro, are constrained by geographic location and hence integration options are more centralized. Some RE resources are variable and have limited predictability. Others have lower energy densities and different technical specifications from solid, liquid and gaseous fossil fuels. Such RE resource characteristics can constrain their ease of integration and invoke additional system costs, particularly when reaching higher shares of RE [51].

Due to above fundamental limitations for any renewable energy generation technology and plant it is essential to integrate renewable energy generation options with control and storage such that they complement each other.

Integration of DG causes bi-directional power flow which reduces the capacity of feeder and transmission line. The other benefits of distributed generation include the reduction of power loss, better voltage support, peak shaving and the improvement of overall efficiency, stability and reliability [52].

In a Region where a majority of electricity generation is based on renewable sources is far beyond the horizon, it is clear that the confluence of government policy, utility planning and global demand growth has the potential to increase penetrations of RE substantially on electricity grids worldwide. This shift in generation portfolios will have profound effects on the operation of the grid, which will in turn affect the operation of RE resources themselves as well as the operation of other resources and equipment connected to the grid [53].

## Different types of grid interfaces

Electricity generation using renewable energy resources is often taken place in small scale due to disperse nature of the resources. Good examples are small hydro, solar photovoltaic, biogas, biomass and small wind turbine based electricity generation systems. The size of these generators typically varies from a few hundreds of kilowatts to several megawatts. These small scale electricity generators are generally connected to the grid at the primary or secondary distribution level and are considered Distributed Generation (DG) or Distributed Resources (DR). Distributed resources include both renewable and non-renewable small scale generation as well as energy storage [54,55].

There are different options for producing electricity from renewable energy sources. Consequently, there are several ways of connecting the gained electricity with the existing grid. The potential renewable energy sources are wind, hydro, solar, biomass, photovoltaic cells, bio fuels and geothermic. The electricity generated from these sources is induced by asynchronous or synchronous generators except for photovoltaic cells. This operation creates co-current flows and gets through an inverted rectifier into the power grid [56].

Often, the small scale renewable generators are not directly connected to the grid. The generation technology or the operational characteristics requires the use of some interface between the generator and the utility distribution grid. For example, solar photovoltaic (PV) panels generate DC electricity and therefore, a power electronics based DC to AC converter is required between the grid and the generator. Some technologies such as induction generator based small hydro or wind can be directly connected to the AC grid (Figure 1). However, concerns such as starting transients, energy conversion efficiency and power quality issues make connecting them through a power electronics interface a better choice [57,58].

Table 1 summarizes some of the common types of generation and their preferred interfacing technologies. Figure 2 and Figure 3 present the block diagrams of interfacing technologies for grid integration of solar energy and wind energy respectively.

## The Renewable energies option for climate change and energy security

Climate change is one of the most difficult challenges facing the world today and preventing will necessitate profound changes in the way we produce, distribute and consume energy. Burning fossil fuels such as coal, oil and gas provides about three-quarters of the world's energy.

However, when these same fuels are burned, they emit Greenhouse Gases (GHGs) that are now recognized as being responsible for climate change [59]. These fuels are ubiquitous. Fossil energy has fuelled industrial development, and continues to fuel the global economy. The primary greenhouse gas emitted through fuel combustion is $CO_2$. Land-use and land-use changes, notably deforestation, also involve emissions of $CO_2$ [25].

Policies to increase both energy efficiency and the share of renewable energy resources have been adopted by many countries as means to mitigate climate change and reduce dependence on external energy supplies [60]. With a high share of both wind power and CHP Denmark is one of the frontrunners in the implementation of such policies, and thus serves as valuable national case study of large-scale integration of new energy technologies.

Gas flaring by the oil companies operating in developing countries like Nigeria has raised temperatures and rendered large areas uninhabitable [30]. The use of renewable energy sources will reduce over dependency on the burning of fossil fuel. Moreover, instead of flaring gas, the gases can be converted to methanol and used as fuel for both domestic and industrial use. With good energy efficiency practices and products, the burning of fossil fuel for energy will be greatly minimized.

RE provides a number of opportunities and not only to address climate change mitigation but also addresses sustainable and equitable economic development, energy access, secure energy supply and local environmental and health impacts [31]. Market failures, up-front costs, financial risk, lack of data as well as capacities and public and institutional awareness, perceived social norms and value structures, present infrastructure and current energy market regulation, inappropriate intellectual property laws, trade regulations, lack of amenable policies and programs, lower power of RE and land use

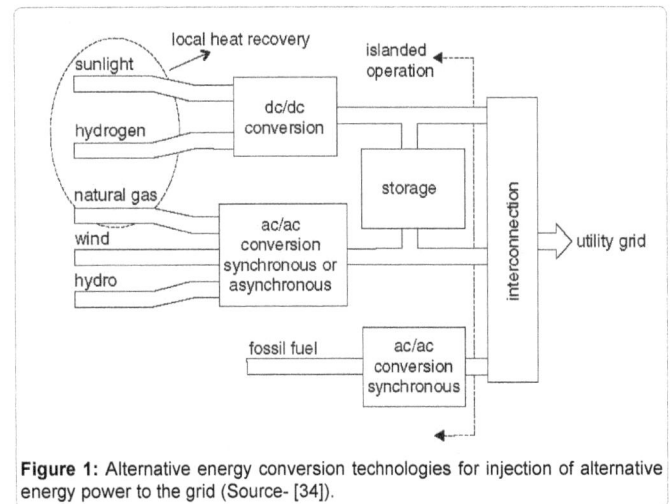

**Figure 1:** Alternative energy conversion technologies for injection of alternative energy power to the grid (Source- [34]).

| Renewable Energy Type | Interfacing Technology |
|---|---|
| Wind Energy | Induction generator/power electronic converter |
| Photovoltaics | Power electronic converter |
| Small hydro power | Synchronous or induction generator, power electronic converter |
| Fuel cells | Power electronic converter |

**Table 1:** Interfacing Technologies (Source [54]).

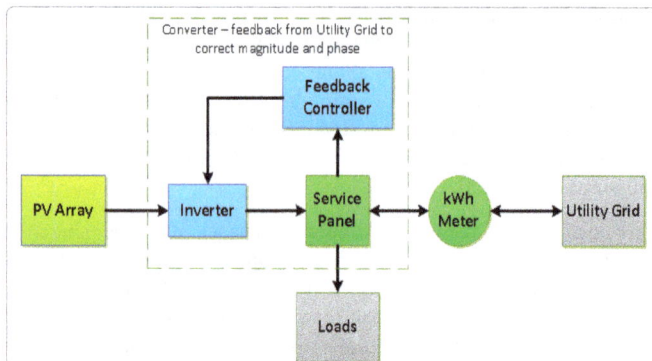

**Figure 2:** Block diagram for grid interface with PV array (Source- [56]).

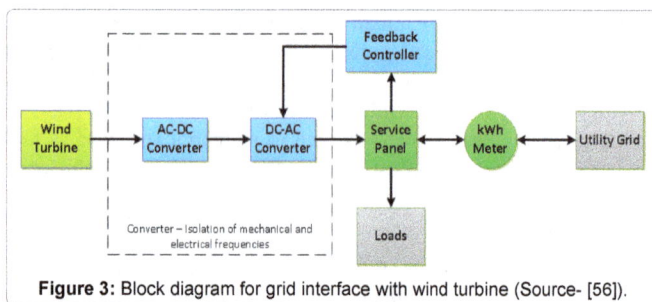

**Figure 3:** Block diagram for grid interface with wind turbine (Source- [56]).

conflicts are amongst existing barriers and issues to expanding the use of RE.

The deployment of renewable energy would result in significant energy security and economic benefits [33]. For example, renewable energy resources have a significant potential for energy efficiency, this would reduce energy import dependency to bring both energy security and economic benefits. For small-sized distributed generators (especially prosumers) with renewable resources, such as PV panels and wind turbines due to their vicinity and integration to the end-users, along with the smart grids implementations, it would greatly decrease the scales and consequences of blackouts. Furthermore, the self-healing features of smart grids would also accelerate the recovery and restoration of the system.

## Challenges for integrating renewable energy into electricity industries

The existing power system heavily relies on power plants that have a controllable power output. Natural gas plants, for example, can be turned on and off as needed, or their output can be increased or decreased to balance changes in power supply. Renewable energy sources, for example wind and solar power are not as controllable as conventional generation resources and they thus present some challenges to the operation of the power grid [60].

Introducing a significant number of renewable energy resources into electricity industries adds new challenges to restructuring in addition to the particular challenges associated with renewable energy resources themselves. The key challenges facing the increased use of wind and solar power are [61]:

• The variable, non-storable nature of renewable energy forms such as wind and solar energy, leads to a need for accurate forecasting and a need to define appropriate boundaries to autonomous decision-making by renewable energy generators for both operation and investment.

• The novel nature of some renewable energy generator technologies, such as wind turbines and photovoltaic systems, leads to uncertainties in their technical performance, particularly during abnormal power system operating conditions when power system security may be at risk.

• The small size of some renewable energy generator installations, such as photovoltaic systems, leads to a rapid increase in the number of supply-side decision-makers and a need to develop appropriate commercial contracts and technical requirements for generator connection to distribution networks, in contrast to the more mature arrangement for generators connected to transmission networks.

• While still in the development phase, renewable energy technologies will continue to require policy support. The challenge exists to provide financial support in a way that encourages the most cost efficient development of the technologies.

• The use of renewable energy in the context of autonomous single-users or small rural communities may raise community social, technical and financial resource questions, as well technical challenges associated with remote locations and long equipment supply chains.

• There are several aspects to be considered in order to integrate RES into traditional networks. However, there are two parameters that have high impact on the integration of RES plants in the network: the selection of the size (rated capacity) and the installation's location of such plants.

For these reasons, the regulatory framework and market rules for a restructured electricity industry may have to evolve to accommodate high levels of renewable energy penetration. To achieve effective outcomes, these issues must be addressed in a consistent manner, at all levels of decision making from the high-level, long time-scale governance level to the technically specific, short term power system operating level.

## Technical issues for grid integration in the light of renewable energy technologies

The technical issues associated with renewable energy compatibility relate to the ability of renewable energy equipment to function effectively as part of the electricity industry as it exists today. Equipment must meet engineering requirements with respect to voltage, frequency, waveform purity, ability to rapidly isolate faulty equipment from the rest of the industry and reasonable ability to withstand abnormal operating conditions (fault ride through) [40].

The technical assessments expressed about concerns for utilization of PV- DG integration into the grid in some countries leading in utilization of PV electricity are as follow [62]:

• Harmonic emission by inverters was considered a present and future concern for high penetration of PV.

• Voltage regulation was considered a big concern for weak grids with high PV penetration. Different regulations allowed overvoltage limits of 5% to 6% by PV plants.

• Network protection was considered a big concern as there is lack of direct control on DGs by the DNSP.

• Unintentional islanding due to high penetration of PV-DG was considered a matter of concern.

The properties of wind turbine generators may increase the power quality (PQ) related problems such as voltage fluctuation, harmonics, voltage unbalance. By integrating wind turbine into the grid can introduce PQ disturbances such as [63]:

• Flickers which are commonly due to rapid changes in the load or the switching operations in the system.

• Steady-state voltage level influence.

• Response to grid disturbances/faults (Stability).

• Wind power may affect the power flow direction in the network and can cause transmission capacity problem.

## Smart grid system

Smart Grid Network is an intelligent, managed, controlled and ultimately self-healing electric distribution network capable of closely matching supply with demand while improving efficiency and reliability. Sensors and control devices on the grid, combined with integrated high speed communications and advanced analytic software, provide utilities with actionable intelligence reports and information [64]. A Smart Electric Grid can identify where electricity is lost or where the system is not in balance or optimized. Such optimization can save 3% or more of overall electric demand without requiring any change in consumer behavior [65].

Smart technologies comprise of the following components: (i) Smart meters to quantify the energy consumption and the power quality of each electrical equipment connected to it and allow the user to take a decision on the best way to consume power and what equipment to consider as regard to power quality, (ii) Artificial intelligent monitoring equipment that has the ability to pre-detect fault and monitor power quality on the network (iii) Real- time transmission power flow monitoring equipment and power electronics that can limit the waveform distortion either from the generator or the load and correct the waveform deformity. Figure 4 shows a pictorial view of Smart Grid structure, from Generation to the Consumers.

Today's electric grid system was designed to operate as a vertical structure consisting of generation, transmission, and distribution and supported with controls and devices to maintain reliability, stability, and efficiency. However, the grid system is facing new challenges including the penetration of renewable energy resources (RER) in the legacy system, rapid technological change and different types of market players and end users [42]. Several factors that contribute to the inability of today's grid to efficiently meet the demand for reliable power supply can be overcome by smart grid system.

The potential promise of the smart grid includes improved reliability and power quality, reduction in peak demand, reduction in transmission congestion costs, potential for increased energy efficiency, environmental benefits gained by increased asset utilization, increased security, ability to accommodate more renewable energy, and increased durability and ease of repair in response to malicious attacks or adverse natural events.

It is now obvious that there is an urgent need to transform the current electricity system to meet future energy demand and reduce Greenhouse Gas Emissions. So we need a modern, smart and intelligent grid which can respond to all the challenges presented by today's grid system.

## Managing Renewable Energy in a Grid

The increasing number of renewable energy sources and distributed

generators requires new strategies for the operation and management of the electricity grid in order to maintain or even to improve the power-supply reliability and quality. The renewable energy sources such as solar, wind, small hydro power etc. has accelerated the transition towards greener energy sources. Keeping in view the aforementioned challenges of integrating renewable energy sources to the grid, some of the possible ways for renewable energy utilization in a grid are [65]:

• The power balance using RES can be carried out by integrating RES with energy storage unit. The benefits of battery energy storage system (BESS) are classified based on end – users as: Transmission level uses, System level uses, ISO Market uses.

• The power-electronic technology plays an important role in distributed generation and in integration of renewable energy sources into the electrical grid, and rapidly expanding as these applications become more integrated with the grid- based systems. During the last few years, power electronics has undergone a fast evolution, due to two factors, the development of fast semiconductor switches that are capable of switching quickly and handling high powers and introduction of real- time computer controllers that can implement advanced and complex control algorithms. These factors have led to the development of cost- effective and grid-friendly converters. The performance of power electronic systems, especially in terms of efficiency and power density, has been continuously improved by the intensive research and advancements in circuit topologies, control schemes, semiconductors, passive components, digital signal processors, and system integration technologies.

• Intermittence of power generation from the RES can be controlled by generating the power from distributing RES to larger geographical area in small units instead of large unit concentrating in one area.

• In case of irrigation load, the load is fed during the night time or off peak load time and this is fed by conventional grid. On other hand power generated by RES like solar PV is generated during day time so we can use this power for irrigation purposes instead of storing the energy for later time which increases the cost of the overall system. Using the solar water pumping for irrigation gives very high efficiency approximate 80% to 90% and the cost of solar water pumping is much lesser than the induction motor pumping type.

• In large solar PV plant output power is fluctuating during the whole day and this power is fed to the grid, continuously fluctuating power gives rise to the security concern to the grid for making stable grid. Solar PV plant owner have to install the different type of storage system which gives additional cost to the plant owner. Once the storage system is fully charged then this storage elements gives no profit to

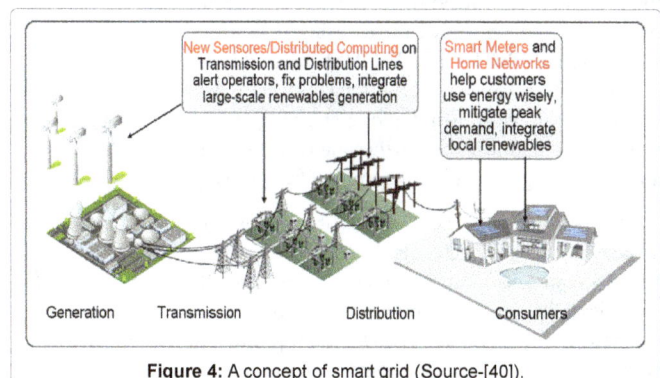

**Figure 4:** A concept of smart grid (Source-[40]).

the system owner. Therefore solar based water pumping system may be installed instead of storage system.

## Strengthening integration of renewable energy into electrical power systems

Partially dispatch able renewable sources pose greater challenges to electric power system operators. In essence these sources of generation cannot be fully controlled (dispatched) since they reflect the time-varying nature of the resource. The main way in which they can be controlled is through reduction of the output. This is in contrast to dispatch able generation that can be controlled by increasing or reducing fuel supply [64].

Solar PV penetration levels remain quite limited despite high growth rates of installed capacity in certain countries. For example, in Germany where active programs of PV installation have been successful, about 10 GW of PV were installed by the end of 2009, producing 1.1% (6.6 TWh or 23.76 PJ) of German electrical energy in 2009 [66]. There is concern that severe grid disturbances with strong frequency deviations can be worsened by large amounts of PV systems [63]. Due to this, the German guideline for the connection to medium-voltage networks requires a defined frequency/ power drop for frequencies above 50.2 Hz [65]. Protection systems in distribution grids also have to be adapted to ensure safety [61]. In general, these adaptations and guidelines indicate that it is important that solar PV become a more active participant in electrical networks [63].

Challenging situations for system balancing caused by high ramp rates for wind power production during storms when individual wind power plant production levels can drop from rated power to zero over a short time span, due to wind turbines cutting out have been reported.

The presence of wind and sunlight are both temporally and spatially outside human control, integrating wind and solar generation resources into the electricity grid involves managing other controllable operations that may affect many other parts of the grid, including conventional generation. These operations and activities occur along a multitude of time scales, from seconds to years, and include new dispatch strategies for ramp able generation resources, load management, provision of ancillary services for frequency and voltage control, expansion of transmission capacity, utilization of energy storage technologies, and linking of grid operator dispatch planning with weather and resource forecasting [66].

**Application of new transmission technologies:** The following transmission technologies can be applied to overcome the challenges of integration of RE into power grid.

☐    Higher voltage level AC transmission (UHVAC): Ultra high voltage AC transmission (UHVAC) is suitable for transmitting power from on-shore RE plants using overhead lines. Research and application of 1000 kV UHVAC transmission as a desirable technology to meet the need for large scale, long distance power transmission from the large coal, hydro, wind and solar energy bases are embarked upon in China to connect the northern and western regions to the central and eastern regions with huge and still fast growing electricity demand.

☐    More flexible AC transmissions (FACTS): Based on advanced power electronic technologies and innovative designs, FACTS equipment can be applied to improve the capacity, stability and flexibility of AC transmission, making it more capable of transmitting large-capacity RE. For example, thyristor controlled series compensators (TCSCs) can be installed in transmission lines to reduce electrical distance, increase damping and mitigate system oscillation; SVC, STATCOM and controllable shunt reactors (CSRs) can be shunt installed on substation buses to solve the reactive power compensation and voltage control problems which are common in RE integration due to their output fluctuation. SVCs or STATCOMs may also be used to improve the performance of RE power plants to meet integration requirements on reactive power and voltage control, while keeping the design of RE generators relatively simple.

☐    Higher voltage level DC transmission (UHVDC): Ultra high voltage DC (UHVDC) transmission is a conventional HVDC transmission technology that is relatively mature and has long been used for long-distance, large-capacity power transmission without midway drop points, as well as for the interconnection of asynchronous power networks. Compared to AC transmission, it has advantages such as lower loss, lower line cost, narrower corridor and rapid power control capabilities [67]. Like AC transmission, DC transmission is also progressing in the direction of ultra-high voltage levels for larger capacity and longer distance power delivery. Again, China is leading in the application of ultra-high voltage DC (UHVDC) transmission.

However, there are still some problems in the use of CSC-HVDC or UHVDC to transmit RE. For example, when HVDC lines are used to transmit only wind power to load centers (Figure 5), not only the low utilization rate problem occurred, but also the minimum start up power of the HVDC lines and problems in frequency stability and voltage stability are encountered and these require advanced technology.

With voltage support from the local AC grid (Figure 6), the stability problems can be mitigated.

☐    More flexible DC transmission: From VSC-HVDC to MTDC and DC grids the major advantages of VSC-HVDC as compared to conventional CSC-HVDC make it not only suitable for application in RE integration, but also more convenient to form multi-terminal DC (MTDC). Three or more converter stations are linked to each other with DC lines, each interacting with an AC grid, which facilitates flexible multi-grid interconnection and even DC grids. These will be useful in future RE integration where multiple resource sites and multiple receiving ends are involved.

In order to connect remote offshore wind power plants to a grid, VSC-HVDC technology is preferred. In order to tap large quantities of offshore wind power from the North Sea, a transnational offshore grid based on multi-terminal VSC- HVDC has been proposed [60] (Figure 7).

**Energy storage and its applications in electricity grid**: Power schedule and dispatch can be made possible with the adoption of energy storage. It allows intermittent power to be harvested at the time of excess and redistributed during scarcity. With this technology, degree of intermittency can be reduced and integration flexibility is enhanced, therefore the contribution from REs can be increased.

Energy storage, due to its tremendous range of uses and configurations, may assist RE integration in number of ways. These uses include, inter alia, matching generation to loads through time shifting; balancing the grid through ancillary services, load following, and load levelling; managing uncertainty in RE generation through reserves; and smoothing output from individual RE plants [65].

Possible energy storage technologies for RE integration include: Mechanical (e.g Pumped hydro, compressed air, fly - wheel), Electromagnetic (e.g Super-capacitors), Chemical (e.g Fossil fuel, biomass), Thermal (e.g Heat pump), and Electrochemical (e.g

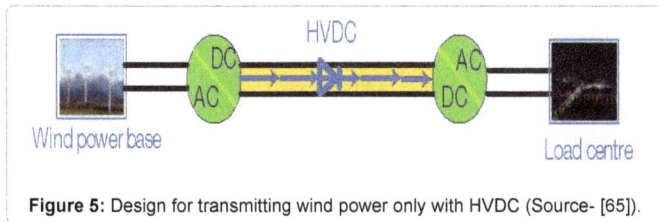

**Figure 5:** Design for transmitting wind power only with HVDC (Source- [65]).

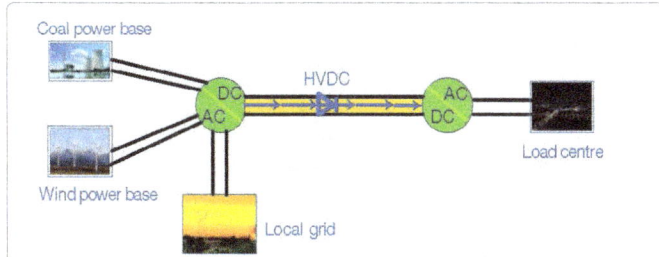

**Figure 6:** Wind-fire bundling design with HVDC transmission and local grid (Source- [65]).

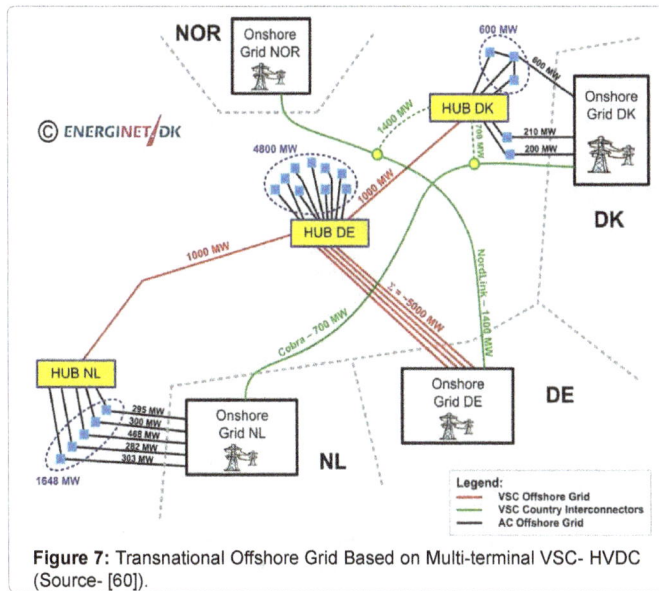

**Figure 7:** Transnational Offshore Grid Based on Multi-terminal VSC- HVDC (Source- [60]).

Batteries). The tremendous application range of storage is shown in Figure 8.

The suitability of electrical energy storage (EES) resource for a particular discharge time-frame is determined by its power density and energy density. Power density refers to the energy storage technology's ability to provide instantaneous power. A higher power density indicates that the technology can discharge large amounts of power on demand. Energy density refers to the ability of the technology to provide continuous energy over a period of time. A high energy density indicates that the technology can discharge energy for long periods. Generally, energy storage technologies with the highest power densities tend to have the lower energy densities; they can discharge enormous amounts of power, but only for a short time. Likewise, technologies with the highest energy densities tend to have lower power densities; they can discharge energy for a long time, but cannot provide massive amounts of power immediately. This quality gives rise to a division of electrical energy storage technologies into categories based on discharge times. These classifications are useful in conceptualizing how

many roles energy storage device can play with respect to renewable integration [67].

Short discharge time resources discharge for seconds or minutes, and have energy-to power ratio (kWh/kW) of less than 1. Examples include double layer capacitors (DLCs), superconducting magnetic energy storage (SMES), and flywheels (FES). These resources can provide instantaneous frequency regulation services to the grid that mitigate the impact of RE's uncontrollable variability.

Medium discharge time resources discharge for minutes to hours, and have energy-to power ratio of between 1 and 10. This category is dominated by batteries, namely lead acid (LA), lithium ion (Li-ion), and sodium sulphur (NaS), though flywheels may also be used. Medium discharge time resources are useful for power quality and reliability, power balancing and load following, reserves, consumer-side time-shifting, and generation-side output smoothing. .

Medium-to-long discharge time resources discharge for hours to days, and have energy-to power ratios of between 5 and 30. They include pumped hydro storage (PHS), compressed air energy storage (CAES), and redox flow batteries (RFBs). RFBs are particularly flexible in their design, as designers may independently scale the battery's power density and energy density by adjusting the size of the cell stacks or the volume of electrolytes, respectively. Technologies in this category are useful primarily for load-following and time-shifting, and can assist RE integration by hedging against weather uncertainties and solving diurnal mismatch of wind generation and peak loads.

Long discharge time resources may discharge for days to months, and have energy-to-power ratios of over 10. They include hydrogen and synthetic natural gas (SNG). Technologies in this category are thought to be useful for seasonal time shifting and due to their expense and inefficiency will likely cause deployment only when RE penetrations are very large. For example, large amounts of solar power on the grid will produce large amounts of energy in the summer months, but significantly less in the winter. Storing excess generation in the summer as hydrogen or SNG and converting it back to electricity in the winter would allow a time-shift of generation from one season to the next. Such technologies can assist RE integration in the long term by deferring the need for transmission expansion and interconnection that arises due to the location dependency of renewable resources.

Table 2 describes various grid-side roles of energy storage and their relevance to large capacity RE integration challenges, along with some examples of EES technologies currently in use. These examples are

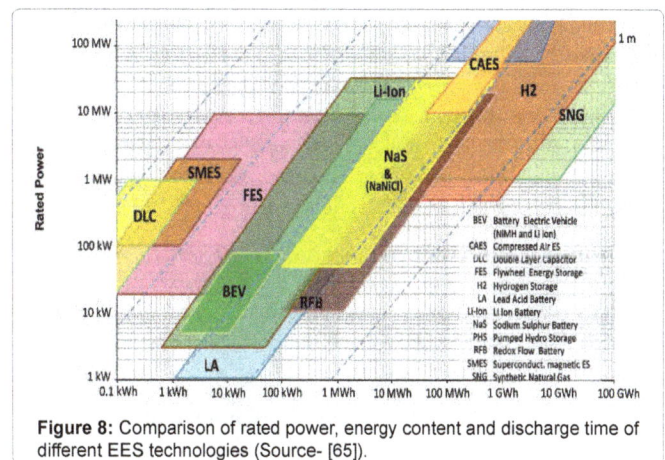

**Figure 8:** Comparison of rated power, energy content and discharge time of different EES technologies (Source- [65]).

| Role | Time Scale(s) | Description | Benefits to RE integration | Examples of EES technologies |
|---|---|---|---|---|
| Time-shifting/ Arbitrage/ Load-leveling | Hours to days | EES allows storage of off peak energy & release during high demand period. | A solution to diurnal generation cycles that do not match load cycles | NAS batteries, CAES, PHS, RFB |
| Seasonal Shifting | Months | EES stores energy for months at a time, releasing it at times of the year when RE output is typically lower | Allows use of Renewably – generated energy year-round, reducing reliance on traditional generation in seasons with e.g low sunlight | Hydrogen, SNG |
| Loading following/ Ramping | Minutes to hours | EES follows hourly changes in demand throughout the day | May mitigate partial unpredictability in RE output during critical load times | Batteries, Flywheels, PHS, CAES, RFB |
| Power quality& Stability | < 1 second | Provision of reactive power to the grid to handle voltage spikes, sags and harmonics | Mitigates voltage instability and harmonies caused or exacerbated by uncontrollable variability of RE generation. | LA batteries, NAS, batteries, flywheels, RFB |
| Frequency regulation | Seconds to minutes | A fast response increase or decrease in energy output to stabilize frequency. | Mitigates uncontrollable moment – to moment variability in RE generation output. | Li – ion batteries, NAS batteries, flywheels, PHS |
| Spinning Reserves | ~ 10 minutes | A fast response increase or decrease in energy output to cover a contingency e.g generator failure | Mitigates partial unpredictability of RE generation output, providing (or removing) energy when the RE resource does not perform as expected | PHS, flywheels, batteries |
| Supplemented Reserves | Minutes to hours | A slower response resource that comes online to replace a spinning reserve. | Provides firm power in the event of a severe and long lasting drop in RE output. Use for RE integration is expected to be in frequent and low value | PHS |
| Efficient use of Transmission Network | Minutes to hours | EES can help grid operator defer transmission system upgrades through time – shifting and more efficient operating reserves. | Reduced transmission costs, mitigates location dependency challenges of RE generation | Li - ion |
| Isolated grid Support | Seconds to hours | EES can assist in the integration of RE on small power grids. | Time shifting and power quality applications to mitigate variability and Unpredictability of RE generation | LA batteries |
| Emergency Power Supply/ Black Start | Minutes to hours | EES may be used to re-start the power system in the event of catastrophic failure | No specific benefit accrues to RE integration, but storage resources may nonetheless provide black start capability to the grid | LA batteries |

**Table 2:** Grid-Side Roles of Electric Energy Storage (EES) System (Source [68]).

impressionistic: the suitability of an EES technology for a particular use is highly context-dependent and will vary according to the needs of the grid operator and the specific design of the EES.

# Conclusion

In this paper, renewable energy resources availability, utilization and grid integration of RES have been presented.

The study shows that, integrating more large capacity RE into the grid brings variability and uncertainty. At the same time, there will continue to be unexpected disturbances stemming from load variation, grid faults and conventional generation outages. Worldwide studies and experience in recent years have shown that new technical solutions are needed to address this conjunction of difficulties. The new solutions will include new technologies, methods and practices, applied in order to provide more flexibility and improve the efficiency of power systems, constantly balancing generation and load. Only this will make the power systems reliable and maintain security of supply, i.e. avoid any interruption in the supply of power.

Furthermore, in order to address the increased variability and uncertainty brought about by integrating higher levels of large-capacity RE, the power system must become more flexible so as to maintain a constant balance between generation and load.

Power system flexibility can be achieved from the generation side (both RE generation and conventional generation), from the load side, and through electric energy storage (EES) acting as either generation or load. It can be better exploited if system operating technologies and practices are improved, and based on control shared over wider geographic areas with the support of transmission expansion.

The technologies to store the excess electricity and control the different processes in the grid with adequate communication ports is one of the basic requirements for a better integration of alternative energy sources. If these conditions are achieved, the integration of alternative energy sources can help to stabilize the current grid in perturbations and do not aggravate the situation.

RE generation can be made more predictable, controllable and dispatch able, or in other words more grid-friendly, by improving the design, operation and modeling technology at the generating unit, plant and plant cluster level. Higher-voltage level transmission, the power electronics based FACTS and DC transmission technologies are paving the way for the transmission expansion needed for accommodating more large capacity RE generation. Based on these technologies, development of probabilistic transmission planning methods is possible for power security and stability.

Demand response, supported by new smart grid, smart building and smart home technologies, is a promising source of power system flexibility in the future, but it is still in its infancy. The rate at which it will mature and be widely applied depends heavily on an understanding of customer behaviour underlying the load demand, as well as on institutional and commercial innovations.

## References

1. http://www.undp.org/seed/eap/activities/wea/drafts-frame.html

2. Agbetuyi AF, Awelewa AA, Adoghe AU , Awosope COA (2013) Technical challenges in connecting wind energy converter to the grid', Int J Renew Sustain Energ 2: 90-92.

3. Albadi MH, El-Saadany EF (2010) Overview of wind power intermittency impacts on power systems. Electr Power Syst Res 80: 627 - 632.

4. Oyedepo SO (2012) Energy and sustainable development in Nigeria: the way forward. Energ Sustain Soc 2: 2– 17.

5. Jain BC (1987) Rural energy centres based on renewable—case study on an effective and viable alternative. IEEE Trans Energy Convers 2: 329–335.

6. Ramakumar R (1983) Renewable energy sources and developing countries. IEEE Trans Power Appar Syst 102: 502–510.

7. Ramakumar R (1996) Energizing rural areas of development countries using IRES. IEEE Trans Energy Conversion 1536–1541.

8. Segurado R, Krajac˘ic´ G, Duic´ N, Alves L (2011) Increasing the penetration of renewable energy resources in S. Vicente Cape, Verde. Appl Energy 88): 466–472.

9. Bernal-Agustin JL, Dufo-Lopez R (2006) Economical and environmental analysis of grid connected photovoltaic systems in Spain. Renew Energ 31: 1107–1128.

10. MNRE (Ministry of New and Renewable Energy India) (2004) Annual Report. Baseline in grid connected renewable energy projects. Technical report, Government of India, New Delhi.

11. Benitez LE, Benitez PC, van Kooten GC (2008) The economics of wind power with energy storage. Energ Economic 30: 1973–1989.

12. Dursun B, Alboyaci B (2010) The contribution of wind-hydro pumped storage systems in meeting Turkey's electric energy demand. Renew Sustain Energ Rev 14: 1979-1988.

13. Black M, Strbac G (2006) Value of storage in providing balancing services for electricity generation systems with high wind penetration. J Power Sourc 162: 949-953.

14. Black M, Strbac G (2007) Value of bulk energy storage for managing wind power Fluctuations. IEEE Transact Energ Convers 22: 197-205.

15. IEA (2010) World Energy Outlook 2010. Int Energ Agency, Paris, France: 736.

16. Modi V, McDade S, Lallement D, Saghir J (2005) Energy Services for the Millennium Development Goals. Energy Sector Management Assistance Programme. United Nations Development Programme, UN Millennium Project, and World Bank, New York, NY, USA.

17. International Energy Agency (IEA) (2011) World energy outlook 2011 Int Energ Agency, Paris, France.

18. http://www.ren21.net/Portals/97/documents/GSR/GSR2011_Master18.pdf

19. http://www.iea.org/Textbase/npsum/ETP2012SUM.pdf

20. Cosentino V, Favuzza S, Graditi G, Ippolito MG, Massaro F, et al. (2012) Smart renewable generation for an islanded system. Technical and economic issues of future scenario. Energy 39: 196-204.

21. Alonso M, Amaris H, Alvarez-Ortega C (2012) Integration of renewable energy sources in smart grids by means of evolutionary optimization algorithms. Expert Sys Appl 39: 5513-5522.

22. Al-Ali AR, El-Hag A, Bahadiri M, Harbaji M, Ali El Haj Y (2012) Smart Home Renewable Energy Management System. Energy Procedia 12: 120-126.

23. European Commission (2010) Joint Research Centre, Institute for Energy (2010). Renew energ snapshots.

24. The U.S. Energy Information Administration (EIA) (2011) International energy outlook 2011.

25. International energy agency (IEA) (2002) Beyond Kyoto: energy dynamics and climate stabilization. Paris: OECD/IEA.

26. Lund H (1999) A green energy plan for Denmark—job creation as a strategy to implement both economic growth and a $CO_2$ reduction. Environ Resour Economic 14: 431–439.

27. Gross R (2004) Technologies and innovation for system change in the UK: status, prospects and system requirements of some leading renewable energy options. Energy Policy 32: 1905–1919.

28. Toke D (2005) Explaining wind power planning outcomes: some findings from a study in England and Wales. Energy Policy 33: 1527–1539.

29. Lund H (2007) Renewable energy strategies for sustainable development. Energy 32: 912–919.

30. Awosika LF (1995) Impacts of global climate change and sea level rise on coastal resources and energy development in Nigeria. In: Umolu J.C, editor. Global climate change: impact on energy development. Nigeria: DAMTECH Nigeria Limited.

31. Oyedepo SO (2012) On energy for sustainable development in Nigeria. Renew Sustain Energ Rev 16: 2583– 2598.

32. Roy N, Mahmud M, Pota H (2011) Impact of high wind penetration on the voltage profile of distribution systems. North American Power Symposium (NAPS), Aug: 1–6.

33. Kumar A, Gao W(2008) Voltage profile improvement and line loss reduction with distributed generation in deregulated electricity markets. TENCON, IEEE Region 10 Conference, Hyderabad, India.

34. Farret FA, Simoes MG (2006) Integration of Alternative Sources of Energy. John Wiley & Sons, Inc., Hoboken, New Jersey.

35. Ahmed PNJ, Haidar, Mohd S (2011) Optimal configuration assessment of renewable energy in Malaysia. Renew Energ 36: 881-888.

36. Alain Liébard, Nahon C (2011) Worldwide electricity production from renewable energy sources.

37. Bhattacharya SC, Abdul Salam P, Runqing H, Somashekar HI, Racelis DA, et al. (2005) An assessment of the potential for non-plantation biomass resources in selected Asian countries for 2010. Biomass Bioenerg 29: 153-166.

38. Phuangpornpitak N, Tia S (2011) Feasibility Study of Wind Farms Under the Thai Very Small Scale Renewable Energy Power Producer (VSPP) Program. Energ Procedian 9: 159-170.

39. Alagoz BB, Kaygusuz A, Karabiber A (2012) A user-mode distributed energy management architecture for smart grid applications. Energy 44: 167-177.

40. IEA (2006) Energy for cooking in developing countries, World Energy Outlook 2006. International Energy Agency, Paris, France.

41. Ackermann T, Morthorst P (2005) Economic aspects of wind power in power systems, Wind Power in Power Systems. T. Ackermann (edn), John Wiley & Sons, New York, NY, USA: 383-410.

42. Weisser D, Garcia RS (2005) Instantaneous wind energy penetration in isolated electricity grids: concepts and review. Renew Energ 30: 1299-1308.

43. http://www.intechopen.com/books/renewable-energy/grid-integration-of-renewable-energy-systems

44. DiPippo R (2008) Geothermal Power Plants: Principles, Applications, Case Studies and Environmental Impact. (2nd edn) Elsevier Ltd., London, UK.

45. Li ZS, Zhang GQ, Li DM, Zhou J, Li LJ, et al. (2007) Application and development of solar energy in building industry and its prospects in China. Energ Policy 35: 4121-4127.

46. Liu YQ, Kokko A (2010) Wind power in China: Policy and development challenges. Energ Policy 38: 5520-5529.

47. Moura PS, de Almeida AT (2010) The role of demand-side management in the grid integration of wind power. Appl Energ 87: 2581-2588.

48. Soder L, Hofmann L, Orths A, Holttinen H, Wan YH (2007) Experience from wind integration in some high penetration areas. IEEE Transact Energ Convers, 22: 4-12.

49. www.un.org/esa/population/publications/wup2007/2007WUP_Highlights_web.pdf.

50. Droege P, Radzi A, Carlisle N, Lechtenbohmer S (2010) 100% Renewable Energy and Beyond for Cities. Report, HafenCity University Hamburg and World Future Council Foundation, Hamburg, Germany.

51. IEA (2008) Energy Technology Perspectives 2008, Scenarios and Strategies to 2050. Int Energ Agency, Paris, France: 646.

52. Sambo AS, Garba B, Zarma IH, Gaji MM (2003) Electricity Generation and the Present Challenges in the Nigerian Power Sector. Energ Resource Rev 4: 7-10.

53. El-Sharkawi MA (2009) Electric Energy – An Introduction. CRC Press, Taylor & Francis LLC, Oxford, UK: 472.

54. Breuer W, Hartmann V, Povh D, Retzmann D, Teltsch E (2004) Application of HVDC for large power system interconnections. Int Council on Large Elect Sys, Paris, France.

55. Billinton R, Allan RN (1988) Concepts of power system reliability evaluation. Int J Elect Power Energ Systems, 10: 139-141.

56. Tony- Burton DS, Jenkins N, Bossanyi E (2001) Wind Energy Handbook, John Wiley & Sons Ltd.

57. Arif MT (2012) Investigation of energy storage required for various location in Australia. Central Regional Engineering Conference 2012, Engineers Australia, Queensland, Australia.

58. Willis HL, Scott WG (2000) Distributed Power Generation, Planning and Evaluation. Marcel Dekker Inc.

59. Venkat P, Saadat M (2009) Smart Grid-Leveraging Intelligent Communications to Transform the Power Infrastructure. Cisco Systems, Inc.

60. Von Dollen D (2009) Report to NIST on the Smart Grid Interoperability Standards Roadmap.

61. Chuong TT (2008) Voltage stability Investigation of grid connected wind farm. World Academy of Science and Technology.

62. BMU (2010) Renewable Energy Sources in Figures, German Federal Ministry for the Environment, Nature Conservation and Nuclear Safety (BMU), Berlin, Germany.

63. Strunz K, Louie H (2009) Cache energy control for storage: Power system integration and education based on analogies derived from computer engineering. IEEE Transactions on Power Systems, 24: 12-19.

64. BDEW (2008) Technical Conditions for Connection to the Medium-Voltage Network. German Association of Energy and Water Industries, Berlin, Germany.

65. Schafer N, Degner T, Jager J, Teil T, Shustov A (2010) Adaptive protection system for distribution networks with distributed energy resources. 10th International Conference on Developments in Power System Protection, Manchester, UK.

66. Caamano-Martin E, Laukamp H, Jantsch M, Erge T, Thornycroft J et al. (2008). Interaction between photovoltaic distributed generation and electricity networks. Progress in Photovoltaics 16: 629-643.

67. Ueda Y, Kurokawa K, Tanabe T, Kitamura K, Sugihara H (2008) Analysis results of output power loss due to the grid voltage rise in grid-connected photovoltaic power generation systems. IEEE Transactions on Industrial Electronics, 55: 2744-2751.

68. Hara R, Kita H, Tanabe T, Sugihara H, Kuwayama A, et al. (2009) Testing the technologies: Demonstration grid-connected photovoltaic projects in Japan. IEEE Power & Energy Magazine 7: 77-85

# 15

# Thermodynamic Analysis of a Combined Brayton and Rankine Cycle based on Wind Turbine

**Hossein Sheykhlou***

*Department of Mechanical Engineering, Urmia University, Iran*

## Abstract

This paper presents the thermodynamic study of a heat and power system which combines an organic Rankine cycle and a gas turbine (GT) cycle. The required power of the compressor in the GT cycle and pump in the Rankine cycle is provided by the WT which according to the amount of cycle required power. For analysis of the cycle, a simulation has been performed using R123 as the working fluid in the Rankine cycle and air and combustion products in the GT cycle. In this end, effect of various parameters such as wind speed, the angular speed of WT, and the gas turbine inlet temperature as well as the compressor pressure ratio, gas turbine isentropic efficiency, condenser temperature and compressor isentropic efficiency on the total thermal efficiency and total exergy efficiency is calculated and analyzed. Thermal efficiency and exergy efficiency of 21.31% and 23.54% is obtained. Also, it is observed that the greatest exergy destruction occurs in the combustion chamber.

**Keywords:** Wind turbine; Combined cycle; Rankine/Brayton; Exergy; First and second laws

## Introduction

Excessive use of fossil fuels has led to more environmental problems such as global worming atmospheric pollution, ozone depletion. Use of renewable energy sources such as wind energy, solar energy rather than fossil fuels can prevent of global warming. Renewable energy is abundant and its technologies are well established to provide complete security of energy supply [1]. Combined cycles are an important tools for energy production which are used in various forms. Combined cycles are flexible, reliable, economical and environmental protection. Several researchers have investigated the performance of power systems using these various heat sources and various power drivers [2-8]. Khaliq et al. [9] analyzed the reheat combined Brayton/Rankine power cycle. They used the second-law approach for the thermodynamic analysis of cycle and investigated the effect of parameters such as pressure ratio, cycle temperature ratio, number of reheats and cycle pressure-drop on the combined cycle performance. Their simulation results showed that the greatest exergy destruction occur in the combustion chamber. Wang et al. [10] proposed the new combined power and ejector–absorption refrigeration cycle that could produce both power output and refrigeration output simultaneously. Khaljani et al. [11] proposed the heat and power combined cycle combines a gas turbine (GT) and an ORC through a single-pressure heat recovery steam generator (HRSG). Their simulation results showed that the increase in pressure ratio and isentropic efficiency of air compressor and gas turbine efficiency improves thermodynamic performance of the system. Also, the most exergy destruction rate takes place in the combustion chamber, and after that in heat recovery steam generator and gas turbine. Maeero et al. [12] analyzed and optimized the second low processes of combined triple power plants. The combined triple power plants includes of the Brayton cycle (gas-based) and two Rankine cycles (steam and ammonia-based). The results of the analysis showed that the most exergy destruction occurs in the heat exchanger. Also, their study showed that the use of feed water heaters increases the efficiency and with increases in ambient temperature, the exergy efficiency decreases. Rabbani et al. [13] analyzed the combined system that coupled a Wind Turbine (WT) with a combined cycle. Required energy of cycle was provided by the wind energy which according to the amount of cycle required energy. Their analysis showed that increasing

the combustion temperature reduces the critical velocity and mass flow rate. Increases in wind speed reduce both energy and exergy efficiency of the overall system. Baskut et al. [14] analyzed the exergy processes of wind turbine power plants. Exergy efficiency of the power plant found to be between 0% and 68.20% and the values of exergy efficiencies of the WTPP were different for different power factor value. Zhu et al. [15] analyzed the energy and exergy processes of a bottoming Rankine cycle for engine exhaust heat recovery. The results showed that working fluid properties, evaporating pressure and superheating temperature are the main factors influencing the system design and performances, the distributions of exergy destruction are varied with working fluid categories and system design constraints. Also, the results showed that with the increasing of the evaporating pressure, the internal exergy destruction of the evaporator decreases, and the external exergy destruction increases. Most working fluids do not have the optimal evaporating pressure due to the relatively high exhaust gas temperature. Ozgener et al. [16] analyzed the exergy and reliability of wind turbine systems. The results showed that exergy efficiency changes between 0% and 48.7% at different wind speeds, considering pressure differences between state points. The research done on the combined cycle can be divided into four categories:

- Examining the main drivers of the combined system and the feasibility of using different energy sources, especially low temperature heat sources.

- Study of types of working fluid that can be used in combined system and cooling various technologies in combined system.

- Investigation of exergy, energy laws, and mass conservation in system.

***Corresponding author:** Hossein Sheykhlou, Department of Mechanical Engineering, Technical Education Faculty, Urmia University, Urmia, West Azerbaijan 57561-15311, Iran, E-mail: chavoshihossein154@yahoo.com

•     Optimization of combined systems.

The present study analyzes a system that couples a Wind Turbine (WT) with a Brayton/Rankine combined power cycle. The aim of the current study is to determine the performance of the combined system through examining the first and second laws of thermodynamics and utilization of wind power to provide cycle required work. The exhaust gases released from Gas Turbine are high temperatures and low pressures, these high temperatures gases drive the Rankine cycle. A working fluid in organic Rankine cycle machine plays a key role. It determines the performance and the economics of the plant [17]. Liu [18] showed that thermal efficiency for various working fluids was a minorant of the critical temperature. Hung [19] showed that wet fluids are unsuitable for ORC systems due to hydrogen bond in wet fluids. Dry and isentropic fluids do not contain liquid droplets at the turbine outlet. This is mainly due to disastrous impact of liquid droplets in turbine exhaust on the performance of turbine in the process of expansion causing the turbine blades wear off and may reduce the turbine efficiency. Rankine cycle employs an organic fluid such as refrigerant, the refrigerant used in the ORC cycle is R123. The working fluid used the Brayton cycle is air that behaves as an ideal gas in throughout the cycle.

## Wind Turbine System

Wind turbines convert the kinetic energy of wind in to the mechanical power and the mechanical power transmitted in the power generation system by the shaft. Wind turbines are mounted on tall towers to receive the most possible energy. AWT with a diameter of 100 m is chosen for the present analysis (Figure 1).

The available power can be determined form the amount of air passing through the rotor of wind turbine per unit time. Mass flow rate of the air stream touching the rotor surface ($A = \pi R^2$) was estimated using Eq. (1). Taking $T_0 = 25°C$ and $p_0 = 0.101$ $MPa$ as reference temperature and pressure of environment and the density of air is $\rho = 1.18$ kg/m$^3$, its mass flow rate is

$$m = \rho \times A \times V_r \tag{1}$$

where $R$ is the radius of the rotor and $V_r$ is wind speed.

Exergy of kinetic energy is found using the Eq. (2).

$$ke = \frac{V_r^2}{2} \tag{2}$$

Available power is found using the Eq. (3).

$$W_a = m \cdot ke \tag{3}$$

$$\frac{p_1}{\rho} + \frac{V_1^2}{2} = \frac{p_2}{\rho} + \frac{V_2^2}{2} \tag{4}$$

Where $V_1$ is the upstream wind velocity and $P_1$ is the upstream pressure at the entrance of the rotor blades and $V_2$ is the downstream wind velocity and $P_2$ is the downstream pressure at the exit of the rotor blades.

The exit velocity can be determined by using Eq. (5).

$$V_2 = \frac{\text{angular speed of turbine}(\omega) \times R}{V_1} \tag{5}$$

where $R$ is the radius of the rotor and $\omega$ is measured in radian per

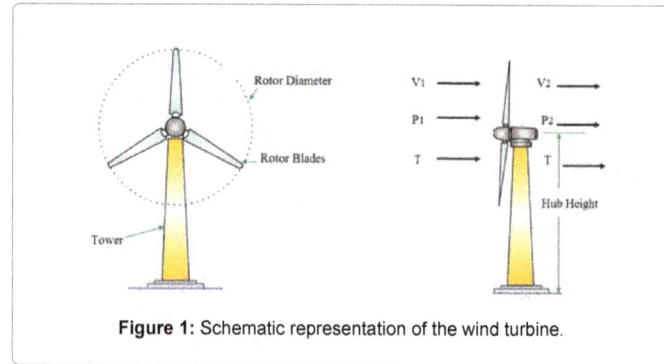

**Figure 1:** Schematic representation of the wind turbine.

second.

$C_p$ is the fraction of upstream wind power captured by the rotor blades. $C_p$ is often called the Betz limit.

Other names for this quantity are the power coefficient of the rotor or rotor efficiency. The power coefficient is not a static value. It varies with tip speed ratio of the wind turbine. The real world is well below the Betz limit with values of 0.35-0.45 common even in best designed wind turbines [20]. Maximum value of $C_p$ as 0.5926 according to Betz criterion. The power coefficient is given by Eq. (6). In this study, electrical equipment and mechanic equipment losses were assumed to be $\eta_{alternator} = 0.98$ and $\eta_{mechanic} = 0.97$, respectively.

$$C_P = \frac{W_e}{\eta_{alternator} \times \eta_{mechanic} \times .5 \times \rho \times 3.14 \times R^2 \times V_r^3} \tag{6}$$

The useful work is found using the Eq. (7).

$$W_u = (p_1 - p_2)\frac{m}{\rho} \tag{7}$$

The exergetic efficiency of a wind turbine is defined as a measure of how well the stream exergy of the fluid is converted in to useful turbine work output or inverter work output. The exergy efficiency is found by using the Eq. (8).

$$\varepsilon = \frac{W_e}{W_u} = \frac{W_e}{E_1 - E_2} \tag{8}$$

## System Description and Assumptions

A schematic of the proposed cogeneration cycle is shown in Figure 2 that includes a cycle of gas turbine (GT) and a cycle of organic Rankine cycle (ORC). WT is used to supply power to the compressor in the GT cycle and pump fluid through a Rankine cycle. The power generation capacity of combined cycle is 26 MW. The Brayton cycle components consist of a combustion chamber, air compressor and GT. The Rankine cycle components consists of a pump, steam turbine, condenser and heat recovery steam generator. Valves of $V_1$ and $V_2$ are used in the system that controls the power penetration to the combined power plant. In the case of the high penetration system, WT produces more power than the required power of compressor and pump. So in this case, the wind speed is above the critical speed and the flow valve $V_2$ is opened and $V_1$ is closed up to additional power is stored in storage unit. In the case of the low penetration system, WT produces less power than the required power of compressor and pump. So in this case, the wind speed is under the critical speed and the flow valve $V_1$ is opened and

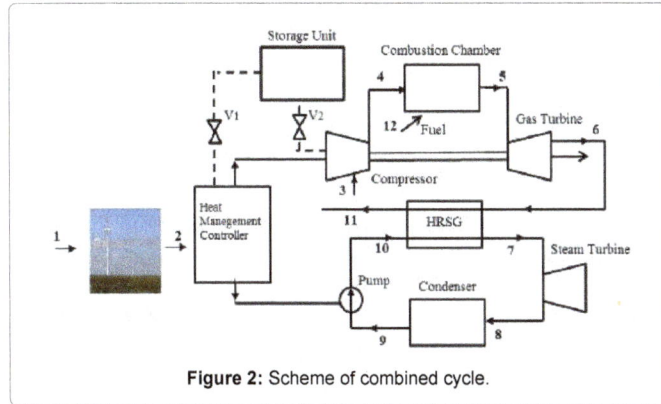

**Figure 2:** Scheme of combined cycle.

| Environment temperature | 25°C |
|---|---|
| Environment pressure | 0.1013 MPa |
| Steam turbine inlet pressure | 0.65 MPa |
| Compressor isentropic efficiency | 0.8 |
| Gas turbine isentropic efficiency | 0.85 |
| Steam turbine isentropic efficiency | 0.82 |
| Pump isentropic efficiency | 0.75 |
| Steam turbine inlet temperature | 130.15°C |
| Pump inlet temperature | 30.15°C |
| Compressor inlet temperature | 25.15°C |
| Compressor inlet pressure | 0.1013 MPa |
| Compressor outlet pressure | 1 MPa |
| Net power output | 26 MW |

**Table 1:** Basic assumptions for the simulation of combined cycle.

$V_2$ is closed up to additional power stored in storage unit be injected in the cycle.

According to Figure 2, the ambient air at point 3 with pressure of 0.101 $MPa$ and temperature of 298.15 K is compressed in air compressor. Then the compressed air flows is entered into combustion chamber. Fuel is injected in the combustion chamber at a pressure of 1.2 $MPa$. Output stream of combustion chamber with a temperature of 1100°C is expanded in the gas turbine and produces net power of 26MW. The exhaust of the Brayton cycle at high temperature and low pressure is used to drive a Rankine cycle. Working fluid in a saturated liquid phase is pumped to high pressure. After being heated in the internal heat recovery steam generator is entered the turbine to produce power and is expanded to the condenser and is condensed to saturated liquid phase.

## Thermodynamic Analysis

The assumptions made for the analysis of systems of expression are:

- The system has reached steady state.

- Pressure drop in the system's components is ignored:

$$p_4 = p_5; \ p_6 = p_{11}; \ p_{10} = p_7; \ p_8 = p_9 \qquad (9)$$

- Kinetic and potential energy and frictional losses are neglected.

- System components are Adiabatic.

- Condenser exit state are saturated liquid;

$$x_9 = 1 \ T_8 = T_9 \qquad (10)$$

- The ambient temperature and pressure are constant ($T_0 = 25°C$ and $P_0 = 100$ kPa).

- Air is treated as an ideal gas with a molar composition of 21% oxygen and 79% nitrogen.

The principle of mass conservation for the various components of the cycle can be written as follows:

$$\sum m_{in} - \sum m_{out} \qquad (11)$$

The calculations are carried out based on the basic assumptions, which are listed in Table 1. Using the law of environmental protection, theory of energy and exergy balance cycles, the balance equations of each component for enthalpy, energy, entropy and exergy are written as follows.

- Compressor

$$W_C = m_3 c_p \left( T_4 - T_3 \right) \qquad (12)$$

$$\eta_C = \frac{(h_{4.is} - h_3)}{(h_4 - h_3)} \qquad (13)$$

- Combustion Chamber

$$-.002\lambda LHV_{ch4} + h_a + \lambda h_f - (1+\lambda) h_p = 0 \qquad (14)$$

Where $\lambda$ is fuel-air ratio and LHV (MJ/Kg) is lower heating value.

$$m_4 h_4 + m_{12} h_{12} = m_5 h_5 \qquad (15)$$

- Gas Turbine

$$W_{GT} = m_5 c_p \left( T_5 - T_6 \right) \qquad (16)$$

$$\eta_{GT} = \frac{(h_5 - h_6)}{(h_5 - h_{6.is})} \qquad (17)$$

- Heat recovery steam generator

$$m_{10} h_{10} + m_6 h_8 = m_7 h_7 + m_{11} h_{11} \qquad (18)$$

- Pump

$$W_P = m_{10} \left( h_{10} - h_9 \right) = \frac{m_{10} v_{10} \left( p_{10} - p_9 \right)}{\eta_p} \qquad (19)$$

- Steam Turbine

$$W_{ST} = m_7 \left( h_7 - h_8 \right) \qquad (20)$$

$$\eta_{ST} = \frac{(h_7 - h_8)}{(h_7 - h_{8.is})} \qquad (21)$$

- Condenser

$$Q_{Cond} = m_8 \left( h_8 - h_9 \right) \tag{22}$$

- The net power input

$$W_{net,input} = W_C + W_P \tag{23}$$

-The net power output

$$W_{net,output} = W_{GT} + W_{ST} - W_C - W_P \tag{24}$$

- The net power input

$$W_{net,input} = \eta_m \varphi \sum_{I=1}^{n} W_u \tag{25}$$

Where Betz limit $\varphi = 0.4$ and n is number of wind turbine and $W_u$ is the output useful work of wind turbine and motor efficiency.

Thermal efficiency of the combined system is defined as the ratio of useful output (specific work output from the cycle and the heat extracted in the condenser) to the input energy (specific work input to the cycle and heat entered to the cycle in the combustion chamber). Performance of the system is shown based on the first law of thermodynamics.

$$e_i = \left( h_i - h_0 \right) - t_0 \left( s_i - s_0 \right) \tag{26}$$

where HHV is higher heating value.

Exergy analysis determines the system performance based on exergy, which is defined as the maximum possible reversible work obtainable in bringing the state of the system to equilibrium with that of the environment, and the evaluation is based on the second law of thermodynamics, because the second law considers not only quantity but also the quality of energy. Taking $T_0$ and $p_0$ as reference temperature and pressure of environment, thermal losses in each of the system components were assumed negligible. Exergy at each point of the cycle is calculated as follows by considering the following assumptions:

$$e_i = \left( h_i - h_0 \right) - t_0 \left( s_i - s_0 \right) \tag{27}$$

$$E_i = m_i . e_i \tag{28}$$

By forgoing the kinetic and potential exergies, the total exergy of fuel (The mixture of $N_2$, $H_2O$, $CO_2$, $O_2$ gases) can be expressed as [21]:

$$ex_{12} = h(T,P) - h_0 - T_0 \left( s(T,P) - s_0 \right) + \sum x_k e_k^{CH} + RT_0 \sum x_K lnx_k \tag{29}$$

Where $e^{CH}$ is chemical exergy per mole of gas k, $x_K$ is the mole fraction of gas k in the environmental gas phase and R is universal gas constant. Exergy efficiency is defined as the output exergy (net exergy work output from the cycle) to the input exergy (net exergy work input to the cycle and exergy entered in the fuel injection):

$$\eta_{ex} = \frac{w_{net,output}}{W_{net,input} + E_{12}} \tag{30}$$

Exergy destruction in each component of the combined cycle is calculated as follows,

$$I = \sum m_{in} e_{in} - \sum m_{out} e_{out} \pm W \tag{31}$$

The total exergy destruction is equal to the summation of the exergy destruction by each of its components.

| State | T (°C) | P (kPa) | h (kj/kg) | s (kg/kg.k) | m (kg/s) | E (KW) |
|---|---|---|---|---|---|---|
| 1 | 25 | 101.3 | 298.4 | 5.7 | - | - |
| 2 | 25 | 157.7 | 298.4 | 5.7 | - | - |
| 3 | 25 | 101.3 | 298.4 | 5.7 | 115.3 | 0 |
| 4 | 371 | 1000 | 647.5 | 7.377 | 115.3 | 22399 |
| 5 | 1100 | 1000 | 1378 | 8.065 | 117.3 | 84493 |
| 6 | 537 | 101.3 | 813.7 | 8.229 | 117.3 | 12554 |
| 7 | 130 | 650 | 471.6 | 1.773 | 18.34 | 30971 |
| 8 | 30 | 260 | 456.8 | 1.831 | 18.34 | 30402 |
| 9 | 30 | 109.7 | 231.4 | 1.109 | 18.34 | 30221 |
| 10 | 30.31 | 650 | 231.9 | 1.109 | 18.34 | 30228 |
| 11 | 514 | 101.3 | 779.4 | 7.909 | 117.3 | 12047 |
| 12 | 298.15 | 1200 | 298.4 | 5.7 | 2 | 85000 |

**Table 2:** Results of simulation for the combined cycle.

| | |
|---|---|
| Pump work (KW) | 9.105 |
| Gas turbine work (KW) | 65869 |
| Net work (KW) | 26000 |
| Steam turbine work (KW) | 273.1 |
| Compressor work (KW) | 40133 |
| Thermal efficiency (%) | 21.31 |
| Exergy efficiency (%) | 23.54 |

**Table 3:** Performance of the combined cycle.

## Results and Discussion

Parametric analysis is carried out to evaluate the effects of various design parameters such as wind speed, angular speed of WT , compressor pressure ratio, compressor isentropic efficiency, gas turbine inlet temperature, gas turbine isentropic efficiency and condenser temperature on the performance of cycle. When one specific parameter is studied, other parameters are kept constant. Table 2 shows the thermodynamic properties such as enthalpy and entropy as well as the mass flow rate and exergy rate at each point of the combined cycle at typical working conditions. The mass flow rate and exergy rate in the wind turbine is variable and depends on the wind speed changes. Table 3 shows the performance of the Brayton/Rankine combined cycle at typical working conditions. Thermal efficiency and exergy efficiency have been obtained respectively 21.3% and 23.5% with existence of wind turbines as the supplier of power of combined system whereas thermal efficiency and exergy efficiency have been obtained respectively 36% and 48% without the wind turbines. Table 4 shows exergy destruction in each component of the combined cycle. The largest exergy destruction occurs in the Combustion Chamber.

Figure 3 shows the effect of wind speed on the WT exergy efficiency and the useful work of WT. According to Eq. (8), exergy efficiency of WT is equal to the ratio of power at inverter output to useful power from WT. With an increase in the wind speed, useful power from WT increases and exergy efficiency of WT decreases. Figure 4 shows the effect of wind speed on the total exergy and thermal efficiencies. With an increase in the wind speed, the mass flow rate of the wind turbine increases. Thus the output wok of the WT and input work to the combined cycle increase. According to Eqs. (26), (30), with the increase of input work to the combined cycle, the total exergy and thermal efficiencies reduce. Figure 5 shows the effect of the angular speed of WT on the total exergy and thermal efficiencies. With an increase in the angular speed of WT, both efficiencies increase.

Figure 6 shows the effect of the compressor pressure ratio on the total exergy and thermal efficiencies and compressor work. With the

| | |
|---|---|
| Compressor (KW) | 5082 |
| Gas turbine (KW) | 5684 |
| Steam turbine (KW) | 295.4 |
| Condenser (KW) | 108.8 |
| Combustion chamber (KW) | 22905 |
| Pump (KW) | 2.219 |
| HRSG (KW) | 21933 |

**Table 4:** Exergy destruction in each component of the combined cycle.

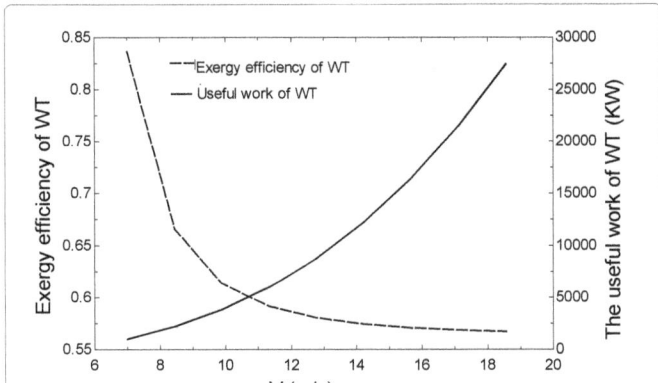

**Figure 3:** Effect of the wind speed on the exergy efficiency of WT and the useful work of WT.

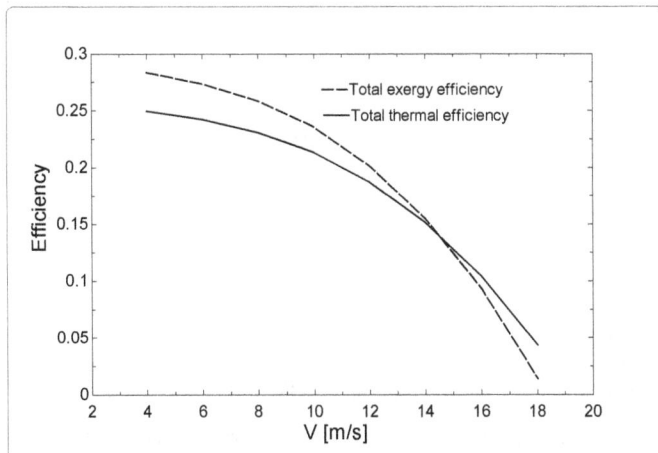

**Figure 4:** Effect of the wind speed on the total exergy and thermal efficiencies.

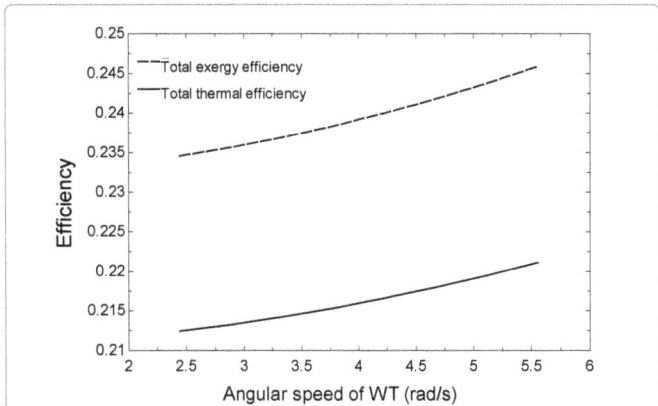

**Figure 5:** Effect of the angular speed of WT on the total exergy and thermal efficiencies.

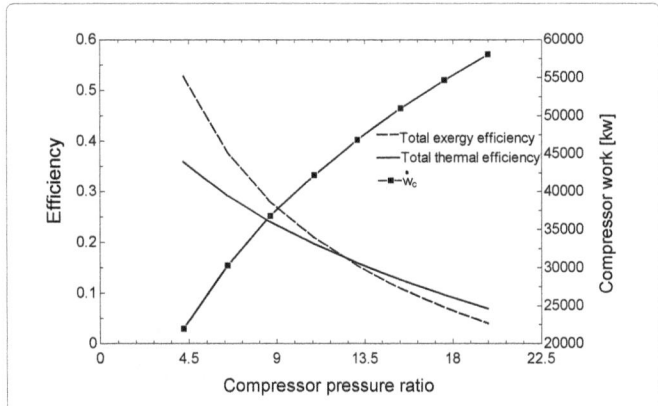

**Figure 6:** Effect of the compressor pressure ratio on the total efficiencies and compressor work.

increase in compressor pressure ratio, the enthalpy of the outlet of the compressor increases and the air flow rate decreases and the flow of fuel consumption also increases. Net power of gas turbine cycle is constant. Thus, work of air compressor increases and by increasing the fuel consumption, input exergy to the combined cycle increase. Reduction of the air flow rate makes decreasing in the heat transferred to ORC. Thus, flow rate of working fluid and work of ORC reduce. By impact of the above factors, the total exergy and thermal efficiencies decrease. Figure 7 shows the effect of the isentropic air compressor efficiency on the total thermal and exergy efficiencies and compressor work. By increasing the isentropic air compressor efficiency, the enthalpy of the outlet of the compressor decreases and the enthalpy of the input remains constant. Since the net power of gas turbine cycle is constant. Thus, the air flow rate increases and the flow of fuel consumption decreases and input exergy to the combined cycle also decreases. These changes

makes the flow rate increasing of working fluid and Rankine cycle work increasing and increasing in the heat transferred to ORC. According to these parameters, the total exergy and thermal efficiencies increase with increasing in the isentropic air compressor efficiency.

Figure 8 shows the effect of the condenser temperature on the total exergy and thermal efficiencies and heat transfer in condenser by increasing condenser temperature, exergy efficiency does not change much but thermal efficiency reduces. With increasing condenser temperature, outlet enthalpy of the condenser increases and input enthalpy of the condenser and flow rate of working fluid of ORC cycle remain constant. Therefore, heat transfer in condenser reduce. According to Eq. (26), thermal efficiency reduce by raising the temperature of the condenser. Figure 9 shows the effect of GT isentropic efficiency on the total exergy and thermal efficiencies. Increasing the isentropic efficiency of gas turbine leads to increase both thermal and exergy efficiencies of the combined cycle. By increasing isentropic efficiency of the gas turbine, the enthalpy of the outlet of the gas turbine decreases and since net power of gas turbine cycle is constant, the air flow rate increases and fuel flow rate reduces and input exergy to the system decreases. The energy balance at the HRSG makes flow rate increasing of working fluid and increasing in the network of ORC. With these changes, the total exergy and thermal efficiencies increase. Figure 10 shows the effect of GT inlet temperature on the total exergy and thermal efficiencies and gas turbine work. Inlet temperature of gas turbine has a significant impact on the total exergy and thermal efficiencies. By increasing inlet

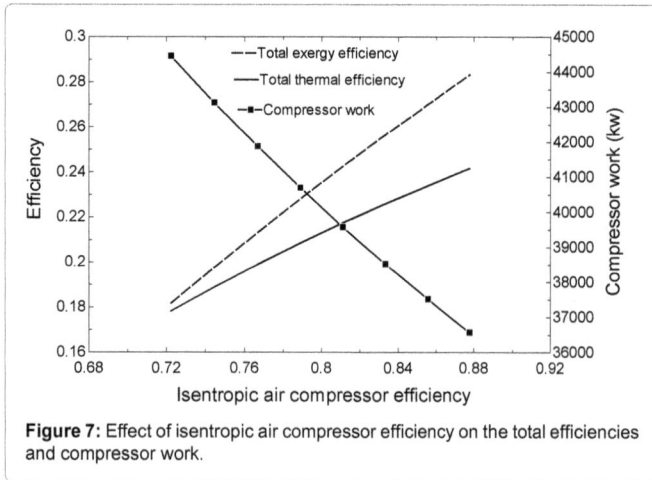

**Figure 7:** Effect of isentropic air compressor efficiency on the total efficiencies and compressor work.

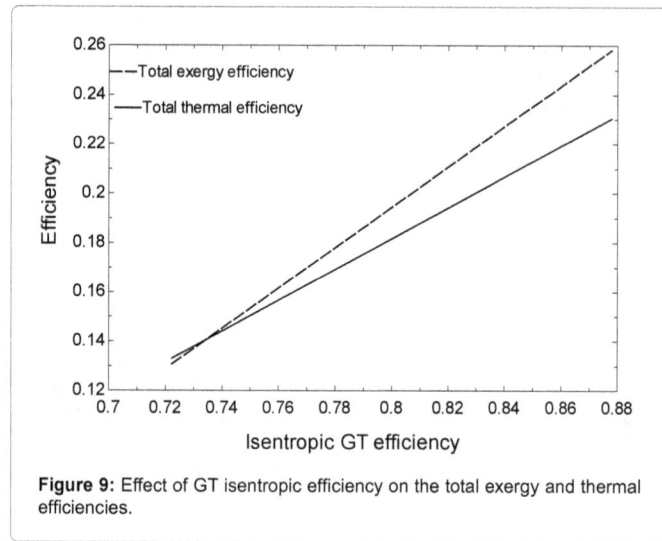

**Figure 8:** Effect of condenser temperature on the total efficiencies and heat transfer in condenser.

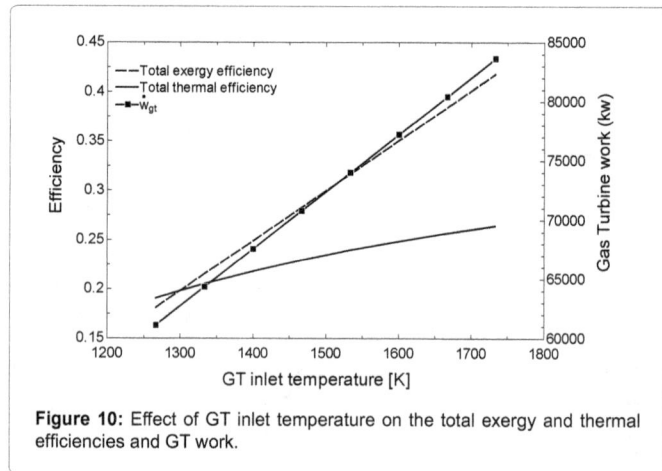

**Figure 9:** Effect of GT isentropic efficiency on the total exergy and thermal efficiencies.

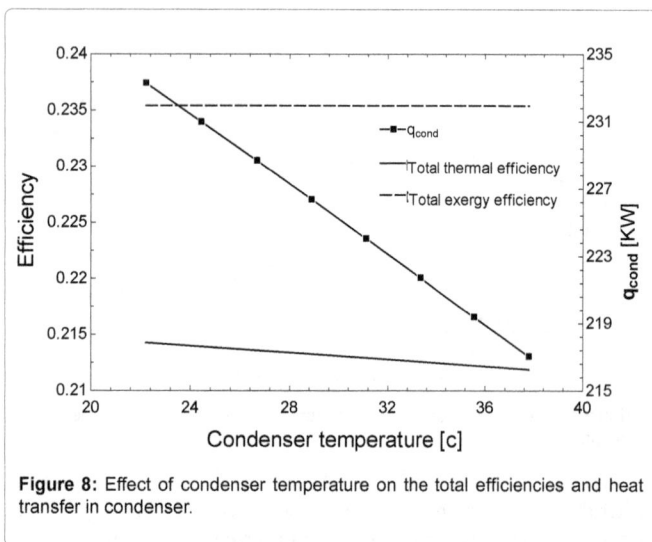

**Figure 10:** Effect of GT inlet temperature on the total exergy and thermal efficiencies and GT work.

temperature of the gas turbine, the enthalpy of the inlet and outlet points of gas turbine and the air flow rate increase and input fuel to the cycle reduces. Thus, work of gas turbine increases and input exergy to the cycle decreases. With these changes, the total exergy and thermal efficiencies increase.

## Conclusion

In this paper, a comprehensive study on a system that couples a Wind Turbine (WT) with a combined heat and power cycle from thermodynamic point of view was investigated with considering two objective functions of first and second law efficiency of the system. The proposed heat and power combined system in this study includes Wind Turbines to supply the power of combined cycle, a gas turbine cycle of 26 MW power and an ORC to produce more power. Adding ORC to the system can produce about 273.1 kW additional power from waste heat recovery of exhaust gases of the GT cycle for the considered base operating conditions. The wind power is used to drive the pump and compressor and if required, additional power is stored by the storage unit that enters to the system in the low wind speed.

· The parametric analysis results of the base case show that the increase in isentropic efficiencies of air compressor and gas turbine and gas turbine inlet temperature improves thermodynamic performance

of the system.

· An increasing wind speed and the compressor pressure ratio reduce both energy and exergy efficiencies of the overall system.

· Exergy analysis showed that the highest exergy destruction occurs in the combustion chamber and exergy destruction is significant in the compressor and gas turbine and the pump of the organic Rankine cycle has the least exergy destruction.

## References

1. Wrixon GT, Rooney ME, Palz W (2000) Renewable energy 2000. Springer-Verlag.

2. Gu W, Weng Y, Wang Y, Zheng B (2009) Theoretical and experimental investigation of an organic Rankine cycle for a waste heat recovery system. J Power Energy 223:523-533.

3. Wei D, Lu X, Lu Z, Gu J (2007) Performance analysis and optimization of organic Rankine cycle (ORC) for waste heat recovery. Energy Convers Manage 48: 1113-1119.

4. Chen H, Goswami DY, Stefanakos EK (2010) A review of thermodynamic cycles and working fluids for the conversion of low-grade heat. Renew Sustain Energy Rev 14: 3059-3067.

5. Roy P, Désilets M, Galanis N, Nesreddine H, Cayer E (2010) Thermodynamic analysis of a power cycle using a low-temperature source and a binary NH3-H2O mixture as working fluid. Int J Therm Sci 49: 48-58.

6. Yamada N, Minami T, Mohamad MNA (2011) Fundamental experiment of pump less Rankine-type cycle for low-temperature heat recovery. Energy 36: 1010-1017.

7. Baik YJ, Kim M, Chang KC, Kim SJ (2011) Power-based performance comparison between carbon dioxide and R125 transcritical cycles for a low-grade heat source. Appl Energy 88: 892-898.

8. Rashidi MM, Bég OA, Parsa AB, Nazari F (2011) Analysis and optimization of a trans critical power cycle with regenerator using artificial neural networks and genetic algorithms. J Power Energy 225: 701-717.

9. Khaliq A, Kaushik SC (2004) Second-law based thermodynamic analysis of Brayton/Rankine combined power cycle with reheat. Applied Energy 78: 179-197.

10. Wang J, Dai Y, Zhang T, Ma S (2009) Parametric analysis for a new combined power and ejector–absorption refrigeration cycle. Energy 34: 1587-1593.

11. Khaljani M, Saray RK, Bahlouli K (2015) Comprehensive analysis of energy, exergy and exergo-economic of cogeneration of heat and power in a combined gas turbine and organic Rankine cycle. Energy Convers Manag 97: 154-165.

12. Marrero IO, Lefsaker AM, Razani A, Kim KJ (2002) Second law analysis and optimization of a combined triple power cycle. Energy Convers Manag 43: 557-573.

13. Rabbani M, Dincer I, Naterer GF (2012) Thermodynamic assessment of a wind turbine based combined cycle. Energy 44: 321-328.

14. Baskut O, Ozgener O, Ozgener L (2011) Second law analysis of wind turbine power plants: Cesme, Izmir example. Energy 36: 2535-2542.

15. Zhu S, Deng K, Qu S (2013) Energy and exergy analyses of a bottoming Rankine cycle for engine exhaust heat recovery. Energy 58: 448-457.

16. Ozgener O, Ozgener L (2007) Exergy and reliability analysis of wind turbine systems: A case study. Renew Sustain Energy Rev 11: 1811-1826.

17. Bertrand T, Sylvain Q, Sébastien D, George P, Vincent L (2010) Economic Optimization of Small Scale Organic Rankine Cycles.

18. Liu BT, Chien KH, Wang CC (2004) Effect of working fluids on organic Rankine cycle for waste heat recovery. Energy 29: 1207-1217.

19. Hung T (2001) Waste heat recovery of organic Rankine cycle using dry fluids. Energy Convers Manage 42: 539-553.

20. Manyonge AW, Ochieng RM, Onyango FN, Shichikha JM (2012) Mathematical Modelling of Wind Turbine in a Wind Energy Conversion System: Power Coefficient Analysis. Appl Mathematical Sciences 91: 4527-4536.

21. Bejan A, Tsatsaronis G, Moran M (1996) Thermal design and optimization. John Wiley & Sons, Inc.

# 16

# Loss of Load Probability of a Power System

**Vijayamohanan Pillai N***

*Centre for Development Studies, Prasanth Nagar, Ulloor, Kerala, India*

## Abstract

By virtue of the vital nature of electric power, both to our economic and personal well being, a power system is expected to supply electrical energy as economically as possible, and with a high degree of quality and reliability. The developed countries in general place higher reliability standards on the performance of electricity supply. However, there has been no significant study in the context of the Indian power sector to analyze reliability in terms of loss of load probability; the technical appraisal of the State power systems in general is confined to examining the plant load factor (PLF) as a measure of capacity utilization only. The present study is a modest attempt to evaluate the reliability of the Kerala power system in India in the framework of a theory-informed methodology.

**Keywords:** Electrical power; Electricity supply

## Introduction

It goes without saying that electricity is vital to the well-being in general and hence a power system is expected to supply energy as economically as possible, as well as with a high degree of quality and reliability. Reliability in its broad sense refers to the probability that a component or system comprising components is able to perform its intended function satisfactorily during a specified period of time under normal operating conditions. Thus the reliability assessment of a power system is mainly concerned with its capability, which is related to the existence and availability of sufficient facilities to satisfy customer load. The basic facilities of a system are in the three sectors of its function, viz., generation, transmission and distribution, which are usually vertically integrated. Electric power produced at the generation end is carried to the consumers via transmission and distribution facilities. In this paper our focus is only on the generation sector.

A modern power system is very large and complex, composed of n power generating stations, where power is generated from fuels (fossil or nuclear) or by hydroelectric stations. Each generating station or plant consists of M plant units or generators, each with a rated capacity. Each of the N stations has an installed capacity Ki megawatts (mw), which is the sum of the rated capacities of its M units, and the system installed capacity to supply power is the sum of the installed capacities of all the stations. In the case of a hydropower system, each power station has usually associated with it a big reservoir behind a dam that supplies hydraulic power to drive each of the M generators.

A power system is unique in that its product is one that must be generated the instant its service is demanded. Another significant characteristic is that the demand for electricity varies greatly at random according to the time of the day and the season of the year. Therefore a power system is designed to supply instantaneously the power demanded by consumers. However, failures in the system do occur when demand exceeds supply as in the case of any other goods and services.

Demand can exceed supply for two main reasons. One is the random deviations of the demand from its expected level such that a very high peak demand exceeds the installed capacity of the system. Capacity of a power system is in general determined after taking due considerations of such unforeseen fluctuations in demand. This is affected by means of reserve or standby capacity over and above the expected peak period demand that is to be met.

Shortage may still occur, even if the load is not far from its expectation; a high demand that does not exceed the installed capacity of the system can exceed the available capacity at that moment. This is due to generator de-ratings, scheduled preventive maintenance and forced outages of generators. Generator de-ratings result from equipment problems and changes in operating conditions, and are a function of the age of the equipment. Outage refers to a certain state of a unit when it becomes unavailable to perform its intended function due to some event directly associated with it. An outage may be either a scheduled one or a forced one. Scheduled outage (or maintenance outage) is a planned event, whereby a component/unit is deliberately taken out of service at a chosen time for preventive maintenance or overhaul or repair; this is to keep the generating units in proper running condition. Forced outage, on the other hand, results when a unit falls out of service due solely to random events such as breakdown, malfunction of equipment, etc.

In the case of a hydropower system, besides these two scenarios, shortage can still occur if the hydraulic power in any storage is not sufficient to turn the concerned generator. The plant unit is then shut down, and the system capacity falls accordingly.

A modest attempt is made in this paper to evaluate the reliability of the Kerala power system. Following a detailed discussion of the methodology used in this study, the maximum likelihood estimates of availability and forced outage rates as well as loss of load probability measures are calculated for the 10 hydropower plants of Kerala.

## Loss of Load Probability: Theory

### Availability and outage measures

In a Markov process, the life history of a repairable electric power system component during its useful life period is represented by a two-state model, the two possible states being labeled 'up' or 'functioning' and 'down' or 'unavailable', denoted by 1 and 0 respectively. Thus when

*Corresponding author: Vijayamohanan Pillai N, Associate Professor, Centre for Development Studies, Prasanth Nagar, Ulloor, Kerala, India- 695011; E-mail: vijayamohan@cds.ac.in

the component fails, it is said to undergo a transition from the up to the down state, and conversely, when repairs are over, it is said to return from the down to the up state. This idea then facilitates to interpret the concept of reliability in terms of the fraction of total time the component remains in the up state. The length of functioning period is also referred to as the time-to-failure, and that of the period under repair as the downtime.

The probabilistic approach to power system reliability analysis views the system as a stochastic process evolving over time. At any moment the system may change from one state to another because of events such as component outages or planned maintenance. Corresponding to a pair of states, say $(i, j)$, there is a conditional probability of transition from the state $i$ to the state $j$.

Suppose the performance of a power plant is continuously monitored to record the sequence of failures and repairs during sustained operation in order to assess its performance. During each failure-repair cycle, the time to failure (when the plant is in upstate) and the time to repair (when the plant is in down state) are recorded. The number of failures per unit of time is known as the failure (or hazard) rate, and the number of repairs per unit of time, the repair rate. The reliability of a power plant is often measured in terms of two availability indices, viz., instantaneous availability, A(t), and steady-state (long-run) availability, A(∞). The former refers to the probability that the power plant is available for operation at any time (t) and the latter to its availability for large values of t, that is, in long run. Thus,

$A(t) = $ Prob(available at time $t$), and

$$A\left(\infty\right) = \lim_{t\to\infty} A(t)$$

The first step in an availability study is to specify certain probability models for the two variables, time-to-failure, denoted by X and time-to-repair, denoted by Y. The second step is the derivation of the availability indices, which in general are the functions of the parameters of the statistical models specified for X and Y.

Usually the failure and repair rates are assumed to be constant; this leads to the assumption that the time-to-failure and the time-to-repair variables follow exponential distribution. The exponential distribution is one of the two (the other being the geometric distribution) unique distributions with the memory less or no-ageing property. That is, future lifetime of a component remains the same irrespective of its previous use, if its lifetime distribution is exponential.

Thus we assume that the time-to-failure, X, is an exponential variable with parameter $\lambda$, so that its density function, viz., failure (hazard) density function, f(x), is given by

$$f\left(x\right) = \frac{1}{\lambda}\exp\left(\frac{-x}{\lambda}\right), \text{ for } x > 0$$

The parameter $1/\lambda$ is the constant failure (hazard) rate. For an exponential distribution of the above form, the mean is given by $\lambda$. Hence the Mean-Time-To-Failure (MTTF) of the power plant is equal to $\lambda$; this is also known as the expected survival time. The probability of a plant surviving at time $t$ in a constant failure rate environment, i.e., its survival function, denoted by R(t), is then obtained by integrating the failure density function, f(x), and is given by $R(t) = \exp(-x/\lambda)$. The complement of this survival probability is the probability of failure in time $t$, given by $1-\exp(-x/\lambda)$.

Similarly we assume an exponential model with parameter $\mu$ for

the time-to-repair variable Y, so that the density function of Y, viz., the repair density function, g(y), is $g\left(y\right) = \frac{1}{\mu}\exp\left(\frac{-y}{\mu}\right)$, for $y > 0$. In this model, $1/\mu$ is the constant repair rate and its reciprocal, , is the mean down (repair) time (MDT) or the expected outage time. The sum of MTTF and MDT is termed the mean-time-between-failures (MTBF) or cycle time.

Shooman [1] and Gnedenko, Belyayev and Solovyev [2] have shown that for the above exponential models, the instantaneous availability of a power plant is

$$A\left(t\right) = \frac{\lambda}{(\lambda+\mu)} + \frac{\mu}{(\lambda+\mu)}\exp\left\{-(\frac{1}{\lambda}+\frac{1}{\mu})t\right\}.$$

The steady-state availability is obtained by taking the limit of $A(t)$ as $t$ approaches infinity. This gives

$$A(\infty) = \frac{\lambda}{(\lambda+\mu)} = \frac{MTTF}{MTBF}.$$

Corresponding to these availability measures, we can also define two down-state probabilities, instantaneous forced outage, denoted by $R(t)$ and steady-state forced outage, denoted by $R(\infty)$ [3,4]. Thus the instantaneous forced outage rate of a plant is

$$R\left(t\right) = \frac{\mu}{(\lambda+\mu)} + \frac{\lambda}{(\lambda+\mu)}\exp\left\{-(\frac{1}{\lambda}+\frac{1}{\mu})t\right\},$$

and the long-run (steady-state) forced outage is

$$R(\infty) = \frac{\mu}{\lambda+\mu} = \frac{MDT}{MTBF}.$$

Now if we let $P_{ij}(t)$, $(i, j = 0,1)$ be the probability of the transition of state from $i$ to $j$ in a small interval of time $t$, where 1 denotes 'up' and 0, 'down' state in a Markov chain, it can be shown (ibid.) that the instantaneous availability and instantaneous forced outage rate, as obtained above, are nothing but the same state transition probabilities, $P_{11}(t)$ and $P_{00}(t)$ respectively. That is, $P_{11}(t) = A(t)$ and $P_{00}(t) = R(t)$. This gives us the remaining two transition probabilities (from 'up' to 'down' state and from 'down' to 'up' state) of the Markov chain:

$$P_{01}\left(t\right) = 1 - P_{00}\left(t\right) = \frac{\lambda}{\lambda+\mu} - \frac{\lambda}{\lambda+\mu}\exp\left\{-\left(\frac{1}{\lambda}+\frac{1}{\mu}\right)t\right\}$$

$$P_{10}\left(t\right) = 1 - P_{11}\left(t\right) = \frac{\mu}{\lambda+\mu} - \frac{\mu}{\lambda+\mu}\exp\left\{-\left(\frac{1}{\lambda}+\frac{1}{\mu}\right)t\right\}$$

It is significant to note that the initial state probabilities obtained for $t = 1$ are nothing but the state transition probabilities, $P_{ij}$. That is $P_{01}(t = 1) = P_{01}$; $P_{00}(t = 1) = P_{00}$; $P_{10}(t = 1) = P_{10}$; and $P_{11}(t = 1) = P_{11}$.

When $t \to \infty$, these probabilities are known as limiting state probabilities that give the steady-state (or stationary or long-run) probabilities of the Markov chain:

$$\lim_{t\to\infty} P_{00}\left(t\right) = P_{00}(\infty) = \lim_{t\to\infty} P_{10}\left(t\right) = P_{10}(\infty) = R = \mu/(l+\mu),$$

gives the forced outage rate (FOR) as defined earlier, and

$$\lim_{t\to\infty} P_{11}\left(t\right) = P_{11}(\infty) = \lim_{t\to\infty} P_{01}(t) = P_{01}(\infty) = A = \lambda/(\lambda+\mu).$$

is the availability rate.

Now it can be shown that $R = \frac{\mu}{\lambda+\mu} = \frac{1/\lambda}{1/\lambda+1/\mu} = \frac{P_{10}}{P_{10}+P_{01}}$, and

$$A = \frac{\lambda}{\lambda + \mu} = \frac{1/\mu}{1/\lambda + 1/\mu} = \frac{P_{01}}{P_{10} + P_{01}}$$

where $P_{ij} = P_{ij}(1)$, $(i, j = 0,1)$, as specified above.

From this it follows that $P_{10} = 1/\lambda$, and $P_{01} = 1/\mu$, with $P_{00} = 1 - P_{01}$ and $P_{11} = 1 - P_{10}$, where $1/\lambda$ is the failure rate and $1/\mu$ is the constant repair rate. That is, we are now able to estimate all the four state transition probabilities simply by using the MTTF and MDT. This is an important result.

## Maximum likelihood estimators

In practice, however, the parameters $\lambda$ and $\mu$ of the exponential models assumed for the time-to-failure ($X$) and time-to-repair ($Y$) are usually unknown. Thus the availability and outage indices are also unknown for most practical problems. Hence we need to estimate these measures from a sample of values on $X$ and $Y$. Note that both $A(t)$ and $A(\infty)$, as well as $F(t)$ and $F(\infty)$, are functions of $\lambda$ and $\mu$, the parameters of the exponential models assumed for $X$ and $Y$. We can, therefore, obtain the maximum likelihood estimates of these measures by substituting the maximum likelihood estimators (MLE) of $\lambda$ and $\mu$ in the above results [5].

To calculate the maximum likelihood estimators (MLE) of $\lambda$ and $\mu$, we observe the power plant unit through $n$ failure-repair cycles, and collect the data on $T$ time-to-failure ($x_1, x_2, \ldots, x_T$) and $T$ time-to-repair ($y_1, y_2, \ldots, y_T$). Actually the data sets are two independent exponential samples.

The maximum likelihood procedure as developed in Kendall and Stuart [6] gives the following estimators:

the MLE of MTTF $\hat{\lambda} = \Sigma x_i / T = \bar{x}$, and

the MLE of MDT ($\mu$): $\hat{\mu} = \Sigma y_i / T = \bar{y}$.

Then the maximum likelihood estimators of availability and outage are obtained by substituting $\lambda$ and $\mu$ into the above results.

Thus the MLE of availability, $\hat{A} = \dfrac{\sum x_i}{\sum x_i + \sum y_i}$.

Thus the MLE of outage, $\hat{R} = \dfrac{\sum y_i}{\sum x_i + \sum y_i}$.

The steady-state (long-run) forced outage is generally known as forced outage rate (FOR), computed as a ratio of the unit's average down-time to the total available time, say, 720 hours a month; that is,

FOR = average forced outage hours/available hours.

The availability measure is then obtained as A = 1 − FOR.

## Mobility

A measure of (what we call) 'propensity to down' (mobility) is given by

$$D = \sum_{i=1}^{k} \sum_{j=1}^{k} P_i P_j \; |i - j|, \text{ where } k \text{ is the number of states of nature,}$$

$P_i$ is the long run probability and $P_{ij}$ the transition probability, In the case of $k = 2$, with $i, j = 0, 1$, we have

$$P_1 = \frac{P_{01}}{P_{10} + P_{01}}, \quad P_0 = \frac{P_{10}}{P_{10} + P_{01}} \text{ and hence } D = \frac{2 P_{01} P_{10}}{P_{10} + P_{01}}.$$

$D$ varies between 0 for immobility (least propensity to down) and 1 for extreme propensity to down.

## Capacity outage distribution

The next step in the generation reliability model is to combine the capacity and availability of the individual units to estimate expected available generation capacity in the system. Thus we obtain a capacity model, in which each generating unit is represented by its *nominal* capacity $k_j$ and its FOR, $R_j$, $j = 1 \ldots N$. Note that for each of the $N$ units of the generating station, the *expected available* capacity $k_j^A$, $j = 1 \ldots N$, is a random variable that can take the value 0 with probability $R_j$ and the value $k_j$ with probability $A_j = 1 - R_j$ as shown below:

$$k_i^A (k_i, R_i) \begin{cases} (k_j, A_j = 1 - R_j), \text{ if unit is available;} \\ \\ (0, R_j), \text{ if unit is in outage.} \end{cases}$$

Then the expected available capacity of a plant unit $j$ is $k_j^A = k_j A_j$ and the expected total generating capacity available at the plant level is: $K^A = \sum_{j}^{N} k_j^A$.

Note that the available capacity at both the unit $k_j^A$ and the plant level $K^A$ is a random variable; and the units fail and get repaired independently of such events of other units. These conditions help us obtain the probability distribution of $K^A$ by combining the independent individual probabilities of $k_j^A$. This in turn gives us a discrete (available) capacity distribution $K^A = (K_i, R_i)$, $i = 1, \ldots, 2^N$. The available capacity states takes on $2^N$ values, equal to the number of combinations of up and down units (due to forced outages) in an $N$-unit system. Each capacity state represents an outage event with one or more units unavailable. This capacity probability distribution is tabulated and referred to as the *capacity outage probability table*.

The capacity of the $i$th state, $Ki$, with $M$ available units and $N - M$ failed units is the sum of the capacities of the $M$ available units, that is,

$$Ki = K_1 + K_2 + \ldots + K_M$$

Given the outage or availability probabilities, the probability corresponding to each available capacity state can be calculated. Remember that the probability of the simultaneous occurrences of two or more independent events is the product of the respective event probabilities. Thus the probability of the $i$th state is equal to the product of the availabilities $A_i$ of the $M$ available units and the FORs $R_i$ of the $N - M$ out-of-service units, that is:

$$P_i = A_1 A_2 \ldots A_M R_1 R_2 \ldots R_{N-M}.$$

For illustration, below we give the capacity outage probability tables for a 2-unit and 3-unit plants and their generalization: (Case 1 and 2)

In general,

Plant availability (capacity state probability, $P_i$)

| Capacity state | | | Plant availability |
|---|---|---|---|
| | Unit 1 | Unit 2 | |
| All up | Up | Up | $A_1 A_2$ |
| 1 up, 1 down | Up | Down | $A_1 R_2 +$ |
| | Down | Up | $A_2 R_1 =$ |
| | | | $A_1 R_2 + A_2 R_1$ |
| All Down | Down | Down | $R_1 R_2$ |

Note: $R_j = 1 - A_j$ is the FOR and $A_j$ is the steady state availability of unit $j$.

**Case 1:** Case of a 2-unit Plant.

| Capacity state | | | | Plant availability |
|---|---|---|---|---|
| | Unit 1 | Unit 2 | Unit 3 | |
| All up | Up | Up | Up | $A_1A_2A_3$ |
| 2 up, 1 down | Up | Up | Down | $A_1A_2R_3$ + |
| | Up | Down | Up | $A_1R_2A_3$ + |
| | Down | Up | Up | $R_1A_2A_3$ = |
| | | | | $A_1A_2R_3+A_1R_2A_3+R_1A_2A_3$ |
| 1 Up, 2 Down | Up | Down | Down | $A_1R_2R_3$ + |
| | Down | Up | Down | $R_1A_2R_3$ + |
| | Down | Down | Up | $R_1R_2A_3$ = |
| | | | | $A_1R_2R_3+R_1A_2R_3+R_1R_2A_3$ |
| All Down | Down | Down | Down | $R_1R_2R_3$ |

Note: $R_j = 1 - A_j$ is the FOR and $A_j$ is the steady state availability of unit $j$.

**Case 2:** Case of a 3-unit Plant.

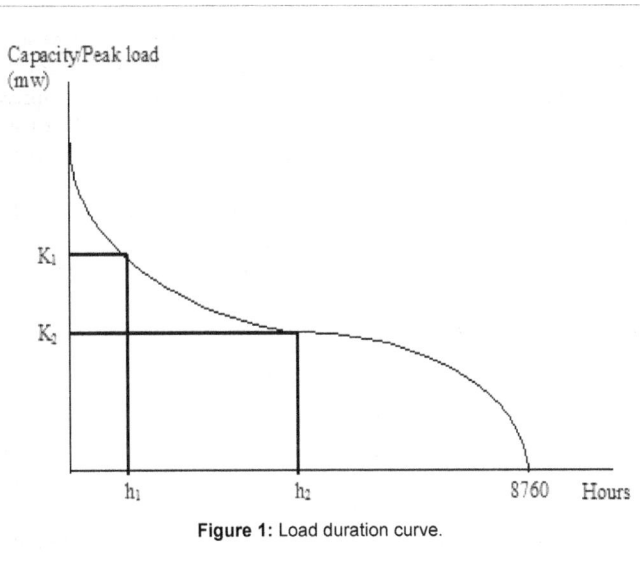

**Figure 1:** Load duration curve.

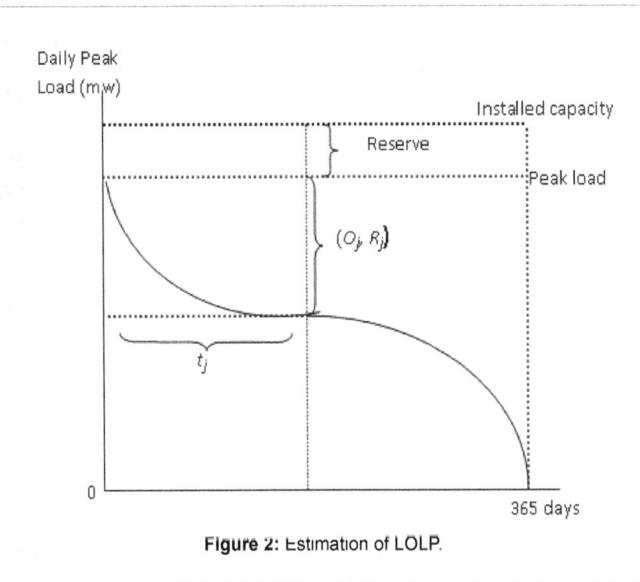

**Figure 2:** Estimation of LOLP.

when all plant units are up = $\Pi\, A_j$ for all $j$.

when all plant units are down = $\Pi\, (1 - A_j)$ for all $j$.

for a 2-unit plant, when 1 unit is up and 1 unit down =

$$\sum_{i\neq j} A_i(1-A_j)\,;\, i,j=1,2.$$

for a 3-unit plant, when 2 units are up and 1 unit down =

$$\sum_{i\neq j\neq k} A_iA_j(1-A_k)\,;\, i,j,k=1,2,3.$$

for a 3-unit plant, when 2 units are down and 1 unit up =

$$\sum_{i\neq j\neq k} A_i(1-A_j)(1-A_k)\,;\, i,j,k=1,2,3.$$

## Loss of load probability (LOLP)

The unreliability of a system in this context is viewed as its inability to meet the daily peak load. A loss of load occurs whenever the system load exceeds the available generating capacity. The overall probability that there will be a shortage of power (loss of load) is called the Loss-of-Load Probability or LOLP. It is usually expressed in terms of days per year, hours per day or as a percentage of time. When expressed as the expected accumulated amount of time during which a shortage of power is experienced, the measure is more correctly referred to as the loss of load expectation (LOLE). The LOLP measure was first introduced by [7].

By combining the availability of each capacity state with the system load duration curve (LDC), we obtain the expected risk of loss of load. A load duration curve is defined as a function whose abscissa specifies the width of the time interval, usually the number of hours in a year, during which customer's (peak) demand for power (D) equals or exceeds the associated level of available capacity ($K^A$) given on the ordinate (Fig. 1). Thus it shows the time duration for which a capacity outage would cause a loss of load (D ≥ K). By normalizing the time variable as a proportion of the total, the value at any point on the abscissa can be taken as the (cumulative) probability that the corresponding load will be equaled or exceeded (D ≥ K). When the daily peak load curve is used, the value of LOLE is in days for the period of study, usually days per year. Because of its monotonicity and continuity, the function can be inverted to obtain the proportion of this time interval. This inverse function can in turn be interpreted as the complementary cumulative density function (i.e., the distribution function) of the customer's demand (Figure 1).

The LOLP can be used to measure loss-of-load risk as illustrated in Figure 2 with a daily peak load curve. $Oj$ is the magnitude of the $j$th outage in the system, $Rj$ is the probability of a capacity outage of magnitude $Oj$, and $tj$ is the number of days that an outage of magnitude $Oj$ would cause a loss of load in the system. Note that capacity outages less than the reserve will not lead to a loss of load; a particular capacity outage greater than the reserve contributes to the overall risk by the amount $(P_j\, t_j)$. Then the system LOLP for the period is:

$$LOLP = \sum_j P_jt_j.$$

Now how to estimate the outage duration, $t_j$?

Suppose the customer's daily maximum demand on a power system over one year can be represented by a normal distribution with mean $\eta$ and standard deviation $\sigma$. Then the proportion of time during which a capacity outage would cause a loss of load (i.e., $D \geq K^A$) is given by

$$\text{Prob}(D \geq K^A) = 1-\text{Prob}(D < K^A) = 1-\varphi(z),$$

where $z = (K^A - )/\sigma$, and $\varphi(z)$ is the area under the standard normal curve given by

$$\phi(z) = (\sqrt{2\pi})^{-1}\int_{-\infty}^{z} \exp(-x^2/2)dx,$$

| Power plants | Unit Capacity (No x mw) | Installed Capacity (mw) | Average Designed Generation Potential (ADGP) | | Storage capacity | |
|---|---|---|---|---|---|---|
| | | | mu | % to IC | mu | % to ADGP |
| Pallivasal | 3x5+3x7.5 | 37.5 | 284.7 | 86.67 | 79.54 | 27.94 |
| Sengulam | 4 x 12 | 48 | 182.2 | 43.33 | 49.61 | 27.23 |
| Neraiamangalam | 3 x 15 | 45 | 237.4 | 60.22 | 67.58 | 28.47 |
| Panniar | 2 x 15 | 30 | 157.7 | 60.01 | 45.47 | 28.83 |
| Poringalakuthu | 4 x 8 | 32 | 171.7 | 61.25 | 63.43 | 36.94 |
| Sholayar | 3 x 18 | 54 | 233 | 49.26 | 99.47 | 42.69 |
| Sabarigiri | 6 x 50 | 300 | 1337.7 | 50.90 | 770.32 | 57.59 |
| Kuttiyadi | 3 x 25 | 75 | 268.1 | 40.81 | 41.46 | 15.46 |
| Idukki | 6 x 130 | 780 | 2397.6 | 35.09 | 2147.88 | 89.58 |
| Idamalayar | 2 x 37.5 | 75 | 380.2 | 57.87 | 254.45 | 66.93 |
| Total | | 1476.5 | 5650.3 | 43.69 | 3619.21 | 64.05 |

**Table 1**: Characteristics of the 10 Hydropower Stations.

| Plant | Units | Monthly MDT (hours) | FOR | Availability |
|---|---|---|---|---|
| Pallivasal | 1 | 39.97 | 0.0555 | 0.9445 |
| | 2 | 23.23 | 0.0323 | 0.9677 |
| | 3 | 72.53 | 0.1007 | 0.8993 |
| | 4 | 74.76 | 0.1038 | 0.8962 |
| | 5 | 40.43 | 0.0562 | 0.9438 |
| | 6 | 29.09 | 0.0404 | 0.9596 |
| Sengulam | 1 | 49.59 | 0.0689 | 0.9311 |
| | 2 | 8.07 | 0.0112 | 0.9888 |
| | 3 | 40.05 | 0.0556 | 0.9444 |
| | 4 | 207.72 | 0.2885 | 0.7115 |
| Neraiamangalam | 1 | 11.24 | 0.0156 | 0.9844 |
| | 2 | 15.48 | 0.0215 | 0.9785 |
| | 3 | 26.39 | 0.0367 | 0.9633 |
| Panniar | 1 | 153.43 | 0.2131 | 0.7869 |
| | 2 | 95.87 | 0.1332 | 0.8668 |
| Poringalakuthu | 1 | 55.69 | 0.0773 | 0.9227 |
| | 2 | 20.83 | 0.0289 | 0.9711 |
| | 3 | 14.18 | 0.0197 | 0.9803 |
| | 4 | 14.10 | 0.0196 | 0.9804 |
| Sholayar | 1 | 173.58 | 0.2411 | 0.7589 |
| | 2 | 127.19 | 0.1767 | 0.8233 |
| | 3 | 79.39 | 0.1103 | 0.8897 |
| Sabarigiri | 1 | 7.58 | 0.0105 | 0.9895 |
| | 2 | 37.59 | 0.0522 | 0.9478 |
| | 3 | 33.96 | 0.0472 | 0.9528 |
| | 4 | 41.12 | 0.0571 | 0.9429 |
| | 5 | 48.27 | 0.0670 | 0.9330 |
| | 6 | 56.67 | 0.0787 | 0.9213 |
| Kuttiyadi | 1 | 1.52 | 0.0021 | 0.9979 |
| | 2 | 1.09 | 0.0015 | 0.9985 |
| | 3 | 0.58 | 0.0008 | 0.9992 |
| Idukki | 1 | 12.44 | 0.0173 | 0.9827 |
| | 2 | 45.02 | 0.0625 | 0.9375 |
| | 3 | 25.29 | 0.0351 | 0.9649 |
| | 4 | 9.82 | 0.0136 | 0.9864 |
| | 5 | 24.82 | 0.0345 | 0.9655 |
| | 6 | 14.83 | 0.0206 | 0.9794 |
| Idamalayar | 1 | 31.17 | 0.0433 | 0.9567 |
| | 2 | 38.96 | 0.0541 | 0.9459 |
| Total | | 46.24 | 0.0642 | 0.9358 |

**Table 2**: Long run Availability and Forced Outage Rates.

which can be read off a standard statistical table. $K^A$ here denotes the available capacity in a certain capacity state; thus we obtain the 'outage duration' that is, the proportion of time during which a forced outage results in a loss of load ($D \geq K^A$) in each of the possible capacity states. The relative contribution of this outage to the overall system loss of load is then obtained by multiplying the availability of a certain capacity state by the corresponding proportion of time that available capacity level is equaled or exceeded. The total LOLP is the sum of all contributions due to the different capacity outages. Multiplying the LOLP by 365 then gives the expected cumulative number of days in a year when loss of load is experienced due to forced outage.

## LOLP of the Kerala Hydro-Power System

Till the mid-1980s, Kerala (India) had a predominantly hydroelectric power system; with the increased dependency on energy import, the hydro-thermal mix has come down, and now hovers around 27:73. However, if we consider installed capacity, the system is still predominantly hydel. There are 44 power generating stations (both in the public and private sectors), including 34 hydel, seven thermal and three wind, with an installed capacity of 2880 MW; out of which the Kerala State Electricity Board (KSEB) owns one wind, two thermal, 30 hydel stations (and shares a national thermal station), accounting for 90.45 percent of the total installed capacity. The present study considers only the 10 old hydropower stations of the State, in view of the availability of sufficiently large time series data. These power stations in the descending order of age (the last plant, Idamalayar, was commissioned in 1987) with their important characteristics are given in Table 1. These ten power stations, with an installed capacity (IC) of 1476.5 megawatt (mw), accounts for about 74 percent of the total own hydel IC (2008.65 mw) and 66 percent of the total own IC (2245.28 mw) of the Kerala power system (Table 1).

Since the early 1980s, Kerala has been suffering from severe capacity shortage in the power sector. Even by 1984-85, the State had an installed capacity of only 1011.5 mw against an estimated demand of 1122 mw. During the two decades from 1976, Kerala's installed capacity in the power sector was growing at an exponential rate of only 3 per cent per annum against 6 per cent of the maximum demand, which in effect was restricted in many ways by power shortage. Only since the late 1990s has there been some perceptible addition to the IC.

That a hydropower system is at the mercy of the vagaries of the monsoon is a foregone conclusion, especially with an insufficient storage capacity. Once the monsoon goes dry, close on the heels follow severe power shortages as was the case in the 1980s and thereafter in

Kerala; power cut/load shedding has become the rule of the day since 1982-83, even with very large import of thermal power, often more than 60 percent of the total available power.

## Availability and forced outage rates

The relevant data for a period of 33 years from 1978-79 (to 2011-12) were collected from the Kerala State Electricity Board (KSEB); the data on plant unit capacity, hours of operation and forced outage are from the KSEB's annual publication, viz., 'System Operations' and the data on daily maximum demand on each of the 10 power stations for three years from 2006-07 were collected from the (unpublished) records from the KSEB head office (*Vydhyuthi Bhavan*) in Trvandrum. In the case of Idukki II stage (3 units) and Idamalayar, commissioned during 1985-87 period, the data are from this period onwards.

It should be noted here that a study of this dimension (theory-informed methodology) is the first of its kind in the case of most of the State Electricity Boards in India, especially the KSEB; the technical

| Plant | Units | Transition Probabilities | | | | Propensity to down |
|---|---|---|---|---|---|---|
| | | $P_{00}$ | $P_{01}$ | $P_{11}$ | $P_{10}$ | |
| Pallivasal | 1 | 0.975 | 0.025 | 0.9985 | 0.0015 | 0.00274 |
| | 2 | 0.958 | 0.042 | 0.9986 | 0.0014 | 0.00272 |
| | 3 | 0.986 | 0.014 | 0.9985 | 0.0015 | 0.00276 |
| | 4 | 0.987 | 0.013 | 0.9985 | 0.0015 | 0.00276 |
| | 5 | 0.976 | 0.024 | 0.9985 | 0.0015 | 0.00274 |
| | 6 | 0.966 | 0.034 | 0.9986 | 0.0014 | 0.00273 |
| Sengulam | 1 | 0.980 | 0.020 | 0.9985 | 0.0015 | 0.00275 |
| | 2 | 0.884 | 0.116 | 0.9987 | 0.0013 | 0.00261 |
| | 3 | 0.975 | 0.025 | 0.9985 | 0.0015 | 0.00274 |
| | 4 | 0.995 | 0.005 | 0.9981 | 0.0019 | 0.00277 |
| Neraiamangalam | 1 | 0.915 | 0.085 | 0.9987 | 0.0013 | 0.00266 |
| | 2 | 0.937 | 0.063 | 0.9986 | 0.0014 | 0.00269 |
| | 3 | 0.963 | 0.037 | 0.9986 | 0.0014 | 0.00272 |
| Panniar | 1 | 0.994 | 0.006 | 0.9982 | 0.0018 | 0.00277 |
| | 2 | 0.990 | 0.010 | 0.9984 | 0.0016 | 0.00276 |
| Poringalakuthu | 1 | 0.982 | 0.018 | 0.9985 | 0.0015 | 0.00275 |
| | 2 | 0.953 | 0.047 | 0.9986 | 0.0014 | 0.00271 |
| | 3 | 0.932 | 0.068 | 0.9986 | 0.0014 | 0.00268 |
| | 4 | 0.932 | 0.068 | 0.9986 | 0.0014 | 0.00268 |
| Sholayar | 1 | 0.994 | 0.006 | 0.9982 | 0.0018 | 0.00277 |
| | 2 | 0.992 | 0.008 | 0.9983 | 0.0017 | 0.00276 |
| | 3 | 0.987 | 0.013 | 0.9984 | 0.0016 | 0.00276 |
| Sabarigiri | 1 | 0.876 | 0.124 | 0.9987 | 0.0013 | 0.00260 |
| | 2 | 0.974 | 0.026 | 0.9986 | 0.0014 | 0.00274 |
| | 3 | 0.971 | 0.029 | 0.9986 | 0.0014 | 0.00274 |
| | 4 | 0.976 | 0.024 | 0.9985 | 0.0015 | 0.00274 |
| | 5 | 0.980 | 0.020 | 0.9985 | 0.0015 | 0.00275 |
| | 6 | 0.983 | 0.017 | 0.9985 | 0.0015 | 0.00275 |
| Kuttiyadi | 1 | 0.518 | 0.482 | 0.99898 | 0.0010 | 0.00203 |
| | 2 | 0.401 | 0.599 | 0.99909 | 0.0009 | 0.00182 |
| | 3 | 0.178 | 0.822 | 0.99934 | 0.0007 | 0.00132 |
| Idukki | 1 | 0.923 | 0.077 | 0.9986 | 0.0014 | 0.00267 |
| | 2 | 0.978 | 0.022 | 0.9985 | 0.0015 | 0.00275 |
| | 3 | 0.961 | 0.039 | 0.9986 | 0.0014 | 0.00272 |
| | 4 | 0.903 | 0.097 | 0.9987 | 0.0013 | 0.00264 |
| | 5 | 0.961 | 0.039 | 0.9986 | 0.0014 | 0.00272 |
| | 6 | 0.935 | 0.065 | 0.9986 | 0.0014 | 0.00268 |
| Idamalayar | 1 | 0.968 | 0.032 | 0.9986 | 0.0014 | 0.00273 |
| | 2 | 0.975 | 0.025 | 0.9986 | 0.0014 | 0.00274 |
| Total | | 0.979 | 0.021 | 0.9985 | 0.0015 | 0.00275 |

**Table 3**: Transition Probabilities and Propensity to Down.

appraisal of these power systems in general is confined to examining the plant load factor (PLF) as a measure of capacity utilization only. It goes without saying that PLF is by no means directly comparable with LOLP, as the two methodologies totally differ from each other; this precludes us from attempting at any comparison with the official measure.

The estimated mean-down-time (MDT; forced outage time), measured in hours per month, of each of the units of the 10 hydro power generating stations are given in Table 2; with a cycle time (MTBF) of 720 hours in general, we derive the corresponding forced outage and availability rates.

The long run availability is the highest, coming very close to unity, for (all the units of) Kuttiadi power plant. Neriamangalam, Poringalkuthu, Idukki, Sabarigiri and Idamalayar plants also have higher availability for all their units (above 90 percent). All but two (units III and IV) of the 6 units of Pallivasal also have higher availability; Sengulam also follows suit with one unit (IV) having the highest FOR of 0.29. The remaining two plants, Panniar and Sholayar, bear the whole brunt of higher FOR. Sholayar unit I has the second highest FOR of 0.24; Panniar unit I follows with only 79 percent of availability.

It is worth finding that an average outage of more than one month in one year occurred in the case of 8 out of the total 39 plant units; Sengulam unit IV has the fate of having the maximum mean outage of a cumulative period of more than 3 months a year and Sholayar unit I, nearly 3 months; the credit of having the minimum mean outage of less than one day in one year goes to the 3 units of Kuttiadi, with 7-18 hours a year only (Table 2).

In the case of Kuttiadi, instantaneous availability readily collapses on the steady-state one, owing to the least MDT (or FOR). For all other plant units, the long run evolves through time limit. For Sabarigiri unit I and Sengulam unit II, it takes a cumulative period of nearly one month to reach the steady state; other units take much more time. In the case of Sengulam unit IV and Sholayar unit I, with very high FOR, the long-run is evolved across a cumulative period of more than 3 months.

Note that minimum mean outage time does not necessarily mean higher mean operating period, as the case of Kuttiadi clearly shows. Even though Kuttiadi has the highest availability (and the least FOR) as per definition, its service period, when accumulated, amounts on an average to only about 7 months a year; that is, all the three units of this plant remain shut down for a cumulative mean period of about 5 months for scheduled maintenance and/or for want of sufficient water in the reservoir. Panniar and Sholayar with lower levels of service time (MTTF) are also shut down for about 3 – 4 months. The shutdown period of other plants in general accumulate up to 1 – 3 months a year.

Averaging all the data, we find that the whole system has an annual MDT of 46.24 hours per month per unit, with a FOR of 6.4 percent and availability of about 93.6 percent. We can also find that a cumulative time of about 1 month is required for translating the 'instant' into the long-run for the system. However, a 6 percent FOR in the face of capacity shortage imposes a heavy tax on the system. Assuming an annual average generation of 6100 million units (mu; average generation of the last 6 years from 2001-02), with sufficient hydraulic power capacity, this level of FOR implies a potential energy shortage to the tune of $6100 \times 0.0642/0.9358 \approx 420$ mu a year. At average revenue of Rs. 3.25 per unit as at present, this represents a financial loss to the system of Rs. 1365 million a year. Moreover, the potential energy (lost) of 420 mu is equivalent to about 80 mw installed capacity at 60 per cent load factor. A zero FOR in this case then implies that it could dispense with the investment requirement of adding about 80 mw capacity to the system, saving immensely in capital costs and working expenses. Note that this saving is in addition to the gain in sales revenue.

Table 3 reports the state transition probabilities along with the measures of propensity to down. The initial instantaneous availability

($P_{11}$, when $t = 1$) for all the plant units is much higher, close to unity, with very low measure of D, the propensity to down; so is the initial instantaneous forced outage rate, but less than $P_{11}$: it takes some time for repair. The system average follows suit.

| Plant | Capacity state | Nominal Capacity mw | Availability | Available Capacity mw $k_j^A$ | Standard Normal Variate $zi$ | Outage duration $t_j$ | LOLP of Capacity State |
|---|---|---|---|---|---|---|---|
| (1) | (2) | (3) | (4) | (5) = (3) x (4) | (6) | (7) | (8) = (4) x (7) |
| Pallivasal | All 6 units up | 37.5 | 0.6671 | 25.02 | 2.32 | 0.010 | 0.007 |
| | 5 up, 1 down | 32.5 | 0.1362 | 4.426 | 1.46 | 0.072 | 0.010 |
| | | 30 | 0.1451 | 4.352 | 1.03 | 0.152 | 0.022 |
| | 4 up, 2 down | 27.5 | 0.00819 | 0.225 | 0.60 | 0.274 | 0.002 |
| | | 25 | 0.02962 | 0.7404 | 0.17 | 0.433 | 0.013 |
| | | 22.5 | 0.00953 | 0.214 | -0.26 | 0.603 | 0.006 |
| | 3 up, 3 down | 22.5 | 0.000146 | 0.00329 | -0.26 | 0.603 | 0.000 |
| | | 20 | 0.00178 | 0.03562 | -0.69 | 0.755 | 0.001 |
| | | 17.5 | 0.00194 | 0.03403 | -1.12 | 0.869 | 0.002 |
| | | 15 | 0.000194 | 0.002905 | -1.55 | 0.939 | 0.000 |
| | 2 up, 4 down | 15 | 0.0000318 | 0.000478 | -1.55 | 0.939 | 0.000 |
| | | 12.5 | 0.000117 | 0.00146 | -1.98 | 0.976 | 0.000 |
| | | 10 | 3.953E-05 | 0.000395 | -2.41 | 0.992 | 0.000 |
| | 1 up, 4 down | 7.5 | 2.090E-06 | 0.0000157 | -2.84 | 0.998 | 0.000 |
| | | 5 | 2.377E-06 | 0.0000119 | -3.27 | 0.999 | 0.000 |
| | All 6 units down | 0 | 4.250E-08 | 0 | -4.12 | 1.000 | 0.000 |
| Sengulam | All 4 units up | 48 | 0.6186 | 29.69 | 2.62 | 0.004 | 0.003 |
| | 3 up, 1 down | 36 | 0.3401 | 12.24 | 0.57 | 0.284 | 0.097 |
| | 2 up, 2 down | 24 | 0.0398 | 0.955 | -1.49 | 0.932 | 0.037 |
| | 1 up, 3 down | 12 | 0.00150 | 0.0180 | -3.54 | 1.000 | 0.002 |
| | All 4 units down | 0 | 1.239E-05 | 0 | -5.59 | 1.000 | 0.000 |
| Neraiamangalam | All 3 units up | 45 | 0.9279 | 41.76 | 0.90 | 0.184 | 0.171 |
| | 2 up, 1 down | 30 | 0.0704 | 2.11 | -1.47 | 0.929 | 0.065 |
| | 1 up, 2 down | 15 | 0.00166 | 0.0249 | -3.84 | 1.000 | 0.002 |
| | All 3 units down | 0 | 1.230E-05 | 0 | -6.21 | 1.000 | 0.000 |
| Panniar | All 2 units up | 30 | 0.6821 | 20.46 | 0.96 | 0.169 | 0.115 |
| | 1 up, 1 down | 15 | 0.2895 | 4.34 | -1.01 | 0.844 | 0.244 |
| | All 2 units down | 0 | 0.0284 | 0 | -2.97 | 0.999 | 0.028 |
| Poringalakuthu | All 4 units up | 32 | 0.8611 | 27.56 | 0.87 | 0.192 | 0.165 |
| | 3 up, 1 down | 24 | 0.1323 | 3.18 | -0.66 | 0.745 | 0.099 |
| | 2 up, 2 down | 16 | 0.00642 | 0.103 | -2.19 | 0.986 | 0.006 |
| | 1 up, 3 down | 8 | 0.00013 | 0.001003 | -3.72 | 1.000 | 0.000 |
| | All 4 units down | 0 | 8.630E-07 | 0 | -5.25 | 1.000 | 0.000 |
| Sholayar | All 3 units up | 54 | 0.5560 | 30.02 | 3.15 | 0.001 | 0.000 |
| | 2 up, 1 down | 36 | 0.3648 | 13.13 | -0.14 | 0.556 | 0.203 |
| | 1 up, 2 down | 18 | 0.0746 | 1.34 | -3.43 | 1.000 | 0.075 |
| | All 3 units down | 0 | 0.00470 | 0 | -6.71 | 1.000 | 0.005 |
| Sabarigiri | All 6 units up | 300 | 0.7242 | 217.26 | 1.77 | 0.038 | 0.028 |
| | 5 up, 1 down | 250 | 0.2412 | 60.30 | -0.01 | 0.504 | 0.122 |
| | 4 up, 2 down | 200 | 0.0323 | 6.46 | -1.79 | 0.963 | 0.031 |
| | 3 up, 3 down | 150 | 0.00220 | 0.330 | -3.57 | 1.000 | 0.002 |
| | 2 up, 4 down | 100 | 7.916E-05 | 0.00792 | -5.35 | 1.000 | 0.000 |
| | 1 up, 5 down | 50 | 1.363E-06 | 0.0000681 | -7.14 | 1.000 | 0.000 |
| | All 6 units down | 0 | 7.811E-09 | 0 | -8.92 | 1.000 | 0.000 |
| Kuttiyadi | All 3 units up | 75 | 0.9956 | 74.67 | 0.69 | 0.245 | 0.244 |
| | 2 up, 1 down | 50 | 0.00442 | 0.221 | -0.30 | 0.618 | 0.003 |
| | 1 up, 2 down | 25 | 6.109E-06 | 0.000153 | -1.30 | 0.903 | 0.000 |
| | All 3 units down | 0 | 2.573E-09 | 0 | -2.29 | 0.989 | 0.000 |
| Idukki | All 6 units up | 780 | 0.8291 | 646.72 | 2.54 | 0.006 | 0.005 |
| | 5 up, 1 down | 650 | 0.1586 | 103.07 | 0.42 | 0.337 | 0.053 |
| | 4 up, 2 down | 520 | 0.0118 | 6.162 | -1.70 | 0.955 | 0.011 |

| | | | | | | | |
|---|---|---|---|---|---|---|---|
| | 3 up, 3 down | 390 | 0.000446 | 0.174 | -3.81 | 1.000 | 0.000 |
| | 2 up, 4 down | 260 | 8.934E-06 | 0.002323 | -5.93 | 1.000 | 0.000 |
| | 1 up, 5 down | 130 | 9.084E-08 | 0.000012 | -8.05 | 1.000 | 0.000 |
| | All 6 units down | 0 | 3.674E-10 | 0 | -10.17 | 1.000 | 0.000 |
| Idamalayar | All 2 units up | 75 | 0.9049 | 67.87 | 1.71 | 0.044 | 0.039 |
| | 1 up, 1 down | 37.5 | 0.0927 | 3.48 | -1.55 | 0.939 | 0.087 |
| | All 2 units down | 0 | 0.00234 | 0 | -4.82 | 1.000 | 0.002 |

**Table 4**: Estimation of LOLP by Capacity States.

| Plant | Daily maximum demand (mw) | | Loss of load | | Expected available capacity | |
|---|---|---|---|---|---|---|
| | Mean | SD | Probability | Days/Year | mw | % to IC |
| Pallivasal | 24.01 | 5.82 | 0.063 | 22.95 | 25.02 | 66.71 |
| Sengulam | 32.69 | 5.85 | 0.138 | 50.38 | 29.69 | 61.86 |
| Neraiamangalam | 39.32 | 6.34 | 0.238 | 86.83 | 41.76 | 92.79 |
| Panniar | 25.66 | 7.62 | 0.388 | 141.46 | 20.46 | 68.21 |
| Poringalakuthu | 27.44 | 5.22 | 0.271 | 98.75 | 27.56 | 86.11 |
| Sholayar | 36.77 | 5.48 | 0.282 | 103.07 | 30.02 | 55.60 |
| Sabarigiri | 261.23 | 31.60 | 0.183 | 66.71 | 217.26 | 72.42 |
| Kuttiyadi | 57.61 | 25.12 | 0.247 | 90.06 | 74.67 | 99.56 |
| Idukki | 624.12 | 61.39 | 0.070 | 25.49 | 646.72 | 82.91 |
| Idamalayar | 55.33 | 11.49 | 0.129 | 47.06 | 67.87 | 90.49 |
| System | 1184.19 | | 0.20 | 73.28 | 1181.03 | 79.99 |

**Table 5**: Loss of Load Probability and Expected Available Capacity.

## Capacity-outage probability and LOLP

The first step in the estimation of LOLP is to find out the available capacity state probabilities. Table 4 reports the distributed levels of available capacity with the corresponding probability of occurrence in accordance with the different combinations of up and down (due to forced outages) units of each of the 10 power stations, estimated as per the section on 'Capacity outage distribution'. Note that the available capacity probabilities of all the states add up to unity. Given the availability and the nominal capacity of each unit, we can find the available capacity ($k_j^A$) corresponding to each state.

Pallivasal, as shown in Table 4, has different levels of available capacity and availability in each of the possible capacity states due to different unit capacities – it has 3 units of 5 mw each and another 3 of 7.5 mw each. Thus in the '1-unit-down' capacity state, we have two levels of available capacity, depending on the capacity of the unit that goes down; if a 5 mw unit fails, the available capacity will be 32.5 mw and in the other case, 30 mw. Also note that in two capacity states ('3 units up', and '2 units up'), the same level of available capacity (15 mw) is obtained; in the '3-units-up' state, it may so happen that all the 3 units of 5 mw each may be in operation (with the other 3 units of 7.5 mw each in outage) and in the next '2-units-up' state, 2 of the 3 units of 7.5 mw each may be in service. All other plants have same-capacity units and hence each capacity state has a unique level of available capacity and probability.

As is already evident, Kuttiadi has the highest availability (almost nearing unity) of maximum capacity (when all units are up). Only 3 plants have an all-units-up availability of more than 90 per cent (Neriamangalam, Kuttiadi and Idamalayar), and 5 plants, of more than 80 per cent (including Poringalkuth and Idukki). Sholayar is the only plant with an all-up availability of less than 60 per cent (Table 4).

The second step in the estimation of LOLP is to bring in the load duration curve (LDC) and derive from it the complementary distribution function of customers' demand. This we accomplish by assuming that the customers' daily maximum demand on the Kerala power system follows a normal distribution. Thus data on the daily maximum demand on each of the 10 power stations for three years from 2001-02 were averaged to avoid variability; and then the respective mean and standard deviation were estimated (Table 5). The maximum demand on Kuttiadi powerhouse is the most variable (coefficient of variation: 43.6%; due to seasonal operation necessitated by insufficient storage) and that on Idukki, the least (coefficient of variation: 9.8%).

Now using these parameters (the mean daily maximum demand and standard deviation), the expected available capacity ($k_j^A$) in each possible state is transformed into its corresponding standard normal variate, $z_j$, and the associated area under the normal curve, $\varphi(z_j)$, is found from a standard statistical table. Then $1 - \varphi(z_j)$ represents the (cumulative) proportion of the outage duration, *i.e.*, the proportion of time during which the load equals or exceeds the available capacity, determined by forced outages in a certain capacity state (Table 4). Thus, in the case of Panniar, about 16.9 percent of the time the maximum demand is likely to equal or exceed the available capacity when all the units are in operation; or, in other words, in the 'all-units-up' capacity state of Panniar, about 17 percent of the time a forced outage is likely to result in a loss of load. It increases to 84 percent in case any one unit falls down.

The proportion of non-supply duration during which loss of load is caused by different capacity outages in the case of all the 10 power plants are given in the penultimate column of Table 4. This outage duration, when all the units are in operation, is negligible for only 6 plants – Pallivasal, Sengulam, Sholayar, Idukki, Sabarigiri and Idamalayar. About 25 percent of the time a loss of load is experienced in the case of Kuttiadi even when all the units are in operating condition. For Poringalkuthu it is about 19 percent, and for Neriamangalam and Panniar, about 18 and 17 percent respectively. Obviously, the factors determining the extent of the non-supply duration are the capacity-demand gap and the variability (standard deviation) of the demand distribution. The smaller the capacity-demand gap, the larger will be the non-supply duration. A surplus capacity coupled with low demand variability or a deficit capacity with high demand variability results in a short non-supply duration. For Kuttiadi the major influencing factor is obviously the higher demand variability, whereas in the case of Neriamangalam and Poringalkuthu, the smaller capacity-demand gap appears to be the main culprit for larger non-supply duration. Both the factors seem to act on Panniar.

If one unit is thrown out of service, demand is likely to exceed for more than 50 per cent of the time in the case of as many as 7 plants and for more than 80 per cent of the time in the case of 3 plants – Neriamangalam, Panniar and Idamalayar. If the available capacity is only one-half of the installed capacity, then demand tends to exceed it for more than 80 – 90 per cent of the time in general.

## LOLP

The expected loss of load in each capacity state is calculated by multiplying the outage duration by the respective availability in that state, given in the last column of Table 4. Summing this over all the capacity states of a plant yields the measure of LOLP; the estimates of LOLP, both as a proportion of time and in terms of number of days a year, for all the 10 plants are given in Table 5. For example, a LOLP of 0.39 for Panniar means that on the whole about 39 percent of the time a loss of load is expected due to forced outages in the case of Panniar. On an annual basis, the expected loss of load is 141.5 days in one year, the expected accumulated amount of time during which demand equals or exceeds the available capacity causing a loss of load due to forced outages; this is the maximum among the 10 plants, followed a little afar by Sholayar (103 days), and Poringalkuthu (99 days). Evidently, the major determinants of this measure are the distribution of availability and non-supply duration in the capacity states. Thus, for example, in the case of Panniar, larger non-supply durations, coupled with the associated, not so small, availability of the respective capacity states contributed to its higher LOLP; that is, the relative contribution to LOLP of larger non-supply duration of lower capacity states is significantly high in this case. On the other hand, for Kuttiadi, the relative contribution to LOLP of larger non-supply durations is negligibly smaller. The minimum LOLP is enjoyed by Pallivasal (23 days a year) and Idukki (25 days a year). A simple average of the LOLPs of all the 10 plants gives the system LOLP of 0.20 or 73.3 days a year with a coefficient of variation of 51.3 percent.

Table 5 also reports the expected available capacity of the 10 plants (when all units are up; also see column (5) of Table 4); as many as 5 plants have available capacity less than 80 percent of the installed capacity: Pallivasal, Sengulam, Panniar, Sholayar and Sabarigiri. Note

that in these cases, the available capacity is barely sufficient to meet the peak load. For the system as a whole, only about 80 percent of the capacity is expected to be available, again not up to the system peak load.

## Conclusion

The vital nature of electric power, both to our economic and personal well being, has prompted the developed countries to place higher reliability standards on the performance of electricity supply. For example, most of the U.S. electric power utilities are designed on the technical assumption that the total accumulated time of supply interruptions (forced outages) should be no more than 1 day in 10 years [8]. This evidently appears to be a very strict design criterion even for developed countries. Some studies have in fact shown this 1-day-in-10-years reliability target as economically unjustified, and that it could reasonably be reduced without adversely affecting the economy [9]. Though the reliability performance of an under-developed electricity supply system such as Kerala's is by no means comparable with that of the developed countries, the estimates of LOLP reported here seem on all counts to be stupendously higher. That the expected cumulative outage time of the power generating system in Kerala amounts to 73 days a year is a shocking revelation of the kind of service rendered.

### References

1. Shooman ML (1968) Probabilistic Reliability: An Engineering Approach. (2nd edn.) McGraw-Hill, New York.

2. Gnedenko BV, Belyayev YK, Solovyev AD (1969) Mathematical Method of Reliability Theory. ( 1st edn.) Academic Press, New York.

3. Vijayamohanan PN (1992) Seasonal Time-of-Day Pricing of Electricity under Uncertainty – A Marginalist Approach to Kerala Power System. PhD Thesis, University of Madras, Chennai.

4. Vijayamohanan PN (2014) An Inquiry into the Distributional Properties of Reliability Rate. Amer J Theoreticd Appl Statistic 3: 197-201.

5. Zehna P (1966) Invariance of Maximum Likelihood Estimation. Annal Mathematic Statistic, 37: 744.

6. Kendall MG, Stuart A (1967) The Advanced Theory of Statistics. (2nd edn) Hafner, New York.

7. Calabrese G (1947) Generating Reserve Capacity Determined by the Probability Method. AIEE Transactions 66: 1439-1450.

8. Vardi, Joseph, Avi-Itzhak, Benjamin (1981) Electric Energy Generation: Economics, Reliability and Rates. The MIT Press, Cambridge and London.

9. Telson ML (1975) The Economics of Alternative Levels of Reliability for Electric Power Generation System. The Bell J Economic 6: 679-94.

# 17

# An Appropriate Extreme Value Distribution for the Annual Extreme Gust Winds Speed

**Banafsheh Abolpour[1], Bahador Abolpour[2,3,4]\*, Hosein Bakhshi[5] and Mohsen Yaghobi[2]**

[1]Department of Civil Engineering, Science and Research Branch, Islamic Azad University, Iran
[2]Department of Chemical Engineering, Shahid Bahonar University of Kerman, Iran
[3]Department of Aerospace Engineering, Payame Noor University, Iran
[4]Department of Computer Engineering, Payame Noor University, Iran
[5]Department of Civil Engineering, Hakim Sabzevari University, Iran

**Abstract**

In this study, an extreme value distribution of the gust wind speeds is obtained in a large selected area of Iran. The generalized Pareto distribution is used to find out the type of wind speed distribution. The three parameters of the generalized extreme value distribution function are reduced to either type I Gumbel, type II Frechet or type III reverse Weibull distribution function for the annual extreme gust wind speeds. It is obtained that, the annual extreme gust wind speeds at 102 stations have a reverse Weibull function distribution. It is also obtained that, type I Gumbel extreme value function is the best model for many of the studied stations.

**Keywords:** Annual extreme gust wind; Extreme value distribution; Gumbel; Frechet; Reverse Weibull

## Introduction

Today, wind power is an important dynamically energy source. Using the wind as a source of energy has been increased [1]. A set of wind turbines in a same location products the electrical power. To maximize this generated power, these locations are built far from shores or in open fields far away from buildings and trees. The suitability of these locations must be obtained using a set of data that approves the wind speed and direction in these locations.

It is observed that, extreme wind speeds are physically bounded. Weibull and reverse Weibull distributions are used for annual the extreme winds [2-5]. It is indicated that, annual fastest-mile wind speeds have the reverse Weibull distribution. A two-parameter generalized Pareto distribution is usable for analyzing the extreme gust wind speeds instead of all Type I distributions. The solution range of the tail-length parameter $c$ of the generalized Pareto distribution may indicates the extreme events for $c$ value approaching zero the Type I Gumbel, for $c>0$ the Type II Frechet and for $c<0$ the Type III reverse Weibull are suitable extreme value functions. More details have been presented in a previous study in USA [6].

Wind load on a structure is a function of various parameters such as wind speed and direction, geometry of the structure and local topography [7,8]. First, design and wind pressures on the external surfaces of the structure are calculated. Then these pressures are converted to the load impact. The amount of wind pressures on the structures in Iran is calculated using the following equations [9]:

$$q = 0.005V^2 \tag{1}$$

$$p = C_e C_q q \tag{2}$$

Where $p$ and $q$ are the design and wind pressures ($dN/m^2$), respectively. $V$ is the basic wind speed (km/h) and $C_q$ and $C_e$ are the shape and speed variation coefficients, respectively. Analyzing the wind speed data indicates that, the maximum annual wind speed at particular locations follows Type I extreme value distribution. This type is the most frequently used model for wind speed [2,10]. The probability distribution of the wind load (as a function of $V$) is a valuable parameter for the structural design of buildings. Nevertheless, this parameter is not necessarily for Type I distribution, because the wind pressure is a function of $V^2$ (instead of $V$). Since some of parameters are random in nature, it is difficult to determine the wind load distribution. Previous studies showed the uncertainty of the wind load and low values of $c$ in the cumulative distribution function (CDF), consequently. Therefore, wind load can be represented by a Type I distribution [2,4,5,11] for close to zero values of $c$.

## Procedure

Considering $W(t)$ as a series of wind speed at a specific site and $V$ as the extreme gusts associated with $W(t)$, the generation of $V$ distribution is not known. However, $V$ can be estimated using its tail quantile probability of a suggestion contingent cumulative distribution function (CDF), $F(v)$, (for $V \leq v$, $d>0$, $(1+c\frac{v}{d})>0$ and $V=W(t) - u$ where $u$ is a sufficiently large threshold of $W(t)$):

$$F(v) = 1 - (1 + c\frac{v}{d})^{-\frac{1}{c}} \tag{3}$$

Where $v$ is a contingent investigation of $V$ and $c$ and $d$ are the shape and scale parameters, respectively. Equation 3 is the generalized pareto distribution (GPD) and will considered an extreme gust of wind as the basis for the review. Element $(1+c\frac{v}{d})^{-\frac{1}{c}}$ on the right side of this Equation is known as the generalized Pareto. Many researchers used the GPD to estimate annual wind speed extreme [4,5,11,12].

## Generalized Distribution of Values

### Wide distribution type I (Gumbel)

There is strong evidence in scientific literature, which advocates

---

**\*Corresponding author:** Bahador Abolpour, Department of Chemical Engineering, Faculty of Engineering, ShahidBahonar University of Kerman, Post Code 76175, Kerman, Iran, E-mail: bahadorabolpor1364@yahoo.com

the use of the Gumbel distributions to fit extremes events [13-18]. Wide distribution of values as its name implies is useful to describe the probabilistic nature [19]. CDF of this variable for $-\infty \leq x \leq \infty$ is:

$$F_X(x) = \exp\left\{-\left[\exp(-\alpha(x-u))\right]\right\} \tag{4}$$

Where $u$ and $\alpha$ are parameters of the distribution. The mean and standard deviation can be calculated using the following limits:

$$\mu_x \approx u + \frac{0.577}{\alpha} \tag{5}$$

$$\sigma_x \approx \frac{1.282}{\alpha} \tag{6}$$

Therefore, if the mean and standard deviation are determined, equations 5 and 6 can be changed and the corresponding values for the distribution parameters can be obtained from these equations.

## Wide distribution type II

Sometimes the best estimate of the distribution of the maximum load on a structure can be provided by Type II [20,21]. CDF of this variable for $0 < x \leq \infty$ is:

$$F_X(x) = k \exp(-\frac{u}{x}) \tag{7}$$

Where $u$ and $k$ are parameters of the distribution and $c = \frac{1}{k}$ for c>0. The mean and standard deviation can be calculated using the following limits:

$$\mu_X = u\Gamma(1-\frac{1}{k}) \text{ for k>1} \tag{8}$$

$$\sigma_X^2 = u^2\left[(1-\frac{2}{k}) - \Gamma^2(1-\frac{1}{k})\right] \text{ for k>2} \tag{9}$$

Note that, the coefficient of variation, $V_x$ is a function of $k$ and is calculated elsewhere [20].

## Wide distribution type III (Weibull)

This distribution is defined by three parameters. There are different functions for the largest and smallest values [22]. For $x \leq w$ CDF can be defined as follow:

$$F_X(x) = \exp\left[-(\frac{w-x}{w-u})k\right] \tag{10}$$

Where $w$, $u$ and $k$ are parameters of the distribution and $c = -\frac{1}{k}$ for c<0. Mean and variance also are calculated using the following equations:

$$\mu_X = w - (w-u)\Gamma(1+\frac{1}{k}) \tag{11}$$

$$\sigma_X^2 = (w-u)^2\left[\Gamma(1+\frac{2}{k}) - \Gamma^2(1+\frac{1}{k})\right] \tag{12}$$

For $x \geq \varepsilon$ CDF is defined as follow:

$$F_X(x) = 1 - \exp\left[-(\frac{x-\varepsilon}{u-\varepsilon})k\right] \tag{13}$$

Where $\varepsilon$, $u$ and $k$ are parameters of the distribution. Mean and variance for this range of values is defined as follows:

$$\mu_X = \varepsilon + (u+\varepsilon)\Gamma(1+\frac{1}{k}) \tag{14}$$

$$\sigma_X^2 = (u-\varepsilon)^2\left[\Gamma(1+\frac{2}{k}) - \Gamma^2(1+\frac{1}{k})\right] \tag{15}$$

Gamma distribution is defined as follow:

$$\Gamma(k) = \int_0^\infty e^{-u}u^{k-1}du = (k-1)! \tag{16}$$

If the GPD assumption were correct, the plot of the cumulative mean exceedance (CME) should follow a straight line. Therefore, $c$ and $d$ can be obtained from the characteristics of this line.

## Data Preparation

Wind speed data was obtained from the center of Iran weather databases. This database included 287 weather stations. 181 of these stations had data ranged from 5 to 54 years. It was felt that hurricanes and tornadoes were worthy of separate consideration. Thus, the stations had been affected by hurricanes and tornadoes were deleted. A minimum record length of 15 years was chosen to allow an adequate amount of wind data to be analyzed in this study. Because of climatic characteristics, a smaller sample may not adequately represent all possible wind patterns. Therefore, stations with less than 15 years of record were removed from the database. Finally, 109 meteorological stations were selected for analysis.

## Evaluate the Information and Forms

Figures 1 and 2 show the mean, standard deviation and maximum values of the annual extreme gust wind speed at all 109 stations in Iran. The parameters $c$ and $d$ were quantified using the CME method. The annual series of median wind speed, $V_{med}$, provided satisfactory results [5]. It is assumed that the extreme events occur in succession.

As a result, pieces of nonlinear CME can be removed. Thus the parameter $c$ will obtain with a more accurate. These findings, is identical with the results by other researchers [23,24]. The extreme gust wind $v$ is plotted against the reduced variate $W$, by both Gumbel and reverse Weibull distributions in Figures 3-5.

$$W_i = \frac{x_i - \mu_{x_i}}{\sigma_{x_i}} \tag{17}$$

As shown in Figures 6 and 7, Type III reverse Weibull distribution, are most suitable for delineating the annual extreme gust wind speeds at the 109 stations; So the inverse Weibull distribution for 109 stations, is the basic representative. The result is in agreement with the conclusion suggested by other studies for severe winds. However, analysis of data of $V_{RN}$ (wind speed estimates using the reverse Weibull distribution) and $V_{GN}$ (wind speed estimates using the Gumbel distribution) present in these figures show the opposite results. It is observed that $V_{RN}$ and $V_{GN}$, for gust winds with great intervals in many of stations, Type I Gumbel distribution has a higher accuracy than Type III distribution. The results of about 102 stations for both Types I and III distributions are approved. Some examples in Figures 3-5 are presented.

## Conclusion

Preliminary annual extreme gust wind speed distribution in the selected 109 stations in Iran was investigated. Based on calculations using CME, it is obtained that the annual extreme gust wind speeds at 102 stations have a reverse Weibull function distribution. However, the results of data analysis and graphic curves showed that Type I Gumbel extreme value function is the best model for many of the studied stations. Wind speed predictions based on Gumble distribution may be

Figure 1: Annual gust wind mean speeds and standard deviation of these values at 109 meteorological stations of Iran.

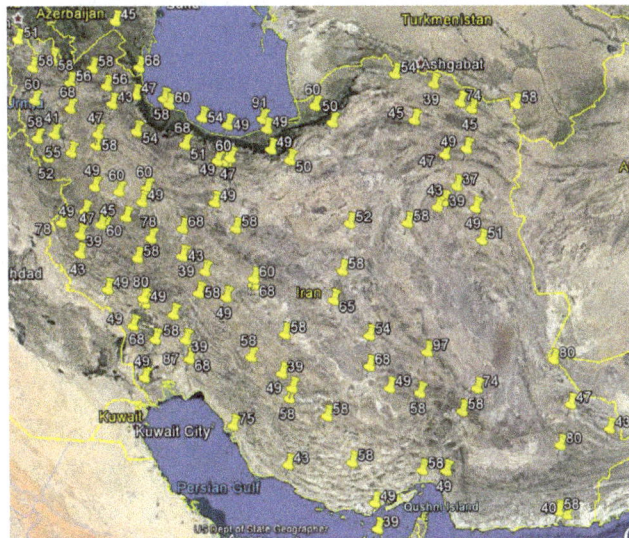

Figure 2: Maximum values of annual gust wind speeds at 109 meteorological stations of Iran.

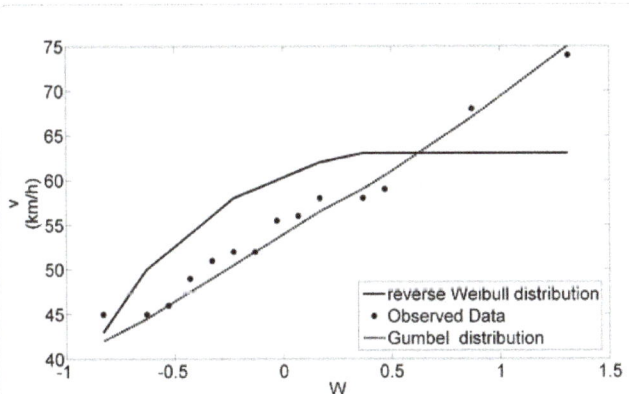

Figure 3: Gumbel and reverse Weibull distribution applied for annual extreme gust wind speeds in Kerman, Rafsanjan station (1973-1987).

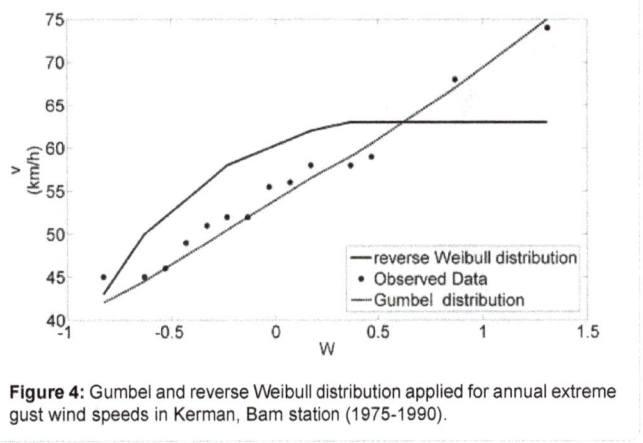

Figure 4: Gumbel and reverse Weibull distribution applied for annual extreme gust wind speeds in Kerman, Bam station (1975-1990).

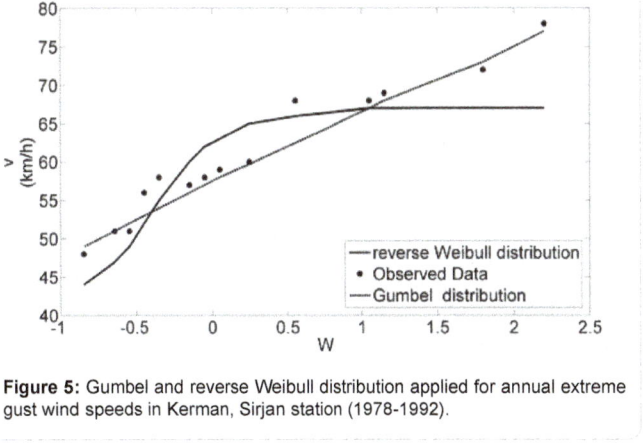

Figure 5: Gumbel and reverse Weibull distribution applied for annual extreme gust wind speeds in Kerman, Sirjan station (1978-1992).

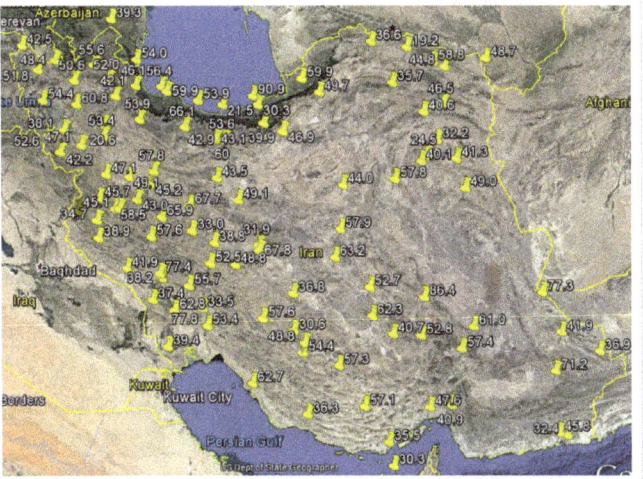

Figure 6: Estimated gust wind speed estimates based on reverse Weibull distribution at 109 meteorological stations of Iran.

unsuitable for the most of time intervals. However, taking into account the time intervals, for a maximum duration of 54 years in 109 selected stations, distributed type I in the modeling extreme gust wind speed, at 102 stations have represented better results.

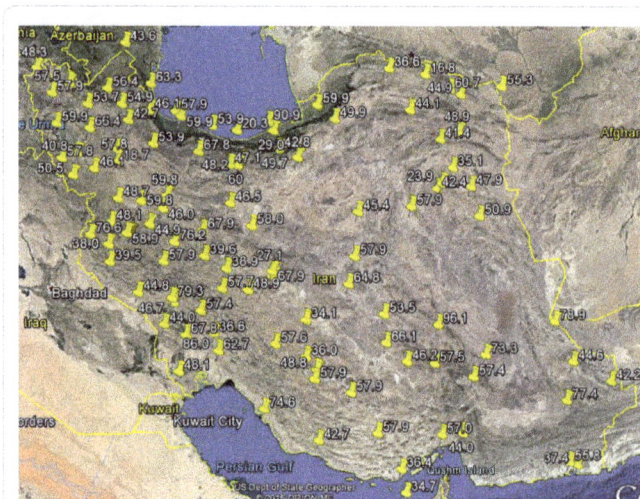

**Figure 7:** Estimated gust wind speed estimates based on Gumbel distribution at 109 meteorological stations of Iran.

## References

1.  World Wind Energy Association (2009) World wind energy report 2008.

2.  Cheng EDH, Chiu ANL (1994) Short-record-based extreme wind simulation. J Nat Inst of Stand Technol 99: 391-397.

3.  Cheng EDH, Chiu ANL (1995) Regional design wind speed estimation. Proceedings of the 9th International Conference on Wind Engineering.

4.  Gross J, Heckert A, Lechner J, Simiu E (1994) Extreme value theory and applications. (Simiu Edition), Kluwer Academic Publishers, Dordrecht, Netherlands.

5.  Simiu E, Heckert N (1996) Extreme wind distribution tails: A peak over threshold approach. J Struct Eng 122: 539-547.

6.  Cheng E, Yeung C (2002) Generalized extreme gust wind speeds distributions. J Wind Engineer Indust Aerodyn 90: 1657-1669.

7.  American Society of Civil Engineers (2010) Minimum design loads for buildings and other structures. ASCE. pp: 424.

8.  Huang S, Li R, Li QS (2013) Numerical simulation on fluid-structure interaction of wind around super-tall building at high Reynolds number conditions. Struct Engineer Mech Int J 46: 2.

9.  Ministry of Housing and Urban Development (2009) National building regulations, Iran.

10. Pinto JG, Neuhaus CP, Kruger A, Kerschgens M (2009) Assessment of the wind gust estimate method in mesoscale modeling of storm events over West Germany. Meteorologische Zeitschrift 18: 495-506.

11. Gross JL, Heckert NA, Lechner JA, Simiu E (1995) A study of optimal extreme wind estimation procedures. Proceedings of 9th International Conference on Wind Engineering, New Delhi, India. pp: 69-80.

12. Heckert N, Simiu E, Whalen T (1998) Estimates of hurricane wind speeds by "peak over threshold" method. J Struct Engineer 124: 445-449.

13. Garciano LE, Koike T (2007) A proposed typhoon resistant design of a wind turbine tower in the philippines. Doboku Gakkai Ronbunshuu F 63: 181-189.

14. Jagger TH, Elsner JB (2006) Climatology models for extreme hurricane winds near the United States. J Climate 19: 3220-3236.

15. Sanabria LA, Cechet RP (2007) A statistical model of severe winds. Geosci Austr Record.

16. Sorensen JD, Nielsen SRK (2007) Extreme wind turbine response during peration. J Phys Conf Ser 75: 012074.

17. Langreder W, Hojstrup J (2007) Going to extremes: A parametric study on peak-over-threshold and other methods. European Wind Energy Conference, Milan, Italy.

18. Kunz M, Mohr S, Rauthe M, Lux R, Kottmeier Ch (2010) Assessment of extreme wind speeds from Regional Climate Models-Part 1: Estimation of return values and their evaluation. Nat Hazard Earth Sys Sci 10: 907-922.

19. Hosking J, Wallis J (1987) Parameter and quantile estimation for the generalized pareto distribution. Technometrics 29: 339-349.

20. Smith R (1990) Handbook of applicable mathematics. Wiley, New York.

21. Van Den Brink HW, Konnen GP (2008) The statistical distribution of meteorological outliers. Geophys Res Lett 35: 1-5.

22. Johnson NL, Kotz S, Balakrshnan N (1995) Continuous univariate distributions (2nd Edn). Wiley, New York, USA.

23. Holmes J, Moriarty W (1999) Application of the generalized pareto distribution to extreme value analysis in wind engineering. J Wind Eng Indust Aerodyn 83: 1-10.

24. Holmes J (2000) Private communication.

# Assessment of Wind Energy Potential for Small Communities in South-South Nigeria: Case Study of Koluama, Bayelsa State

**Akintomide Afolayan Akinsanola[1,2]\*, Kehinde Olufunso Ogunjobi[2], Akintayo T Abolude[1], Stefano C Sarris[1] and Kehinde O Ladipo[2]**

[1]*School of Energy and Environment, City University of Hong Kong, Hong SAR, China*

[2]*Department of Meteorology and Climate Science, Federal University of Technology Akure, Nigeria*

## Abstract

Although the concept of wind energy potential assessments has matured considerably, there is only limited application and adoption in regions of energy crisis where electricity demand far exceeds supply. For Nigeria, seeking alternate sources of energy to meet its energy demand is essential and must be met in a sustainable practice. This study analyzed the electricity generation potential from wind at Koluama, Bayelsa State, Nigeria using a combination of 10-m monthly mean wind speed and direction data (1984-2013) and five year daily wind speed data (2009-2013). The data were subjected to different statistical tests and also compared with the two-parameter Weibull probability density function. Maximum mean day of year (DOY) wind speed recorded was 5.25 m/s and minimum wind speed was 0.92 m/s, while seasonal mean wind speed during the dry months (DJF) is estimated to be 4.05 m/s and 4.32 m/s during the wet months of June, July August and September (JJAS) for the 30-year period considered. Wind power density (WPD) ranged from 82 W/m² to 145 W/m² in November and August respectively.

Lastly, small scale wind-to-electricity power generation was assessed using six (6) practical wind turbines. The AV 928 turbine had the maximum energy yield, despite relatively low capacity factor of less than 10%.

**Keywords:** Wind variability; Wind power potential; Capacity factor; Turbine output

## Introduction

Despite abundant fossil fuel resources available both globally and in Nigeria, the electricity demand of the Nigerian population is still much larger than the present supply [1]. This coupled with the threat of foreseen fossil fuel depletion, changing climate, air pollution and oil price volatility [2-4], implies that alternative approaches to electricity generations are been constantly considered. These alternatives approach are majorly renewables because they offer a sustainable solution and are quite immune to energy security concerns [4,5]. Energy security concerns can be broadly viewed under the indices of availability, affordability and resilience [5]. The energy mix has witnessed an increasing share of renewables towards attaining sustainability, better air quality and mitigating climate change [6-10]. Installed capacity of wind energy has increased with previous studies highlighting the wind energy growth regime [7-14]. Furthermore, [15] stated that "wind energy assessment is a topic of both scientific interest and an issue of relevance with ecological, economic and political implications". While wind energy has been continually exploited by nations with large wind resources [15-20], there exists limited evidence of such studies in Nigeria where electricity supply is still below demand [1]. Residential power supply in small communities and rural environments have been severely impacted by this crisis thereby posing a challenge to comfort, health and productivity of residents, culminating in high poverty levels and low Gross Domestic Product (GDP).

This study seeks to investigate the wind energy resources in Koluama, Bayelsa State, Nigeria (4.47°N; 5.77°E; altitude 6.1m; air density 1.225 kg/m³) using 30 years historical data to establish the trend and variability of wind speed and direction, estimate wind power density and potential via a range of turbines and make recommendations based on results. The results from this study will provide an analysis of the possible impact of local meteorology on the adoption and use of wind energy for powering small communities both in Nigeria and regions where a transition to renewables is required or electricity supply is short of demand.

## Study Area

The study area as shown in Figure 1 is Koluama, situated in Bayelsa State within the eastern region of Niger Delta, Nigeria. Geographically, it is located between latitude 04°47`N and longitude 5°77`E. The location features a tropical monsoon climate with a lengthy and heavy rainy seasons and a very short dry season from December to January. The Harmattan, which climatically influences many cities in West Africa is less pronounced over the region. Koluama's heaviest rainfall occurs during September with an average of 367 mm of rain and the lowest rainfall occurs during December with an average rainfall of 20 mm. Temperature throughout the year varies from 25 to 28°C. The vegetation is mainly mangrove and salt water swamps, but a major part had largely been destroyed by oil exploration [21,22].

## Data and Methodology

### Data

Thirty years monthly mean wind speed and direction data at a height 10 m above sea level from (1984-2013) and five years (2009-2013) daily wind data were assessed from the archive of Nigeria Meteorological Agency (NIMET) for the study area. The data were recorded continuously using cup-generator anemometer. These dataset

**\*Corresponding author:** Akintomide Afolayan Akinsanola, Department of Meteorology and Climate Science, Federal University of Technology Akure, Ondo 340001, Nigeria

E-mail: mictomi@yahoo.com; aakinsano2-c@my.cityu.edu.hk

**Figure 1:** Location of Koluama, Bayelsa State, Nigeria (Source: https://google.com/maps).

were then analyzed to determine the monthly, seasonal and yearly wind resource potentials for power generation.

## Methods

In previous studies, various statistical distributions exist for describing and analyzing wind resource data. Some of these include normal and lognormal, Rayleigh and Weibull probability distributions to mention a few [23-25]. However, of these statistical methods, the Weibull distribution has been found to be accurate and adequate in analyzing and interpreting the situation of measured wind speed and in predicting the characteristics of prevailing wind profile over a place [26-28]. Thus, in this study, the Weibull two parameter probability density function (PDF) was employed in carrying out the analyses of over the study area and is mathematically expressed as:

$$f(v) = \left(\frac{k}{c}\right)\left(\frac{v}{c}\right)^{k-1} \exp\left[-\left(\frac{v}{c}\right)^k\right] \qquad (1)$$

Where k is the Weibull shape parameter, c is the scale parameter and $f(v)$ is the probability of observing wind speed v (m/s).

The Weibull Cumulative Density Function (CDF) corresponding to the PDF is given as:

$$F(v) = 1 - \exp\left[-\left(\frac{v}{c}\right)^k\right] \qquad (2)$$

Where F (v) is the cumulative distribution function of observing wind speed v.

The mean value of the wind speed $V_{weibull}$ and standard deviations σ for the Weibull distribution as defined in terms of the Weibull parameter k and c are given as:

$$V_{weibull} = c\Gamma\left(1 + \frac{1}{k}\right) \qquad (3)$$

and

$$\sigma = \sqrt{c^2\left\{\Gamma\left(1 + \frac{2}{k}\right) - \left[\Gamma\left(1 + \frac{1}{k}\right)\right]^2\right\}} \qquad (4)$$

Where $\Gamma(x)$ is the gamma function of (x)

**Evaluation of wind power density (WPD):** The WPD evaluation can be carried out in two forms. One based on available power in the wind as captured by the wind conversion system and estimated directly from the wind speed v (m/s) and the other based on the Weibull two-parameter method. These two approaches are given as

$$P(v) = \frac{1}{2}\rho A v^3 \qquad (5)$$

$$p(v) = \frac{p(v)}{A} = \frac{1}{2}\rho c^3 \Gamma\left(1 + \frac{3}{k}\right) \qquad (6)$$

Where, P (v) is the wind power (W), p (v) is the wind power density (W/m²), ρ is the air density (kg/m³) at the site and A is the swept area of the rotor blade (m²).

However, to simulate the electrical power output of a model wind turbine requires using:

$$P_e = \begin{cases} 0 & V < V_C \\ P_{eR}\dfrac{V^K - V_C^K}{V_R^K - V_C^K} & V_C \leq V < V_R \\ P_{eR} & V_R \leq v \leq V_F \\ 0 & v > V_F \end{cases} \qquad (7)$$

Where $P_{eR}$ is the rated electrical power, $V_c$ is the cut-in wind speed, $V_R$ is the rated wind speed and $V_F$ is the cut-out speed respectively of the model wind turbine.

Also, the average power output (Pe, ave) from a turbine corresponding to the total energy production and related to the total income/cost analysis was evaluated from:

$$P_{e, ave} = \left\{\frac{e^{\left(\frac{V_c}{c}\right)k} - e^{-\left(\frac{V_R}{c}\right)^k}}{\left(\frac{V_R}{c}\right)^k - \left(\frac{V_c}{c}\right)^k} - e^{\left(\frac{V_F}{c}\right)k}\right\} \qquad (8)$$

The capacity factor (CF) which is the ratio of average power output to turbine rated/maximum output [2] is evaluated from

$$CF = \frac{P_{e, ave}}{P_{eR}} \qquad (9)$$

Since the standard height for most turbines is 80m, the wind speed over Koluama was estimated to a height of 80 m using

$$V_{ref} = V_{10}\left(\frac{h_{80}}{h_{10}}\right)^{á} \qquad (10)$$

Where $V_{ref} = V_{80}$ = wind speed at 80 m, $V_{10}$ = wind speed at 10-m height, $h_{80}$ = 80 m height, $h_{10}$ = 10 m height and α = wind shear coefficient for the sites = 0.147 [26]. The mathematical expressions in equations 1-10 can also be found in previous available data [14-20,27,28].

## Results and Discussion

### Mean climatology

The wind climatology over the study area for the years 2009 to

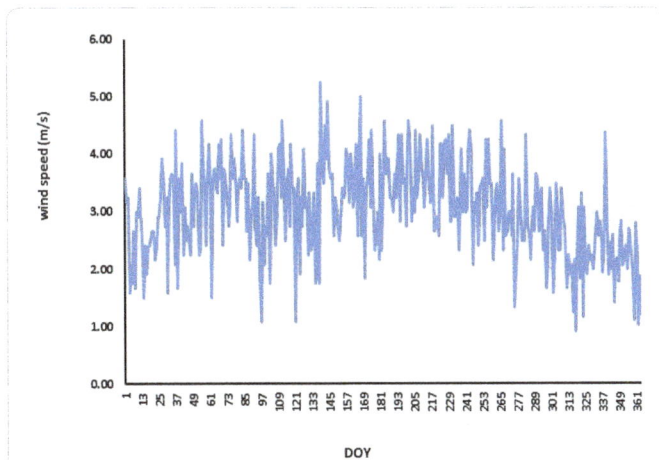

**Figure 2:** Day of Year average wind speed (m/s) at Koluama for 2009-2013.

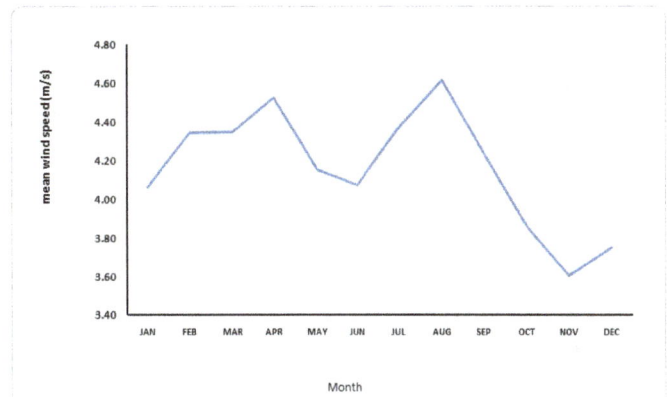

**Figure 3:** Monthly mean climatology of wind speeds from 1984-2013.

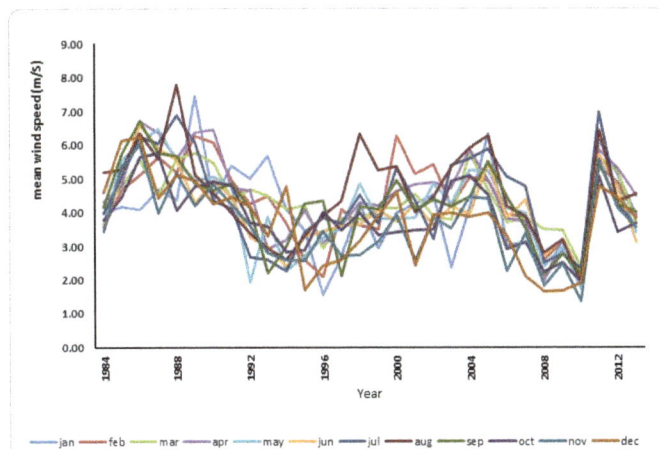

**Figure 4:** Climatological plot of yearly mean wind speed for the period 1984-2013.

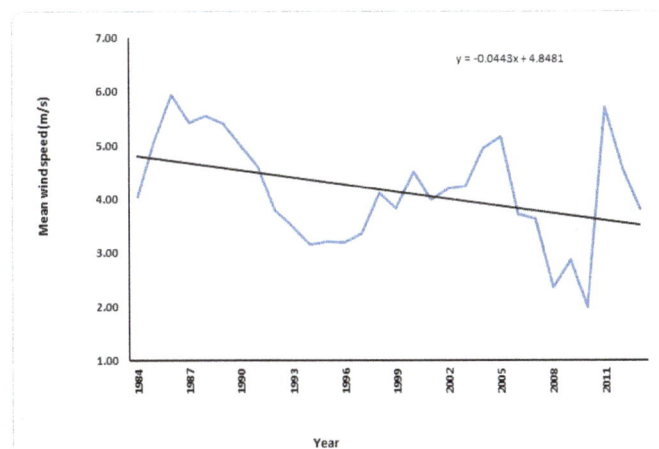

**Figure 5:** Monthly variation of mean wind speed for the 30 years period under consideration.

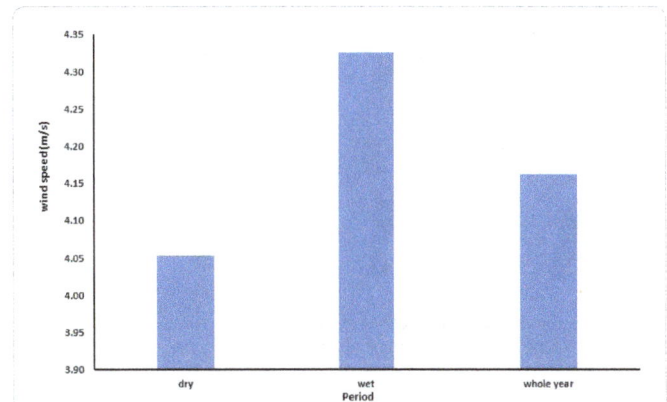

**Figure 6:** Plot of dry months (DJF), wet months (JJAS), and the whole-year (January-December) of mean wind speed for Koluama.

2013 was investigated and shown in Figure 2 using average day of the year (DOY) wind speed (m/s). For the period considered, maximum mean DOY wind speed was approximately 5.25 m/s recorded on Julian day 139 while the minimum was 0.92 m/s recorded on Julian day 319. On the overall, mean wind speed distribution at the study site can be broadly described as 3.03 ± 0.82 m/s with a range of 4.33 m/s.

Considering the thirty years monthly wind speed climatology, a decline in monthly mean wind was observed over the study area as shown in Figure 3. Also, a decline in yearly mean wind speed measurements from 1984-2013 was observed in Figure 4. The peak value of 5.95 m/s was observed in 1986, a low of 1.98 m/s in 2010 and a decreasing trend in wind speed at the rate of -0.044 m/s per year is evident over the study area. Furthermore, the monthly variation of mean wind speed over the study area is presented in Figure 5. The highest observed value of mean wind speed is 4.62 m/s in August and the lowest of 3.60 m/s in November. Additionally, the mean wind speed during the dry months (December, January and February) and wet season (June, July, August and September) from 1984 to 2013 is presented in Figure 6. The average wind speed varies from 4.05 m/s in the dry months to 4.32 m/s in the wet months. This stronger wind regime during the wet season is expected have impact on rainfall and other weather related events considering Koluama's proximity to the Atlantic Ocean.

The prevailing wind direction at Koluama for January-December is shown in Figure 7. The dominant wind direction over the study area is southerly and south westerly winds throughout the year. It is important to note that there were few cases of strong northerly wind during the dry months of December and January. These observed wind directions over the study area is partly modulated by sea breezes from the Atlantic Ocean, while the seasonal wind reversal (monsoon) is responsible for the dominant south westerly wind in the wet season. However, during

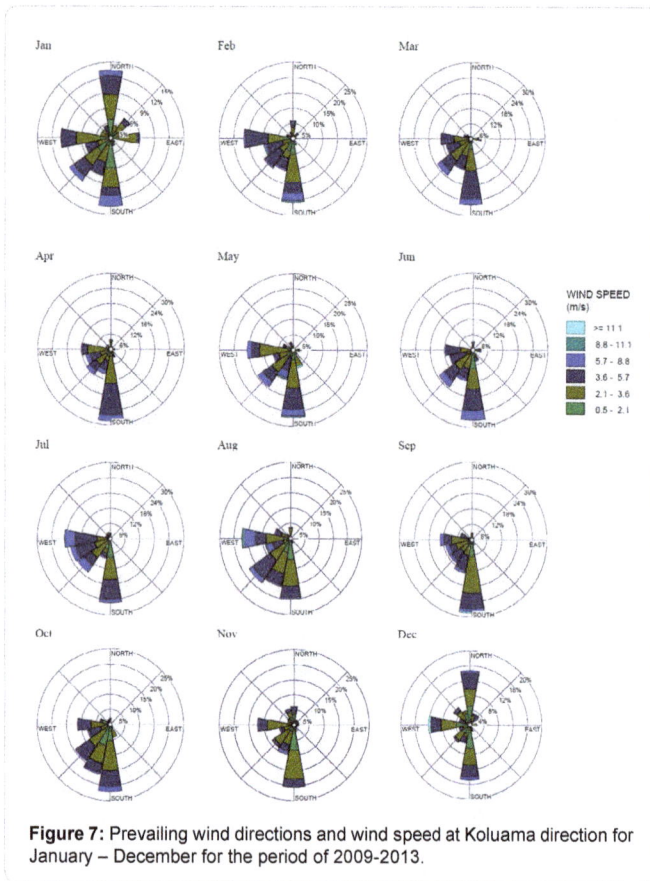

**Figure 7:** Prevailing wind directions and wind speed at Koluama direction for January – December for the period of 2009-2013.

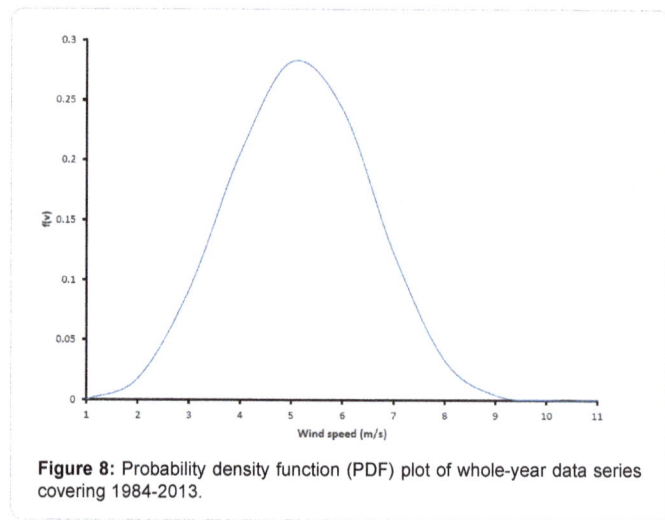

**Figure 8:** Probability density function (PDF) plot of whole-year data series covering 1984-2013.

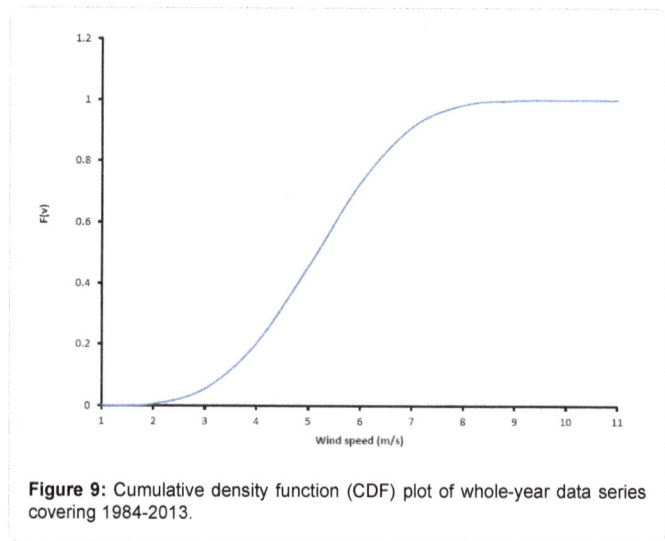

**Figure 9:** Cumulative density function (CDF) plot of whole-year data series covering 1984-2013.

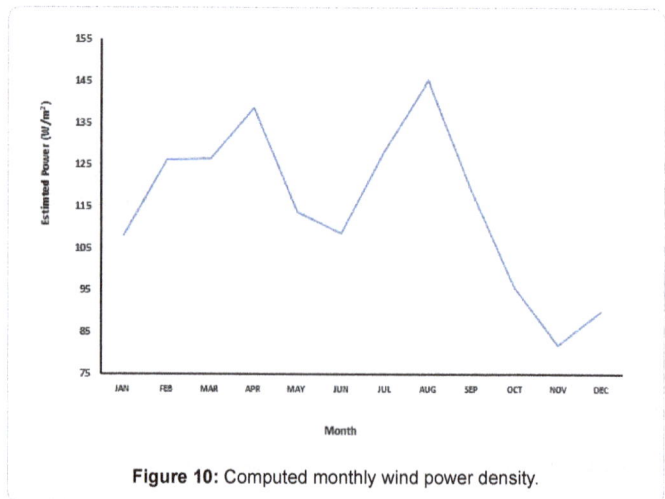

**Figure 10:** Computed monthly wind power density.

| Period | $V_m$ (m/s) | C | K | Weibull (m/s) | σ actual (m/s) | σ Weibull (m/s) |
|--------|-------------|-----|-----|---------------|-----------------|------------------|
| Jan | 4.065 | 4.529 | 3.328 | 4.085 | 1.343 | 1.346 |
| Feb | 4.342 | 4.82 | 3.576 | 4.442 | 1.168 | 1.348 |
| Mar | 4.348 | 4.827 | 3.581 | 4.248 | 0.849 | 1.348 |
| Apr | 4.523 | 5.009 | 3.738 | 4.523 | 1.232 | 1.349 |
| May | 4.153 | 4.623 | 3.407 | 4.253 | 1.232 | 1.347 |
| Jun | 4.073 | 4.539 | 3.336 | 4.173 | 1.069 | 1.346 |
| Jul | 4.376 | 4.856 | 3.606 | 4.296 | 1.365 | 1.348 |
| Aug | 4.615 | 5.105 | 3.821 | 4.715 | 1.339 | 1.35 |
| Sep | 4.235 | 4.708 | 3.48 | 4.235 | 1.156 | 1.347 |
| Oct | 3.856 | 4.309 | 3.143 | 3.896 | 0.971 | 1.344 |
| Nov | 3.605 | 4.041 | 2.921 | 3.705 | 1.118 | 1.342 |
| Dec | 3.752 | 4.199 | 3.051 | 3.852 | 1.25 | 1.343 |
| Dry | 4.053 | 4.517 | 3.318 | 4.123 | 1.254 | 1.346 |
| Wet | 4.325 | 4.802 | 3.56 | 4.385 | 1.232 | 1.348 |
| Whole year | 4.162 | 4.632 | 3.415 | 4.192 | 1.174 | 1.347 |

**Table 1:** Weibull results and estimation parameters using 30 years wind speed data.

the dry months of December and January, the north easterly trade wind flows towards the ocean as a result of the monsoon retreat, which explains the observed strong northerly wind in the dry months.

## Wind energy potential assessment

Using wind speed data from 1984 to 2013, Table 1 depicts results of the Weibull statistical analysis and the standard deviation for the predicted Weibull parameters k and c, used for the for corresponding monthly PDF and CDF plots presented in Figures 8 and 9. The probability distribution function of wind speed is essential in evaluating the availability of wind power at a site. It also permits the selection of appropriate wind machines for exploiting the wind energy. The 30 year average k and c Weibull parameters obtained were used in equation 1 to obtain probability distribution for the different wind speeds from 1 m/s to 11 m/s (incremental of 1 m/s) which is then plotted in Figure 8 and the cumulative probability plotted in Figure 9. These figures reveal that up to 70% of the data series ranged from about 2.0 to 5.8 m/s. The most probable wind speed at the site is about 4.2 m/s while wind speed greater than 8.0 m/s shows very low probability.

Figure 10 depicts the estimated monthly wind power, generated at Koluama. Two peaks occurred in April (139.79 W/m²) and August (145.48 W/m²) and a low in November (81.94 W/m²). Also, two decreasing wind power trends were observed: i) April through June and ii) August through November. Similarly, two increasing wind power trends: i) June through August and ii) November through April. Furthermore, because the majority of the available turbine hub heights are at 80 m, the wind profile characteristics were estimated for this height using equation 10. The speeds at this new height were then employed for Weibull re-analyses. Thus at heights above 10 m, the economic viability of wind energy at the site is best investigated using equation 10. The wind speed profile between heights 10 to 100 m is presented in Figure 11.

## Adapting real wind turbine to Koluama

Previous studies have evidently shown that wind turbine installation is capital intensive [27-29]. Therefore it is important to assess turbine

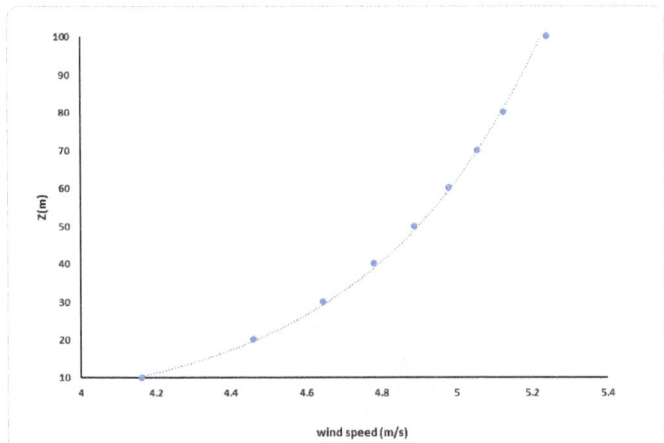

**Figure 12:** Capacity factors of generation using the six wind turbines.

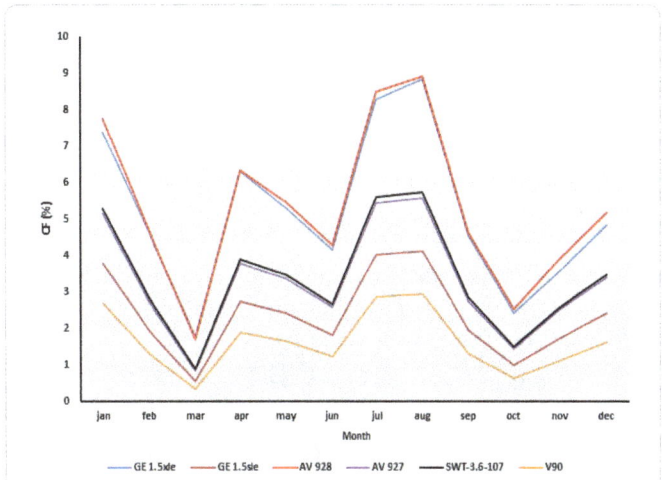

**Figure 13:** Power output (kW) of six wind turbines using mean wind speed at 80 m.

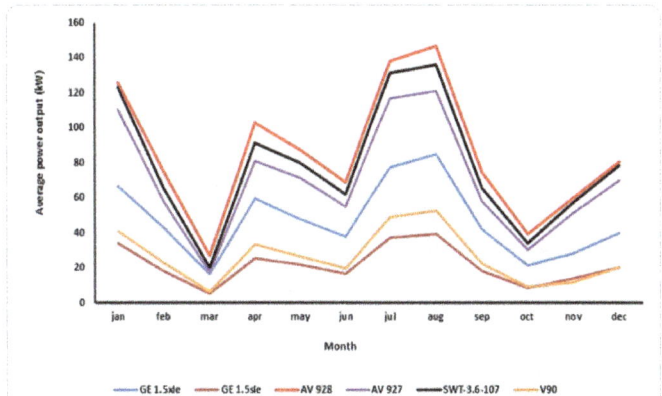

**Figure 11:** The vertical wind shear profile from 10 m to 100 m height at the study site.

parameter relative to the site's wind profile. This entails estimating the amount of electrical power that a particular wind turbine will likely generate and thus, the estimated Capacity Factor (CF) is a pointer to the turbine's generation capacity. Based on the above, we assessed six different turbines with technical parameters given in Table 2. Equations 7-9 were then used to evaluate the power output from each of the turbines. Figure 12 depicts the estimated capacity factor of generation for six turbines: GE 1.5xle, GE 1.5sle, AV 928, AV 927, SWT-3.6-107 and V 90. Capacity factors (in %) of generation peaked in August with approximate values of 9, 4, 9, 6, 6 and 3 respectively while been lowest in March. Turbines GE 1.5xle and AV 928 had the highest capacity factors of generation at the study site. The monthly average power output (kW) for the six turbines: GE 1.5xle, GE 1.5sle, AV 928, AV 927, SWT-3.6-107 and V 90 were 47, 22, 86, 70, 71 and 26 respectively as seen in Figure 13. This result indicates that the AV 928 had the highest energy production (kW) of the turbines considered.

## Conclusion

This study assessed the wind energy potential for power generation over Koluama, Bayelsa State, Nigeria. Thirty years of mean wind data at 10 m height were assessed and analyzed. Also, the data was subjected to Weibull two-parameter and other wide range of statistical analyses. The results revealed the following;

i.    Monthly mean wind speeds peaked in August at 4.62 m/s and

| Wind turbine | $V_c$ (m/s) | VF (m/s) | VR (m/s) | PeR (kW) | Hub Height (m) | Rotor diameter (m) |
|---|---|---|---|---|---|---|
| GE 1.5sle | 3.5 | 25 | 14 | 1500 | 65/80 | 77 |
| GE 1.5xle | 3.5 | 20 | 11.5 | 1500 | 80 | 82.5 |
| AV 928 | 3 | 25 | 11.6 | 2500 | 80 | 93.2 |
| AV 927 | 3 | 25 | 13.1 | 3300 | 60/80 | 93.2 |
| V 90 | 4 | 25 | 15 | 3000 | 80 | 90 |
| SWT-3.6.107 | 3 | 25 | 13 | 3600 | 80 | 107 |

**Table 2:** Technical details of the wind turbine model.

were at a low in November at 3.60 m/s. A secondary peak was further observed in April at 4.52 m/s. Some seasonal variability was observed between the dry (DJF) and wet (JJAS) months mean wind speeds, with lower mean wind speed of 4.05 m/s in DJF and a higher value of 4.32 m/s during the wet JJAS period;

ii.      Maximum (minimum) recorded annual mean wind speeds was 5.95 m/s (1.98 m/s) in the year 1986 and 2010 respectively. The wind speed exhibited a decreasing trend of -0.044 m/s per year;

iii.      Annual prevailing wind direction over the study area was majorly southerly, south-westerly and westerly. This may be attributed to the proximity of the Atlantic ocean;

iv.      Maximum estimated wind power density (145.48 W/m$^2$) was observed in August and at a low in November with 81.95 W/m$^2$;

v.      Turbine monthly power output peaked in August and was least in March and October. The AV 928 turbine had the maximum energy output followed by the SWT-3.6-107. However, the capacity factor was relatively low (~10%) and this raises concern about the economic viability of large scale wind power project in the study site.

## Acknowledgement

We are grateful to the Nigeria Meteorological Agency (NIMET) for the provision of the ground observation data.

## References

1. Oyedepo SO (2012) On energy for sustainable development in Nigeria. Renew Sustain Energy Rev 16: 2583-2598.

2. Dunning CM, Turner AG, Brayshaw DJ (2015) The impact of monsoon intraseasonal variability on renewable power generation in India. Environ Res Lett 10: 064002.

3. Reddy BS, Assenza GB, Assenza D, Hasselmann F (2009) Energy efficiency and climate change: Conserving power for a sustainable future. SAGE Publications India Pvt Ltd, New Delhi.

4. Alagappan L, Orans R, Woo CK (2011) What drives renewable energy development? Energy Policy 39: 5099-5104.

5. Valentine SV (2011) Emerging symbiosis: Renewable energy and energy security. Renew Sustain Energy Rev 15: 4572-4578.

6. PortmanME, Duff JA, Koppel J, Reisert J, Higgins ME (2009) Offshore wind energy development in the exclusive economic zone: legal and policy supports and impediments in Germany and the US. Energy Policy 37: 3596-3607.

7. Liu Y, Kokko A (2010) Wind power in China: Policy and development challenges. Energy Policy 38: 5520-5529.

8. Meyer NI (2007) Learning from wind energy policy in the EU: Lessons from Denmark, Sweden and Spain. European Environment 17: 347-362.

9. Ackerman T, Soder L (2000) Wind energy technology and current States: A review. Renew Sustain Energy Rev 4: 315-374.

10. Kenisarin M, Karsli VM, Caglar M (2006) Wind power engineering in the world and perspectives of its development in Turkey. Renew Sustain Energy Rev 10: 341-369.

11. Faulin J, Lera F, Pintor JM, García J (2006) The outlook for renewable energy in Navarre: An economic profile. Energy Policy 34: 2201-2216.

12. Ely CR, Brayshaw DJ, Methven J, Cox J, Pearce O (2013) Implications of the North Atlantic Oscillation for a UK-Norway renewable power system. Energy Policy 62: 1420-1427.

13. Brayshaw DJ, Troccoli A, Fordham R, Methven J (2011) The impact of large scale atmospheric circulation patterns on wind power generation and its potential predictability: A case study over the UK. Renew Energy 36: 2087-2096.

14. Crawford RH (2009) Life cycle energy and greenhouse emissions analysis of wind turbines and the effect of size on energy yield. Renew Sustain Energy Rev 13: 2653-2660.

15. García-Bustamante E, Gonzálezn Rouco JF, Jiménez PA, Navarro J, Montávez JP (2009) A comparison of methodologies for monthly wind energy estimation. Wind Energy 12: 640-659.

16. Weisser D (2003) A wind energy analysis of Grenada: An estimation using the "Weibull" density function. Renew Energy 28: 1803-1812.

17. Shu Z, Li QS, Chan PW (2015) Investigation of offshore wind energy potential in Hong Kong based on Weibull distribution function. Appl Energy 156: 362-373.

18. Celik AN (2003) Weibull representative compressed wind speed data for energy and performance calculations of wind energy systems. Renew Energy 44: 3057-3072.

19. Bustamante EG, González-Rouco JF, Jiménez PA, Navarro J, Montávez JP (2008) The influence of the Weibull assumption in monthly wind energy estimation. Wind Energy 11: 483-502.

20. Celik AN (2003) Assessing the suitability of wind speed probability distribution functions on wind power density. Renew Energy 28: 1563-1574.

21. Olowoyo DN (2011) Physicochemical characteristics of rainwater quality of Warri axis of Delta state in Western Niger Delta region of Nigeria. J Environ Chem Ecotoxicol 3: 320-322.

22. Akinsanola AA, Ogunjobi KO (2014) Analysis of rainfall and temperature variability over Nigeria. Glob J Hum-Soc Sci B 14: 11-40.

23. Ajayi OO, Fagbenle RO, Katende J, Okeniyi JO (2011) Availability of wind energy resource potential for power generation of Jos, Nigeria. Frontiers Energy 5: 376-385.

24. Carta JA, Ramirez P, Velazquez S (2009) A review of wind speed probability distributions used in wind energy analysis: Case studies in the Canary Islands: case studies in the Canary Islands. Renew Sustain Energy Rev 13: 933-955.

25. Akpinar EK, Akpinar S (2005) An assessment on seasonal analysis of wind energy characteristics and wind turbine characteristics. Energy Convers Manage 46: 1848-2167.

26. Fadare DA (2008) Statistical analysis of wind energy potential in Ibadan, Nigeria, Based on Weibull Distribution Function. The Pacific J Sci Technol 9: 110-119.

27. Lynn PA (2012) Onshore and offshore wind energy: An introduction. John Wiley & Sons, US.

28. Keyhani A, Ghasemi-Varnamkhasti M, Khanali M, Abbaszadeh R (2010) An assessment of wind energy potential as a power generation source in the capital of Iran, Tehran. Energy 35: 188-201.

29. Fagbenle RO, Katende J, Ajayi OO, Okeniyi JO (2011) Assessment of wind energy potential of two sites in North-East, Nigeria. Renew Energy 36: 1277-1283.

# An Experimental Appraisal on the Efficacy of MWCNT-H2O Nanofluid on the Performance of Solar Parabolic Trough Collector

**Harwinder Singh\* and Pushpendra Singh**

*Department of Mechanical, Production and Industrial Engineering, Delhi Technological University, India*

### Abstract

An application of MWCNT nanoparticles and distilled water was used to prepare the nanofluid and this type of MWCNT based absorbing medium was found to be highly efficient in investigation of the performance of solar parabolic trough collector due to better thermo physical properties (i.e. thermal conductivity) acquired by the MWCNT based nanofluid. In present research study author decided to take volume concentration 0.01% and 0.02% and high quality surfactant Triton X-100 was used to enhance the dispersion quality of nanoparticles in conventional fluid. The test were performed under different volume flow rate conditions of nanofluid i.e. 160 L/h and 100 L/h. Experimental results show that with an incremental change in volume concentration from 0.01% to 0.02%, there is a substantial increment in efficiency of parabolic collector but observed only at 160 L/h.

**Keywords:** Parabolic trough collector; MWCNT nanofluid; Triton X-100 surfactant; Collector performance testing

## Abbreviations

$Q_u$: Useful heat gain (Watt); $Q_u$: Useful heat gain (Watt); $\dot{m}$: Mass flow rate (Kg/s); $C_{nf}$: Specific heat of MWCNT nanofluid $\left[\frac{J}{kg-k}\right]$; C: Specific heat of base fluid $\left[\frac{J}{kg-k}\right]$; $t$: Time interval (half an hour); $D_i$: Internal diameter (m); $U_i$: Overall heat loss coefficient; F: Collector efficiency factor; $F_R$: Collector heat removal factor; $G_T$: Total solar intensity (W/m²) $\left[\frac{W}{m-k}\right]$; $T_{max}$: Maximum temperature (K); $T_{mini}$: Minimum temperature (K); Tout: Outlet temperature (k); Tin: Inlet temperature (k); W: Width of collector (m); L: Length of collector (m); T: Total experimental duration; $D_o$: Outer diameter (m); C: Concentration ratio; $F_R$: Collector heat removal factor; $h_r$: Convective heat loss coefficient; $K_{nf}$: Thermal conductivity $\left[\frac{W}{m-k}\right]$; S: Absorbed heat flux

## Greek symbols

$\varphi_p$: Weight fraction of MWCNT nano particles in nano fluid; $\rho_{nf}$: Density of MWCNT nanofluid $\left[\frac{kg}{m^3}\right]$; $\rho$: Density of base fluid $\left[\frac{kg}{m^3}\right]$; $\rho_{np}$: Density of nano particles $\left[\frac{kg}{m^3}\right]$; $\mu_{nf}$: Dynamic viscosity of MWCNT nanofluid $\left[\frac{Kg}{m-sec}\right]$; $\mu$: Dynamic Viscosity of base fluid $\left[\frac{Kg}{m-sec}\right]$; $v_{nf}$: Kinematic viscosity of MWCNT nanofluid $\left[\frac{m^2}{sec}\right]$; $v$: Kinematic viscosity of base fluid $\left[\frac{m^2}{sec}\right]$; $E_i$: Instantaneous energy production; $\eta th$: Thermal efficiency; $\eta_{ot}$: Overall thermal efficiency

## Introduction

Parabolic trough collector is a prominent way to convert solar radiations into solar thermal energy and transfer this heat or thermal energy to working fluid for purpose of electric power generation. These days solar energy devices are in use widely and enhancement in performance of solar device are very necessary due to purpose of decrease down the effect of environmental pollutants released from conventional methods. From the last two decades scientists gave effort to improve the performance of solar parabolic trough collector and thermal storage systems for achievement of maximum power and there was a performance booster comes after the discovery of nanoparticles. Application of nanoparticles in conventional fluid also become a new approach to enhance the thermo physical properties of working fluid and among other nanoparticles, MWCNTs possess better thermal, mechanical and optical characteristics and MWCNTs based nanofluid

as a working fluid has an capability to enhance the outcome of solar thermal devices. Suspension of metallic and non metallic particles in base fluid is simply known by nanofluid and this term is originated and investigated by Haddad and it has also been seen that nanofluid attain higher dispersion quality as comparison to microfluid [1]. Due to hydrophobic nature, MWCNT nanoparticles have poor dispersion quality in base fluid and stability of nanoparticles in base fluid can be increased with the help of surfactant, which has both hydrophobic and hydrophilic functional groups [2]. Davis et al. evaluate the shear thinning behavior in the viscosity of CNT nanofluid and they found that viscosity of CNTs based nanofluids is function of concentration of nanoparticles in base fluid, He also concluded that with increase in concentration of CNTs, interactions between nanotubes with each other increases and which results in movement between tubes will be stopped [3]. Ding et al. study about the heat transfer process with nano fluid containing CNTs and results concluded that carbon nano tubes enhance the heat convection coefficient as comparison to total enhancement in thermal conductivity. The reason behind more enhancements in heat convection coefficient is high aspect ratio of using CNTs [4]. Lotfi et al. studied experimentally that heat transfer can be enhanced due to presence of MWCNT nanoparticles in water as comparison to simple water and enhanced heat transfer due to MWCNT and water based nano fluids used in horizontal shell and heat exchanger applications [5]. Yousefi et al. evaluate the effect of MWCNT nanofluid on the efficiency of flat plate collector with different mass flow rate of nanofluid 0.0167 to 0.05 kg/s and also with decided weight fraction of CNTs was 0.2% and 0.4%, he concluded an substantial

\*Corresponding author: Harwinder Singh, Department of Mechanical, Production and Industrial Engineering, Delhi Technological University, Delhi College of Engineering, Shahbad Daulatpur, Main Bawana Road, Delhi 110042, India
E-mail: harrymehrok14@gmail.com

increase in efficiency with surfactant at 0.2% MWCNT nanofluid, while an incremental change in efficiency was observed at 0.4% MWCNT nanofluid without surfactant [6]. Kasaeian et al. conducted an experimental study on solar trough collector with the application of MWCNTat decided volume concentration 0.2% to 0.3% in mineral oil and he concluded that 4-5% and 6-7% enhancement in efficiency with MWCNT and mineral oil based nanofluid as comparison to pure oil [7]. Yousefi et al. studied experimentally that effect of $Al_2O_3$ nanofluid on flat plate collector with different mass flow rates 1, 2 and 3Lit/min and he concluded that 28.3% enhancement in efficiency at 0.2% weight fraction of nanoparticles along with 15.63% efficiency enhancement with the application of surfactant Triton X-100 due to enhancement in heat transfer [8].

## Experimentation & Data Findings

### Nanomaterial

In this experimental study high class MWCNT nanoparticles (97% purity) with 20-40nm in diameter were obtained from Nano Green Technologies LLP (India). The Triton X-100 was used to achieve high quality dispersion of MWCNT in distilled water as base fluid for investigation and it is non-ionic natural surfactant (Table 1).

The SEM (Scanning Electron Microscopy) image of MWCNT nanoparticles produced by secondary electron at different resolution and magnification is shown in (Figures 1 and 2).

### Preparation of nanofluid

MWCNT with 0.01% and 0.02% volume concentration used in distilled water and Triton X-100 surfactant was used in sufficient amount to avoid aggregation and instability between nanotubes, which results in better dispersion behavior. BRANSON 3510 Sonication device followed by magnetic stirrer was used for homogeneous mixing of MWCNT particles in distilled water. Sonication time also affect to dispersion behavior and corresponding thermal properties of carbon nanotubes and after going through several literature study in this field, the soniaction time was decided 45 minutes for mixture amount of 2 liters. Surfactant Triton X-100 due to its non ionic nature showed better dispersion quality for MWCNTs based suspension among other surfactants. Proper dispersion of carbon nanotubes in base fluid is not easy to maintain so that surfactant like Triton X-100 is necessary for better dispersion. It has been seen that Triton X-100 has acquired benzene ring in structure and absorb to graphitic surface in very strong manner due to π-π stacking type interactions [6]. In this experimental study Triton X-100 is used almost same in amount as calculated for MWCNT in base fluid after going through many research discussions. Surfactant is used to bring single phase in solution used as working fluid and fig showed MWCNT based nanofluid contain Triton X-100 with it for proper suspension of MWCNTs throughout experimental span (Figure 3).

### Experimental methodology

The parabolic trough collector was experimentally tested at Thapar University (Punjab). The parabolic trough collector has a copper receiver tube in which working fluid is flowing and gets heated at outlet. Temperatures measure at inlet and outlet through thermocouples and flow in piping and receiver was forced convection due to electric pump with 18W capacity used at inlet side. Collector system also has a storage tank with certain 8L capacity and ball valve was used at inlet side after pump to control the volume flow rate of working nanofluid in solar concentrating collector system. Storage tank and piping system was fully insulated through glass wool and aluminium foil insulation to prevent heat loss from the solar system. Total solar heat flux throughout the day was measured by solar power meter (Tenmars TM-207) and also flowing wind speed was measured by CFM/CMM vane anemometer (PRECISE AM804). Temperatures at inlet and outlet was measured after half an hour as decided before initializing the experimental work and experimental readings were taken from forenoon 9:30 am to afternoon 3:00 pm according to Indian standard time (Table 2 and Figure 4).

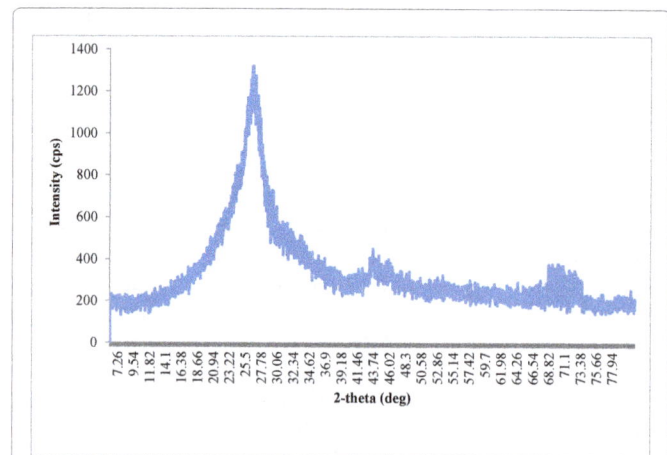

**Figure 1:** XRD image of MWCNT nanoparticles.

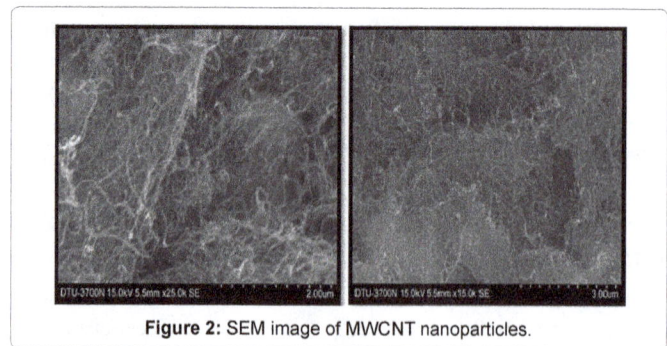

**Figure 2:** SEM image of MWCNT nanoparticles.

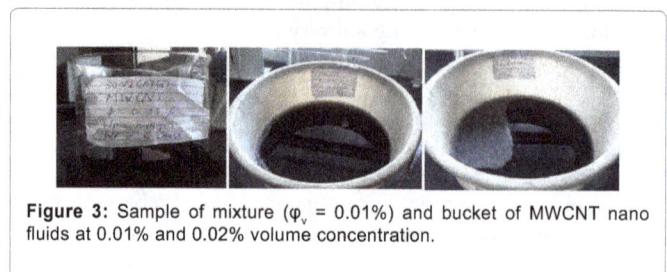

**Figure 3:** Sample of mixture ($\varphi_v$ = 0.01%) and bucket of MWCNT nano fluids at 0.01% and 0.02% volume concentration.

| Item Description (MWCNTs) | |
|---|---|
| Purity | > 97% |
| Length of Nanotubes | 1-10 micrometer |
| No. of Walls | 3-15 |
| Density | 0.15-0.35g/cm³ |
| Surface Area | 350 m²/g [9] |
| Specific Heat | 630 J/Kg-k [9] |
| Thermal conductivity | 1500 W/m-k [9] |

**Table 1**: Properties of MWCNT nanoparticles.

| Length of collector | 1.2 m |
|---|---|
| Breadth of collector | 0.915 m |
| Aperture area | 1.0188 m² |
| Rim angle | 90˚ |
| Focal length | 0.3 m |
| Inside diameter of receiver tube | 0.027 m |
| Outside diameter of receiver tube | 0.028 m |
| Inside diameter of glass cover | 0.064 m |
| Outside diameter of glass cover | 0.066 m |
| Concentration ratio | 9.66 |

**Table 2:** The specification of parabolic trough collector [10-13].

**Figure 4:** Parabolic trough collector (location: Thapar university).

## Performance Testing of Parabolic Solar Collector

Thermal steady state analysis was employed to evaluate the performance of solar parabolic collector and further assume piping and storage system was fully insulated. Various experiments were performed on solar collector through different volume flow rate and at certain weight fraction of nanoparticles in distilled water i.e. 6L. Data related to performance of solar parabolic collector was evaluated through efficiency and also useful heat gain as discussed below:

$$Q_u = mC_{nf}\left(T_{out} - T_{in}\right) \tag{1}$$

and

$$Q_u = F_R\left(W - D_O\right) L\left[S - \frac{U_l}{C}(T_{fi} - T_a)\right] \tag{2}$$

Here $\dot{m}$ is mass flow rate and $C_{nf}$ is specific heat of nanofluid, which is calculated as follow:

$$c_{nf} = \frac{[(1 - \varphi_p)\rho_f c_f + \varphi_p \rho_{np} c_{np}]}{\rho_{nf}} \tag{3}$$

Here '$c_f$' & '$c_{np}$' is specific heat of base fluid (water) and nanopartcles (MWCNT). Further '$\varphi$' is volume concentration of nanopartcles. Density and viscosity of mixture can be calculated through given equations:

$$\rho_{nf} = \left(1 - \varphi_p\right)\rho_f + \varphi p \rho_{np} \tag{4}$$

$$\mu_{nf} \quad \mu_f \,/\, 1 - \varphi_p^{\;2.5} \tag{5}$$

Here '$\rho_{nf}$' & '$\rho_{np}$' is the density of nanofluid and nanoparticles. Instantaneous energy production is directly proportional to useful heat gain and is described as below:

$$E_i = \frac{Q_u}{G_T R_b W\ L} \tag{6}$$

Here '$G_T$' is total solar intensity W/m² and '$R_b$' is bond resistance is taken as constant. Further thermal and overall thermal efficiency of solar parabolic trough collector is discussed in following equations:

$$\eta_{ot} = \frac{mCnf(Tmax - Tmini)}{A_{aper} G_{avg} T} \tag{7}$$

Here '$\eta th$' is thermal efficiency of parabolic collector and further $G_T$ and t is total solar intensity (W/m²) and time interval (half an hour).

$$\eta_{ot} = \frac{mCnf(Tmax - Tmini)}{A_{aper} G_{avg} T} \tag{8}$$

Here '$\eta_{ot}$' is overall thermal efficiency of parabolic collector and further $G_{avg}$ and '$T$' is average solar intensity (W/m²) and total test time period for experimental work.

Also collector efficiency factor (F) and heat removal factor ($F_R$) of collector system is discussed below:

$$F = \frac{D_i h_f}{D_i h_f + D_O U_l} \tag{9}$$

$$F_R = \frac{\dot{m}C_P}{\eth D_O L U_l}(1 - exp\ (-\frac{F\pi D_O U_l L}{\dot{m}C_P})) \tag{10}$$

This equation almost matches with Hottel-Whillier-Bliss equation of flat plate collector. Here 'h' is convective heat transfer coefficient $\left[h = N_U \times \frac{K}{D_i}\right]$ & '$F_R$' is an important design parameter because it is measure of the thermal resistance comes in the path of absorbed solar radiation in reaching the collector fluid. In equation '$F_R U_l$' is a negative efficiency parameter and it has negative effect on useful heat gain further effect encountered on instantaneous efficiency of collector, which is defined by the ratio of useful heat gain to the incident radiation coming on the solar collector.

### Thermophysical properties of MWCNT nanofluid and water

Thermal properties like thermal conductivity and viscosity of water was calculated through various experimental test runs on KD2 Pro conductivity meter and kinematic viscometer with different temperature. Experimentally measured density of water was found almost equivalent to standard density of water, therefore standard value of density was considered for research work. All experimental and standard results were used to calculate the thermophysical properties of MWCNT based nanofluid for both 0.01% and 0.02% weight fraction (Tables 3 and 4).

## Results and Discussion

### MWCNT based nanofluid used as working fluid

In this present study nanofluid was prepared at 0.01% and 0.02% of MWCNT in distilled water as base fluid with the application of Triton X-100 surfactant in appropriate amount. Prepared nanofluids

| Thermal conductivity (K) | $1000 \dfrac{kg}{m^3}$ |
| --- | --- |
| Density (ρ) | $1000 \dfrac{kg}{m^3}$ |
| Dynamic viscosity (μ) | $0.854*10^{-3} \dfrac{Kg}{m-sec}$ |
| Kinematic viscosity (ν) | $0.854*10^{-6} \dfrac{m^2}{sec}$ |
| Specific heat (C_p) | $4.187 \dfrac{KJ}{kg-k}$ |

**Table 3:** Thermophysical properties of water.

| Thermo physical Properties | Mixture I (φ_p = 0.01%) (MWCNT+ Distilled Water) | Mixture II (φ_p = 0.02%) (MWCNT + Distilled water) |
| --- | --- | --- |
| Thermal Conductivity ($K_{nf}$) | $0.617369817 \dfrac{W}{m-k}$ | $0.617369817 \dfrac{W}{m-k}$ |
| Dynamic viscosity ($\mu_{nf}$) | $0.000854213 \dfrac{Kg}{m-sec}$ | $0.000854427 \dfrac{Kg}{m-sec}$ |
| Kinematic viscosity ($v_{nf}$) | $0.854*10^{-6} \dfrac{m^2}{sec}$ | $0.854*10^{-6} \dfrac{m^2}{sec}$ |
| Specific heat ($C_{Pnf}$) | $4186.91 \dfrac{J}{kg-k}$ | $4186.82 \dfrac{J}{kg-k}$ |
| Density ($\rho_{nf}$) | $999.925 \dfrac{kg}{m^3}$ | $999.85 \dfrac{kg}{m^3}$ |

**Table 4:** Calculated thermo physical properties of MWCNT nanofluid.

| Different volume flow rate | F | $F_R$ | $F_R U_l$ |
| --- | --- | --- | --- |
| 160 L/h | 0.9754369 | 0.97186 | 12.9063 |
| 100 L/h | 0.8586555 | 0.85422 | 11.3440 |

**Table 5:** $F_R U_l$ for parabolic trough collector with 0.01% MWCNT nanofluid.

| Different volume flow rate | F | $F_R$ | $F_R U_l$ |
| --- | --- | --- | --- |
| 160 L/h | 0.9754437 | 0.97178 | 12.9052 |
| 100 L/h | 0.8586919 | 0.85426 | 11.3467 |

**Table 6:** $F_R U_l$ for parabolic trough collector with 0.02% MWCNT nanofluid.

as working fluid was flowing through collector receiver tube at different volume flow rates. It has been seen that overall thermal efficiency outcomes from 0.02% weight fraction MWCNT nanofluid at 160 L/h was 5.45% and higher than as comparison to results found at different fraction and with different flow rates. Figure showed that thermal efficiency of 0.02% weight fraction MWCNT nanofluid at 160 L/h was 12.63% measured, which is greater than other results of thermal efficiency from nanofluid at different weight fraction and volume flow rates. Further this experimental study also include heat losses in collector and it has been seen that $F_R U_l$ has a negative effect on instantaneous efficiency and useful heat gain, further calculated values of $F_R U_l$ in case of MWCNT nanofluid at various flow rates are shown in (Tables 5 and 6). Surfactant Triton X-100 is a non-ionic and high foaming surfactant, which reduces heat transfer b/w water and nanotubes. Surfactant mixed at higher amount with MWCNT nanofluid has also considerable negative effects on performance of solar collector [6]. Overall thermal efficiency of 0.02% MWCNT based nanofluid at 100 L/h showed poor results as comparison to other results outcomes from various experiments conducted

through MWCNT nanofluid and decrement in thermal efficiency can be due to higher viscosity of fluid and corresponding pressure drop at 100 L/h. It has also been seen that enhancement in thermal conductivity is dependent upon bulk temperature of nanofluid; Therefore Incremental change in mass flow rate has a considerable effect on bulk temperature and thermal conductivity of MWCNT nanofluid [6]. Further results of Thermal efficiency along with instantaneous energy production are shown graphically as below for different 0.01% & 0.02% volume concentration and at different decided flow rates 160 L/h and 100 L/h (Figures 5-8).

### Water as working fluid

Water (base fluid) was used as working fluid in solar parabolic trough collector. Experimental study was done during 9: 00 am to 3: 00 pm and data related to inlet and outlet temperature, temperature difference, useful heat gain and efficiency of collector was measured at various flow rates. (Figures 9 and 10) showed graphical variations in thermal efficiency and instantaneous energy production of collector through water at 160 L/h and 100 L/h. Figure 10 showed maximum thermal efficiency was 7.28% measured during the time interval 11: 00-11: 30 am for water at 160 L/h and further maximum thermal efficiency at 100 L/h was 6.39% measured during the time interval 10: 30-11: 00 am. $F_R U_l$ is a negative efficiency parameter, which account an effect on performance of solar collector as discussed before and (Table 7) showed $F_R U_l$ for water used as working fluid in solar collector device.

Water showed higher value of '$F_R$' at 160L/h as comparison to '$F_R$' at 100 L/h. basically a heat removal factor is defined by the heat lost from the collector system and collector efficiency factor is completely opposite to heat removal factor, it means that how much heat absorbed by the collector system and denoted by 'F'. Thermal losses from the receiver tube can calculate through loss coefficient '$U_L$' and it depends upon area of receiver tube. Collector efficiency factor and loss coefficient can be calculated from similar expression as described for flat plate collector case. Parabolic trough collector is a type of concentrating collector and used to produce high temperature, which means that thermal radiations are important for evaluation of thermal losses and are temperature dependent.

### Effect of inlet temperature and mass flow rate

Inlet temperature of fluid has considerable effect on collector performance, when inlet temperature of fluid is increasing results in surface temperature of absorber tube and convective losses from absorber tube are also increases. These losses are increases continuously with change in day time and have a negative effect on collector performance or instantaneous efficiency as shown in graphical results of MWCNT nanofluid and water. Mass flow rate of fluid also showed great effect on system performance because of

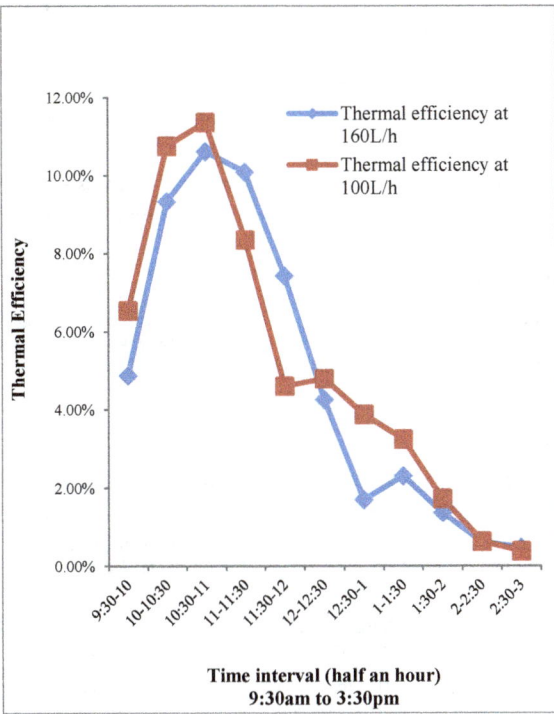

**Figure 5:** Thermal efficiency of MWCNT nanofluid at 160L/h & 100L/h with 0.01%.

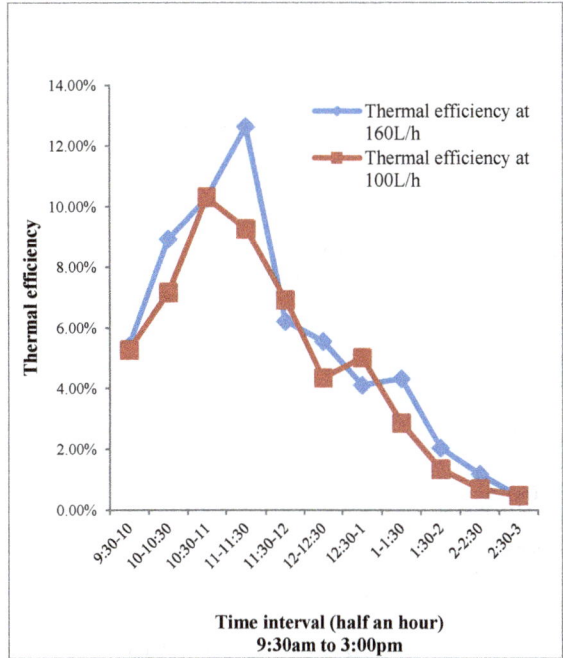

**Figure 7:** Thermal efficiency of MWCNT nanofluid at 160L/h & 100L/h with 0.02%.

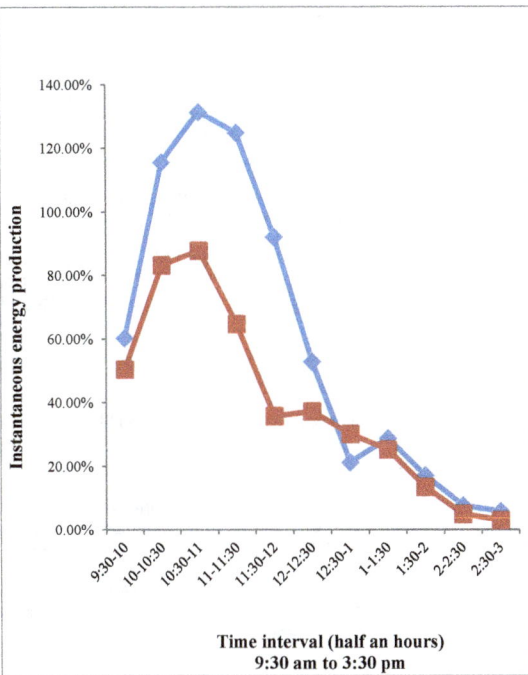

**Figure 6:** Instantaneous energy production of MWCNT nanofluid at 160L/h & 100L/h with 0.01%.

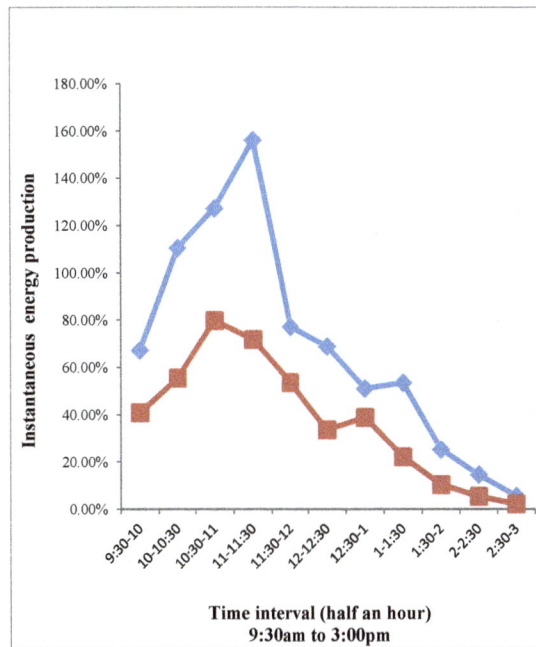

**Figure 8:** Instantaneous energy production of MWCNT nanofluid at 160L/h & 100L/h with 0.02%.

increasing coefficient of heat transfer increasing, which results in incremental change occur in collector efficiency factor and collector heat removal factor also increased as shown in Tables 5 and 6 for MWCNT nanofluid and also same behavior shown in Table 7 for water.

## Conclusion

In this experimental study effect of MWCNT nanofluid on solar parabolic trough collector performance was investigated. The effect of mass flow rate of MWCNT nanofluid mixture containing Triton X-100 at different weight fraction 0.01% and 0.02% was studied. The results showed that 0.02% MWCNT nanofluid possess highest value

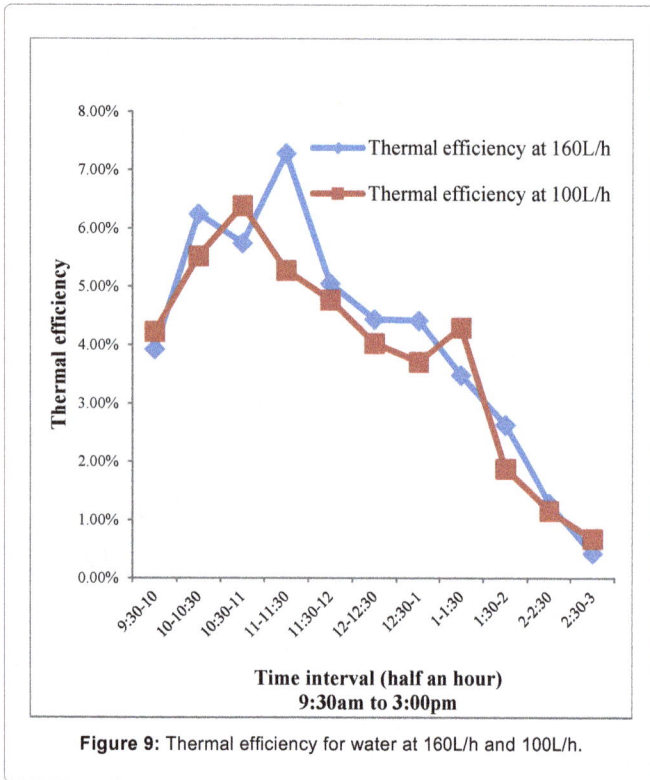

**Figure 9:** Thermal efficiency for water at 160L/h and 100L/h.

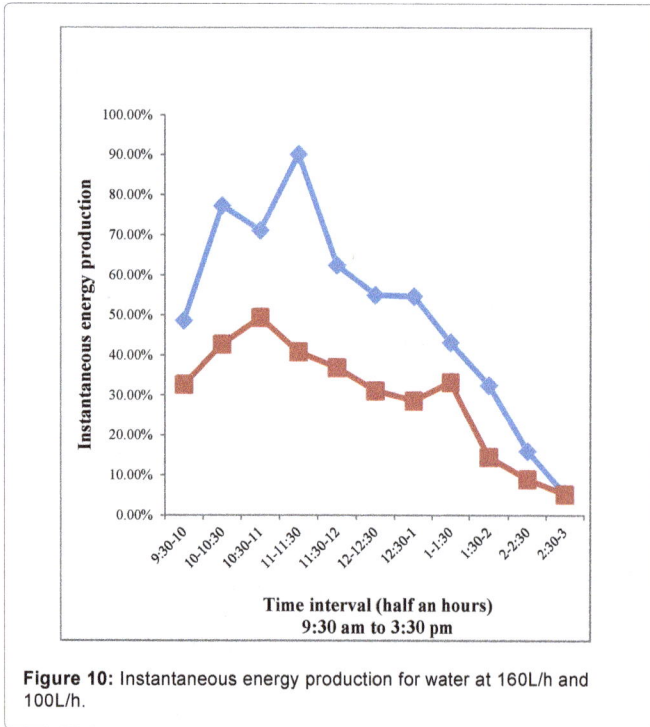

**Figure 10:** Instantaneous energy production for water at 160L/h and 100L/h.

| Different volume flow rate | F | $F_R$ | $F_R U_l$ |
|---|---|---|---|
| 160 L/h | 0.975429002 | 0.97185 | 12.9061 |
| 100 L/h | 0.858619195 | 0.85419 | 11.3436 |

**Table 7:** $F_R U_l$ for parabolic trough collector with water as working fluid.

increased by increasing the mass flow rate of fluid was measured in this experimental study, which results in incremental change occur in efficiency.

## Acknowledgement

I am very thankful to Mr. Kundan lal, Assitant professor of Thapar University for allow me to conduct the experimental work on prototype of PTC system and I am also very grateful for his guidance and support enabled throughout my experimental procedure

## References

1. Haddad Z, Abid C, Oztop HF, Mataoui A (2004) A review on how the researchers prepare their nanofluids. Int J Thermal Sci 76: 168-189.

2. Rastogi R, Kaushal R, Tripathi SK, Sharma AL, Kaur I, et al. (2008) Comparative study of carbon nanotube dispersion using surfactants. J Colloid Interface Sci 328: 421-428.

3. Davis VA, Ericson LM, Parra-Vasquez ANG, Fan H, Wang Y, et al. (2004) Phase behaviour and rheology of SWNTs in superacids. Macromolecules 37: 154-160.

4. Ding Y, Alias H, Wen D, Williams RA (2006) Heat transfer of aqueous suspension of carbon nanotubes (CNT nanofluids). Int J Heat Mass Transfer 49: 240-250.

5. Lotfi R, Rashidi AM, Amrollahi A (2012) Experimental study on the transfer enhancement of MWNT-water nanofluid in a shell and tube heat exchanger. International Communication in Heat Mass Transfer 39: 108-111.

6. Yousefi T, Veisy F, Shojaeizadeh E, Zinadini S (2012) An experimental investigation on the effect of MWCNT-H$_2$O nanofluid on the efficiency of flat-plate solar collectors. Exp Thermal Fluid Sci 39: 207-212.

7. Kasaeian A, Daviran S, Azarian RD, Rashidi A (2015) Performance evaluation and nano fluid using capability study of a solar parabolic trough collector. Energy Convers Manag 89: 368-375.

8. Yousefi T, Veisy F, Shojaeizadeh E, Zinadini S (2012) An experimental investigation on the effect of Al$_2$O$_3$-H$_2$O nanofluid on the efficiency of flat-plate solar collectors. Renewable Energy 39: 293-298.

9. Akhavan-Behabadi MA, Shahidi M, Aligoodarz MR (2015) An experimental study on heat transfer and pressure drop of MWCNT-water nano-fluid inside horizontal coiled wire inserted tube. Int Communication Heat Mass Transfer 63: 62-72.

10. Sukhatme SP, Nayak JK (2012) Solar Energy principles of thermal collection and storage. Tata McGraw-Hill Education.

11. Shaker M, Birgersson E, Mujumdar AS (2014) Extended Maxwell model for the thermal conductivity of nanofluids that accounts for nonlocal heat transfer. Int J Thermal Sci 84: 260-266.

12. Esfe MH, Saedodin S (2014) An experimental investigation and new correlation of viscosity of ZnO-EG nanofluid at various temperatures and at various temperatures and different solid fractions. Exp Thermal Fluid Sci 55: 1-5.

13. Sunil K, Kundan L, Sumeet S (2014) Performance evaluation of a nanofluid based parabolic solar collector – an experimental study. Proceedings of Twelveth IRF International Conference.

of thermal, overall thermal and instantaneous energy production among other concentration of nanofluid and water and further also at different flow rates. 0.02% MWCNT nanofluid at 160 L/h showed lowest amount of $F_R U_l$ and highest value for collector efficiency factor among other concentration of fluid. Temperature difference

# Building Integrated Solar Thermal (BIST) Technologies and Their Applications: A Review of Structural Design and Architectural Integration

Xingxing Zhang[*1], Jingchun Shen[1], Llewellyn Tang[*1], Tong Yang[1], Liang Xia[1], Zehui Hong[1], Luying Wang[1], Yupeng Wu[2], Yong Shi[1], Peng Xu[3] and Shengchun Liu[4]

[1]Department of Architecture and Built Environment, University of Nottingham, Ningbo, China

[2]Department of Architecture and Built Environment, University of Nottingham, UK

[3]Beijing Key Lab of Heating, Gas Supply, Ventilating and Air Conditioning Engineering, Beijing University of Civil Engineering and Architecture, China

[4]Key Laboratory of Refrigeration Technology, Tianjin University of Commerce, China

## Abstract

Solar energy has enormous potential to meet the majority of present world energy demand by effective integration with local building components. One of the most promising technologies is building integrated solar thermal (BIST) technology. This paper presents a review of the available literature covering various types of BIST technologies and their applications in terms of structural design and architectural integration. The review covers detailed description of BIST systems using air, hydraulic (water/heat pipe/refrigerant) and phase changing materials (PCM) as the working medium. The fundamental structure of BIST and the various specific structures of available BIST in the literature are described. Design criteria and practical operation conditions of BIST systems are illustrated. The state of pilot projects is also fully depicted. Current barriers and future development opportunities are therefore concluded. Based on the thorough review, it is clear that BIST is very promising devices with considerable energy saving prospective and building integration feasibility. This review shall facilitate the development of solar driven service for buildings and help the corresponding saving in fossil fuel consumption and the reduction in carbon emission.

**Keywords:** Building integrated solar thermal; Design criteria; Operation; Application

## Introduction

In order to achieve the global carbon emission target, the high fraction of locally available renewable energy sources in energy mix will become necessary in addition to a significantly reduced energy demand. Solar energy is one of the most important renewable sources locally available for use in building heating, cooling, hot water supply and power production. Truly building integrated solar thermal (BIST) systems can be a potential solution towards the enhanced energy efficiency and reduced operational cost in contemporary built environment.

According to the vision plan issued by European Solar Thermal Technology Platform (ESTTP), by 2030 up to 50 % of the low and medium temperature heat will be delivered through solar thermal [1]. However currently, the solar thermal systems are mostly applied to generate hot water in small-scale plants. And when it comes to applications in space heating, large-scale plants in urban heating networks, the insufficient suitable-and-oriented roof of most buildings may dictate solar thermal implementation. For a wide market penetration, it is therefore necessary to develop new solar collectors with feasibility to be integrated with building components. Such requirement opens up a large-and-new market segment for the BIST system, especially for district or city-level energy supply in the future.

BIST is defined as the "multifunctional energy facade" that differs from conventional solar panels in that it offers a wide range of solutions in architectural design features (i.e., colour, texture, and shape), exceptional applicability and safety in construction, as well as additional energy production. It has flexible functions of buildings' heating/cooling, hot water supply, power generation and simultaneously improvement of the insulation and overall appearance of buildings. This facade based BIST technologies would boost the building energy efficiency and literally turn the envelope into an independent energy plant, creating the possibility of solar-thermal deployment in high-rise buildings.

## Working Principle of Typical BIST System

The typical BIST system is schematically shown in Figure 1. The system normally comprises a group of modular BIST collectors that receive the solar irradiation and convert it into heat energy, whereas the heating/cooling circuits could be further based on the integration of a heat pump cycle, a package of absorption chiller, a modular thermal storage and a system controller. In case of some unsatisfied weather conditions, a backup/auxiliary heating system (e.g., boiler) is also integrated to guarantee the normal operation of system.

In the typical BIST system, the overall energy source is derived from solar heat, which is completely absorbed by the modular BIST collectors. This part of heat is then transferred into the circulated working medium and transported to the preliminary heat storage unit, within which heat transfer between the heat pump refrigerant and the circulating working medium will occur. This interaction will decrease the temperature of circulating medium, which enables the circulating medium absorbing heat in the facades for next circumstance.

Meanwhile in the heat pump cycle (compressor-condenser-

---

**\*Corresponding author:** Zhang X and Tang L, Department of Architecture and Built Environment, University of Nottingham Ningbo, China, E-mail: Xingxing.Zhang@nottingham.edu.cn; Llewellyn.Tang@nottingham.edu.cn

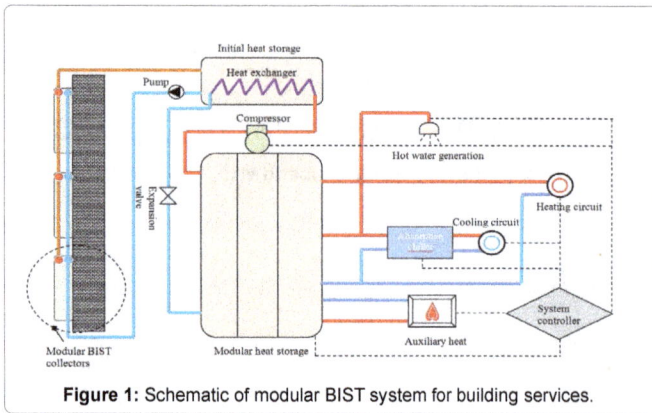

**Figure 1:** Schematic of modular BIST system for building services.

expansion valve-evaporator), the liquid refrigerant will be vaporized in the heat exchanger, which, driven by the compressor, will be subsequently converted into higher-temperature-and-pressure, supersaturated vapour, and further releases heat energy into the tank water via the coil exchanger (condenser of the heat pump cycle), leading to the temperature rise of the tank water. Also, the heat transfer process within the coil exchanger will result in condensation of the supersaturated vapour, which will be downgraded into lower-temperature-and-pressure liquid refrigerant after passing through the expansion valve. This refrigerant will undergo the evaporation process within the heat exchanger in the initial heat storage again, thus completing the heat pump operation. When the water temperature in the tank accumulates to a certain level, i.e., 45ºC, then water can be directly supplied for utilization or under-floor heating system. For the cooling purpose, an additional appliance of absorption chillers should be coupled with.

## Category of BIST Technologies

The BISTs can be classified into air-, hydraulic- (water/heat pipe/refrigerant) and PCM-based types according to the heat transfer medium. Air based type is characterized by lower cost, but lower efficiency due to the air's relatively lower thermal mass. This system usually uses the collected solar heat to pre-heat the intake air for the purpose of building ventilation and space heating. Hydraulic-based BISTs are most commonly used building integrated solar thermal devices that enable the effective collection of the striking solar radiation and conversion of it into the heat for the purpose of hot water production and space heating. The PCM-based type is usually operated in combination with air, water or other hydraulic measures that enable storing parts of the collected heat during the solar-radiation-rich period, and releasing them to the passing fluids (air, water, or others) during the solar-radiation-poor period, in order to achieve a longer period of BIST operation. In this aspect, the heat transfer medium based classification was adopted and mainly illustrated as follows:

## Air-based BIST technology

Air-based solar thermal systems use air as the working fluid for absorbing and transferring solar energy. It can directly heat a room or pre-heat the air passing through a heat recovery ventilator or an air coil of an air-source heat pump. It is a promising solar thermal technology with the main advantages of: anti-freezing and anti-boiling operation, non-corrosive medium property, and low cost and simple structure. Therefore, no damage caused by leakage, stagnation condition and frost problems needs to be dealt with, offering a possibility of reliable and

cost-effective solutions even at a low irradiation level.

But air has a relatively low heat capacity, resulting in higher mass or volumetric flows and poor thermal removal effectiveness for BIST systems. In other words, more occupancy space is necessary in building components to fit air-handling equipment (ducts and fans) compared to that in the hydraulic-based system. And higher parasitic power consumption and acoustic problems are also worthy of attention. Generally, air-based BIST system could be simply delivered from single channel and double channels, as shown in Figure 2. The air-based solar thermal facade could be formulated by incorporating an air gap between the back surface of glazing covers (or PV panels, external sheet, construction mass) and the building fabric (facade, glazing or roof). In practical application, air based BIST in space heating systems is usually operated with fixed airflow rates, thus the outlet temperature varies along with the change of solar irradiation in a day. This is because if it runs at a fixed outlet temperature by varying the flow rate, both heat removal factor and collector performance would be low [2-4]. When air circulation is combined within photovoltaic (PV) modules, effectiveness for PV cooling would be very low once the air temperature is above 20 °C [3,4].

## Water-based BIST technology

Water is the most suitable heat transfer medium for solar thermal technology owing to its high thermal capacity and thermal conductivity, and low viscosity and cost. Besides, it allows easy storage of solar heat gains, and is suitable for direct domestic hot water production and indirect space heating. However, water is corrosive in nature (especially at high temperature) as well as freezing and scaling based, which poses a challenge in the design tubing and plumbing. Though glycol/water mixture has been widely adapted to lower freezing risk, it is still worthy of attention with water pressure difference at different BIST levels (heights). And more measures should be taken into consideration of an envelope structure, accessibility and 8 protection of water leakage. Figure 3 demonstrates different water-based BIST structures.

The water-based BIST can be mainly divided into two modes

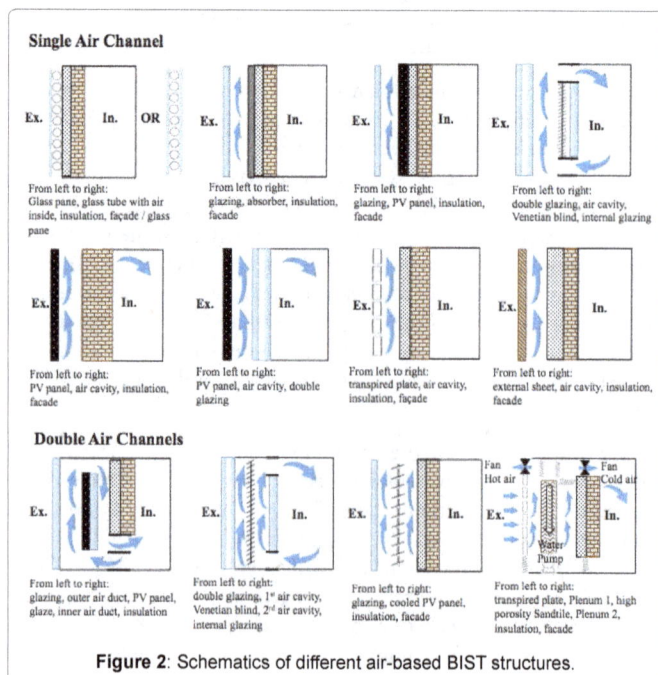

**Figure 2:** Schematics of different air-based BIST structures.

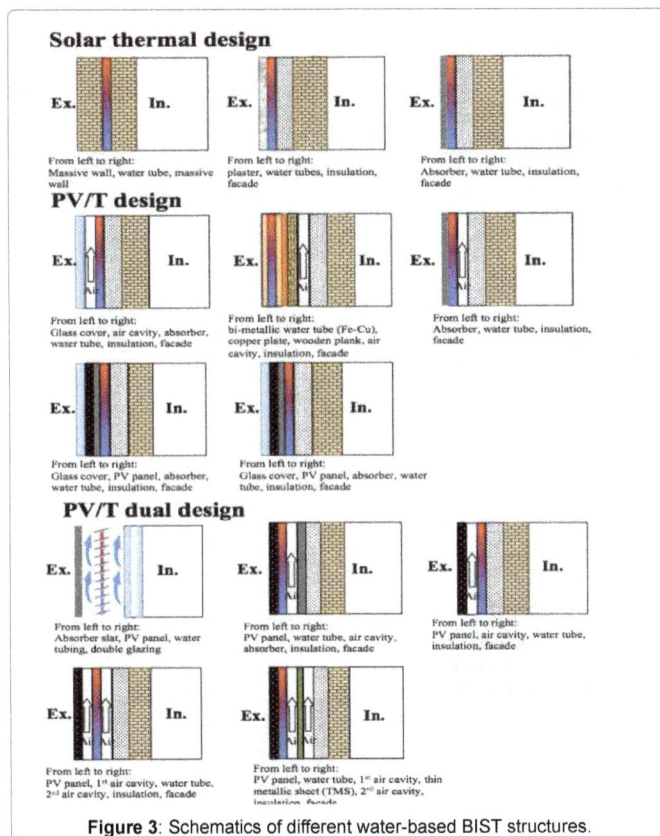

**Figure 3**: Schematics of different water-based BIST structures.

of arrangement: single water channel and multiple flow channels (integrated air and water flow). The former structural arrangements are evolved from the mechanism of the passive solar heating, which often provide the solutions to hot water purpose. The latter structural arrangements are normally formulated by incorporating additional air gap between the PV module (absorber, water flow tubes) and the building fabric. As a result, air could be delivered from above, below or on both sides of the water tubes. With the help of electrical or mechanical equipment, these designs can provide the solutions to the inversed thermo-siphon phenomena and the uncertainty of heat transfer. In some cases, the PV module acts as the thermal absorber (named PV/thermal design) and the water flow channels could be round, rectangle and other irregular geometries. The water-based PV/thermal designs take advantages of both electricity production and heat extraction. At present, the technical drawback of the traditional PV panels lies in the relatively low electrical efficiency in the range 6 to 18%. PVs' electrical efficiency varies in an inversely linear trend with the PV cells' surface temperature, leading to around 0.5% efficiency declining per degree rise in cells' temperature. The water-based PV/Thermal designs are therefore developed to control the temperature of the PV cells, increase the corresponding electrical efficiency during practical operation, and meanwhile make advanced utilization of the waste PV heat [5,6]. Usually, the multiple water flow channel structure is designed for the end-users who have multiple demands in hot air, space heating, agriculture/herb drying or increased ventilation, and electricity generation.

## Refrigerant-based BIST technology

Compared to water, refrigerant has properties of lower boiling/freezing, lower viscosity, and higher thermal capacity, which therefore enables to efficiently transfer a larger amount of heat even with a small amount. In the indirect heat-pump based BIST system, the thermal collector also works as a heat-pump evaporator, which could respond more quickly to the absorption of solar heat even in the poor weather conditions than other BIST types [6-8]. In terms of the refrigerant types, Chlorofluorocarbons (CFC) are the most commonly-applied heat-transfer fluids due to the properties of stability, non-inflammability and non-corrosivity. But CFC has high Ozone Depletion Potential (ODP) and Global Warming Potential (GWP). As a result, the hydrochlorofluorocarbons (HCFC) take the place. Recently, natural fluids, such as propane (R-290), butane (R-600), isobutane (R-600a), propylene (R-600), ammonia (R-717) and carbon dioxide (R-744) have been considered as long-term heat transfer fluid with great environmentally behaviour at a very low or near zero ODP and GWP [3].

In terms of refrigerant types, Chlorofluorocarbon (CFC) refrigerants are more commonly used as heat transfer fluids owning to properties of stability, non-inflammability, and non-corrodility. But CFC has high Ozone Depletion Potential (ODP) and Global Warming Potential (GWP). As a result, the hydrochlorofluorocarbons (HCFC) then takes the place. Recently, natural fluids, such as propane (R-290), butane (R-600), isobutane (R-600a), propylene (R-600), ammonia (R-717) and carbon dioxide (R-744) have been considered as long-term heat transfer fluid with great environmentally behaviour at a very low or near zero ODP and GWP [3].

Refrigerant as the working fluid is more prevalent in the application of PV/thermal system coupled with the heat pump [8]. From the simulation results, it can be found that a lower evaporation temperature was associated with a higher system efficiency because a lower evaporation temperature leads to a lower surface temperature of absorbers for solar cells to override the limitation of electricity generating efficiency, meanwhile the heated refrigerant can be further re-utilized for thermal production with increasing the coefficient of performance (COP) of heat pumps [8].

The heat pipe as a core ensures the advantages as one-directional, two-phase thermosyphons, very low hydraulic resistance, constant liquid flow and isothermal heat absorbing surface. The heat pipe technology has significant development in the evacuated tube collector (ETC) with two basic types of single tube and double tubes. Because of the major drawback of a fragile glass cover, the sole utilization of heat pipe has been introduced to alleviate this problem [9]. And it has been currently brought forward as an individual heat transfer component into the design of BIST systems.

Because of the special characteristics of heat pipe technology, it has been identified that the heat-pipe based BIST system presents a great potential with attractive merits of light weight, easy assembling/installation, high versatility, good scalability, and excellent adaptability to the building design [10]. Currently, there are mainly three types of heat pipes applied in BIST systems, including conventional straight heat pipe, loop heat pipe, and micro-channel heat pipe [11-13] as shown in Figure 4. In practice, as a potential heat transfer medium applied in BIST systems, the heat pipe has to overcome the limitations in on-site assembly requiring higher vacuum degree, and manufacture cost owning to labour intensive processing, meanwhile the difficulties in maintenance and replacement.

## PCM-based facade BIST technology

Thermal energy can be generally classified two phases of sensible and latent heat. PCM, because of its higher energy storage density and

| BIST Type | Advantages | Disadvantages |
|---|---|---|
| Air | Anti-freezing or anti-boiling, and non-corrosive; <br> Low cost; <br> Simple structure; | Low heat capacity; <br> Potential leakage and noise; <br> Lower efficiency; <br> Large mass or volume; <br> High in heat loss. |
| Water | Nontoxic; <br> High in specific heat; <br> Cost effective; <br> Perform well in cold climates; <br> Almost constant energy collected; <br> Smaller storage volume; | Potential mineral deposits; <br> Possible leakage, freezing, corrosion and overheating; <br> Unstable heat removal effectiveness. |
| Refrigerant | Small fluid volume; <br> Quick response to solar heat under different weather conditions; <br> Lower vaporization latent heat; <br> Stable performance; <br> High efficiency; <br> Re-utilization | High cost; <br> Unbalanced liquid distribution; <br> Need to consider environmental behaivor <br> Requirement of recharging refrigerant. |
| Heat pipe | Compact and super high heat exchange ability; <br> Low in hydraulic and thermal resistances; <br> Constant liquid flow <br> Isothermal heat absorbing surface <br> Versatility, scalability, and adaptability of the design <br> Small weight <br> Easy assembling and installing | High cost; <br> High degree in vacuum processing; <br> Difficulty in maintenance and replacement |
| PCM | Improvement in thermal comfort and building envelop; <br> Diversity in building integration; | Difficult to operate; <br> Complex behaviour; <br> Diverse affection factors; <br> Sensitive heat injection; |

Table 1: The characteristic comparison of BIST in terms of heat transfer medium

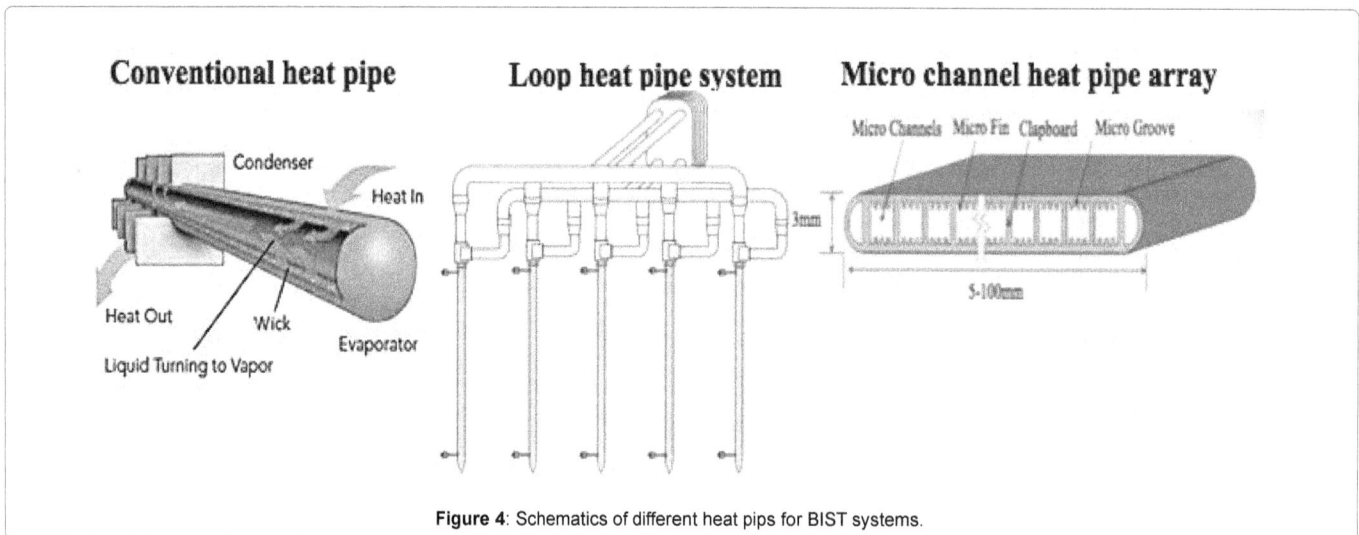

Figure 4: Schematics of different heat pips for BIST systems.

thermostatic fusion, is particularly attractive in the BIST application due to the two-phase heat transfer process [14]. In recent years, a new technique has been proposed to use the PCM slurry as pumpable heat transfer fluids and as heat storage systems [15-17]. Through microencapsulating and isolating PCM from its surroundings and the carrier fluid, the PCM is less likely to hamper the heat transfer process [18]. Microencapsulated PCM slurry, semi-clathrate hydrate slurry, shape-stabilized PCM slurry and PCM emulsions are the common approaches to form the heat transfer fluids in a range of melting temperature from 0°C to 20 °C [19]. These potentials allow the increase of system energy efficiency, and the reduction of pipe size and collector area, moreover save pumping power consumption due to less quantity of

heat transfer fluids [19]. But till now, more applications of PCM [20-25] have been found in the BIST system for the function of thermal storage with the characteristics of a great thermal energy storage capacity, a high heat transfer property and a positive phase change temperatures. Although the heat transfer of PCM is quite complicated, it has broad prospects for the practical application in the future. Therefore, it still needs more intensive investigation for its function as a heat transfer fluid.

### General comparison of different SFT technologies

In general, a comprehensive feature including the advantages and disadvantages of each heat transfer medium based BIST technologies

| Schematic structure | Description |
|---|---|
| **Wall-based BIST** | |
| <br>1 Cover rail<br>x glass fleece | "AKS Doma Flex" is a glazed flat plate system and its distinct feature locates in the aluminum absorbers that equipped with a highly selective coating and laser-welded copper pipes that guarantee optimum performance. The wood or aluminum collector frame provides a very high level of freedom in both size and shape as either the facade solutions or the high durable and stable installation for building [31] |
| <br>1 Cover rail<br>2 Glass fixing profile<br>3 Casing<br>4 Glazing<br>5 Edge bond<br>6 Absorber<br>7 Rear panel<br>8 Thermal insulation | "H+S MegaSlate II" is a glazed flat plate system characterized by both the thickness of 38mm and gluing technique that eases the integration into the building skin and provide a new level of freedom in the field of glazed flat plates. The glazing is glued to the collector structure with the same technique used to glue double-glazing, and the gap is filled with argon gas, which helps reduce heat losses. But no freedom is available in module shape and size, nor in absorber or glazing surface texture/colour [32]. |
| | The unglazed BIST system is composed of a cost effective facade collector coupled to a reversible heat pump for heating/cooling of high-rise buildings. It works as a low temperature solar collector as well as an atmospheric heat exchanger and a night time heat-dissipater to boost the heating/cooling efficiency of the system. The active wall element is a low temperature unglazed solar collector with a capillary mat embedded in the finishing layer of the external insulation of the building. This cost effective solution allows installing the external insulation and the active facade in the same step. A glycol solution circulates in the capillary tubes to transfer heat to the heated room via the storage and the heat pump [33]. |
| <br>1. Insulating layer<br>2. Coated metal sheet<br>3. Corrugated pipe<br>4. Connection pipe<br>5. Post profile<br>6. Copper line<br>7. Metal cover<br>8. Glazing cover<br>9. Connection absorber<br>10. Absorber plate | The building integrated solar thermal facade system is fully integrated into an aluminium glass facade and represents the building shell. This kind of BIST consists of a typical glass, thermal absorber, soft connections, pipes with the mullion, insulation and thermal metal panel [34]. |
| <br>facade of the building<br>solar collector<br>aluminum heat pipes | The aluminium profiled heat pipe solar collector consists with the extruded aluminium alloy with heat pipes of original cross-sectional profile and wide fins and longitudinal grooves. The flat absorber plate can be composed of several fins, and the opposite end of heat pipe serves as a heat sink surface. When it's integrated with building, number of heat pipes and their length is optional and can be replaced anytime [35]. |

In this BIST system, macro-encapsulated PCM panels are installed inside the air chamber of a ventilated facade. It has multiple operation modes as mechanical or natural ventilation mode. The use of the PCM not only increases the solar energy absorption capacity during winter but it uses as a cold storage system during warm periods to improve the thermal behaviour of the whole building [36].

The thermal absorber [37] is only 5mm wide and is made up by two parallel thin flat-plate metal sheets, one of which is extruded by machinery mould to formulate arrays of mini corrugations, while the other sheet remains smooth for attaching the building wall. A laser-welding technology is applied to join them together, forming up the built-in turbulent flow channels. Such unique compact structure engenders not only high heat transfer capacity but also convenience in rapid assembly and installation. If several absorbers are connected together as a complete larger area of building façade, the question of the connecting arrangement will arise. Therefore, a flexible connection arrangement scheme is proposed through two sets of inlet and outlet around the four corners of the flat plate panel. Basically, there are three kind of arrangements as parallel, series and combined.

### Window-based BIST

This new glazing integrated transparent BIST based on low-cost window technology will allow heat generation, visual contact to the exterior and provide solar and glare control in the same time. In summer, this BIST will be used as a heat source for solar cooling systems. The approach is to integrate apertures with angular selective transmittance into the absorber of a solar thermal collector, which is integrated in the transparent part of the facade. These apertures will selectively shield the direct irradiation of the sun (coming from directions with higher solar altitude angles) while retaining visibility through the window horizontally or downwards [38].

This a new vacuum tube solar air collector for facade integration in high-rise buildings bases on many advantages of the heat transfer medium, air like in common building installations. The vacuum insulation of the tubes is responsible for a good thermal efficiency at high operating temperatures and low irradiation. The heat transfer medium, air has an intrinsic fail-safe behaviour without leakage, frost, and stagnation problems, which offers the possibility of reliable, efficient and cost-effective solutions [39].

### Balcony-based BIST

This facade-based solar water heating system employs a modular panel incorporating a unique loop heat pipe (LHP). The outdoor part is a modular panel which receive solar irradiation and convert it into the heat carried by the vaporised heat pipe working fluid, whereas the indoor part is the combination of a flat-plate heat exchanger, a hot water tank, a circulating pump and water piping connections, which raises the temperature of the circulated water by absorbing heat from the heat pipe vapour, resulting in the condensation of the vapour at the exchanger [40].

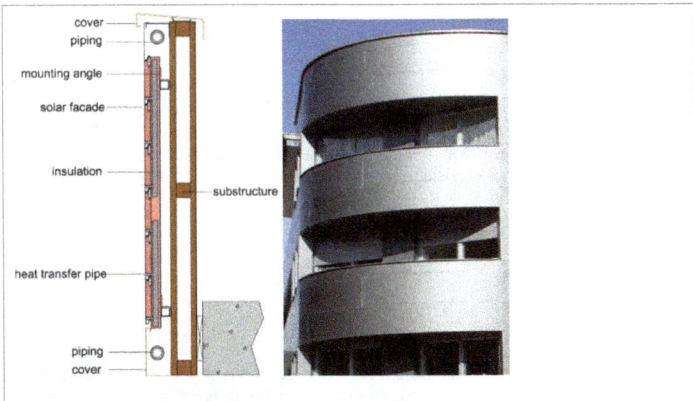

WAF's solar balcony is a special version of the BIST. With this design, the facade elements are directly integrated into the front of the balcony. All pipes are placed behind special cover plates. In this way, the visual appearance of the front of the balcony is not compromised [26].

## Sunshield-based BIST

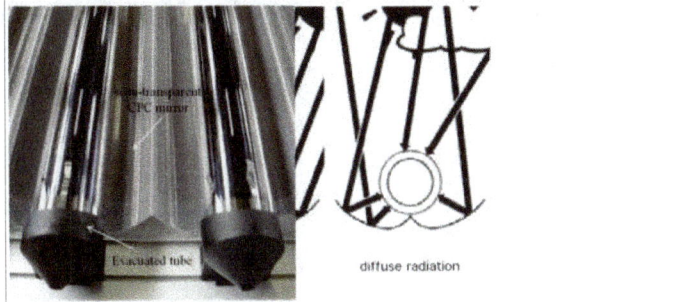

Ritter XL applies the evacuated tubes with semi-transparent Compound Parabolic Concentrating (CPC) mirrors integrated into facade elements for heating and sun shading. It makes up with the aluminium frame, a borosilicate glass, selective absorber, a CPC mirror, the stainless steel and copper pipes. Because of the absorber form of the evacuated tube, the number and size of the collector is optional and flexible to be customised for solar shading design for building [41].

## Roof-based BIST

"VarioSol E Collector System" is a glazed flat plate system using a unique SKYTECH absorber technology, which is made by roll-folding technology connects absorber sheet and tube over the entire circumference leading an optimised heat transfer to the fluid inside. The thin collectors can be simply mounted onto flat roof rather than modified original building structure, providing an optimum solution for building renovation [42].

"Genersys" is a flat plate collector consisting of a one-piece forged metal casing, where solar glass is fixed by means of a frame made from non-corrosive aluminium profile. The stamped Al-Mg sheet absorber fins with high-selective conversion layer span the copper pipe meander. The series connected collectors ensures the good options for the building integration as construction or integration elements into sloping roof, stand on flat roof and building façade [43].

"COBRA Evo." is a kind of flat-plate collectors. Then principal design highlights are the highly selectively coated copper absorber plate, which converts most of heat radiation from the sun to the absorber plate, as well as the consistently lasered copper serpentine coil, which makes both the arches and the manifolds thermally active. In virtue of these technologies, the system provides excellent operating performance even under unreasonable climate conditions. In addition, there is a large range of mounting accessories for roof-mounted, stand on flat roof and building façade [44].

| | "QUICK STEP Solar Thermie" is an innovative, low efficiency, unglazed system for roofs. If the system is integrated into the standard Rheinzink QUICK STEP roof covering system, the module look exactly like the traditional non active ones without any field positioning and dimensioning issues [45]. |
| --- | --- |
| | "Atmova roof title" is a copper tile or facade component. The working principle is that collecting thermal energy from the ambient air, wind, and rain as well as any available solar radiation and then fed via the pipe system with its heat-transporting fluid to the heat pump. After that the heat pump converts this energy into usable heat for heating and the hot water supply that flows directly into a hot water storage tank and which can be used for immediate consumption. The high-energy yield (500 W/m$^2$ and COP> 3.5) with the long lifetime of the system guarantee the above-average economic benefit of the system [46]. |
| | "TECU Solar System" is an innovative integrated copper roof system. The heats transfer fluid in the copper path of collector, capturing the solar heat to produce thermal energy. Because the whole collector is positioned integrally with roof system, therefore it is fully protected from external intrusions and completely invisible [47]. |
| | "SolTech Sigma" is a specially developed tile more transparent active energy roofs. Underneath the glass tiles, the specially developed liquid based absorber modules harvest the energy from the sun, and the generated solar energy ends up in a specific storage tank that is connected to the building's central heating system [48]. |

**Table 2:** Prototypes of different innovative BIST technologies.

can be summarized in Table 1. And further selections of these BIST technologies should be carried out depending on different application scenarios.

## BIST Structural Design in Terms of Architectural Element

The basic structure of BIST is originally derived from the conventional solar thermal collectors, as shown in Figure 5, which includes three fundamental configurations as glazed or unglazed flat-plate and evacuate tube shapes. The flat-plate BIST usually compromises a glazing cover (optional), a fluid-cooling thermal absorber, an insulation layer (optional) and the supporting enclosure, whilst the cylinder BIST normally consists of evacuated tubes with fluid-cooling

thermal absorbers inside. The overall BIST system also has a storage tank and additional mechanical devices, such as fans, pumps, complex controllers or other auxiliary devices to redistribute solar energy.

However, the specific structures of the BIST systems are various when they serve as different building elements in practice. This section provides the generally discussion about the typical BIST designs in terms of the architectural element.

### Typical BIST structures in terms of architectural elements

**Wall-based BIST technology:** Figure 6 gives a schematic structure of a sample wall-based BIST, mainly composed with a series of weatherboarding, a connection metal sheet, a group of copper heat-

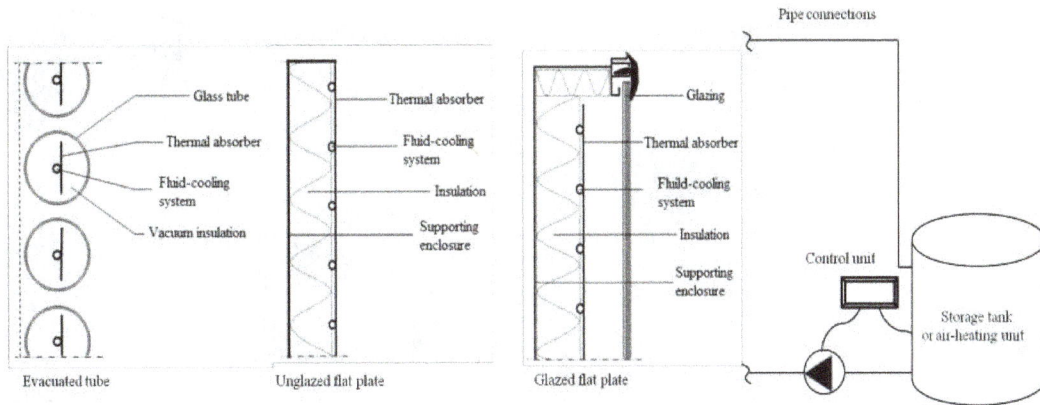

Figure 5: Integration constrains for BIST (non-exhaustive).

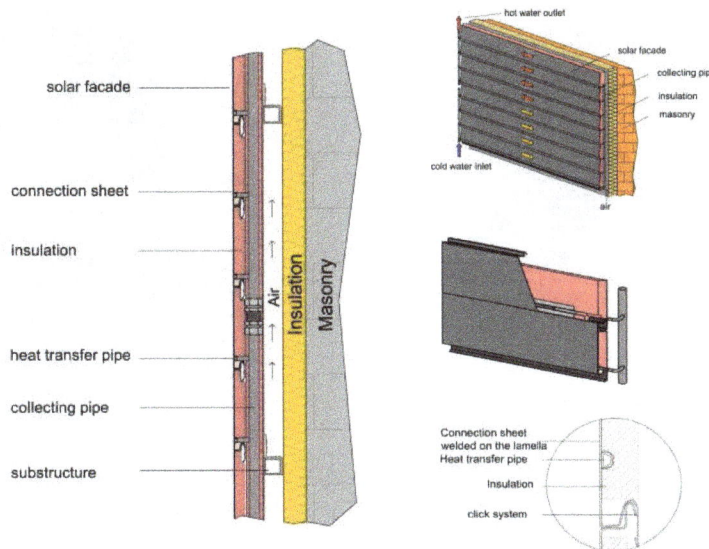

Figure 6: Schematics of BISTs as different building envelope parts with typical examples.

transfer pipe and an insulation layer. The connection metal sheet and the copper pipe are welded into a piece of weatherboarding, which ensures optimum heat transfer performance. A polyurethane (PUR) insulation is usually used at the back of panel to minimise convective heat losses, especially PUR foam is selected to avoid air cavities between the pipe and the insulation. In addition, mineral wool around 50mm in thickness is used for collecting pipes insulation. The coating of the weatherboarding consists of a special solar varnish with an excellent absorption and emission, being extremely weather resistant as well. All the weatherboarding pieces are assembled via the mounting clicks, enabling an easy self-assembly approach. The working principle of such BIST is as same as the conventional solar thermal collectors. Solar energy is optimally converted into thermal energy owing to the selective solar varnish coating. A heat transfer medium, such as a water-glycol mixture, passes through the solar circuit behind the facade elements to transfer heat directly from the absorber. The absorbed heat is further transferred to the heat storage system by a heat exchanger.

Window-based BIST technology: A sample of the semi-transparent solar thermal window is illustrated in Figure 7. It is a double glazed unit fixed in the window frames with multifunction as glazing system for insulation, natural lighting, domestic hot water and heating and/or air conditioning [27]. This solar thermal window comprises two layers of glass panes, a certain volume of argon filling, a group of ⊠-shape profiled thermal absorbers, a copper serpentine pipe with working fluid and aluminium strips. During operation, the solar flux through such window is divided mainly into three parts according to the seasons: (1) directly intercepted by the front part of solar absorber lamellas, which is almost constant all over the year; (2) second part is reflected by the reflectors strips on the back side of the horizontal lamella; (3) final part corresponds to other solar radiation passing on through the glazing unit. The obtained energy is then transferred by water circulation to a thermal exchanger in a storage tank and the hydraulic connections fitted in the frames. Because the solar window composes itself directly into the building envelop, it contributes as well to the thermal and phonic insulation. Apart from the thermal contributions, it can also favours on passive solar heat gain and day lighting supplies in winter

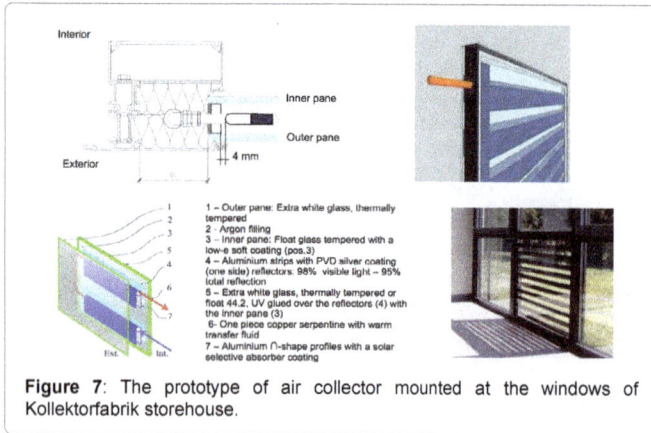

**Figure 7**: The prototype of air collector mounted at the windows of Kollektorfabrik storehouse.

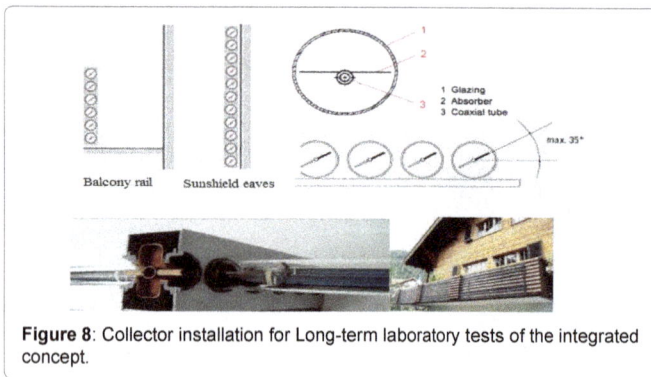

**Figure 8**: Collector installation for Long-term laboratory tests of the integrated concept.

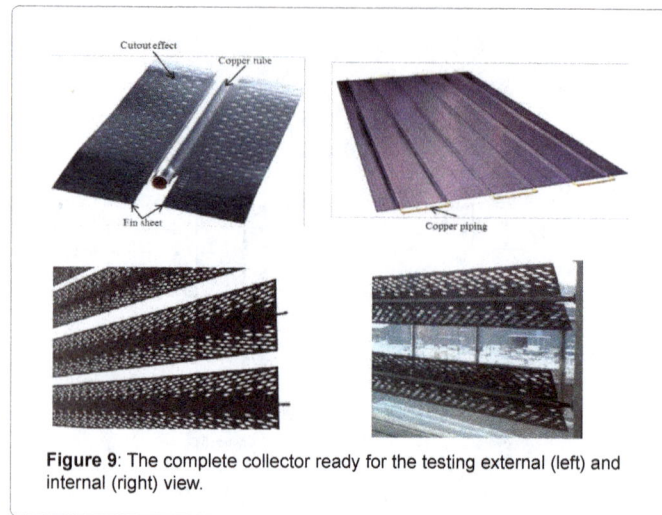

**Figure 9**: The complete collector ready for the testing external (left) and internal (right) view.

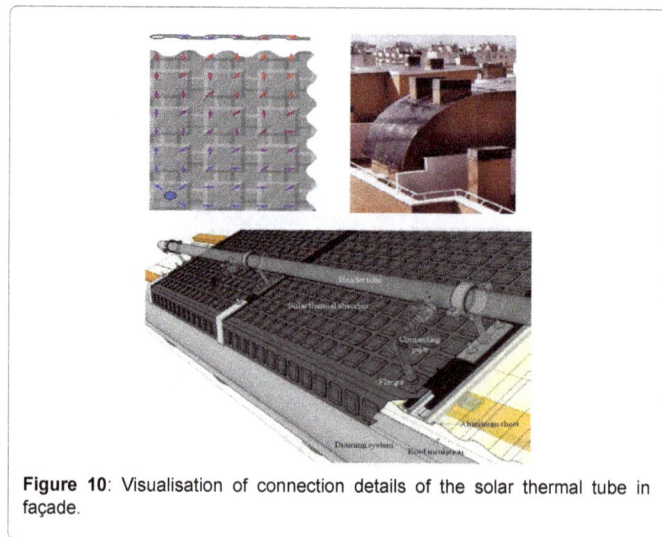

**Figure 10**: Visualisation of connection details of the solar thermal tube in façade.

whereas solar shading and day lighting diffuse in summer.

Balcony -based BIST technology: Figure 8 presents a group of evacuated tubes in a parallel module. Each tube is composed of individual glass tube, a black or black-tube thermal absorber with fluid-cooling system, surrounded by a vacuum interior space. The high-level vacuum insulation minimises the heat losses from the system, enabling such BIST a thermally effective solution. The standard arrangement is a field of several glass tubes with manifold tubes at top and bottom. The tubes are standardized products with easy joining while the number of paralleled tubes can be flexible according to the heat demand or construction size. But the structure and external appearance of the evacuated tubes limit their integration possibilities into the envelope to some extent, but applications like balcony rails or eaves, corridors are promising [28].

**Sunshield-based BIST technology:** Figure 9 shows a kind of solar thermal absorber strip that is made of aluminium with a metallically joined copper tube. Resulting from a combination of highly specialised surface, good fin efficiency and excellent heat-transfer properties, this BIST has a high performance. The special rhombic shape of pipe allows a turbulent flow and increased heat-transferring area to increase heat transfer to the heating medium. The flexible size makes the absorbers in either serial or parallel flow as desire. This kind of BIST has specific applications in the Nordic regions within glass, in southern hemisphere without glass [29].

**Roof-based BIST technology**: Figure 10 shows a sample of the unglazed BIST roof system. Two corrugated stainless steel sheets are welded together to form a thermal absorber with the cushion structure

upon the rubber roof insulation. The peculiar structure also enables various types of integration on all types of roof profiles, even curved ones. Because of the selective black surface, the roof-based BIST possesses high absorption capacity and low emissivity equal to the flat plate collectors but with a reduced price, and the operating temperatures are between 30 and 50 °C. Compared to previous mentioned BIST, roof-based one has a perfect waterproof quality owing to the self-draining aluminum profiles beneath the solar thermal absorber, which is used to evacuate any possible water infiltration and fix the corrugated BIST on the roof. Its applications covers large-scale hot water pre-heating, swimming pool heating, preheating of heat pumps systems and seasonal storage [30].

## Overview of different BIST structures in terms of architectural elements

An overview of the structural design of currently available BIST systems is presented in Table 2 in terms of different architectural elements.

## Design Criteria for BIST Technology

### Design standards

Generally, the building envelope should provide protection from external conditions, such as solar irradiation, temperature, humidity, precipitation and wind, in order to achieve a pleasant indoor thermal comfort. Therefore, it is vital to take consideration of technical issues, i.e. efficiency, effectiveness, safety, durability and flexibility, together with constructive and formal issues at the early design stage of a BIST system [49].

Regarding as the construction component, there are statutory instruments, directives and standards for the facade application. In European, the currently available standards are: (1) the *Regulation 305/2011* that defines the essential requirements of building construction products placed on the market for an economically reasonable life cycle; (2) *European Technical Approval Guidelines (ETAG)* that seriously address the technical requirements for building components; (3) *Construction Products Directive (CPD) and Construction Products Regulation (CPR)* that give the upper and lower limit figures relating to the construction products; (4) *Directive on the Energy Performance of Buildings (EPBD)* that specifies the detailed requirements for building, with extensive impacts to selection and application of a range of construction products. On the whole, BIST is fundamental to possess the construction, hydraulic and hygiene characteristics (listed in Figure 11). In terms of the function, shielding, comfort maintaining and communication availability are the items to be addressed; In terms of the construction concern, protection from external intrusions, load-bearing capability, prevention of thermal bridge and moisture condensation, mechanical stress, stability, energy efficiency, thermal reservation, safety in use, provision to material volume expansion, as well as fitting with other envelop materials are the factors to be considered; In terms of hydraulic concern, prevention from the water leakage, and balance to the hydraulic pressure difference should be considered; And in terms of hygienic concern, health, environmental adaptability should be the issues to be addressed [50]. The common BIST components are relatively heavier than the conventional facade components, therefore risk assessment and management should be particularly addressed for achieving the safe installation. The standards addressing the BIST safety issues are: (1) the *Micro-generation Installation Standard MIS 3001 (Issue 1.5)* and (2) *EST CE131, Solar water heating systems – guidance for professionals – conventional models* [51].

### Architectural consideration

A great deal of work has been thoroughly conducted in the technical assessment of BIST systems, such as configuration design, absorber material, paint and coatings, and connection methods etc. [4]. However, the considerations of functional, constructive and formal requirement play a marginal role in the BIST design. It can be briefly articulated into wall-, window-, balcony- and roof-based envelope parts with typical schematic structures illustrated in Figure 12. The advantages and disadvantages of various BIST applications, in terms of functional, constructive and aesthetic aspects, are listed in Table 3.

On basis of the analyses in Table 3, when looking the features for each BIST type from functional, constructive and aesthetic points of view, some conclusions can be made as below. Firstly, the opaque facade is usually composed of multi-layers with functions of external protection and insulation. Such features exactly offset the limitations of flat-plate BIST, which is less flexible in translucency and module

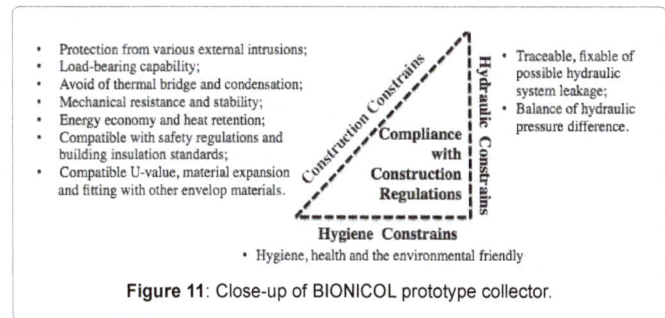

**Figure 11**: Close-up of BIONICOL prototype collector.

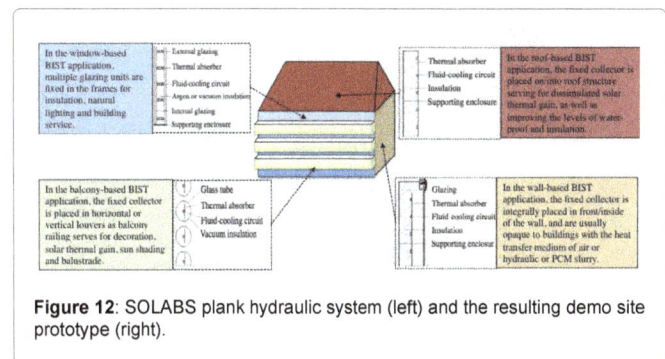

**Figure 12**: SOLABS plank hydraulic system (left) and the resulting demo site prototype (right).

thickness. Therefore, the wall-based BIST application is especially common in a renovation project as a cladding element with further insulation and protections of weather and mechanical stress. Secondly, the transparent and translucent window-based application concerns more on daylight transmission, outdoor visual relation and partial sun shading. In such case, the light-weighted glazed/unglazed collector or evacuated tubes are recommended to integrate with glazing in an alternating or interlaced pattern to have partial sun shading, or create a dummy effect. Thirdly, the evacuated tube shows more promising applications in the balcony-based integration for its lightweight, higher efficiency and convenience in assembly and pipe connections. Finally, roof has the great popularity for the installation of BIST systems. It gives the superiority in dissimulating solar collector, higher solar thermal yield, and convenient installation methods.

## Operation Conditions of BIST Systems

The proper control of operational conditions, such as temperature and flow rates, is usually an important factor for the performance enhancement of BIST systems [52]. Solar radiation is not the significant affecting factor for the efficiency of air based solar collectors [53], whilst the airflow distribution and the upward heat loss caused by the ambient wind and the approach velocity have considered closely affecting the operation performance of air-based BIST system. Gunnewiek et al. [54] studied the airflow distribution on the face of an unglazed solar thermal collector in a computational fluid dynamics (CFD) model at a constant wind speed of 5 m/s. The wind was found to reinforce the factors producing outflows, and the recommended minimum average suction velocity required to avoid such outflow were given for four operating conditions. Fleck et al. [55] experimentally studied the effects of the ambient wind on working performance in regard of wind direction, speed and fluctuation intensity on an unglazed solar collector. The measurements indicated that a high magnitude of turbulence dominated the near wall, and the maximum collector

| Envelop part | Advantages | Disadvantages |
|---|---|---|
| **Wall-based** | • An economical and high efficient solution to embed an active solar collector in the finishing layer of the external insulation;<br>• Lessen climate disturbance of original facade unit;<br>• Improving thermal insulation of the building;<br>• Offset against the cost of conventional facade.<br>• Most solar collecting area;<br>• Simplify in the piping arrangement;<br>• Well suited for superimpose in an existing building;<br>• High grade of pre-fabrication possible; | • Renewable energy components require high demands in self quality, material expansions compatible and installation due to totally external exposure;<br>• More costs for outdoors' cleaning and maintenance;<br>• Risk of condensation and thermal frost within insulation;<br>• Cold bridge and acoustic problems at a penetration hole;<br>• Additional imposed staticandwind that require additional fixed structures. |
| **Window-based** | • If the renewable energy component is placed within a cavity of glazing unite, it has no reduced life expectancy;<br>• Regulating the visual relations inside/outside and the supply of fresh air / daylight and passive solar gains;<br>• High grade of pre-fabrication possible; | • Low light transmission through renewable energy components;<br>• Additional moveable shading in clear vision area is necessary;<br>• Have risks in reducing life expectancy caused by water leakage, or thermal breakage and expansion under high temperature. |
| **Balcony/ Sunshield-based** | • Optimum overall shading with full room height visions allows an adequate daylight distribution;<br>• Energy output independent for orientation and solar angle for the vacuum tube collector;<br>• Easy cleaning and maintenance;<br>• Making solar energy visible;<br>• Well suited for superimpose in existing building;<br>• High grade of pre-fabrication;<br>• Offset against the cost of shading. | • Renewable energy component requires high demands in both quality and installation due to totally external exposure for both overall performance and safety aspect;<br>• Additional moveable shading in clear vision area is necessary;<br>• Additional support structure is needed. |
| **Roof-based** | • Simplify in the piping arrangement;<br>• Lessen climate disturbance of original roof structure;<br>• Improving the roof's thermal insulation;<br>• Less costs for outdoors' installation, cleaning and maintenance. | • Additional imposed static, wind, snow load that require additional fixed structures and load assessments;<br>• Have risks in reducing life expectancy caused by water leakage or thermal breakage and expansion under high temperature. |

**Table 3:** Characteristics of BIST integration with building.

efficiency occurred at the wind speeds between 1 and 2 m/s instead of non-wind speeds. Kumar et al. [56] conducted a two-year field study focusing on the relation between wind induced convective heat transfer and upward heat losses in unglazed collectors. The estimated wind heat transfer coefficient has been found to correlate to the wind speed by linear regression and power regression. Shukla et al. [57] reported that the increasing in wind velocity (below 2 m/s) increases collector efficiency; but as the wind velocity exceeding 7m/s, effects on efficiency becomes negligible.

For the hydraulic-based BIST system, mass-flow rate is considered as a variable for the extraction of maximum exergy, which can lead to identification of inefficient parts and optimum operating conditions. Although the present control methods that regulate the outlet temperature of the collector field by suitable adjusting the working fluid flow may ensure a smooth operation [58], but the reported experimental results were usually around 3% for the maximum exergetic efficiency. Badescu et al. [52] implemented a direct optimal control method in a detailed collector model with realistic meteorological data. The proposed operation at a properly defined constant mass-flow rate may be necessarily associated to the maximum exergy extraction, but it required a priori knowledge of meteorological data time series. A lot of analyses were presented in a general simulation model for the optimal design and operating parameters of solar collectors [59]. Apart from aforementioned assumptions either assuming a constant overall heat loss coefficient or considering the fluid inlet temperature equals to the ambient temperature, Jafarkazemi et al. [60] both compared and evaluated the energetic and exergetic performance of a flat plate solar collector. Based on the theoretical results, it was obvious that energy and exergy efficiencies have conflicting behaviours in many cases. Designing the system with inlet water temperature approximately 40°C more than the ambient temperature as well as a lower flow rate can enhance the maximum energy and exergy efficiency close to 80% and 8%, whereas thickening back insulation over 5 cm has little effect on both energy and exergy performances.

High temperature is detrimental to the reliability, durability and safety during operation. Normally, well-insulated glazed flat-plate collectors achieve maximum stagnation temperatures of 160–200°C, evacuated tube collectors 200–300°C [60]. Solar overheating problem may occur within BIST systems for a number of reasons, such as oversized capacity, no way of dumping the excess heat, low pitched solar collectors, switching on the electric backup immersion heater for long periods etc. But it's basically because the hot water being produced is far greater than the actual demand [61, 62]. Fan et al. [63] elucidated the relation between operating conditions, such as flow rates and properties of working fluid and heat transfer condition in a glazed solar collector. The decreased flow rate and decreased content of glycol in the glycol/water mixture were discovered to lead a growing risk of overheating in the upper part of the collector panel. The magnitude of the stagnation temperature for any solar collector depends on the solar insolation level and the ambient air temperature [64], therefore many approaches have been tried to avoid the overheating of both solar absorbers and the BIST system. Except for the simply approaches like utilizing high temperature tolerant components, venting or shading absorbers, a more intelligent method is to purposely degrade the optical performance. The relevant techniques are optical switches based on scattering layers, thermotropic layers and temperature varied particle solubility fluids [65]. In order to block the direct light into the collector, a prismatic structure based solar thermal collector was proposed under the principle of switchable total reflection. The experimental study indicated that the maximum temperature inside the collector can be regulated by thermal evaporation or manual regulation of the switching fluid level near the prisms, and there was a strong light reducing effect around noon with less sensitivity

for seasonal changes [66]. On the other side, the enhancement of heat transfer using oscillatory which flows to prevent the overheating phenomenon is another way to degrade thermal performance from a solar thermal system aspect [67]. The preliminary estimations showed a dramatic heat transfer enhancement using oscillatory flows compared with the forced convection of heat by standard unidirectional flows. Furthermore, Harrison et al. [64] proposed a summary and overview of stagnation control from both the collector level and the system level approaches as listed in Table 4. To optimise the performance of the latent heat storage in solar air heating system, Arkar et al. [68] studied the relations among PCM melting temperatures, air flow-rate by the thermal storage of solar air heating system. Based on the transient conduction-dominated numerical model, the optimal air flow-rate of 40m$^3$/h- m$^2$ was suggested.

## BIST Pilot Project Application

### Practical applications

This section summarises the pilot projects within the scope of the successful BIST building integration. The pilot projects are further expected to give an outlook of current and possible future combinations of solar thermal technologies with the passive building design concept towards green building or net zero energy buildings. Table 5 presents a list of representative BIST incorporated projects at a current stage, which is based on various types of buildings. Detailed descriptions are given in each project.

### Critical analysis

The above mentioned BIST devices has created a considerable energy saving prospective, and endorses an innovative solar thermal approach for building integration with efficient and durable devices, variable choices of colour and texture and mature processing techniques. The pilot BIST projects show how the synergic collaboration between manufacturers and researchers can be. Diverse directions with miscellaneous motives have led to a lot of different BIST building integration variations. The core element of a BIST system in the above buildings has exhibited a high energy capacity, and the end users therefore benefit of an attractive pay-back time on their investment.

However, the shortfalls in current BIST systems are primarily:

| Stagnation Control Scheme | Protects Collectors? | Protects System Components? | Fail-safe Operation? | Cost Impact? | Performance or durability Impact? |
|---|---|---|---|---|---|
| Drain-back or drain-down | No | Yes | Yes (must be installed to drain completely if also used for freeze protection) | Pumping power consumption may be increased | Collector loop open to atmosphere may increase corrosion or fouling |
| Control based (e.g., night heat rejection, recirculation etc.) | Yes | Yes | No. (requires active control and pumping) | Not for hardware. (additional pump use & potential energy loss) | May result in available energy being dumped at night |
| Steam Back | Not always | Not always | Yes (careful design and placement of system components required) | Expansion tank may have to be oversized | Potential thermal shock/scalding on restart |
| Collector Venting (Integral to collector) | Yes (if carefully designed) | Yes (if carefully designed) | Yes (for thermally activated versions) | Modest hardware cost | May experience small performance penalty if not carefully implemented |
| Heat Waster on Collector loop | Yes | Yes | Some designs operate passively - others require power or pumps, fans etc. | Significant hardware cost | If powered may require aux. generators or PV |
| Heat pipe control (Evacuated-tube collector) | No | Yes | Yes (for thermally activated versions) | Modest Hardware cost | System may be inoperable for remainder of day |

**Table 4:** Summary and overview of some collector and system level approaches to stagnation control.

| | |
|---|---|
| **Category 1: Family house & apartment cases** | |
| **No. 1 "Home for life" concept house, Aarhus, Denmark** | |
| This project dissimulates BIST within the roof to form a kind of active house with a living area of 190 m². The total solar collecting area of 6.7m² is integrated in the lowest part of the roof surface with an auxiliary heat pump system, which directly supply 50-60% of the yearly household hot water heating demand as well as a supplement for downwards room heating. | Picture source: VKR Holding A/S |
| **No.2 House J, single family house, Nenzing, Austria** | |
| This project exhibits the building integration issue between dummy elements and the glazed flat plate collectors. A dark blue cladding is chosen as finishing to best cope with the imposed dark blue appearance of 12 m² solar collectors mounted on the south facade. Due to the use of similar collector size fitting with the modular rhythm of standard cladding, and the same types of jointing, the homogeneity of colour size and jointing makes two parts of facade component acceptable. | Picture source: Domasolar |
| **No.3 Row dwelling I-Box concept building, Tromsoe, Norway** | |
| The whole design is paid considerable attention to various building materials, forms, and colours of the architectural composition of envelope without an abrupt BIST appearance. Each 5 m² BIST system is whole integrated with each residential units. | Picture source: Viessmann |
| **No.4 Sunny woods, apartment house, Zürich-Höngg, Switzerland** | |
| The whole BIST application is natural and plain in the integration design with simple accessible external connections. The solar-tube collectors act as a transparent baluster for the balcony, while the valuable sunshine can be stored through windows into a stone flooring. | Picture source: solaripedia |
| **No. 5 Jadarhus ISOBO Aktiv, Sandnes, Norway** | |

The BIST looks like mosaic titles inserted in the timber roof of the traditional rafter construction with a total living area of 178 m² and an annual heat demand of 44kWh/m². The solar collectors make up an area of 8 m² as the roof surface for 90% of hot water preparation and 95% of space heating.

Picture source: Jadarhus AS

**No. 6 2-flat penthouse in Vienna, Austria**

As a project of sustainable renovation, this ingenious BIST application with two direct flow vacuum tube collectors of 2.88 m² is installed at the roof terrace area as an innovative, modern, solar pergola. It has multi-functions as hot water production, sun shading and pergola. This project exhibits a special case of building integration from the overall architectural concept.

Picture source:Viessmann

**No. 7 Sun house, Tyrol, Austria**

There are two groups of inclination installations in this project as on the roof with 45° inclination for a maximum yield in the summer, and in the facade with 70° inclination for a maximum yield in the winter. Therefore it shows a good exhibition for both the solar thermal collectors and the photovoltaic modules perfectly adapting into the house.

Picture source: SIKO SOLAR GmbH

**No. 8 KOMBISOL® house, Tyrol, Austria**

This BIST application shows a perfect example of cladding façade by 70m² solar collectors and the thermal mass storage by concrete. The whole solar thermal façade can provide needed energy for hot water preparation and space heating with an additional geothermal heat pump. When there is excess energy from the solar thermal collectors, it can be not only stored in the ground collectors but also in the concrete core of the house, which enables a comfortable and warm climate in winter and also a cooling function in summer.

Picture source: SIKO SOLAR GmbH

**No. 9 Passive house Caldonazzi, Austria**

25% BIST installation is mixed with 75% insulated windows on the south facing façade to realize the energy concept of minimization heat loss from good insulation, passive solar energy and energy from solar collectors. The solar system with 17m² solar collectors is a combined system for both hot water preparation and space heating. It was calculated that the building cost was 1.4 Euro/m² in all.

Picture source: AEE INTEC

**No. 10 Passive house Rudshagen, Oslo, Norway**

This project utilizes the high performance polymer collectors (19.5 m²) in the south facing facade of the "solar house" to satisfy both space heating and domestic hot water preparation, which is 61 kWh/m²/year in total heat demand. This BIST application is benefit from lightweight collectors for easy domestic installation.

Picture source: Aventa AS

**No. 11 Petersbergenstrasse, Austria**

Picture          source:          AEE          INTEC

This project exhibits us a case of prefabricated BIST system to streamline installation. The 50 m² wall based glazed flat plate solar collectors has been erected 10° southwest oriented for DHW preparation and space heating. These pre-manufactured fields are made by fixing the wooden back wall of the collector to the timber frames with only about 10 steel angles for the fixation, which causes almost no effect on thermal bridges.

**Category 2: Commercial & public building cases**

**No. 12 Alpiner Stützpunkt Schiestlhaus of the Austria Tourist Club, Austria**

This project is a special utilization of BIST system for the first large mountain refuge building with the capacity to accommodate 70 people. The specialty lies in the upper south-facing facade designed as an energy-facade system for the generation of thermal energy. Through the heat exchanger, the solar collector transfers heat to three buffer storage tanks with a total capacity of 2,000 litres. Besides, both a rape oil operated unit and a solid fuel range can also load heat into the buffer storage tanks.

Photo: DI Wilhelm Hofbauer/ Treberspurg & Partner Architekten

**No. 13 Retrofitted Office Building in Ljubljana, Slovenia**

This is a detail designed retrofitting project with BIST application. The air heating vacuum tube collectors replace the balustrade on the fifth floor, while the transparent solar thermal collectors are attached to the stairwell. Both collector areas face almost south, the solar collectors 15° towards east and the air-heating tubes 15° towards west. Both components are developed to be a substitute for the building skin as well as the thermally activated building system of 100 m² office space with fluid at temperatures above 35 °C during the heating season.

[55]

**No. 14 HQ, AKS DOMA Solartechnik**

This project realizes the neutral $CO_2$ energy supply for the 470 m² offices and the 1,380 m² production hall through the BIST application. The total 80m² BIST system provides the whole heating through a wall heating and a floor heating operated with very low flow temperature, which offers ideal conditions for the operation of the solar thermal plant.

Picture source: AEE INTEC

**No. 15 Hotel Jezerka, Czech Republic**

This BIST application pays special consideration to the horizontal installation with 236m² solar collectors in the south facing balconies for heating tap water and swimming pools. These collectors were designed as handrails of balconies, or as flat roofs where vertical collectors would be overloaded by wind. The price of each collector was € 364.8 with total solar gain of 120 000 kWh/year

Photo: Thermosolar

**No. 16 Environmental Research Station Schneefernerhaus - UFS, Germany**

This retrofitting project with BIST system is located at Germany's highest mountain from a stone building into a research station. With adding additional thermal insulation, 100m² solar collectors were integrated in the building facade as well for both DHW preparation and space heating. The related water based heat distribution consists with floor heating and radiators, while heat pumps and electricity are set for auxiliary heating. The price of each collector was € 943 with total solar gain of 60 000 kWh/year.

Photo: Fraunhofer ISE (Fraunhofer Institute for Solar Energy Systems)

**No. 17 Granby Hospital, Granby (Quebec ) / Canada**

This application is one of the few air-based BIST systems. The transpired solar fresh air heating system in this hospital is specially designed to satisfy the demand of large mounts of fresh air. The solar system is aesthetically presented with 82 m²-curved facades to provide 8160 m³/h to the space below. The working principle is that the perforated metal absorber draws in heated fresh air off the surface of south-facing walls, where it is then distributed throughout the building as pre-heated ventilation air. Overall, the system owns operating efficiencies up to 70% with total payback's within 5 years with energy savings about 149.1GJ.

Photo: Matrix Energy

**No. 18 Social housing in Paris, France**

This is the first social house building with BIST application. The new multifunctional semitransparent collector encapsulated into a double skin facade, which weights 45kg, offers complete privacy from the passengers commuting nearby, ensures light penetrating into the back of room, as well restricts the noise flow. Besides, the solar panel captures solar energy to produce enough power to meet 40% of the domestic hot water needs providing 44% of domestic hot water needs.

Photo:ROBIN SUN

**No. 19 The Bellona Building, Oslo, Norway**

This building is constructed of in situ concrete with facades in plaster and glass. The installation shapes of solar collectors serve as the self-shading facade and cover large parts of the south-facing wall. There are windows on the inward-facing part and 240 solar collectors on the outward-facing part. Owning to good insulation, excellent window, minimized thermal bridge and low air leakage factor, solar collectors and geothermal heat pump can cover all the heating demand.

Photo: Finn Staale Feldberg

| | |
|---|---|
| **No. 20 School building in Geis, Switzerland** | |
| This is another successful prefabricated BIST application. Because of early design phase intervention, designers paid considerable attention to the facade design, layout, size and the fixed modular dimensions of the solar collectors. The total 63 m² collector field fully respects the rhythm of window openings and the colour of both window frame and concrete bricks, showing a convincing result. | Photo: Schweizer-metallbau |
| **No.21 The centre d'exploitation des Routes Nationales (CeRN), Bursins, Switzerland** | |
| In the south-facing facade, a large area of stainless steel unglazed metal collectors is utilized as multifunctional facade claddings for floor heating, hot water production and an excellent corrosion-resistant building element as well. This BIST application harmonized all the active and non-active panels with stainless steel elements of same dimension and appearance, therefore they fit the modular demands of the building. The active solar panels weigh about 10 kg/m², an important consideration for easy assembly. | Photo: Energie Solaire SA |

(1) most concepts of absorber parts in the BIST are directly inherited from the conventional solar thermal collectors, which exist instability under long term weather exposure, difficulty in both on-site assembly and practical application and complexity in fluid channel structure resulting in the bulk volume and fragile for the BIST application; (2) most BIST design that only functions as a structural cladding element of glass curtain-wall, rooftop, or traditional wall surface; (3) limited considerations are given to the irregular geometry, colour and texture design leading to boring building appealing; (4) lack of building related studies in terms of lighting pollution, acoustic effect, structural load and thermal performance etc. Typical strategies can be assigned to dissimulation into building envelope, special placement and modular building component design. Furthermore, it can be found that the majority of new building projects are solar house or passive house, while those renovation projects provide a new direction of multi-functional transformation instead of sole repairing. More focuses have been put on threatening resource shortage, comfort living environment as well as position architectures themselves in the former niche. It is worthy of mention that as an innovative choice of multifunctional building envelope, BIST has superiority in good insulation, ability of capturing solar heat and high compactness, which fully satisfies the current boom branch of green buildings and zero energy buildings.

## Conclusion

This paper presents an overall review of the currently available BIST systems and their applications by emphasizing on structural design and architectural integration. Different BIST systems in terms of the working medium, i.e., air, hydraulic (water/heat pipe/refrigerant) and PCM, are fully described. The fundamental structure of BIST derived from the conventional solar thermal collectors includes three basic configurations as glazed or unglazed flat-plate and evacuate tube shapes. However, the specific structures of the BIST systems are various when they serve as different building elements in practice. Some design criteria including efficiency, safety, durability and flexibility, and constructive issues at the early design stage of a BIST system are introduced. The practical operation conditions of BIST systems are also discussed. There are many advantages by using the BIST systems, whose aesthetics and building envelope characteristics can match that of existing building products and therefore hasten their adoption to buildings. These systems can also provide substantial savings to the building or home owners from reduced heating, maintenance and repair costs.

Through the comprehensive literature review looking into the BIST structural design and their architectural integrations, the assessment indicates a promising future for considerable energy saving with a potentially broad market. However, there are still barriers existing in those systems: (1) most concepts of absorber parts stayed in the conventional solar thermal system. When applied as a BIST component, it has to face the instability under a long term weather exposure, difficulty in both on-site assembly and practical application, complexity in fluid channel structure resulting in the bulk volume and fragile as multi-function building facade; (2) most BIST design that only functions as a structural cladding element of glass curtain-wall, rooftop, or traditional wall surface; (3) limited considerations are given to the irregular geometry, colour and texture design leading to boring building appealing; (4) lack of building related studies in terms of lighting pollution, acoustic effect, structural load, thermal performance and types of jointing etc.

To break through above limitations and achieve a broader market deployment, new BIST technologies are still desired to emerge urgently. Further appropriate recommendation for the related research development in terms of structural design and architectural integration would be: (1) integration of structural and finish materials together to work as true building material; (2) compulsory structural/rigidity test for the BIST serving as a load bearing structural element; (3) development of light-weight and long-life polymer materials to replace the current promising materials, like metal, glass and ceramic, to minimize loads on existing architectural structure; (4) integration of BIST design into architectural or life-cycle design tools (such as BIM - building information modelling) to quickly assess the appropriate structure and integration method of a BIST system for building.

This review shall facilitate the development of solar driven, distributed (or centralised) service for buildings, which would lead to the corresponding saving in fossil fuel consumption and the reduction in carbon emission.

### Acknowledgement

The authors would acknowledge our sincere appreciation to the financial supports from the Ningbo Natural Science Foundation (2015A610039).

### Reference

1. ESTTP European Solar Thermal Technology Platform (2009) Solar Heating and Cooling for a Sustainable Energy Future in Europe. European Solar Thermal Technology Platform.

2. Kalogirou SA (2004) Solar thermal collectors and applications. Progress in Energy and Combustion Sci 30: 231–295.

3. Shukla R, Sumathy K, Erickson P, Gong J (2013) Recent advances in the solar water heating systems: A review. Renew Sustain Energy Rev: 173–90.

4. Zhang X, Shen J, Lu Y, He W, Xu P, et al. (2015) Active Solar Thermal Facades (ASTFs): From concept, application to research questions. Renew Sustain Energy Rev 50: 32-63.

5. Tripanagnostopoulos Y (2007) Aspects and improvements of hybrid photovoltaic/thermal solar energy systems. Solar energy 81: 1117–1131.

6. Zhang X, Zhao X, Smith S, Xu J, Yu X, et al. (2012) Review of R&D Progress and Practical Application of the Solar Photovoltaic/Thermal (PV/T) Technologies. Renew Sustain Energy Rev 16: 599-617.

7. Chen Xi, Hongxing Y, Lin Lu, et al. (2011) Experimental studies on a ground coupled heat pump with solar thermal collectors for space heating. Energy 36: 5292-5300.

8. Zhao X, Zhang X, Riffat S, et al. (2011) Theoretical investigation of a novel PV/e roof module for heat pump operation. Energy Convers Manag 52: 603–614.

9. Chen K, Oh SJ, Kim NJ, Lee YJ, Chun WG (2010) Fabrication and testing of a non-glass vacuum-tube collector for solar energy utilization. Energy 35: 2674–2680.

10. Zhang X, Zhao X, Shen J, et al. (2014) Dynamic performance of a Solar Photovoltaic/Loop-heat-pipe Heat Pump Water Heating System. Appl Energy 114: 335-352.

11. Boris R, Sergii K, Rostyslav M, Olga A, Andrii R (2013) Solar Collector Based on Heat Pipes for Building Facades. Smart Innovation, Systems and Technologies [Internet]. Berlin, Heidelberg: Springer Berlin Heidelberg: 119–126.

12. He W, Hong X, Zhao X, Zhang X, Shen J, et al. (2015) Operational performance of a novel heat pump assisted solar facade loop-heat-pipe water heating system. Appl Energy 146: 371–382.

13. Deng Y, Zhao Y, Wang W, Quan Z, Wang L, et al. (2013) Experimental investigation of performance for the novel flat plate solar collector with micro-channel heat pipe array (MHPA-FPC). Appl Thermal Engg 54: 440–449.

14. Zhao CY, Lu W, Tian Y (2010) Heat transfer enhancement for thermal energy storage using metal foams embedded within phase change materials (PCMs). Solar energy 84:1402–1412.

15. Delgado M, Lázaro A, Mazo J, Marín JM, Zalba B (2012) Experimental analysis of a microencapsulated PCM slurry as thermal storage system and as heat transfer fluid in laminar flow. Appl Thermal Engg 36: 370–377.

16. Zhang Y, Wang S, Rao Z, Xie J (2011) Experiment on heat storage characteristic of microencapsulated phase change material slurry. Solar Energy Materials and Solar Cells 95:2726–2733.

17. Qiu Z, Zhao X, Peng Li, et al. (2015) Theoretical investigation of the energy performance of a novel MPCM (Microencapsulated Phase Change Material) slurry based PV/T module. Energy 87: 686-698.

18. Alvarado JL, Marsh C, Sohn C, Phetteplace G, Newell T (2007) Thermal performance of microencapsulated phase change material slurry in turbulent flow under constant heat flux. Int J Heat Mass Transfer 50: 1938–1952.

19. Youssef Z, Delahaye A, Huang L, Trinquet F, Fournaison L, et al. (2013) State of the art on phase change material slurries. Energy Convers Manag 65:120–132.

20. Tyagi VV, Buddhi D (2007) PCM thermal storage in buildings: A state of art. Renew and Sustain Energy Rev 11:1146–1166.

21. Koschenz M, Lehmann B (2004) Development of a thermally activated ceiling panel with PCM for application in lightweight and retrofitted buildings. Energy and Buildings 36: 567–578.

22. De Gracia A, Navarro L, Castell A, Ruiz-Pardo Á, Álvárez S, et al. (2013) Experimental study of a ventilated facade with PCM during winter period. Energy and Buildings 58: 324–332.

23. Saman W, Bruno F, Halawa E (2005) Thermal performance of PCM thermal storage unit for a roof integrated solar heating system. Solar energy 78: 341–349.

24. Diarce G, Urresti A, García-Romero A, Delgado A, Erkoreka A, et al. (2013) Ventilated active facades with PCM. Appl Energy 109: 530–537.

25. Rodriguez-Ubinas E, Ruiz-Valero L, Vega S, Neila J (2012) Applications of Phase Change Material in highly energy-efficient houses. Energy and Buildings 50:49–62.

26. http://www.waf.at/

27. http://www.robinsun.com/

28. http://www.schweizer-energie.ch/html/index.html

29. http://en.ssolar.com/ProductApplicationAreas/SunstripStripsforcollectors/tabid/307/Default.aspx

30. http://www.energie-solaire.com/

31. http://www.domasolar.com/

32. http://www.hssolar.ch/

33. http://www.cost-effective-renewables.eu/publications.php?type=brochure

34. http://www.heliopan.info/Galerie/

35. Rassamakin B, Khairnasov S, Zaripov V, Rassamakin A, Alforova O (2013) Aluminum heat pipes applied in solar collectors. Solar Energy 94:145–154.

36. PG. Prototype for transparent thermal collector for window integration (2011) European Community Seventh Framework Programme. Seventh framework programme cooperation - theme 4: D3.1.2.

37. Xu P, Zhang X, Shen J, et al. (2015) Parallel experimental study of a novel super-thin thermal absorber based photovoltaic/thermal (PV/T) system against conventional photovoltaic (PV) system. Energy Reports 1: 30-35.

38. Probst MCM (2008) Architectural integration and design of solar thermal systems. PhD thesis of école polytechnique fédérale de lausanne.

39. http://www.cost-effective-renewables.eu/

40. Wang Z, Duan Z, Zhao X, Chen M (2012) Dynamic performance of a facade-based solar loop heat pipe water heating system. Solar Energy 86:1632–1647.

41. http://ritter-xl-solar.com/en/home/

42. http://www.winklersolar.com/

43. http://www.genersys-solar.com/

44. http://www.soltop.ch/de/home.html

45. http://www.rheinzink.ch/

46. http://www.atmova.ch/vorteile00.html

47. http://www.kme.com/en/product_lines_intended_for_alternative_energies

48. http://soltechenergy.se/

49. http://solarthermalworld.org/sites/gstec/files/standardisation.pdf

50. Roecker C, Munari M, de Chambrier E, Schueler A, Scartezzini JL (2009) Facade Integration of Solar Thermal Collectors: A Breakthrough. Berlin, Heidelberg: Springer Berlin Heidelberg: 337–341.

51. Ridal J, Garvin S, Chambers F, Travers J (2010) Risk assessment of structural impacts on buildings of solar hot water collectors and photovoltaic tiles and panels – final report [Internet].

52. Badescu V (2007) Optimal control of flow in solar collectors for maximum exergy extraction. Int J Heat Mass Transfer 50: 4311–4322.

53. Leon MA, Kumar S (2007) Mathematical modeling and thermal performance analysis of unglazed transpired solar collectors. Solar energy 81: 62–75.

54. Gunnewiek LH, Hollands KGT, Brundrett E (2002) Effect of wind on flow distribution in unglazed transpired-plate collectors. Solar energy 72: 317–325.

55. Fleck BA, Meier RM, Matović MD (2002) A field study of the wind effects on the performance of an unglazed transpired solar collector. Solar energy 73: 209–216.

56. Kumar R, Rosen MA (2011) A critical review of photovoltaic–thermal solar collectors for air heating. Appl Energy 88: 3603–3614.

57. Shukla A, Nkwetta DN, Cho YJ, Stevenson V, Jones P (2012) A state of art review on the performance of transpired solar collector. Renew Sustain Energy Rev 16: 3975–3985.

58. Meaburn A, Hughes FM (1996) A simple predictive controller for use on large scale arrays of parabolic trough collectors. Solar energy 56: 583–595.

59. Farahat S, Sarhaddi F, Ajam H (2009) Exergetic optimization of flat plate solar collectors. Renew Energy 34:1169–1174.

60. Jafarkazemi F, Ahmadifard E (2013) Energetic and exergetic evaluation of flat plate solar collectors. Renew Energy 56:55–63.

61. Akhtar N, Mullick SC (1999) Approximate method for computation of glass cover temperature and top heat-loss coefficient of solar collectors with single glazing. Solar energy 66: 349–354.

62. Mahdjuri F (1999) Solar collector with temperature limitation using shape memory metal. Renew Energy 16: 611–617.

63. Fan J, Shah LJ, Furbo S (2007) Flow distribution in a solar collector panel with horizontally inclined absorber strips. Solar energy 81:1501–1511.

64. Harrison S, Cruickshank CA (2012) A review of strategies for the control of high temperature stagnation in solar collectors and systems. Energy Procedia 30: 793–804.

65. http://www.google.co.uk/patents/US4270517

66. Slaman M, Griessen R (2009) Solar collector overheating protection. Solar energy 83: 982–987.

67. Lambert AA, Cuevas S, del Río JA (2006) Enhanced heat transfer using oscillatory flows in solar collectors. Solar energy 80:1296–1302.

68. Arkar C, Medved S (2015) Optimization of latent heat storage in solar air heating system with vacuum tube air solar collector. Solar energy 19:10–20.

69. Giovanardi A (2012) Integrated solar thermal facade component for building energy retrofit. PhD thesis of Doctoral School in Environmental Engineering, UNIVERSITÀ DEGLI STUDI DI TRENTO in collaboration with EURAC RESEARCH.

70. Payakaruk T, Terdtoon P, Ritthidech S (2000) Correlations to predict heat transfer characteristics of an inclined closed two-phase thermosyphon at normal operating conditions. Appl Thermal Engg 20:781–790.

71. http://www.bionicol.eu/

72. Munari Probst MC, Roecker C (2007) Towards an improved architectural quality of building integrated solar thermal systems (BIST). Solar Energy 81: 1104-1116.

73. Musall I, Weiss T, Voss K, Lenoir A, Donn M, et al. (2011) Net Zero Energy Solar Buildings: An Overview and Analysis on Worldwide Building Projects. IEA Solar Heating & Cooling Programme.

74. Switzerland EPPA (2012) Integrated Concepts - Construction Aspects. European Community Seventh Framework Programme. Seventh framework programme cooperation - theme 4: d4.1.2.

75. http://projects.iea-shc.org/task39/projects/default.aspx

# Performance Analysis and Comparison of Different Photovoltaic Modules Technologies under Different Climatic Conditions in Casablanca

**Elmehdi Karami[1*], Mohamed Rafi[2], Amine Haibaoui[1], Abderraouf Ridah[1], Bouchaib Hartiti[2] and Philippe Thevenin[3]**

[1]*Department of Physics, LIMAT Laboratory, Ben M'sick, Morocco*
[2]*Mohammedia Faculty of Science and Technology, MAC & PAM Laboratory, ANEPMAER Group, Morocco*
[3]*Laboratory Optical Materials, Photonics and Systems, University of Lorraine, Metz, France*

## Abstract

The main goal of this work is to study the performance of silicon-based photovoltaic modules of different technologies (Monocrystalline (c-si), Polycrystalline (p-si) and Amorphous (a-si)) installed on rooftop of the Ben m'sik faculty at Hassan II university, Casablanca, Morocco (Latitude 33°36"N, Longitude 7°36"W). This study is based on daily measurements under various climatic conditions (clear, cloudy and rainy). In order to improve the performance evaluation, the real-time measurements were taken for every five minutes of different climatic parameters (solar irradiation, ambient temperature, module temperature, wind speed and direction) and electrical parameters (power, current and voltage). In fact, we studied the PV array efficiency, the inverter efficiency and the system efficiency. In addition, we performed an evaluation to the PV array, reference and final yields and the performance ratio (PR). The results show maximum values for module efficiency, final efficiency and system efficiency on a clear day for all three technologies due to high irradiation. The maximum values of PR are 72.10%, 91.53% and 86.20%, are obtained on a cloudy day, this is due to the low temperature and the high wind speed. Minimum values of PR, module efficiency, reference efficiency and final efficiency on a rainy day are due to the low sun exposure and the rain which affect the generated energy and stability of PV systems.

**Keywords:** PV performance; PV array efficiency; PR inverter efficiency

## Introduction

The cost increase of conventional energy, the limitation of its resources, the uncertainty on energy supply and global warming have caused renewed interest in installations using solar energy, especially in areas with favourable climatic conditions. Currently, photovoltaic systems are becoming among the most popular renewable energy resources and have numerous applications in various fields. Indeed, the photovoltaic industry has achieved durative development at an annual average rate of 42% since 2009 [1]. In 2015 the total cumulative installations amounted to 242 GWp [2]. Morocco adopts an energy policy to use solar energy, the aim is to reach 6 GW of installed capacity from renewable energy resources by 2020 [3].

The performance of photovoltaic systems is affected by climatic conditions. It is directly affected by solar irradiation and indirectly by operating temperature, which depends on many factors such as ambient temperature, wind speed and direction. The prediction of the performance of these systems is therefore important in several related aspects such as system sizing and control. In addition, this performance data is important for system planning and financing, as well as energy market analysis, especially when these systems are injected into the grid.

Several research projects have been carried out in different parts of the world on the performance and characteristics of the grid connected photovoltaic systems, for example Elkholy, found that low solar irradiation has a significant impact on the energy quality of the Photovoltaic system output [4]. Dabou, presented a study on the effects of climatic conditions on the performance of the grid connected photovoltaic system. The results show that these performances are affected on a cloudy and sandy day due to successive and rapid change of clouds and exposure to sand which affect the generated energy and stability of the photovoltaic system [5], while there are no studies that include experimental results on PV performance in the Casablanca

region and the interaction of these PV with the environment in this region, this paper presents an experimental study with a critical analysis of different PV modules based on silicon for 3 days under variable climatic conditions (clear, cloudy and rainy), the experimental analysis was carried out in order to evaluate the real performance of the selected technologies under real conditions in Casablanca.

In fact, the performance of photovoltaic systems depends on the continuous and unpredictable change of several variables, such as solar irradiation, ambient temperature and wind speed. Therefore, the presence of a meteorological station is essential.

## Materials

### PV array

Our 6 kWp photovoltaic installation (Figure 1) is facing equator, tiletd by 30° and devised into 3 mini-installations using three silicon technologies, of nearly 2 kWp for each one. Each mini-installation is connected to a Sunny Boy inverter. Both polycrystalline and monocrystalline contain 8 "Solar World" modules of 255 watts each, while the amorphous contains 12"Next Power" modules of 155 watts each. The details PV modules specifications are presented in Table 1.

### Inverters

In our installation, we used string inverter architecture (Figure

---

**\*Corresponding author:** Elmehdi Karami, Department of Physics, LIMAT Laboratory, Ben M'sick, Morocco, E-mail: karami.abdelkebir@ gmail.com

Figure 1: View of 6 kWp PV installed in Casablanca.

| Performance under STC condition | SW 255 MONO | SW 255 POLY | NT-155 Amorphous |
|---|---|---|---|
| Maximum power (Pmax) | 255 Wp | 255 Wp | 155 Wp |
| Open circuit voltage (Voc) | 37.8 V | 38 V | 85.5 V |
| Maximum power point voltage (Vmpp) | 31.4 V | 30.9 V | 65.2 V |
| Short circuit current (Isc) | 8.66 A | 8.88 A | 2.56 A |
| Maximum power point current (Impp) | 8.15 A | 8.32 A | 2.38 A |

Table 1: PV modules electrical characteristics.

2), in this architecture an inverter is placed at the end of each chain which aims to increase the number of DC/DC converter which leads to the possibility of extracting the maximum power [6]. The main specifications of the inverter are showing in Table 2.

## Weather station

In order to collect the meteorological data, we installed one of the 20 stations developed in the PROPRE.MA project [7], funded by Research Institute in Solar Energy and New Energies (IRESEN). This station measures the horizontal and 30° tilted solar irradiations, ambient temperature, PV modules temperature and the wind speed and its direction (Figure 3).

The solar sensor used in our metrology system is a polycrystalline silicon module, this solar "Sun Plus 20" is suitable for industrial and professional uses. An anemometer was used to measure the running speed and wind direction. For the measurement of ambient and module temperatures we used four temperature sensors PT100 module, for room temperature the sensor is in direct contact with air, but protected from sun and rain. Its shelter is well ventilated but provides enough against rain. For module temperature, sensors are equipped with a specially insulated attachment system of a better contact with the back of the modules.

The monitoring of the different measurements assured by four PC DUINO (is a mini PC or single board computer platform that runs PC like OS such as Ubuntu and Android ICS). The recorded parameters provide information about the power levels, DC/AC currents and voltages as well as the metrological parameters. The data is recorded with five minutes time step and saved on daily files.

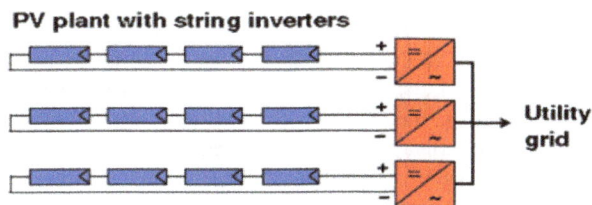

Figure 2: Inverter architecture.

| Inverter | Specification |
|---|---|
| Maximum DC power | 2600 W |
| Maximum DC voltage | 700 V |
| PV - voltage range at MPPT | 175 V-560 V |
| Maximum DC current | 15 A |
| Maximum AC power | 2500 VA |
| Nominal AC power | 2500 W |
| Nominal AC voltage | 230 V/60 Hz |
| Nominal AC current | 14.2 A |
| Maximum efficiency - Euro efficiency | 96.3%-95.3% |

Table 2: Sunny Boy 2500 HF inverter specifications.

Figure 3: View of weather station.

## Methodology

The conversion efficiency value of PV modules is very low. Since their available power depends on environmental conditions such as solar irradiance, temperature and other weather conditions. Performance assessment of PV modules becomes important. In order to investigate the energy performance of the PV modules, some indicators proposed by IEC 61724, NREL and SMA etc., can be used [8]. In this context, AC power, DC power, solar irradiance and surface area are used to calculate the performance of modules.

The instantaneous PV array efficiency ($\eta_{PV}$) is important parameter. The power conversion efficiency for one square meter surface area depends on size of PV modules, DC power ($P_{DC}$) and solar irradiance as given in equation 1 [9].

$$\eta_{PV} = \left( \frac{P_{DC}}{G * A} \right) * 100\% \tag{1}$$

Where;

G is instantaneous solar irradiance (W/m²) and A the area of PV array (m²).

The efficiency of the solar inverters can be calculated based on the value of electrical DC power delivered to the inverters from PV generator ($P_{DC}$) and the AC power obtained from inverters ($P_{AC}$). The instantaneous inverters efficiency ($\eta_{inv}$) is defined as the ratio of output to input power as given in equation (2) [10]:

$$\eta_{inv} = \left(\frac{P_{AC}}{P_{DC}}\right) * 100\% \tag{2}$$

The instantaneous system efficiency:

$$\eta_{syst} = \left(\frac{P_{AC}}{G * A}\right) * 100\% \tag{3}$$

The reference yield ($Y_R$) is the reference time in hour and it is calculated as given in equation 4 [11]:

$$Y_R = \left(\frac{G_t}{G_{STC}}\right) \tag{4}$$

Where;

$G_T$ is the total solar irradiance (kWh/m²) and $G_{STC}$ is the irradiance under standard test condition (1 kW/m²).

The PV array yield ($Y_A$) is the time which PV module operates under STC. This time can be calculated as given in equation 5 [12]:

$$Y_A = \left(\frac{E_{DC}}{P_{PV.rat}}\right) \tag{5}$$

Where;

$E_{DC}$: DC energy output (daily) of PV array (kWh).

$P_{PV.rat}$: PV rated power ($kW_p$).

The final yield ($Y_F$) is defined as the energy output divided to the nameplate power of the photovoltaic generator in STC, the $Y_F$ as given in equation 6 [13]:

$$Y_F = \left(\frac{E_{AC}}{P_{PV.rat}}\right) \tag{6}$$

Where;

$E_{AC}$: AC energy output (daily of inverter (KWh).

The performance ratio is an important performance evaluation of PV systems and it means to a measure of the quality of a PV plant that is independent on environmental parameters. Furthermore, it is stated as percent and describes the relationship between actual and theoretical energy outputs of the PV plant as formulated in equation 7 [14]:

$$PR = \left(\frac{Y_F}{Y_R}\right) \tag{7}$$

The array capture losses ($L_C$) are due to the solar PV array losses and are given by equation 8 [15]:

$$L_C = Y_R - Y_A \tag{8}$$

The system losses ($L_S$) are as result of the inverter losses and are given by equation 9 [15]:

$$L_C = Y_R - Y_A \tag{9}$$

## Results and Discussion

### Solar irradiation and DC/AC power measurements

Figures 4 and 5 shows the variation of solar irradiation, ambient

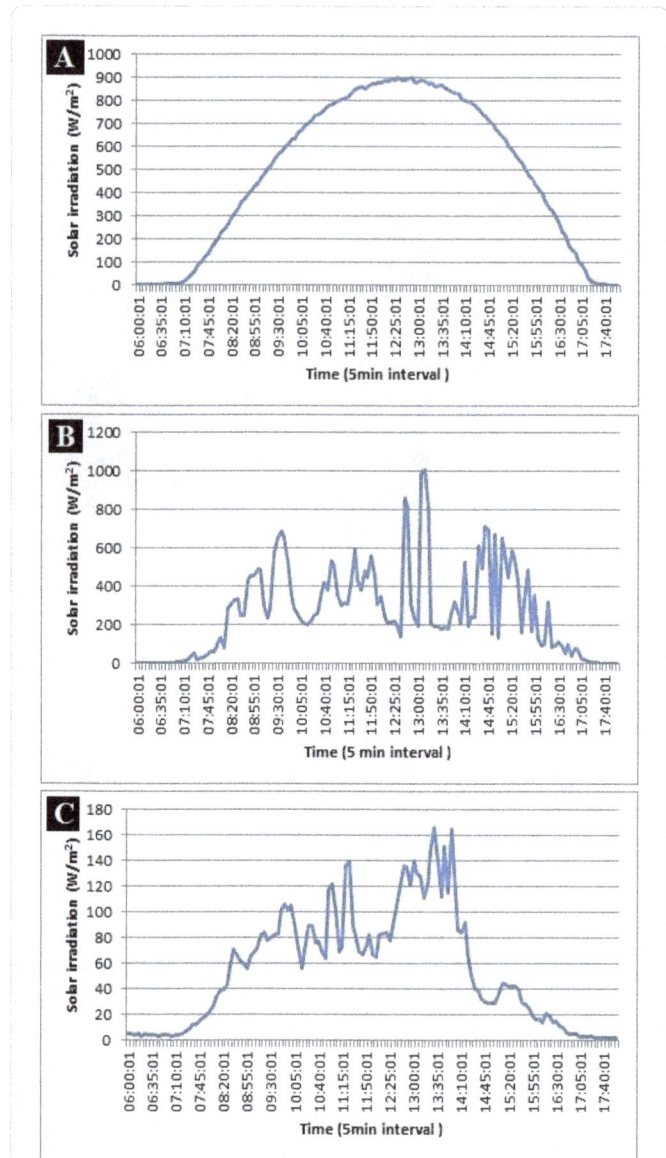

**Figure 4:** Solar irradiation in different days (clear (A), cloudy (B) and rainy (C)).

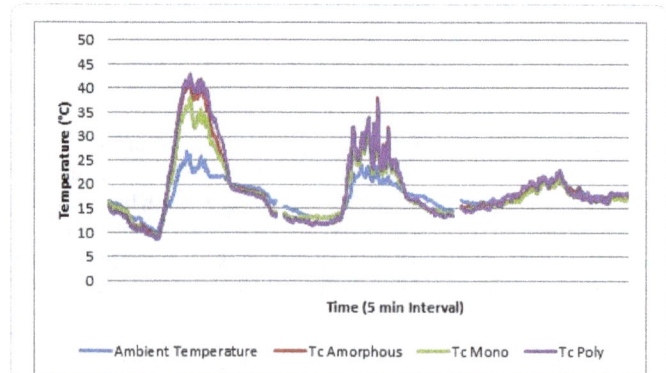

**Figure 5:** Ambient temperature and modules temperature in different days (clear, cloudy and rainy).

temperature and module temperatures during the three days (clear, cloudy and rainy). The clear day is characterized by a maximum irradiation during the noon, this leads to an increase of the ambient temperature and the temperature of the modules. During the other two days we observed that the solar irradiation has an irregular shape and the wind speed is greater in comparison with the clear day, which explains the small difference between the ambient temperature and the temperatures of the modules. This means that the modules cool down fairly quickly with the help of wind speed, these results are coherent with the results reported in Al-Otaibi et al. [16].

The Figures 6 and 7 show the DC powers of the photovoltaic modules and the AC active powers of the inverters. During the clear day the powers are higher and similar to solar irradiations compared to a cloudy and a rainy day where the fluctuations of solar irradiation conducted fluctuations of the system's powers. At relatively high temperatures (clear day noon) the p-si generates more power than the c-si, but at low temperatures (rainy day) le c-si generates more power than the p-si. During the three days, the a-ci generated less power than c-si and p-si, but during the cloudy day which characterized by low irradiation and relatively low temperatures, the a-si become more efficient, this is due to the better performance under diffused irradiation.

The efficiencies of the modules and the photovoltaic systems are shown in Figures 8 and 9. For the three days, it is clear that the efficiencies of the modules and the systems vary inversely with respect to temperature fluctuation for the modules c-si and p-si, in contrast, for the a-si the temperature variation has practically no effect on the module and system efficiency. In the rainy day there is a disconnection of the p-si inverter, which means that there is no sufficient DC power to generate the AC power, this is in good agreement with the results reported in Dabou [5].

Figure 10 shows the variation of the efficiencies of the inverters

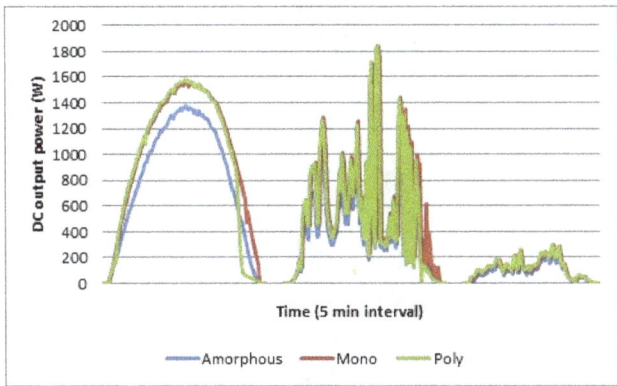

Figure 8: Daily variation of PV array efficiency in different days (clear, cloudy and rainy).

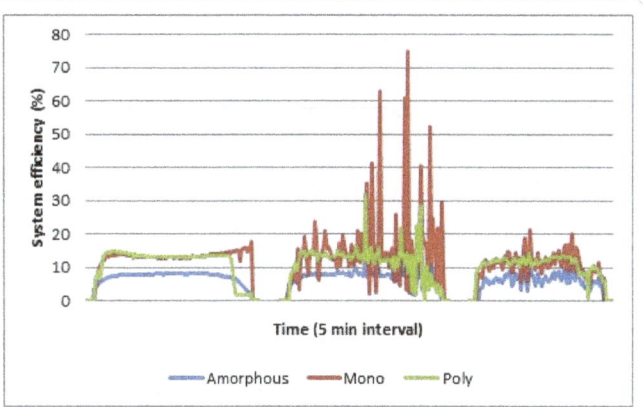

Figure 9: Daily variation of system efficiency in different days (clear, cloudy and rainy).

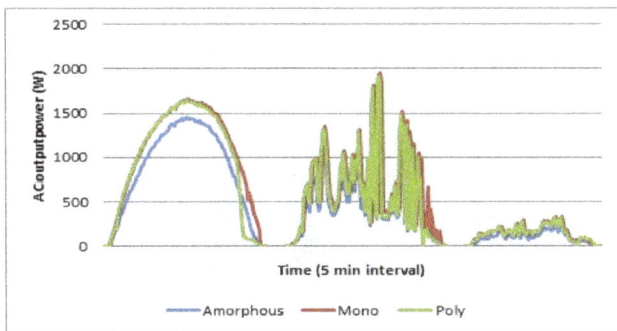

Figure 6: DC output power in different days (clear, cloudy and rainy).

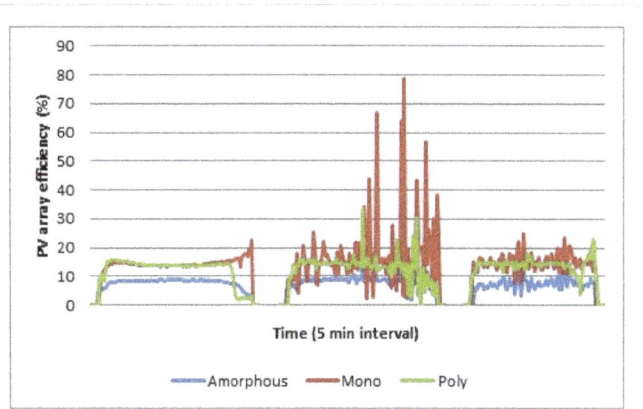

Figure 10: Daily variation of inverters efficiency in different days (clear, cloudy and rainy).

corresponding to each technology. During the clear day, there is a stability of the efficiencies of the inverters around 95%. As for the cloudy and rainy day, the efficiencies of the inverters have an irregular shape and some peaks. This variation is normal because of the rapid changing of the solar irradiation which is caused by the clouds and the rain. In the rainy day there is a significant decrease in the efficiency of the c-si and p-si inverters which is caused by the low input voltage supplying the inverters that were out of the maximum efficiency range of inverters [17].

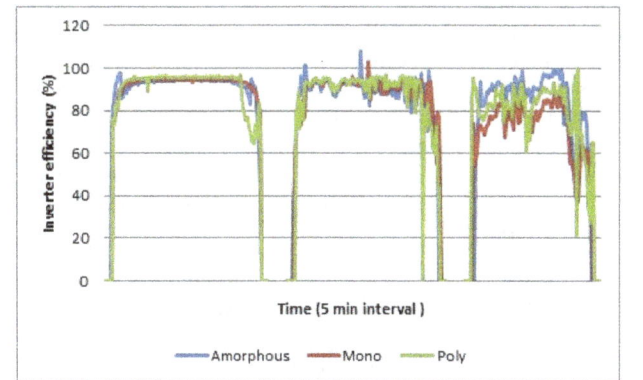

Figure 7: AC output power in different days (clear, cloudy and rainy).

| Day | Solar irradiation (KWh/m²) | Ambient temperature (°C) | Wind speed (m/s) |
|---|---|---|---|
| Clear day 16/11/2016 | 5.85 | 23.95 | 2.7 |
| Cloudy day 19/11/2016 | 3.19 | 21.48 | 4.06 |
| Rainy day 02/11/2016 | 0.63 | 18.02 | 5.2 |

**Table 3:** Daily solar irradiation, average of ambient temperature and wind speed.

| Day | Technologies | PV array efficency (%) | Inverter efficency (%) | System efficency (%) |
|---|---|---|---|---|
| Clear day 6/11/2016 | Amorphous | 8.09 | 93.50 | 7.56 |
| | Mono | 14.01 | 94.14 | 13.19 |
| | Poly | 13.15 | 95.29 | 12.53 |
| Cloudy day19/11/2016 | Amorphous | 8.21 | 92.12 | 7.57 |
| | Mono | 14.56 | 92.64 | 13.49 |
| | Poly | 13.55 | 93.56 | 12.68 |
| Rainy day 02/11/2016 | Amorphous | 6.50 | 90.21 | 5.87 |
| | Mono | 13.22 | 77.62 | 10.26 |
| | Poly | 12.72 | 83.33 | 10.60 |

**Table 4:** Daily PV array efficiency, inverters efficiency and system efficiency in different days for each technology.

| Day | Technologies | Reference yield (KWh/KWp/Day) | Array yield (KWh/KWp/Day) | Final yield (KWh/KWp/Day) | Capture losses (h/Day) | System losses (h/Day) | Performance ratio (%) |
|---|---|---|---|---|---|---|---|
| Clear day 16-11-2016 | Amorphous | 5.85 | 4.50 | 4.21 | 1.35 | 0.24 | 71.96 |
| | Mono | 5.85 | 5.55 | 5.22 | 0.30 | 0.32 | 89.31 |
| | Poly | 5.85 | 5.20 | 4.95 | 0.65 | 0.15 | 84.61 |
| Cloudy day 19-11-2016 | Amorphous | 3.20 | 2.50 | 2.30 | 0.69 | 0.20 | 72.10 |
| | Mono | 3.20 | 3.15 | 2.92 | 0.04 | 0.23 | 91.53 |
| | Poly | 3.20 | 2.95 | 2.75 | 0.24 | 0.20 | 86.20 |
| Rainy day 02-11-2016 | Amorphous | 0.63 | 0.54 | 0.39 | 0.09 | 0.15 | 62.90 |
| | Mono | 0.63 | 0.62 | 0.49 | 0.005 | 0.13 | 77.77 |
| | Poly | 0.63 | 0.59 | 0.50 | 0.04 | 0.09 | 79.36 |

**Table 5:** Daily reference yield, array yield, final yield, capture losses, system losses and performance ratio in different days for each technology.

Table 3 sums the solar radiation, ambient temperature and the wind speed during the three days. The efficiencies of the modules, the inverters and the systems are showing in Table 4. In Tables 1 and 2 we observed that the clear day is characterized by an intense solar irradiation and a mean temperature close to the standard condition temperature (STC) which influences the performances of the photovoltaic module. The results show that the efficiency of the module and that of the system are maximal in the cloudy day for the three technologies due to the low temperature and the high wind speed which affects the module temperature [18].

During the rainy day the efficiencies are minimal for the three technologies due to the low solar irradiation and the rain [19].

## Performances of the PV systems

The reference, the module and the final yields, as well as the capture losses, the systems losses and the performance ratio are shown in Table 5. The results show that maximal values of the module, reference and the final yields for the three technologies are registered during the clear day. The yields of c-si are a bit greater than those of p-ci due to the different coefficients of temperature and power. The a-si has more important capture and system losses in comparison with those of c-si and p-si, this result can be explain to the a-si's greater temperature coefficient. Also the PR varies with solar irradiation [20] and the PR is maximal for the three technologies in the cloudy day owing the fact that the decreasing in losses which is caused by the decrease of module temperature.

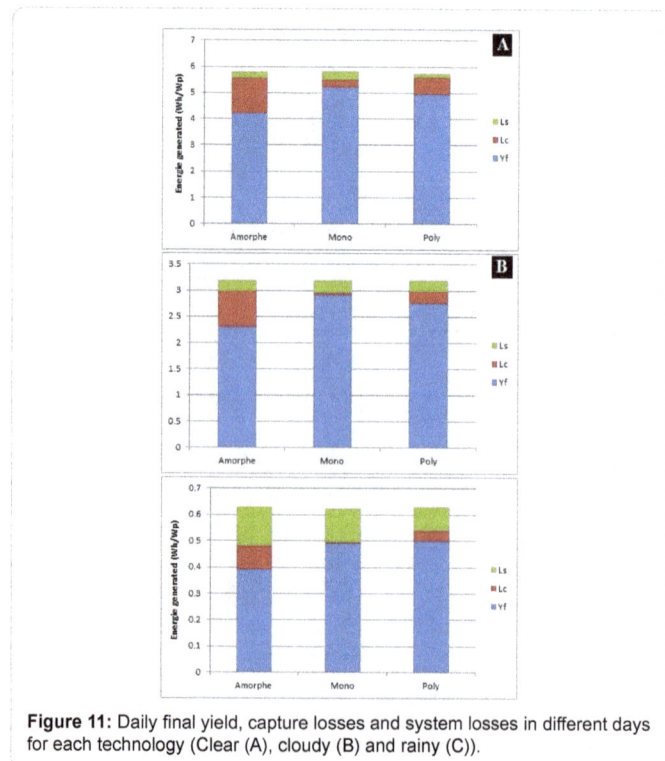

**Figure 11:** Daily final yield, capture losses and system losses in different days for each technology (Clear (A), cloudy (B) and rainy (C)).

Figure 11 (clear (A), cloudy (B) and rainy (C)) shows the final yields, the losses of the captures and the losses of systems for the three technologies during the three days. On the clear day, the three technologies have generated a great amount of energy in comparison with the two other days. The losses of the system for the three technologies in both the cloudy and rainy day increase with decreasing of the inverter efficiency. Also the capture losses of a-ci during the three days have the most important values. In fact, two factors are responsible of the poor effectiveness of the a-si: STEABLER-WRONSKI effect [21] and the low mobility of holes within the material.

## Conclusion

The experimental results show that the three PV technologies have a different behaviour depending on the days. This differences are mainly due to the variations of the spectral component, the weather condition, the installation type, etc. this study on the silicon PV modules of different technologies was performed in order to understand the impact of the different parameters and to evaluate the consequences on the energetic production. The obtained results show that: the efficiency's maximal values for the PV array on a cloudy day are 8.21%, 14.56% and 13.55% for a-si, c-si and p-si respectively. As for the system efficiency we have obtained as maximal values: 7.57%, 13.49% and 12.68% for a-si and p-si respectively, also on a cloudy day. In the clear day, the maximal values of system efficiency are 93.50%, 94.14% and 95.25% for a-si, c-si and p-si respectively. $Y_A$, $Y_F$ and $L_C$ were shown to have maximal values for the three technologies on a clear day. The maximal values of PR are 72.10%, 91.53% and 86.20% for a-si, c-si and p-si respectively, these values were obtained on a clear day. We concluded that the c-si and p-si modules considering their yields are the best performing modules in the three days. However, in an unfavourable weather conditions the a-si modules begin to perform better due to its better handling of the diffused irradiation.

### References

1. Bakos GC (2009) Distributed power generation: A case study of small scale pv power plant in Greece. Appl Energy 86: 1757-1766.

2. Solarbuzz TM, Marketbuzz TM (2010) Annual world solar photovoltaic industry report 2010.

3. Ministry of Energy, Mines, Water and Environment report (2009) National Energy Strategy Horizon 2030.

4. Elkholy A, Fahmy FH, Abou El-Ela AA, Nafeh AE, Spea SR (2016) Experimental evaluation of 8Kw$_p$ grid-connected photovoltaic system in Egypt. J Electr Syst Inform Technol 90: 1-13.

5. Dabou R, Bouchafaa F, Arab AH, Bouraiou A, Draou MD, et al. (2016) Monitoring and performance analysis grid-connected photovoltaic under different condition climatic in south of Algeria. Energy Convers Manage 130: 200-206.

6. Pavan AM, Castellan S, Quaia S, Roitti S, Sulligoi G (2007) Power electronics conditioning systems for industrial photovoltaic fields: Centralized or string inverters? ICCEP 7th International Conference Clean Electrical Power, 2007. pp: 208-214.

7. Aarich N, Erraissi N, Akhsassi M, Lhannaoui A, Mustapha R, et al. (2014) "Propre.Ma" project: Roadmap & preliminary results for grid-connected PV yields maps in Morocco. IEEE Conference IRSEC 2014 - International Renewable and Sustainable Energy Conference.

8. Tina GM, Ventura C, De Fiore S (2011) Sub-hourly irradiance models on the plane of array for photovoltaic energy forecasting applications. Proceedings of the 38th photovoltaic specialists conference.

9. Marion B, Deceglie M, Silverman TJ (2014) Analysis of measured photovoltaic module performance for Florida, Oregon and Colorado locations. Sol Energy 110: 736-744.

10. Piotrowicz M, Maranda W (2013) Report on efficiency of field-installed PV-inverter with focus on radiation variability. 20th International Conference Mixed Design of Integrated Circuits and Systems.

11. Başoğlu ME, Kazdaloğlu A, Erfidan T, Bilgin MZ, Çakır LB (2015) Performance analyzes of different photovoltaic technologies under Izmit, Kocacli climatic conditions. Renew Sustain Energy Rev 52: 357-365.

12. Carr AJ, Pryor TL (2004) A Comparison of the performance of different PV module types in temperate climates. Sol Energy 76: 285-294.

13. Hussin M (2012) Design installation and testing results of 1 kWp amorphous-silicon FS GCPV system at UiTM, Malaysia. IEEE International Conference of Control System, Computing and Engineering.

14. Leloux J, Narvarte L, Trebosc D (2012) Review of the performance of residential PV systems in Belgium. Renew Sustain Energ Rev 16: 178-184.

15. Kymakis E, Kalykakis S, Papazoglou TM (2009) Performance analysis of a grid connected Photovoltaic Park on the island of Crete. Energy Convers Manage 50: 433-438.

16. Al-Otaibi A, Al-Qattan A, Fairouz F, Al-Mulla A (2015) Performance evaluation of photovoltaic systems on Kuwaiti schools rooftop. Energy Convers Manage 95: 110-119.

17. Rodrigo PM, Velázquez R, Fernández EF (2016) DC/AC conversion efficiency of grid-connected photovoltaic inverters in central Mexico. Solar Energy 139: 650-665.

18. Edalati S, Ameri M, Iranmanesh M (2015) Comparative performance investigation of mono- and poly-crystalline silicon photovoltaic modules for use in grid-connected photovoltaic systems in dry climates. Appl Energy 160: 255-265.

19. Humada AM, Hojabri M, Hamada H, Samsuri F, Ahmed MN (2016) Performance evaluation of two PV technologies (c-Si and CIS) for building integrated photovoltaic based on tropical climate condition: A case study in Malaysia. Energy Buildings.

20. Aste N, Pero CD, Leonforte F (2014) PV technologies performance comparison in temperate climates. Sol Energy 109: 1-10.

21. Prasad R, Shenoy SR (1996) Staebler-Wronski effect in hydrogenated amorphous silicon. Physics Lett 196: 85-90.

# Energy Demand Based Procedure for Tilt Angle Optimization of Solar Collectors in Developing Countries

**Samer Yassin Alsadi\* and Yasser Fathi Nassar**

*Department of Electrical Engineering, Palestine Technical University, Palestine*

## Abstract

For efficient performance of photovoltaic (PV) panels and flat-plate solar collectors, one of the most important factors that should be considered is tilt angle. The common approach used by researchers has been to calculate the tilt angle ($S_s$) which maximizes the amount of solar radiation received by the collector. Economically, solar systems must provide a maximum energy to the customer not to collect maximum solar radiation. In some situations, there is a mismatch between them. However, solar harvesters need to be tilted at the correct angle to maximize the performance of the system. In this paper, the average monthly solar fraction of the system (the fraction of energy that is supplied by solar energy) is used as an indicator to find out the optimum tilt angle ($S_f$) for a solar system. This manner is profitable for most developing countries where there is no law governing the exchange of energy between the main provider of electrical energy in the country (in our case the general electrical company) and investors in the solar energy (for example, house owner). Regardless, that the solar radiation in the case optimum tilt angle based upon the maximum solar radiation collection ($S_s$) is 4% greater than that of the offered tilt angle ($S_f$), but we get an improvement in the solar fraction coefficient reached to 0.31%, which is equivalent to a yearly sum of 540 MWh. The solar radiation is calculated using the clear sky ASHRAE model and then multiplied by a magnification factor to meet most of the energy demand. This factor is physically presenting the solar conversion efficiency multiplied by the area of the solar collectors. Having the total monthly energy demand, the monthly solar fraction coefficient can be calculated by dividing the total monthly energy delivered by the solar system by the total monthly energy demand.

**Keywords:** Solar radiation; Optimum tilt angle; Electrical load; Solar fraction coefficient

## Introduction

Solar systems, like any other systems, need to be operated with the maximum possible performance. This can be achieved by proper design, construction, installation and orientation. The orientation of the collector is described by its azimuth and tilt angles. Generally, systems installed in the northern hemisphere are oriented due south and tilted at a certain angle [1]. Accordingly, it is important to determine the optimal tilt angle at which maximum solar radiation is collected. The tracking systems, that follow the direction of the sun on its daily sweep across the sky, allow the maximization of solar radiation incident on the collector's surface. A gain of 40% in solar radiation incident on the collector is achieved if a two axis tracking system is adopted instead of a fixed collector. However, tracking systems are expensive, need energy for their operation and are not always applicable especially for small scale systems [2].

Functionally, solar systems must provide a maximum energy to the customer not to collect maximum solar radiation. We mean by solar system the all system that including solar harvester (thermal solar collector or/and PV panel) and the storage (thermal or electrical). In some situation there is a mismatch between them, when we comparing the daily load curve with the available solar radiation we recognize the lack of harmony between them. This mismatch is depending on the location and the nature of the load. This paper introduces applied solar energy aspects that is optimization of tilt angle for maximum solar energy contribution in the energy-grid and presents a method for calculating it. This method takes into account the monthly energy consumption and the available solar radiation on the site of interest. It was selected two different locations with different loads behaviour for the purpose of comparison where Brack El-Shati is locating in a desert area (Sahara) which characterized by a dry and warm during the day and cold during the night in the winter season and very hot during the

summer season. While Tulkarm is a city locating in a mountainous area belongs to the Mediterranean basin, characterized by moderate climate where the winter season is rainy and warm and in summer it relatively hot with high humidity. Table 1 presents the coordinates of two different sites with different electrical load distribution and different climate [3].

## Theoretical Approach

The followed approach can be summarized in the following steps:

### Calculation of solar radiation

There are many models for estimating solar radiation. Most of them are presented in text books of solar energy [4,5] and most recent researches [6-10], hourly total radiation ($I_{t,s/f}$) on an inclined surface using both tilt angles ($S_s$ - The optimum tilted surface of the collector in order to collect maximum solar radiation and $S_f$ - The optimum tilted surface of the collector in order to achieve maximum fraction coefficient or to provide a maximum solar energy) has been calculated. The model considers the anisotropy diffuse sky model formulated by Hay and Davis [4] and includes components of beam directly from the sun and diffuse irradiation from the circumsolar and the sky dome and beam and diffuse irradiation reflected from the ground. The total

**\*Corresponding author:** Samer Yassin Alsadi, Department of Electrical Engineering, Faculty of Engineering and Technology, Palestine Technical University-Kadoorie, Tulkarm-Palestine, 00970, Palestine
E-mail: samer_sadi@yahoo.com

| Site | Country | Latitude | Longitude | Elevation (m) |
|---|---|---|---|---|
| Brack El-Shati | Libya | 27.53°N | 14.28°E | 334 |
| Tulkarm | Palestine | 32.31°N | 35.03°E | 125 |

**Table 1:** Geographic location of the sites.

solar radiation ($I_{t,s/f}$) - the subscript $s/f$ refers to the slopes $S_s$ and $S_f$ respectively- for an hour as the sum of three components is given as:

$$I_{t,s/f} = (I_b + A_i I_d) R_{b,s/f} + (1 - A_i) I_d \frac{1 + cos S_{s/f}}{2} + (I_b + I_d) \varrho_g \frac{1 - cos S_{s/f}}{2}, \left[\frac{W}{m^2}\right] \quad (1)$$

Where $I_b$ is the hourly beam radiation from the sun on a horizontal surface, $I_d$ is the hourly diffuse radiation parts of the circumsolar and the isotropic on a horizontal surface, so the total diffuse radiation on a horizontal surface will be equal to the sum of these two components, having neglected the horizon brightening diffuse radiation component, according to Hay and Davis anisotropic sky model. $A_i$ is the anisotropic index which is a function of the transmittance of the atmosphere for beam radiation and then $A_i = \frac{I_b}{I_o}$, where $I_o$ is the hourly extraterrestrial radiation on a horizontal surface [9], which equal to:

$$I_o = G_{sc}\left(1 + 0.033 \frac{360n}{365}\right) cos\theta_z, \left[\frac{W}{m^2}\right] \quad (2)$$

Where $G_{sc}$ is the solar constant 1367 W/m², n is denotes to day of the year and $\theta_z$ is the solar zenith angle at the time and day of interest.

The $R_{b,s/f}$ is a geometric factor which presents the ratio of beam radiation on the tilted surface to that on a horizontal surface at any time, $R_{b,s/f} = \frac{cos \theta_{i,s/f}}{cos \theta_z}$ in where $\theta_{i,s/f}$ is the solar incident angle and calculated from the following equation [9], with the corresponding azimuth surface angle $\psi$ and tilt angle $S_{s/f}$:

$$\theta_{i,s/f} = cos^{-1}\left[sin S_{s/f} \; cos\theta_z \; cos(\phi - \psi) + cos S_{s/f} \; sin\theta_z\right] \quad (3)$$

and $\theta_z$, is the solar zenith angle; and $\phi$, is the solar azimuth angle.

$$\theta_z = cos^{-1}\left[sin\delta \; sinL + cos\delta \; cosL \; cosh\right] \quad (4)$$

$$\phi = cos^{-1}\left[\frac{sin\delta \; cosL - cos\delta \; sinL \; cosh}{sin\theta_z}\right] \quad (5)$$

Where $L$ denotes the local latitude, angle $\delta$ is the declination angle and $h$ is the hour angle: $h=15(t_s-12)$ in where $t_s$ presents the solar time and $\varrho_g$ is the ground-reflectivity.

In this paper, the ASHRAE clear-sky model is adopted to estimate the hourly beam normal ($I_{bn}$) and diffuse ($I_d$) solar radiation. The ASHRAE clear-sky model appears to be general enough for the objective of the paper, furthermore, we don't need to any information about the location of interest, except the latitude angle.

The direct beam radiation and sky diffuse are calculated from the following formula [6]:

$$I_{bn} = A e^{\frac{-B}{cos\theta_z}}$$
$$I_b = I_{bn} cos \theta_z \quad (6)$$
$$I_d = C I_{bn}$$

Where A, B and C are constants for every day and are given in Table 2 for the 21st day of each month [6].

### Optimum tilt angle based upon the maximum solar radiation collection

The common approach used by researchers has been to calculate the tilt angle which maximizes the amount of radiation received by the

| Months | A: W.m⁻² | B: Dimensionless | C: Dimensionless |
|---|---|---|---|
| January 21 | 1,230 | 0.142 | 0.058 |
| February 21 | 1,215 | 0.144 | 0.060 |
| March 21 | 1,185 | 0.156 | 0.071 |
| April 21 | 1,135 | 0.180 | 0.097 |
| May 21 | 1,103 | 0.196 | 0.121 |
| June 21 | 1,088 | 0.205 | 0.134 |
| July 21 | 1,085 | 0.207 | 0.136 |
| August 21 | 1,107 | 0.201 | 0.122 |
| September 21 | 1,151 | 0.177 | 0.092 |
| October 21 | 1,192 | 0.160 | 0.073 |
| November 21 | 1,220 | 0.149 | 0.063 |
| December 21 | 1,233 | 0.142 | 0.057 |

**Table 2:** Constants for ASHRAE equations for the 21st day of each month.

collector. Many investigations have been carried out to determine, or at least estimate, the best tilt angle was found as [3]:

$$S_s = 1.5 + 1.35L - 1.069 \times 10^{-2} L^2 \quad (7)$$

Where $L$ is the latitude angle.

### Monthly electrical load of the site

A two-year data of daily electrical load ($Q_L$) in Tulkarm and Brack El-Shati is obtained from Tulkarm Municipal-Electrical department and General Electrical Company of Libya, respectively. The data has been rearranged into the form of monthly load.

### Calculation of the solar fraction coefficient

The performance of a solar system is characterized by the annual solar fraction (the fraction of load supplied by the solar energy). We mean by solar system the all system that including solar harvester (thermal solar collector or/and PV panel) and the storage (thermal or electrical). Economic constrains preclude the establishment of a solar system with 100% fraction coefficient. This coefficient could be determined monthly for more accurate in analysis and it is defined as:

$$f = \frac{Q_L}{\chi H_t} \quad (8)$$

Where $H_t$ is the total monthly solar radiation and $\chi$ is a magnification factor to meet most of the energy demand. This factor is physically presenting the solar conversion efficiency multiplied by the area of the solar collectors. Of course, the value of $\chi$ is depending on all of available solar radiation, load and the fraction coefficient. For realization of the problem we choose a value of 92% for the annual fraction coefficient, therefore magnification factor was found:

$\chi$=61,000 m², for Tulkarm-Palestine and

$\chi$=150,000 m², for El-Shati-Libya    (9)

### Results and Discussion

An MS Excel-sheet has been prepared to estimate the solar radiation incident on a tilted surface using the above mentioned equations. Figure 1 presents contour plots for total monthly solar radiation in kWh/m² incident on a tilted surface as a function of the tilt angle ($S$), for both sites Tulkarm and Brack El-Shati. For stationary solar collectors, the optimum tilt angle ($S_s$) based upon the maximum solar radiation collection was found as:

$S_s$=35° for Tulkarm-Palestine and $S_s$=30° for Brack El-Shati-Libya

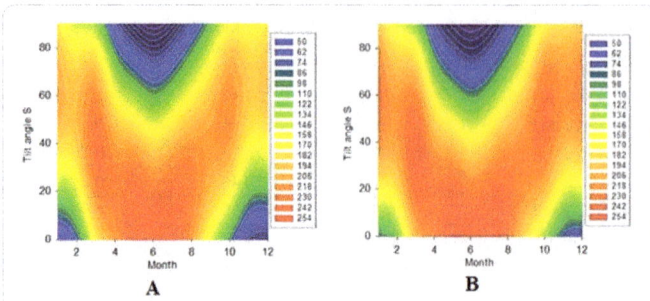

**Figure 1:** Contour plots of total monthly solar radiation (kWh/m²) incident on a tilted surface as a function of the tilt angle $S$; **A.** For Tulkarm-Palestine site; **B.** For Brack El-Shati-Libya site.

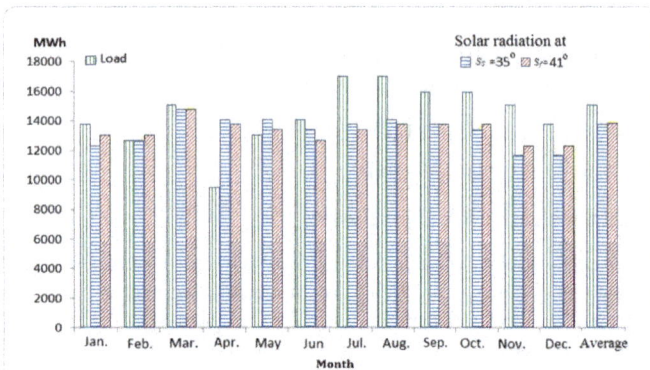

**Figure 2:** Monthly load distribution and the solar radiation on tilted surface, 35° and 41° for Tulkarm-Palestine.

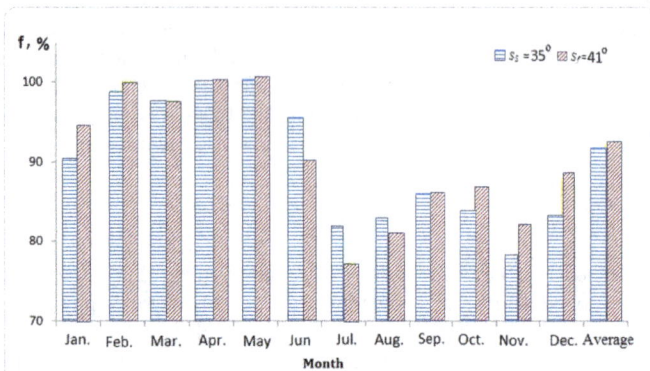

**Figure 3:** The monthly and annual average solar fraction coefficient for tilt angles, 35° and 41° for Tulkarm-Palestine.

Having the electrical load, the solar fraction coefficient has been determined. The optimum tilt angle ($S_f$) will be located according to maximum solar fraction coefficient and it was found as:

$S_f$=41° for Tulkarm-Palestine and

$S_f$=26° for Brack El-Shati-Libya

Figure 2 presents the monthly load distribution and the solar radiation incident on tilted surfaces of 35° and 41° for Tulkarm site. Figure 3 presents the monthly solar fraction and the annual solar fraction coefficient for Tulkarm site. In the same way Figure 4 presents the monthly load distribution and the solar radiation incident on tilted surfaces of 30° and 26° for Brack El-Shati site. Figure 5 presents the monthly solar fraction and the annual solar fraction coefficient for Brack El-Shati site.

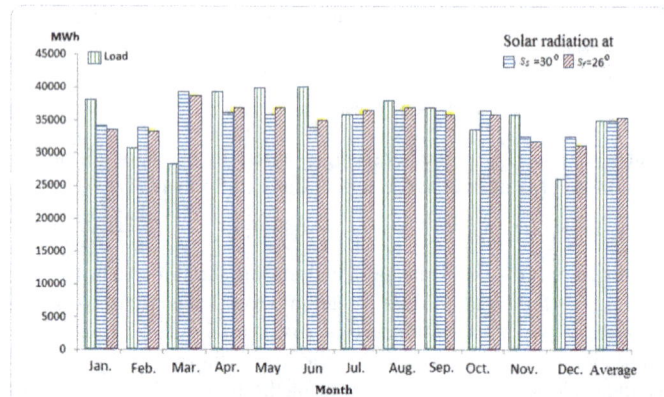

**Figure 4:** Monthly load distribution and the solar radiation on tilted surface, 30° and 26° for Brack El-Shati-Libya.

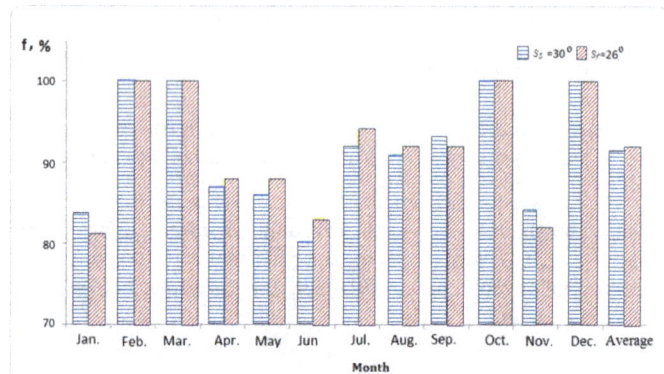

**Figure 5:** The monthly and annual average solar fraction coefficient for tilt angles, 30° and 26° for Brack El-Shati-Libya.

Looking at Figure 2 we find that the total energy from system with optimum tilted surface $S_s$=35° comes close to the load in 3 positions, June, July and August. While with the angle $S_f$=41° the harvested energy comes close to the load in January, February, October, November and December. However the energy from the system matches to the load in the rest of the months (March and September). When the radiation is larger than the load as it is the case in April and May the fraction coefficient is equal to 1, as it indicated in Figure 3.

As a result we find that the annual fraction coefficient for $S_f$=41° is larger than that for $S_s$=35°. From an economical point view putting solar panels at $S_f$=41° more economic benefit of those at $S_s$=35°. The situation is exactly the same for Brack El-Shati site, as it demonstrated in Figures 4 and 5.

The reason of this is that, in the developing countries where there is no law governing the energy exchange between the provider and the consumers, the losses in the collected solar energy is very high, accordingly the term "maximum" losses its meaning, because the maximum and other value may be coming equal in the case of both energy collected by the two angles will have the same solar fraction coefficient of unity.

## Conclusion

Increased solar energy gain justifies changing the tilt angle of solar collectors from $S_s$ to $S_f$. In our case this gain reaches up to 540 MWh yearly which is equivalent to 318 oil barrel and reduced carbon dioxide emission of (410 ton of $CO_2$), in addition of other pollutants reduction.

The present work has studied the optimum tilt angles for solar system by using the monthly solar fraction as an indicator and has reached the following conclusions:

1. There is no explicit function for optimizing the tilt angle of solar collectors, for a specific situation there is an optimum angle, the approach is outlined in this research.

2. The optimum tilt angle of the collector depends on the energy demand behaviour and the magnification factor.

3. In our case study, the optimum tilt angle for the maximum solar fraction for Libya is less than any of that for the maximum solar radiation at the collector by 4°. On the other hand this was found to be 6° greater for Palestine.

4. The authors recommend that further work should be conducted in countries where there is law governing the exchange of energy between the main provider and the investors in order to estimate the economic benefit of the exchange process.

## References

1. Shariah A, Al-Akhras MA, Al-Omari IA (2002) Optimizing the tilt angle of solar collectors. Renew Energy 26: 587-598.

2. Darhmaoui H, Lahjouji D (2013) Latitude based model for tilt angle optimization for solar collectors in the mediterranean region. Energy Procedia 42: 426-435.

3. Asadi SY, Nassar YF, Ali KA (2016) General polynomial for optimizing the tilt angle of flat solar energy harvesters based on ASHRAE clear sky model in mid and high latitudes. Energy Power 6: 29-38.

4. Duffie JA, Beckman WA (2013) Solar engineering of thermal processes (4thedn). John Wiley & Sons, Inc.

5. Nassar YF (2006) Solar energy engineering active applications. Sebha University, Libya.

6. Alsadi S, Nassar YF (2016) Correction of the ASHRAE clear-sky model parameters based on solar radiation measurements in the Arabic countries. Int J Renew Energy Technol Res 5: 1-16.

7. Yeole SV, Awate AU, Sarode PL (2016) A review of advancements in design configurations for optimizing solar flat plate collector effectiveness. Int J Innovative Res Sci Engineering Technol 5: 4894-4902.

8. Kocer A, Yaka IF, Sardogan GT, Gungor A (2015) Effects of tilt angle on flat-plate solar thermal collector systems. Brit J Appl Sci Technol 9: 77-85.

9. Arslanglu N (2016) Optimization of tilt angle for solar collectors: A case study of Bursa, Turkey. Int J Electrical Comput Energetic Electronic Commun Engineering 10: 517-520.

10. Despotovic M, Nedic V (2015) Comparison of optimum tilt angles of solar collectors determined at yearly, seasonal and monthly levels. Energy Convers Manag 97: 121-131.

# 23

# Stored Heat Evaluation in Geothermal Systems: A Case of a Mexican Field

**Aragón-Aguilar Alfonso[1]\*, Izquierdo-Montalvo Georgina[1], López-Blanco Siomara[1] and Gómez-Mendoza Rafael[2]**

[1]Instituto de Investigaciones Eléctricas, Gerencia de Geotermia, Morelos, México
[2]Instituto Mexicano de Tecnología del Agua, Morelos, México

## Abstract

México is rich in renewable energy sources (solar, wind, biomasss, hydropower and geothermal). Nevertheless, the potential of this type of energy has not been fully exploited. Geothermal energy in the country is one of the renewable energies used for electric generation although Hydropower is between renewable energy sources with the highest installed capacity. México ranks fourth, in the world, in installed electric capacity, from geothermal resources. In this work is presented a methodology for stored heat evaluation in the zone central eastern of Los Humeros geothermal field. The wells grouped in this section of the field are non producers, however in the neighboring zone (central western) are producers. We present an analysis, which shows an evolution, in producer wells from two phases toward one phase (steam) in its produced mass. It was determined temperatures distribution in this central zone using data of producers and non producer wells. Moreover, through using isothermal surfaces and establishing temperature bounds of 200, 250 and 300°C were determined net thicknesses in each well with possibility for heat storage. The innovative contribution of this work is focused to rescue non producer wells with high temperature although low permeability and scarce recharge. Considering different scenarios of reservoir properties in the studied zone were determined stored heat and its corresponding evaluation for obtaining electric generation. In determinations, values of specific heat ($c_T$) between 1500 and 2900 [kJ/(m³°C)] and the reservoir temperature, 200°C < ($T_R$) < 300°C, were used. The obtained results are expressed in MW$_T$h and show the feasibility for extending the methodology to other similar fields. Through variation of extraction factor between 0.01 and 0.05, and efficiencies conversion of 0.10 and 0.25, energy in MW$_T$h was determined. The methodology results are useful in taking decisions about feasibility of a project for heat extraction for its commercial exploitation.

**Keywords:** Renewable energies; Geothermal energy; Heat recovery; Production decline; High enthalpy; Dry steam; Hot rock; Recharge; Entrance water

## Introduction

To date, generation electricity in Mexico is mainly based on fossil-fuelled power plants (hydrocarbons and coal), 72.6%, and more than one fifth (22%) on hydroelectric plants. Mexico ranks ninth in the world in crude oil reserves, fourth in natural gas reserves in America and it is also highly rich in renewable energy sources (hydropower, wind, biomass, geothermal and solar). However, the potential of this type of energy has not been fully exploited. Hydropower is the renewable energy source with the highest installed capacity within the country [1].

Renewable energy sources can be defined as sustainable resources available over the long term at a reasonable cost that can be used without negative effects [2]. In México the electric capacity from wind is approximately 2.4%, Geothermal electric capacity represents1.8 %, biomass capacity is 1.2 and the solar potential for electricity is largely untapped, leaving room for great improvements in the future [3,4].

Distribution of installed capacity for electric generation in Mexico from renewable energies, updated till 2014 [5] is given in Table 1.

The main uses of geothermal energy in the world [7] are: 32% for heat pumps; 30% for health resorts and spas; 20% for heating buildings; 8% for greenhouses; 4% for industrial processes; 4% for aquaculture and 2% for any other applications as dry fruits. Electricity generation from geothermal resources reduces damage to environment avoiding the fuels burning and the risks that represent their transportation and storage.

There are 24 countries in the world which generate electricity from geothermal resources, whose total installed capacity is 10898 MW [8]. México is ranked in fourth in installed electric capacity, from

| Energy type | Installed capacity operating MW | Percentage participation |
|---|---|---|
| Hydropower | 11707 | 80.82 |
| Wind | 1289 | 8.46 |
| Geothermal | 958 | 6.67 |
| Biomass* | 645 | 3.82 |
| Solar | 37 | 0.23 |

**Table 1:** Capacity of electric generation in Mexico during 2014, using renewable energies [6].

geothermal resources, after The U.S. (3098 MW), Philippines (1974 MW) and Indonesia (1197 MW).

The electric installed capacity in México, from geothermal resources is 958 MW [9]. To date have been identified more than 400 hot springs along the country [10], and there are four geothermal fields in continuous operation. These fields are: "Cerro Prieto B. C." (720 MW), "Los Azufres, Mich" (188 MW), "Los Humeros Pue", (40 MW) and "Las Tres Vírgenes B.C.S." (10 MW). Thirty seven power plants of several types (condensing, back pressure and binary cycle) between 1.5 and 110 MW operate in these fields, fed by 229 geothermal wells. The

**\*Corresponding author:** Alfonso AA, Instituto de Investigaciones Eléctricas, Gerencia de Geotermia, 113, Col. Palmira, Cuernavaca, Morelos, México, CP 62490, E-mail: aaragon@iie.org.mx

production wells have depths [11] between 600 and 4400 meters and global water-steam ratio is about 1.2.

This work is focused to present the results of the analysis carried out in a section of Los Humeros Mexico, geothermal field, whose stored heat could be extracted by different ways to conventionals. Considering thermodynamic characteristics and petrophysical properties [12] of the rock in the analyzed reservoir section, is presented an evaluation of its stored heat and its probable generation electric capacity. The use of the reservoir heat for power generation represents a great advantage in the solution of global warming by avoiding combustion of fossil fuels, which increases atmospheric $CO_2$. Another advantage is the possibility of power availability in marginal regions [13].

Zones of thermal springs are concentrated mainly along the Mexican volcanic belt although are distributed throughout the country. Most fluids are derived from surface waters that have percolated into the earth along permeable pathways such as faults [14]. The Los Humeros geothermal field is nested inside the plioquaternary volcanic caldera complex with less than 500 ka of age. This complex is located in the eastern part of the Mexican volcanic belt [15]. The location of the field is at the border between Puebla and Veracruz states, approximately 220 km to east of Mexico City, with latitude 19.68°N and longitude 97.45° W [16]. The topographical level of the field varies between 2800 and 2900 masl and the average temperature at the surface [17] between -2°C (in spring) and 15°C (in winter). Figure 1 shows location of the Los Humeros geothermal field in the Mexican Republic.

The drilling operations in Los Humeros geothermal field started since 1981 and to date have been drilled 41 wells, 18 of them are producers and 3 are injectors. The successful results of the wells located at northern area, were the base for new drillings exploration along the field. During exploration stage for expansion of the field were drilled five wells in this central area however none has been producer. In these wells, were found temperatures upper 300°C at depths greater than those located at central western zone, nevertheless a common characteristic is the low permeability in all of them. Due to lack of permeability conditions and that the found temperatures at higher depths in this section of the field, to date no more wells drilled.

Is a natural effect of exploitation, the decline in production parameters, which starts to appear due to wearing in the reservoir energy and duration of the production time. Some of the producer wells of The Los Humeros geothermal field have been showing a quick decline in their produced mass in conjunction with changes in phases of their fluid [18]. It has been observed that fluid gradually changes from two phase toward one phase (steam), increasing therefore its enthalpy [15,19,20]. The lack of fluid in the Los Humeros

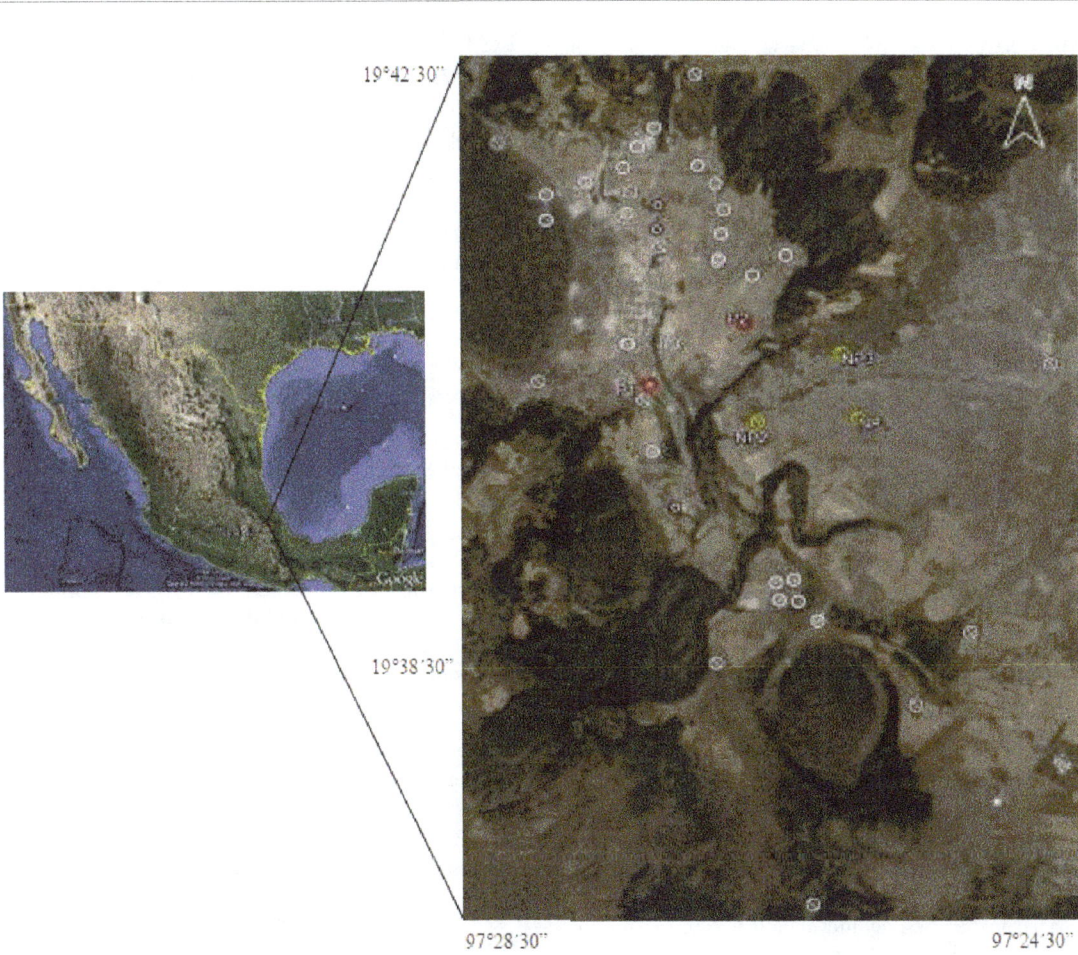

**Figure 1**: Map showing location of the Los Humeros geothermal field into Mexico and analyzed section of the field. The analyzed wells in central section are marked with red, the producers and with yellow the non producers.

reservoir is related to scarce recharge entrance which besides is due to its low permeability. The decline in produced mass and the increase in enthalpy allow assume lack of fluid, nevertheless heat remains stored in the reservoir.

The importance of this research is that through characterization of a zone of Los Humeros geothermal field with high temperature, but low permeability and scarce recharge water entrance, can be rescued wells with thermal characteristics. Moreover it is observed that non-producer wells with low permeability but high temperature are grouped in the central eastern section of this field. In this work is applied the USGS volumetric [21] evaluation method for estimating the probable electrical generation capacity of the reservoir analyzed section.

## Conceptual Background

Characterization and exploitation of petroleum systems are sustained by application of methodologies for analysis reservoirs behavior. From the developed technology, knowledge has been generated of exploration, drilling, exploitation and modeling reservoir which, modified to geothermal reservoirs characteristics, has shown can be applied with successful results [22]. Both systems type (petroleum and geothermal) could be nested in different structural environments; however they are characterized, in general terms, by their boundaries. The reservoir has an impermeable base and, a top that works as a seal layer.

Differences between both systems are types and fluids composition. While in oil systems the mean pressures are in order of 800 bars, in geothermal systems vary between 100 to 200 bars. Mean temperatures in oil systems are in the order of 180°C, in geothermal systems, temperatures vary in the order of 350°C. The recharge due to water influx is a basic factor in both systems. According to the flow regime in the reservoir in some cases appear prematurely breakthroughs due to a displacement not uniform, under these conditions there is a risk for resource effective recovery. This last situation could result in an entrapment into the formation; of oil (in petroleum systems) and; of heat (in geothermal systems).

Both type of reservoirs work during the primary production stage by their own energy, which, decreases according to the formation characteristics and production time. In this work are analyzed prevailing conditions in a section of a geothermal reservoir with a system of low permeability and low recharge water entrance, including the evaluation of its stored heat.

In geothermal systems, the most used techniques, for improving wells productivity and retard their decline trend are among others: chemical stimulations to the rock matrix [23], fracturing by thermal shock [24], hydraulic fracturing [25]. Under controlled conditions the thermal shock has shown successful results through opening fractures near the injection wells [24]. However the successful of any operation to improve productivity depends on the recharge characteristics to the reservoir.

One of the motivations of this research is sustained by high temperatures in the central eastern section of the field which allow assuming heat presence. Therefore its extraction to surface and its successful use constitute a challenge. Los Alamos National laboratory was actively engaged in field testing and demonstration the hot dry rock geothermal energy concept during the period from 1974 through 1995. The tests were carried out in the Fenton Hill hot dry rock site in the Jemez Mountains of north-central New México [26,27]. However after this project ended, a vast amount of information was obtained

concerning the characteristics and performance of confined hot dry rock reservoirs, some of them could be applied in new projects. However, one of the main lessons from this project is the low possibility in the practice to connect two wells through the creation of a hydraulic fracture between both.

In order to improve efficiency system it would be recommended generating a fracture using a defined well and identify their characteristics (fracture length, direction, depth, capacity, thickness, permeability). After knowing the fracture parameters; locate and to drill a second well for intercept this and by this way achieve connection between both wells. Different studies have carried out, related to heat recovery from geothermal reservoirs with low permeability and recharge [28-33].

Numerical simulation about feasible electric energy generation which can be extracted from a unitary rock volume carried out by Sanyal et al. [33]. The study assumes uniform reservoir rock properties including permeability and one of among others obtained results suggest an efficiency volume factor of 26 MWe/km³. The study adds that taking into account this correlation would be necessary 0.19 km³ of rock formation volume for generating 5 MWe.

## Calculation Methodology

The heat conduction is calculated from next expression:
$$q = K_T \left( \frac{\Delta T}{Z} \right) \qquad (1)$$
where q (W/m²) is the heat flow in a squared meter, $\Delta T$ (°C) is the temperature difference between two levels, z (m) is the depth and $K_T$ [W/(m°C)] is the thermal conductivity of the rock.

The term [$\Delta T/z$], in Equation (1), is referring to the rock formation thermal gradient. The thermal conductivity is equivalent to heat flow per second which crosses an area of 1 m², under a thermal gradient of 1 (°C/m) in the flow direction.

The equation called as the volumetric method is used for geothermal reserves estimation, its advantage is a quick applicability for any type of geologic resources. The parameters can be measured or estimated; however, the probable errors could be compensated at least partially [34].

In the volumetric method, the reservoir thermal energy is calculated as [21]:
$$q_R = c_T Ah(1-\phi)(T_R - T_{ref}) \qquad (2)$$
where $q_R$ (kJ) is the reservoir thermal energy, $c_T$ (kJ/(m³°C) is the volumetric specific heat of the system (rock and water), A (m²) is the reservoir area, h (m) is the reservoir thickness, $\phi$ is the porosity in the formation interval, $T_R$ (°C) is the average reservoir temperature, $T_{REF}$ (°C) is the average surface temperature.

Porosity represents void spaces of the rock formation and with permeability and storage are petrophysical properties influencing the underground flow capacity [35]. The void spaces reduce the capacity of heat storage and its transfer, so, the porosity into Equation (2) is a factor decreasing the final value of the estimated thermal energy.

The variables of Equation (2) which are related with reservoir properties provide uncertainty due to the tools accuracy used in their determinations. It was proposed [21] the use of a range of values, between 50 and 150% for these variables in order to calculate a general diagnosis value and establishing evaluation criteria.

## Influence Parameters

Equation for stored thermal energy determination ($q_R$) includes variables which have uncertainty in their determination such as the area (A), thickness (h), porosity ($\varphi$) and average reservoir temperature ($T_R$).

Parameters of main importance in evaluation of heat content in a reservoir volume portion are the temperature, the geometry and thermal properties of rock formation. In this analysis the area value is calculated, taking as boundaries the chosen wells. The area was selected taking into account the productive and thermal characteristics in producer wells and non- producers.

Using measured data along temperature profiles were determined isotherms distributions, across this section, for 200, 250 and 300°C. For thicknesses determination, were considered the logged temperatures, between limits 200°C and 300°C. The result of this analysis is the determination of different lengths in the thickness for each well and determining their thermal interest intervals.

Due to lack of transient pressure test data, were used losses fluid circulation logs during drilling, for qualitative determination of reservoir permeability in each well. These circulation losses profiles were combined with the calculated heating index using two temperature logs taken at the major resting time available in each well. The constructed graphs from fluid lost circulation volumes during drilling of each well are shown in Figure 2. The major volumes of fluid circulation losses were found at shallow depths in each well as can be seen in these graphs. Therefore it can be assumed that losses fluid circulation at shallow depths, are not related with geothermal reservoir.

## Study Area

The surface distribution and location of the wells analyzed in this studied area are shown in Figure 3. The analyzed area shows producer wells (P), and non-producers (NP). Highlights the reservoir heterogeneity due to prevailing contrasting conditions, i.e., in some cases there is a non-producer well, too close to a producer well. However it is feasible, in general terms, to take into account that non-

producer wells are grouped in the eastern section of the analyzed area, as can be seen in Figure 3. For this work were analyzed six wells, three producers and three non-producers.

The production behavior through exploitation time in the producer wells is being monitored by measurements at surface conditions. From observations in produced mass flow rate by the wells, highlights the changes in the steam-water ratio, which results in a decrease of the liquid fraction. In order to analyze production parameters at reservoir conditions, it was necessary to transform the parameters at bottom hole conditions. It was used the WELLSIM simulator program [36-38] with production measurements carried out at surface conditions for obtain these parameters at reservoir conditions.

Temperatures higher than 200°C were measured at least at somewhere of their profile in the involved wells in this study. However it is important to emphasize that horizontal distribution of temperature is non-uniform.

For defining the interest interval in the well, the thickness (h) was determined considering 200°C as the lower limit of temperature. The net thickness, for this study, is determined from the difference between the depth of isotherm 200°C and total depth of each well. Although there are temperature measurements higher than 350°C in some wells, in this work it was evaluated the profitable thickness, assuming limits between 200 and 300°C.

## Results

It was observed that measured values, at long standby times, are nearby to those calculated using the Horner static temperature method (1931). Considering availability of data, were chosen measurements done in the studied wells, with about 24 - 30 hrs of standby times. Figure 4 shows an example of measured temperature profiles at the total depth, losses circulation and heating index (defined as rate of temperature change °C/hr) of one producer well (P1). Figure 5 is an example of temperature profiles logged at total depth, losses fluid circulation and heating index (°C/hr) in a non-producer well (NP3).

For each well its temperature behavior profile was analyzed which combined with other parameters, provides some qualitative idea about the formation permeability. Using temperature data measured with a difference of about 12 hrs between logs, the profiles of heating index were determined for each well. The profile of heating index of wells used as demonstrative cases is shown at right side of Figures 4 and 5. The heating index (°C/hr) reveals the heat entrance rate at the wellbore, after it has been cooled due to drilling fluid. So the peaks in the graph indicate rapidity of heat flow from the reservoir to the well.

The profile of fluid circulation losses during well drilling is shown at the left side of the same Figures 4 and 5. One of the main characteristics identified in this field during the wells drilling is that the field in general showed low volumes of circulation losses during drilling. The major volumes of fluid circulation losses were found at shallow depths in each well as can be seen in these two shown wells. But in all the wells studied it was found similar behavior in volumes of fluid circulation losses during drilling. It is important to emphasize that the major volumes identified at shallow depths in any case were no greater to 50 m³/hr.

Variations of fluid circulation losses measured at deep zones of the well never were more than 20 m³/hr. In some cases were found greater volumes of fluid circulation losses in non- producer wells than in producers. It can be assumed that this behavior be related to the

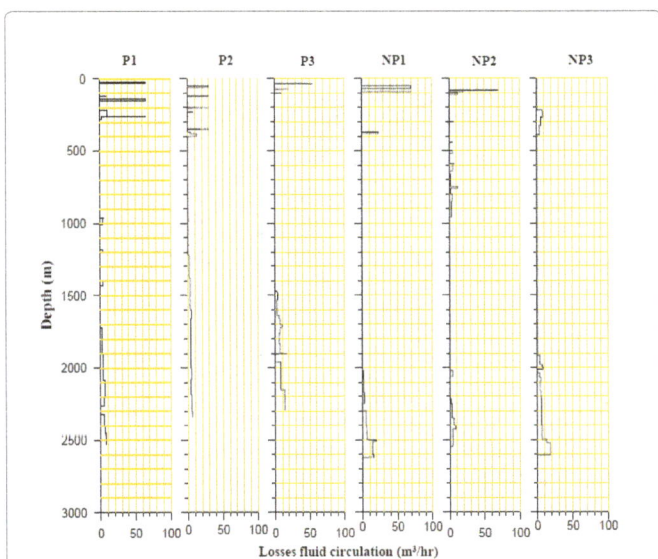

**Figure 2**: Schemes profiles of losses fluid circulation measured during drilling of the six wells analyzed in this work.

**Figure 3:** Location of analyzed wells in studied field section, showing the producers (P) with red marks and non-producers (NP) with yellow marks.

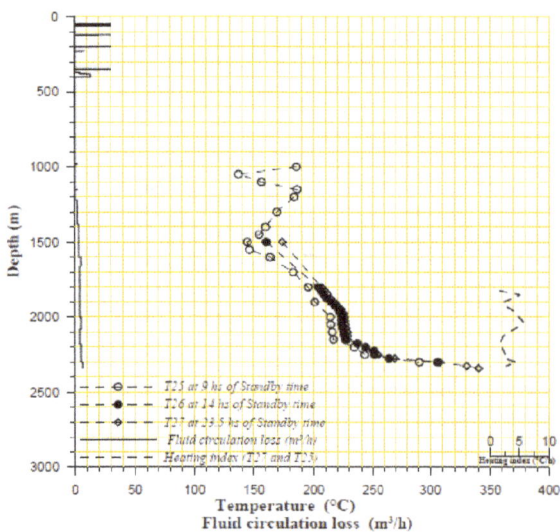

**Figure 4:** Profiles of temperature logged at different standby times and fluid circulation losses during drilling in well P1.

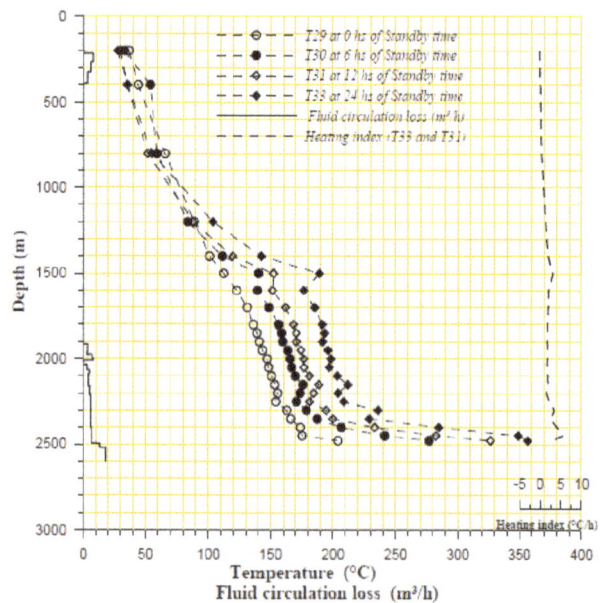

**Figure 5:** Profiles of temperature logged at different standby times and fluid circulation losses during drilling in well NP3.

existence of low permeability at well depth. An important observation is that the measured low volumes of fluid circulation losses are related with its heating index increase as can be seen in Figures 4 and 5.

From the analysis carried out in all the involved wells, we can observe in some of them, a clear increase in the calculated values of heating index. This behavior it was observed in producer wells. Through comparison of profiles behavior of heating index in a producer well (P1) with another non-producer well (NP3), it can to identify the difference in behavior between these wells type. It is important to emphasize that it was identified a good difference in heating index values in producer wells, although low volumes of their circulation losses. However, it is possible to identify changes, in lesser ranges, in the increase of heating index in non-producers wells. Minor changes were observed in the heating index profile of these wells as can be seen in Figure 5. This condition could be explained taking into account that the drilling fluid cools the rock formation, but after standby time and by lack of water entrance, the heat again returns to rock. Through combination of temperature profiles with the heating index, the thickness interval interest for each well were defined, assuming the useful limits between 200°C and 300°C. Table 2 shows location of the depths in the wells for each isotherm, as indicative of thicknesses of interest in the analyzed wells.

Were estimated the isotherms for 200, 250 and 300°C using temperature measured data of each analyzed well. A superficial distribution of temperatures for isotherms of 200 and 300°C in the analyzed section of the field is shown in Figure 6a and b. A scheme for thickness determination is shown in Figure 7 which results from the overlapping of isotherms of 200°C and 300°C.

In the studied reservoir section three different thicknesses were identified. Thickness lengths in the rock formation were determined for 200, 250 and 300°C. The feasible thickness lengths that can be exploited from heat stored are shown in Table 3.

The isotherms distribution in producer wells occur at higher levels than those determined for the non-producers wells. Furthermore, the producer wells are grouped at the west section of the analyzed area, leaving the eastern section, for grouping of the non-producer wells.

The analysis behavior of pressure, temperature and losses circulation during drilling was applied to all the wells involved in this study area, even though in this work only are shown of wells P1 and NP3. Through correlation of temperature profiles with fluid circulation losses, heating index, the interest thickness in each well, its heat storage was determined. Values of the depths of each isotherm, the total depth drilled in the well, and the useful thickness were used for calculating the stored heat in the rock volume of the analyzed section.

The analysis carried out allows assuming existence of temperatures upper to 200 °C in some wells of this analyzed section, therefore are candidates of a research in order to rescue them for using its stored heat. The analyzed total area was determined and using mean thicknesses from interest temperatures in the wells, was estimated the feasible volume for heat storage. The boundaries of this area were assumed to the east by the non-producer wells NP1 and NP3, and to west, the bound is marked by the half-length between the non- producers wells, and its nearby producer. So, we assumed the half of the distance between the NP2 and P1 wells, and the NP3 and P2 wells. The estimated area according to last assumptions resulted in a value of 1.21E06 m².

Through the use of measurements of temperature profiles in the analyzed wells, with Equation (1) were calculated the thermal gradients at different depths along each one of these. Values of thermal gradient were calculated at depths since 1500 m to the total depth of each well, whose results are shown in Table 4.

The stored heat ($q_R$) in the formation volume bounded by the involved wells in this study was determined using Equation (2). Due to uncertainties in measurements reservoir parameters and

| Well | Total depth (m) | Temperature location (depth) | | |
|---|---|---|---|---|
| | | T = 200 °C | T = 250 °C | T = 300 °C |
| P$_1$* | 2340 | 1460 | 1640 | 1820 |
| P$_2$* | 2440 | 1550 | 1750 | 2020 |
| P$_3$* | 2290 | 1300 | 1380 | 1490 |
| NP$_1$* | 2620 | 2500 | 2520 | 2540 |
| NP$_2$* | 2540 | 1620 | 1920 | 2350 |
| NP$_3$* | 2600 | 1440 | 1620 | 2020 |

*The producer wells are called with P, while non-producer wells with NP

**Table 2**: Estimates of the depths for specific temperatures (200, 250 and 300°C) along the analyzed wells, using logged temperatures.

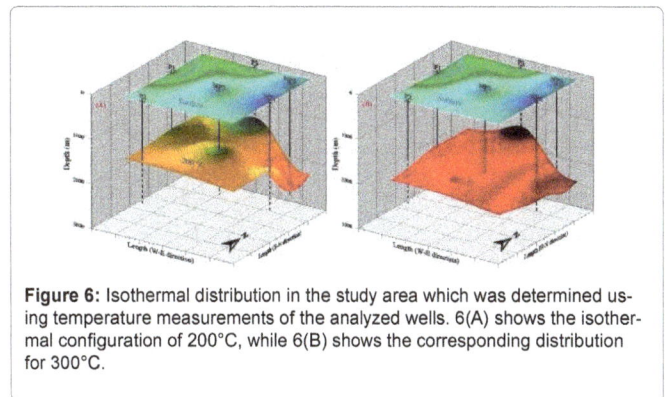

**Figure 6:** Isothermal distribution in the study area which was determined using temperature measurements of the analyzed wells. 6(A) shows the isothermal configuration of 200°C, while 6(B) shows the corresponding distribution for 300°C.

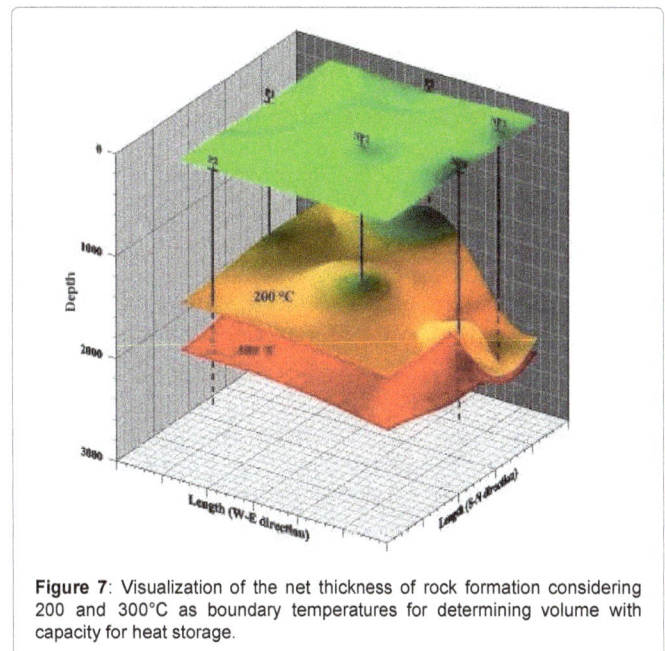

**Figure 7**: Visualization of the net thickness of rock formation considering 200 and 300°C as boundary temperatures for determining volume with capacity for heat storage.

| Well | Thickness (m) | | |
|---|---|---|---|
| | T = 200°C | T = 250°C | T = 300°C |
| *P1 | 880 | 690 | 510 |
| *P2 | 890 | 690 | 410 |
| *P3 | 980 | 900 | 790 |
| NP1 | 110 | 90 | 70 |
| *NP2 | 20 | 620 | 197 |
| NP3 | 1150 | 970 | 570 |

**Table 3**: Available thicknesses resulting from temperature locations for 200, 250 and 300°C in each of the analyzed wells.

| Depth (m) | $\Delta T/\Delta z$ (°C/m) | | | | | |
|---|---|---|---|---|---|---|
| | P1 | P2 | P3 | NP1 | NP2 | NP3 |
| 1500 | 0.06 | 0.04 | -0.02 | 0.04 | 0.11 | 0.51 |
| 1800 | 0.02 | 0.01 | 0.04 | 0.08 | 0.10 | 0.10 |
| 2000 | 0.20 | 0.08 | 0.10 | 0.05 | 0.03 | 0.02 |
| 2100 | 0.15 | 0.12 | 0.18 | 0.04 | 0.08 | 0.02 |
| 2200 | 0.23 | 0.87 | 0.40 | 0.08 | 0.07 | 0.12 |
| 2300 | 0.07 | | 0.40 | 0.04 | 0.21 | 0.12 |
| 2400 | 0.15 | | | 0.08 | 0.65 | 0.26 |
| 2500 | 0.43 | | | 0.12 | 0.28 | 0.24 |
| 2550 | 0.26 | | | 0.12 | 0.10 | 0.18 |

**Table 4**: Calculated thermal gradient profiles along depths of analyzed wells in the study zone.

| $T_R$ °C | C [kJ/ (m³°C)] | $q_R$ (MW$_T$h) | $T_R$ °C | C [kJ/ (m³°C)] | $q_R$ (MW$_T$h) | $T_R$ °C | C [kJ/ (m³°C)] | $q_R$ (MW$_T$h) |
|---|---|---|---|---|---|---|---|---|
| 200 | 1500 | 6.55E+07 | 250 | 1500 | 8.31E+07 | 300 | 1500 | 1.01E+08 |
| | 1700 | 7.42E+07 | | 1700 | 9.42E+07 | | 1700 | 1.14E+08 |
| | 1970 | 8.29E+07 | | 1970 | 1.05E+08 | | 1970 | 1.28E+08 |
| | 2100 | 9.16E+07 | | 2100 | 1.16E+08 | | 2100 | 1.41E+08 |
| | 2300 | 1.00E+08 | | 2300 | 1.27E+08 | | 2300 | 1.55E+08 |
| | 2500 | 1.09E+08 | | 2500 | 1.39E+08 | | 2500 | 1.68E+08 |
| | 2700 | 1.18E+08 | | 2700 | 1.50E+08 | | 2700 | 1.82E+08 |
| | 2900 | 1.27E+08 | | 2900 | 1.61E+08 | | 2900 | 1.95E+08 |

**Table 5**: Estimated values of stored heat in the rock volume of analyzed section, using different specific heats, and temperatures in the rock formation.

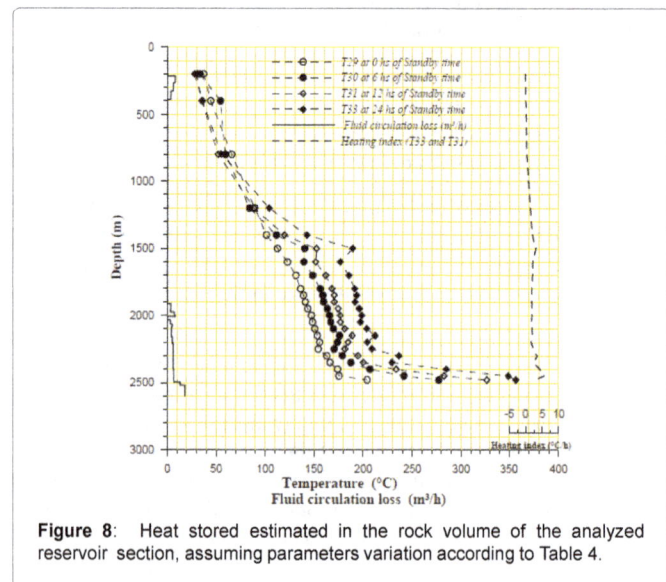

**Figure 8**: Heat stored estimated in the rock volume of the analyzed reservoir section, assuming parameters variation according to Table 4.

rock properties, different values were used taking into account their variation range. So, the used values were: For specific heat ($c_T$) between 1500 and 2900 [kJ/ (m³°C)]; The reservoir temperature, 200 °C < ($T_R$) < 300°C. The surface temperature ($T_{ref}$), was assumed of 15°C, the average reservoir thickness (h) in 800 m and 15% as the mean value for porosity ($\varphi$). Determinations for different reservoir temperatures were carried out for values of 200, 250 and 300°C.

Table 5 shows the results obtained for the heat stored ($q_R$) in the analyzed rock volume, for reservoir temperature cases of 200°C, 250°C and 300°C and specific heat of rock formation between 1500 and 2900 [kJ/(m³°C)].

The graphical results of the estimated stored heat ($q_R$) using Equation (2) are shown in Figure 8. As mentioned before, the main variation in parameters used were: Specific heat between 1500 and 2900 [kJ/(m³ °C] and; temperatures for 200°C, 250°C and 300°C.

For obtaining generation electricity (MWeh) from the stored heat shown previously (Table 5) it was assumed the time life of 30 years for the analyzed system. Furthermore due to uncertainties of involved parameters it has been assumed besides specific heat of the rock formation, the conversion efficiency and factor extraction, according to following:

Case 1: Factor extraction ($R_g$): 0.01, Efficiency conversion (e): 0.10

Case 2: Factor extraction ($R_g$): 0.05, Efficiency conversion (e): 0.25

Taking into account last assumptions was estimated (MW$_e$h), for $T_R$ values of 200°C, 250°C and 300°C and different values of specific heat in the probable rank of the rock formation. Table

6 shows a summary of the results obtained for the analyzed system under the different conditions above mentioned.

The graphical results of the estimated energy in the analyzed zones with variation of specific heat, reservoir temperature, extraction factor and conversion efficiency are shown in Figure 9.

From the graph of Figure 9 it can be seen the influence to use values of $R_g$ and e, for obtaining the marked differences in the estimated energy.

## Discussion

The los Humeros geothermal field is nested into Mexican volcanic system which influences in its heterogeneity for finding producer wells nearby to non-producer wells. The chosen wells are grouped into the studied section with high temperature but low permeability outside the production zone. It was determined that the isotherms distribution in the zone of production wells is located at lesser depths that in the non-producer wells.

The characteristics of wells in the analyzed zone are low permeability

| $T_R$ | C [kJ/(m³°C)] | MWeh Case 1 | MWeh Case 2 | $T_R$ | C [kJ/(m³°C)] | MWeh Case 1 | MWeh Case 2 | $T_R$ | C [kJ/(m³°C)] | MWeh Case 1 | MWeh Case 2 |
|---|---|---|---|---|---|---|---|---|---|---|---|
| | 1500 | 0.249 | 3.114 | | 1500 | 0.316 | 3.955 | | 1500 | 0.384 | 4.796 |
| | 1700 | 0.282 | 3.529 | | 1700 | 0.359 | 4.482 | | 1700 | 0.435 | 5.436 |
| | 1970 | 0.316 | 3.944 | | 1970 | 0.401 | 5.010 | | 1970 | 0.486 | 6.076 |
| 200 | 2100 | 0.349 | 4.359 | 250 | 2100 | 0.443 | 5.537 | 300 | 2100 | 0.537 | 6.715 |
| | 300 | 0.382 | 4.774 | | 2300 | 0.485 | 6.064 | | 2300 | 0.588 | 7.355 |
| | 2500 | 0.415 | 5.189 | | 2500 | 0.527 | 6.592 | | 2500 | 0.640 | 7.994 |
| | 2700 | 0.448 | 5.604 | | 2700 | 0.570 | 7.119 | | 2700 | 0.691 | 8.634 |
| | 2900 | 0.482 | 6.019 | | 2900 | 0.612 | 7.646 | | 2900 | 0.742 | 9.273 |

**Table 6**: Estimated energy (MWe) from the stored heat in the analyzed zone for extraction factors of 0.01 and 0.05, and efficiencies conversion of 0.10 and 0.25.

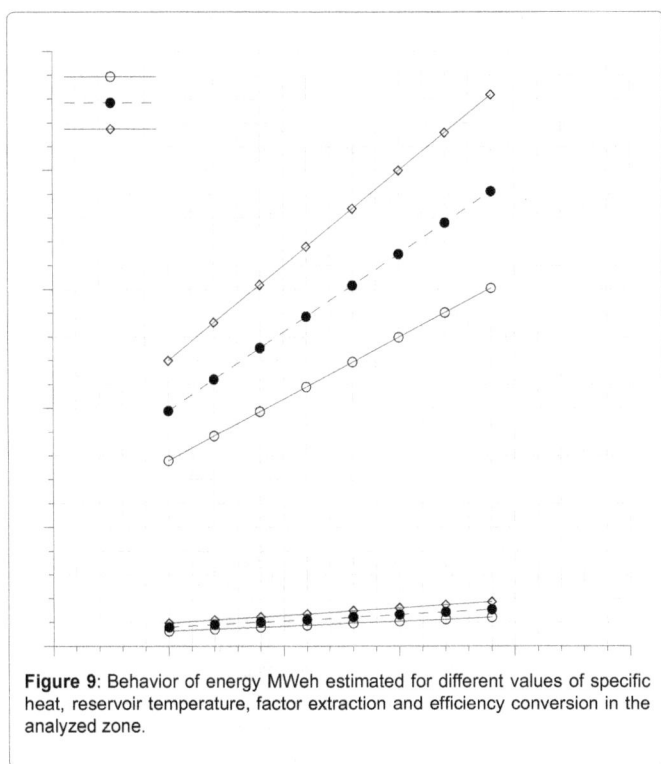

**Figure 9**: Behavior of energy MWeh estimated for different values of specific heat, reservoir temperature, factor extraction and efficiency conversion in the analyzed zone.

and high temperatures at deep conditions. In the rock formation of non producer wells, exists heat that could extracted by means different to the conventional. This study is focused to evaluation of the stored heat in the rock formation, considering useful thicknesses limited by temperatures of 200°C and the total depth of each well. The length between the depth for the 200°C isotherm and the total depth of each well is greater than the length between the isotherm of 300°C and its total depth. The calculated volumes are influenced by the mean values of thicknesses for each well. The calculation of stored heat is a function of the reservoir temperature, the thickness, porosity, the area of the analyzed zone, specific heat. In Table 5 the calculations were done for different reservoir temperatures and specific heat of the rock.

Considering that some variables introduce an uncertainty grade due to methods for their measurement and reservoir heterogeneity by

Brook et al., in this study were used variations in reservoir temperature ($T_R$) and in specific heat (C) of rock formation [21].

The rock volumes estimated using the thicknesses length between 200°C and total depth, are higher than those estimated for 300°C and the total depth (according to Table 2). The differences in variation of parameters values are influence factors in determination of stored heat, as can be distinguished in Table 5 and Figure 8.

The use of Equation (2) implies reservoir variables which involve some uncertainty even in homogenous systems. The uncertainty increases in heterogeneous systems such as the analyzed case. For calculating the stored heat all the variables intervening have an uncertainty grade; the area, the thickness, porosity, reservoir temperature. By this reason it is highly recommended use range of parameters values.

The rock formation thermal properties are influence parameters in the estimated energy, which could to help in taking decisions about the feasibility of a heat recovery project.

## Conclusions

It was found that the production decline is a variable dependent of reservoir properties, exploitation time, recharge water entrance, etc., among other factors.

The analyzed parameters in the studied zone have allowed carry out its characterization for reservoir understanding behavior, and planning its development. The technique used for defining isotherm depths of 200, 250 and 300°C allow to configure thicknesses of the study zone.

In the analyzed zone was determined the stored heat using data of six of its wells (three producers and three non producers). The uncertainty degree of the variables used was solved through values variation. Different values in the specific heat ($c_T$) between 1500 and 2900 [kJ/(m³°C)] and the reservoir temperature, 200°C < ($T_R$) < 300°C, were used. The results obtained are expressed in $MW_Th$ and show the feasibility for extending the methodology to other similar fields.

Making variations in the extraction factor between 0.01 and 0.05, and in the efficiencies conversion between 0.10 and 0.25, energy in MWeh was determined. The obtained results help to sustain technically the feasibility of a project for heat extraction for its commercial exploitation.

One of the contributions of this work is focused to rescue non producer geothermal wells with high temperature, although low permeability and scarce recharge.

## Acknowledgements

The authors express their gratitude to authorities of IIE and CFE (Geothermal Management) by the support in developing this research.

## References

1. Alemán-Nava G, Casiano-Flores V, Cárdenas-Chávez D, Díaz-Chavez R, Scarlat N, et al. (2014) Renewable energy research progress in Mexico: A review. Renew and Sustain Energy Rev 32: 140-153.

2. Charters WWS (2001) Developing markets for renewable energy technologies. Renew Energy 22: 217–222.

3. Bertani R (2007) World Geothermal Generation. Geo-Heat Center Quarterky

4. Bulletin 28: 8-19.

5. http.cfe.gob.mx/es/LaEmpresa/generacionelectricidad

6. SENER, CRE, CFE (2014) US Embassy-Mexico City Renewable energy.

7. Aragón-Aguilar A, Izquierdo-Montalvo G, Arellano-Gómez V (2013) Security regulations in Mexican renewable energies: Case of geothermal projects. Smart Grid Renew Energ 4: 21-31.

8. Lund JW, Freeston D, Boyd TL (2010) Direct utilization of geothermal energy 2010 worldwide review. Proceedings World Geothermal Congress, Bali Indonesia.

9. Bertani R (2012) Geothermal power generation in the world 2005-2010 update report. Geothermics 41: 1-29.

10. Gutiérrez-Negrín L (2012) Update of the geothermal electric potential in Mexico.

11. Geotherm Resourc Council Transact 36: 671-677.

12. Hiriart G, Gutiérrez-Negrín L, Quijano L, Ornelas A, Espindola S, et al. (2011) Evaluación de la Energía Geotérmica en México. Informe para el Banco Interamericano de desarrollo y la Comisión Reguladora de Energía, México: 167.

13. http://www.os.is/gogn/unu-gtp-sc/UNU-GTP-SC-14-05.pdf.

14. Elders W, Izquierdo-Montalvo G, Aragón-Aguilar A, Tovar-Aguado R, Flores- Armenta M (2014) Significance of deep zones of intense bleaching and silicification in the Los Humeros high-temperature geothermal field, México: Evidence of the effects of acid alteration. Geotherm Resourc Council Transact 38: 497–502.

15. Brown DW, DuTeau R (1997) Three principal results from recent Fenton Hill flow testing. Proceedings of the Twenty-First Workshop on Geothermal Reservoir Engineering, Stanford University, Stanford Cal, USA.

16. Lenhardt N, Gotz A (2015) Geothermal reservoir potential of volcanoclastic settings: The valley of Mexico Central, Mexico. Renew Energy 77: 423-429.

17. Martínez-Reyes J, González-Partida E, Perez RJ, Tinoco M, Jorge A (2014) Thermodynamic state of the volcánic caldera and geotermal reservoir of Los Humeros, Puebla, México. Collapse Caldera Workshop, Earth and Environment Science 3, Mexico.

18. Lorenzo-Pulido C (2008) Borehole geophysics and geology of well H-43 Los Humeros Geothernal field, Puebla, México. Geothermal training program, United Nations University, I-S Reykjavik, Iceland: 44.INEGI (2013) Prontuario de información geográfica municipal de los Estados

19. Unidos Mexicanos, Chignautla, Pue. México, Clave Geoestadística: 21054.

20. Arellano V, Barragán RM, Ramírez M, López S, Paredes A, et al. (2015) The response to exploitation of the Los Humeros (México) geothermal reservoir. World Geothermal Congress, Melbourne, Australia.

21. González-Partida E, Tello-Hinojosa E, Verma MP (2001) Interacción agua geotérmica manantiales en el campo geotérmico de Los Humeros Puebla, México Ingeniería Hidráulica en México 16: 185-194.

22. Tello E (2005) Estado de equilibrio soluto mineral y saturación de minerales de alteración en fluidos geotérmicos de alta temperatura. Tesis Doctoral, Facultad de Ingeniería UNAM, México DF.

23. Brook CA, Mariner DR, Mabey Jr. Swanson M, Guffanti LJP, et al. (1978) Hydrothermal Convection Systems with Reservoir Temperatures 90 °C. Ed. Muffler, LPJ, Asses of Geothermal Resources of US, USGS Circular 790.

24. Blodgett L, Slack K (2009) Geothermal 101: Basics of geothermal energy production and use. Geothermal Energy Association, Washington DC, USA: 55.

25. Katagiri K, Ott WK, Nutley BG (1980) Hydraulic fracturing aids geothermal field development. World Oil 191: 75-88.

26. Bodvarsson GS, Tsang CF (2012) Injection and thermal breakthrough in fractured geothermal reservoirs. J Geophysical Res: Solid Earth 87: 1031-1048.

27. Keiiti A, Fehler M, Aamodt RL, Albright JN, Potter RM, et al. (2012) Interpretation of seismic data from hydraulic fracturing experiments at the Fenton Hill, New Mexico, hot dry rock geothermal site. J Geophysical Res: Solid Earth 87: 936–944.

28. Brown D (1995) The US hot dry rock program-20 years of experience in reservoir testing. Proceedings of the world Geothermal Congress, Florence, Italy 4: 2607-2611.

29. Brown D (2009) Hot dry rock geothermal energy: Important lessons from Fenton Hill. Proceedings Thrity-fourth Workshop on Geothermal Reservoir Engineering, Stanford University, Stanford Cal. USA.

30. Kruger P, Karasawa H, Tenma N, Kitano K (2000) Analysis of heat extraction from the Hijiori and Ogachi HDR geothermal resources in Japan. Proceedings World Geothermal Congress, Kyushu-Tohoku Japan: 2677-3682.

31. Buttner G, Huenges E (2003) The heat transfer in the region of the Mauna Kea (Hawaii)—Constraints from borehole temperature measurements and coupled thermo-hydraulic modeling. Tectonophysics 371: 23–40.

32. Erdlac RJ, Armour L, Lee R, Snyder S, Sorensen M, et al. (2007) Ongoing resource assessment of geothermal energy from sedimentary basins in Texas. Proceedings Thirty-Second Workshop on Geothermal Reservoir Engineering, Stanford University, Stanford Cal, USA.

33. DiPippo R (2004) Second Law assessment of binary plants generating power from low-temperature geothermal fluids. Geothermics 33: 565-586.

34. Fridleifsson GO, Elders WA, Thorhallsson S. Albertsson A (2005) The Iceland Deep Drilling Project–A Search for Unconventional (Supercritical) Geothermal Resources. Proceedings World Geothermal Congress, Antalya, Turkey.

35. Sanyal SK, Butler SJ (2009) Feasibility of geothermal power generation from petroleum wells. Geothermal Resources Council Transactions 33: 673–680.

36. Rybach L, Leroy J, Muffler P (1981) Geothermal systems: Principles and case histories. Wiley, New York USA: 359.

37. Garg KS, Combs J (2011) A reexamination of USGS volumetric "Heat in Place" Method. Proceedings Thirty-Sixth Workshop on Geothermal Reservoir Engineering, Stanford University, University Cal, USA.

38. Freeston D, Calum G (1993) Wellbore simulation – Case studies. Proceedings Eighteenth Workshop on Geothermal Reservoir Engineering, Stanford University, University Cal, USA.

# 24

# Computational Examination of Utility Scale Wind Turbine Wake Interactions

**Tyamo Okosun and Chenn Q Zhou\***

*Department of Mechanical Engineering, Purdue University, Center for Innovation through Visualization and Simulation, Purdue University Calumet, USA*

## Abstract

Numerical simulations of small, utility scale wind turbine groupings were performed to determine how wakes generated by upstream turbines affect the performance of the small turbine group as a whole. Specifically, various wind turbine arrangements were simulated to better understand how turbine location influences small group wake interactions. The minimization of power losses due to wake interactions certainly plays a significant role in the optimization of wind farms. Since wind turbines extract kinetic energy from the wind, the air passing through a wind turbine decreases in velocity, and turbines downstream of the initial turbine experience flows of lower energy, resulting in reduced power output. This study proposes two arrangements of turbines that could generate more power by exploiting the momentum of the wind to increase velocity at downstream turbines, while maintaining low wake interactions at the same time. Simulations using Computational Fluid Dynamics are used to obtain results much more quickly than methods requiring wind tunnel models or a large scale experimental test.

**Keywords:** Turbine; Kinetic energy; Upstream and downstream

## Introduction

Wind energy is a growing source of alternative power in many countries around the world. With many advantages, including a cost per kilowatt hour equivalent to that of an average coal power plant (2.5-3.5 ¢/kWh), wind energy has established itself as an affordable and stable competitor in the energy production market [1]. The earliest recorded machines that harnessed the powter of the wind are the grain grinding mills in Persia [2]. When European travelers and crusaders discovered this technology and brought it to Europe, the transfer of concepts eventually led to the trademark "Dutch" windmill designs, often used to draw water from the ground or grind grain. With the advent of electricity, the next logical evolution in wind power was the wind turbine, which directly converts the kinetic energy of the wind to electricity. A major figure in advancing windmills to create wind turbines in 1891 was the Danish scientist Poul La Cour [3]. In the twentieth century, several countries produced variations of the horizontal axis wind turbine (HAWT) in attempts to create more efficient designs. Compared to drag based designs, these lift based turbines are far more efficient. Even as small scale wind power was experiencing a decline during the 1930s, continued advancements were made in larger scale wind energy generation. In 1941, the collaboration between American engineer Palmer C. Putnam and the Morgan Smith Company of York, Pennsylvania produced the largest wind turbine in the world. This turbine, capable of producing 1.25 MW, would hold that record for almost 40 years. Putnam's goal when designing this turbine was to create a turbine that would be able to produce energy at a rate competitive with common power production methods [4]. It is with this same goal that many utility scale wind turbines are designed today. By 2008 the United States had an estimated capacity of 25000 MW of wind energy. By 2030, the U.S. hopes to use wind power for 20% of its electrical generation requirements [5]. Because of this, there is a considerable need for research dedicated towards the improvement of wind turbine and farm efficiency.

Available wind energy is a significant issue in wind farm arrangement. Because a 1.0 m/s decrease (from 10 m/s to 9 m/s) in wind velocity can result in a 25% drop in power output for a given turbine, it is crucial to site wind turbines such that turbines upstream will not adversely affect the overall performance of the farm. The reduction of available wind power within a farm is usually referred to as a wake or

array effect. Since kinetic energy is extracted by each turbine, the air flow exiting a turbine contains less energy for a downstream turbine to extract. By accounting for this effect, a wind farm operator can more accurately decide where to place turbines within a farm to improve energy output. Maintenance costs can also be reduced by accounting for the wake effect, since wakes generated by turbines often induce wind flow outside of ideal design parameters [4].

Wind turbine manufacturers continue to reduce the cost of wind turbines and improve their operating efficiencies, and over the past several decades, the reliability and efficiency of wind turbines have improved greatly. Better wind turbines, in conjunction with a reduction of the overhead costs associated with installing a wind farm and government tax incentives for operations have made wind power more enticing than ever [5]. However, in order to provide electricity at a price that remains competitive with other power production methods, ongoing research within the field of wind energy must be pursued.

Of great importance is the aerodynamic performance of HAWTs. Hansen and Butterfield [6] discussed the research on the aerodynamics of wind turbines and some popular methods of analyzing aerodynamic performance. At the time, the most popular technique for the macro scale analysis of rotors involved Blade Element Momentum (BEM) theory. Due to its ease of use, as well as its accuracy, the BEM method was one of the most widely used methods for examining the flow physics of wind turbine rotors during the end of the twentieth century, and it continued to be the basis for almost all rotor design codes during the past decade [6-9]. While the BEM method provides a ballpark estimation of a rotor's performance, wind tunnel testing is still more accurate at predicting overall power output due to the BEM

**\*Corresponding author:** Zhou CQ, Professor, Department of Mechanical Engineering, Purdue University, Center for Innovation through Visualization and Simulation, Purdue University Calumet, USA, E-mail: czhou@purduecal.edu

method's assumptions regarding complex flow phenomena. Though modifications of the BEM method have been attempted, the inherent assumptions that form the basis of this method's calculations limit its overall accuracy [10].

Combinations of BEM based techniques with the more complex full Navier-Stokes equations are typically more successful than pure modifications of the technique in terms of accuracy. In 2002, Sørensen and Shen [11] developed an aerodynamic model for examining the three dimensional flow field surrounding a wind turbine. The model they developed employs a combination of the three dimensional Navier-Stokes equations and the actuator line technique. Using this technique, they validated the model by performing an analysis of a three bladed 500 kW HAWT. While the BEM methods remain popular for calculating rotor loads and performance, in order to create an accurate depiction of unsteady flow physics, dynamic loading, and other complicated flow physics, more sophisticated methods for examining forces on rotors are needed.

Vortex wake methods have been applied with varying degrees of success to the analysis of wind turbines. Oftentimes, vortex wake methods are implemented into existing analysis models. AeroDyn is currently one of the most popular models for wind turbine design and analysis in the United States. It functions by using a combination of the BEM theories and a simplified variation of the vortex wake methods. However, while fairly accurate, it still lacks the ability to provide complete aerodynamic results, leading to attempts to combine the accuracy of vortex wake models with the speed of AeroDyn [12]. Hugh D. Currin, Frank N. Coton, and Byard Wood [13,14] have developed and validated a new wake model for the analysis of HAWTs. The prescribed vortex wake code *HAWTDAWG*, developed at the University of Glasgow, was linked to AeroDyn and the structural dynamics code FAST in an effort to provide greater accuracy under dynamic flow conditions. The results obtained from this study were compared to experimental Phase VI wind tunnel data, as well as to basic BEM and vortex wake methods built into AeroDyn. *HAWTDAWG* produced results comparable to both the experimental data and the BEM methods during steady, axial flow.

The most accurate numerical method for analysis of fluid flow is directly solving the Navier-Stokes equations governing fluid motion. This is the underlying basis of Computational Fluid Dynamics (CFD). While there are projects geared towards improving currently existing models, with the ever-increasing power of computational systems, using CFD to calculate results based on the Reynolds Averaged Navier-Stokes (RANS) equations has proven to be both accurate and efficient. In 2002, Sørensen et al. [15] completed a RANS simulation of the NREL Phase VI Rotor. The calculated results from this simulation were in good agreement with both BEM methods previously used to model the rotor, as well as with experimental measurements made in the NASA Ames wind tunnel. This simulation also proved that accurate aerodynamic results could be obtained from a 3D CFD simulation. While CFD is more computationally costly than any of the aforementioned methods of analysis, due to advances in modern day computing, CFD is becoming a popular method for analysis when large amounts of detail are required [16].

Currently, there are two major methods of applying CFD for wind turbine analysis. To analyze a given problem, either a commercially available solver can be used or new code can be written and applied. In this project, due to its ease of use and availability, the commercially available CFD package ANSYS Fluent* is used. Various research projects have been conducted using commercial CFD codes, many of which

have performed additional validation studies supporting the accuracy of commercial software. One of the most common commercial CFD codes, ANSYS Fluent* is used in many projects due to its flexibility and accuracy. Amano et al. [17] used Fluent* to analyze and improve the design of a wind turbine rotor blade, resulting in the addition of a swept edge to the blade in an effort to extract greater amounts of power from the wind. Palm et al. [18] used Fluent* to determine basic relations that can be used to predict the power output of a given tidal farm configuration.

In 2007, Wußow et al. [19] used ANSYS Fluent® to model a full-sized wind turbine and determine the aerodynamic flow physics in the wake downstream of the rotor. By using a direct model with a body fitted grid, this simulation aimed to accurately capture the flow physics generated by the interaction of rotating turbine blades with a given wind flow. In this case, the rotor rotational speed was fixed at a value corresponding to the velocity of the oncoming wind at the wind turbine hub height. Although this assumption reduces the computational time required for the simulation, it prevents aerodynamic forces from being accurately modeled on the wind turbine rotor. However, this does not affect the downstream wake aerodynamics, which is the main focus of the project. Using the Large Eddy Simulation (LES) turbulence model, Wußow et al. generated results that matched experimental data closely, albeit representing only the lower range of available data. Thus, simplified CFD simulations can provide accurate results in the small time frames encountered in many industrial settings [20].

Several other commercial CFD software packages exist, including ANSYS CFX. CFX was applied by McStravick et al. [21] to analyze the Eppler 423 airfoil as a wind turbine blade. The domain was designed to contain only one blade with periodic boundary conditions to simulate the full turbine. While these other CFD packages exist and have been applied to the analysis of many types of simulations, the popularity and dependability of ANSYS Fluent* make it ideal for the current study.

## Objectives

The primary objective of this project is to complete numerical simulations of utility scale wind turbines in two major arrangements designed to optimize energy extraction and decrease the effect of wake losses. From these simulations, the flow physics surrounding the rotors of multiple turbines will be examined. Also, interactions between wind turbines within the simulated wind farms will be examined to determine the effect wakes have on the energy production potential of the proposed siting arrangements. Three major wind farm arrangements will be simulated in this project. The first is a common space saving arrangement. By placing wind turbines in small rows, typically groups of three to five turbines, space is conserved. However, by placing turbines this close to one another, large losses may be induced by wakes. Simulations of this geometry will be used as the baseline.

The second arrangement uses groups of triangularly spaced turbines. As wind passes through a turbine, some of the air cascade around the sides of the turbine. The air passing around the turbine is then forced into the freestream air flow, slightly accelerating the fluid passing around the turbine. By locating turbines in triangular groups, with two turbines placed relatively close downstream and to the left and right sides of the upstream turbine, it is hypothesized that the downstream turbines will be able to generate more power.

The third arrangement is essentially a reversed triangular pattern, or a "delta." The delta simulations are designed to determine if an increase in velocity in a single turbine downstream of two turbines results in a larger overall increase in power than seen in the triangular arrangement. The project will also attempt to determine any adverse issues with the triangular and delta arrangements by examining various angles of wind flow, as well as the spacing of turbines within a given area. In all of the aforementioned turbine arrangements, the turbines are placed on a flat surface for simulation, similar to a level field with very few trees, or an offshore wind turbine farm.

## Theory

CFD numerically solves the complex PDEs known as the Navier-Stokes equations, which govern fluid flow, by discretizing them into a simpler system of algebraic equations that can be easily calculated at various points within a fluid domain. A major concern when analyzing any flow field with CFD is the accuracy of the simulation. Not only can computers introduce rounding errors, results generated by CFD can only be as accurate as the physical models developed to represent fluid flow. Nonetheless, CFD has proven to be an extremely accurate tool for predicting fluid flow in an extremely broad spectrum of applications, including the simulation of wind energy aerodynamics [22].

The simulations performed in this study are run as steady state cases using SIMPLE pressure-velocity coupling. The following methods are used for spatial discretization: 1) least squares cell based gradients, 2) Fluent standard pressure discretization, and 3) the second order upwind method for momentum, turbulent kinetic energy, and specific dissipation rate. A brief comparison of the wakes generated by transient simulations with those observed in the steady state model was conducted early on in the project. The differences observed in wake formulation are mostly limited to increased vortex generation and vorticity in the wake, with minor variations in velocity. In order to optimize the computational efficiency of the simulations, the steady state model was selected over transient models. Initialization of all cases was performed using Fluent's hybrid initialization scheme, which solves the Laplace equation to produce an initial velocity field compliant with the boundary domains and cell zone conditions.

Several of the available turbulence models were considered for use in this project, including the Standard k-ε model, the k-ω and k-ω SST models, and more complex models such as the Large Eddy Simulation (LES) and Detached Eddy Simulation (DES) models. The k-ω SST (shear stress transport) model combines the near-wall/vortex accuracy of the k-ω model with the freestream accuracy of the k-ε model. This results in a turbulence approximation model capable of accurately representing the flow physics at the boundary layers surrounding the wind turbine blades, as well as the size, shape, and intensity of the wake as it travels downstream of the turbine.

The LES and DES methods were also examined for use in this project. However, since LES and DES simulations are typically run as transient simulations, they require extremely large computation times. Since the results (size, shape, and position of the turbine wakes) from LES and DES methods differ little from the results from the k-ω SST turbulence model during low turbulence flows, it seems clear that the k-ω SST model is best suited for the simulations in this project [23]. The equations for the k-ω and k-ω SST models are as follows:

Equation for k:

$$\frac{\partial k}{\partial t} + U_j \frac{\partial k}{\partial x_j} = P_k - \beta^* k\omega + \frac{\partial}{\partial x_j}\left[\left(\upsilon + \sigma_k \upsilon_T\right)\frac{\partial k}{\partial x_j}\right]$$

Equation for ω:

$$\frac{\partial \omega}{\partial t} + U_j \frac{\partial \omega}{\partial x_j} = \alpha S^2 - \beta \omega^2 + \frac{\partial}{\partial x_j}\left[\left(\upsilon + \sigma_\omega \upsilon_T\right)\frac{\partial \omega}{\partial x_j}\right] + 2\left(1-F_1\right)\sigma_{\omega 2}\frac{1}{\omega}\frac{\partial k}{\partial x_i}\frac{\partial \omega}{\partial x_i}$$

## Numerical Methodology

The design of the wind turbine in this study is based on GE's 1.5 MW utility scale turbines, shown in Figure 1. Since its properties and dimensions are publicly available, this project is able to simulate utility scale wind turbines under realistic operating conditions without the need for proprietary information.

The size, scale, and general operating conditions of this turbine class are available on GE Energy's website.

The blades are based on the popular NREL S809 airfoil with decreasing chord length. The detailed blade design is shown in Figure 2. The blade is also twisted along its length by 22 degrees. A preprocessing software (Gambit) was used to generate the simulation mesh.

Using this blade design, a three-bladed HAWT is created by attaching the blades to a basic hub and nacelle geometry with a 77 m tower. For each domain, a mesh of roughly 13.5 million cells is generated. The simulations are performed using a rectangular box domain 1.5 km in length, 1 km in height, and 1.5 km in width, with a velocity inlet, pressure outlet, and fixed ground. To simulate the rotor assembly under operating conditions, a cylindrical sliding mesh zone 80 meters in diameter is created surrounding the turbine rotor. This zone is meshed independently of the external flow domain and consists of roughly 1.5 million cells. The cell size near the turbine rotor is specified to be 0.1 meters, expanding at a growth rate of two to 0.5 meters at the outer interfaces. The remaining domain mesh quality is controlled by blocking the domain. The turbine is surrounded by a cube 150 meters per side and the cell size at the turbine tower and nacelle is also specified to be 0.5 meters. The mesh is allowed to expand to a maximum cell

**Figure 1**: GE's 1.5 MW turbines [24].

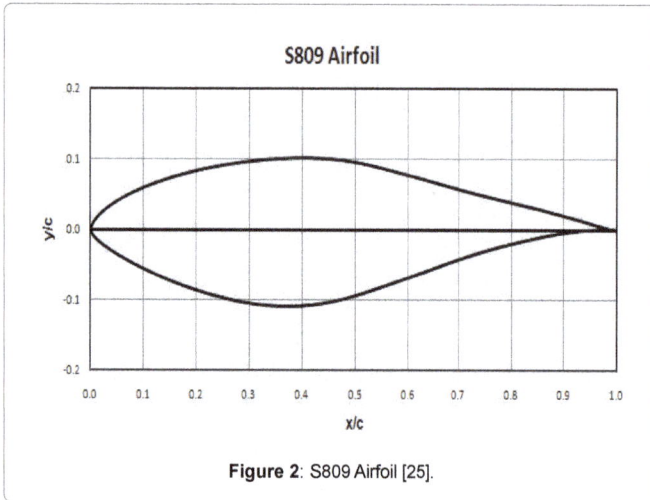

**Figure 2**: S809 Airfoil [25].

size of 3.5 meters within this zone. Any turbines are then surrounded by a flow block 500 meters wide by 200 meters tall, with a maximum cell size of 7 meters traversing the length of the domain from inlet to outlet. Finally, the remaining domain is meshed as a farfield zone with a maximum cell size of 17 meters. Figure 3 showcases the 3D model of the turbine and the flow domain for simulation.

The next step in creating the geometry is to define the boundary conditions. Boundary conditions define various fluid properties at certain locations within the domain. In this case, since a uniform freestream velocity over the wind turbines of 12 m/s is desired, the velocity inlet boundary condition is set at the inlet face of the domain's large boundary box. This inlet can be set to any given velocity, and it can be varied from case to case. The base simulation inlet speed is 12 m/s, with a turbulent intensity of 5%, and a turbulence length of one meter. The domain outlet is set at the back face of the boundary domain and is defined as a pressure outlet with the pressure at one atmosphere.

The walls and ground of the boundary domain are simple no-slip walls. However, in order to remove the boundary layer on the side and top walls during steady state simulations, they are set to have a constant translational velocity equal to that of the freestream. Therefore, in the base case, the side and top walls move at 12 m/s. This eliminates the wall boundary layer and allows for unimpeded freestream flow.

The wind turbine blades, tower, and nacelle are all basic wall boundary conditions with the no-slip condition, and the volumes are set to specific fluid cell zones. Every volume except the cylindrical volumes surrounding the wind turbine rotor assemblies are defined as stationary fluid zones. The cylindrical volumes are each defined as its own individual fluid zone, so that a sliding mesh rotational speed may be applied in Fluent'.

With the wind turbine geometry determined, the turbines are then placed in various positions for analysis. Three different arrangements of three turbines are simulated in this study. The first arrangement is designed to determine the effects of wakes along a row of wind turbines. The turbines are located 400 m apart and are aligned along the direction of wind flow. The distance between these turbines corresponds to approximately 5 rotor diameters, which is typical on a wind farm. This arrangement can be seen in Figure 4 below.

Two other arrangements are simulated in this study: 1) a triangular arrangement and 2) a reversed triangular arrangement referred to as the delta arrangement. Both of these arrangements are designed with

various distances between the turbines in an attempt to determine the impact the upstream turbine wakes have on the intake velocities of the downstream turbines. The base case arrangement for both the triangular and delta setups involves one turbine either 400 m upstream or downstream of two turbines, with the two turbines placed 150 m to either side of the single turbine. In the other cases, the downstream distance between turbines is either increased to 500 m or 600 m and the side to side distance of the two turbines is changed from 100 m to 200 m. Figures 5 and 6 shows an isometric view of both the triangular and delta turbine arrangements.

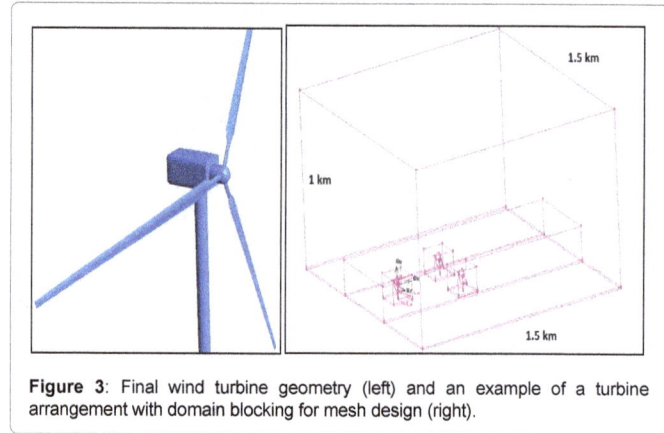

**Figure 3**: Final wind turbine geometry (left) and an example of a turbine arrangement with domain blocking for mesh design (right).

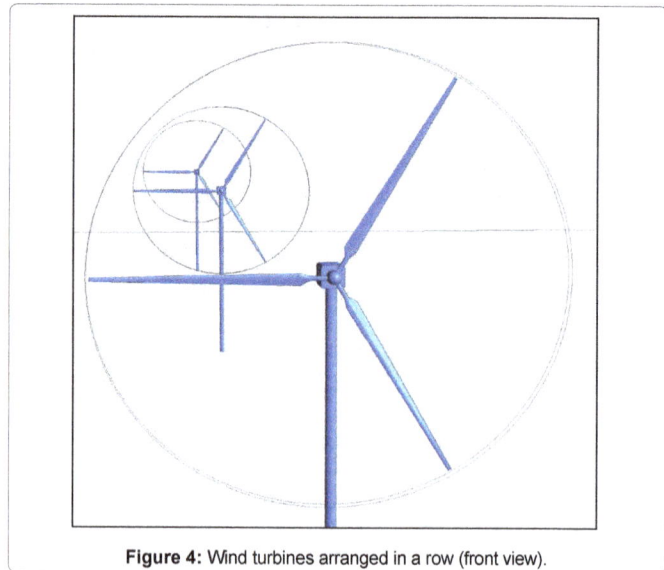

**Figure 4:** Wind turbines arranged in a row (front view).

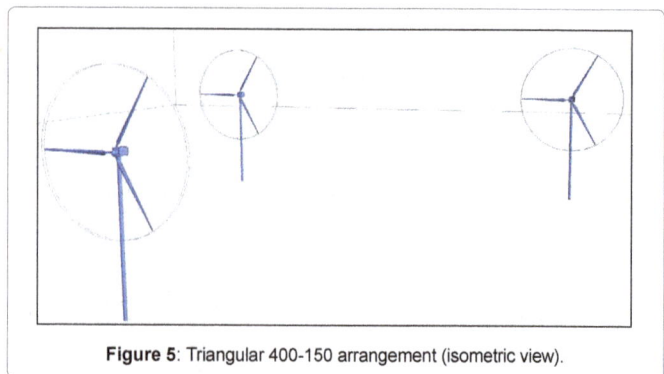

**Figure 5**: Triangular 400-150 arrangement (isometric view).

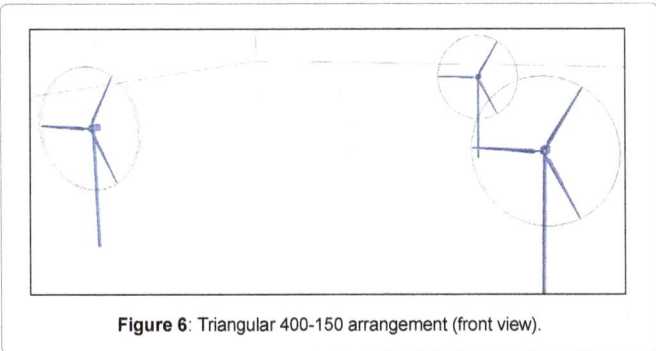

**Figure 6**: Triangular 400-150 arrangement (front view).

In addition to the direct wind flow geometries, four geometries are created to simulate variations in the incoming angle of the wind. This is done by taking the base case of 400 m by 150 m and rotating the turbines to face an incoming wind flow, which has been turned by a given amount. Then all three turbines are rotated by the same amount about the origin such that they face the inlet of the new geometry. Using this modification, simulations can be conducted using the same flow domain, with turbines interacting as if the incoming wind angle had changed. In this study, both the delta and triangular cases have geometries created for a five and ten degree angle of attack.

A grid sensitivity study, using the 400 by 150 m triangular arrangement, found that decreasing the mesh size below 10 million cells prevented solution convergence. The study also determined that increasing the mesh size above 14 million cells yielded data that did not differ much from the data obtained from k-ω SST simulations, while drastically increasing the computational requirements for timely solution convergence.

## Validation

To ensure accurate prediction of airflow surrounding the rotating wind turbine blades, simulations of the 0.66 m chord length S809 wind turbine airfoil experiencing wind flow similar to that at the tip of a wind turbine blade were performed. To achieve validation of the numerical results, a comparison was made with experimental results obtained using a Reynolds number of 2,000,000 corresponding to an inlet velocity of 52.28 m/s. Simulations were conducted at several angles of attack relative to the wind direction. The data is compared to experimental results obtained from the Delft University 1.8 m x 1.25 m low turbulence wind tunnel used by Wolfe and Ochs to validate various CFD models of the S809 airfoil [ 24-26]. This brief comparison ensures that the basic aerodynamic flow phenomena are accurately represented in the CFD simulation. Figure 7 details the comparison between experimental pressure data measured along the airfoil surface and calculated pressure data from the CFD simulation.

In addition, measurements taken by Barthelmie et al. [27] were used to validate the simulated velocity deficits generated by the wind turbine wakes. Barthelmie et al. used a ship-mounted SODAR to measure wind turbine wakes in an offshore farm in Denmark. The speed profile of the operating turbines was measured and compared with meteorological measurements from nearby masts. The farm contained Bonus Mk III 450 kW wind turbines with rotor diameters of 37 meters. In order to validate the CFD model, a single turbine geometry of matched scale was created.

As observed in Tables 1 and 2, the wake velocities calculated by the CFD model in all cases is either within the error range of the sodar measurements or very close to the range. Also noted by Barthelmie et

al. is the fact that wakes meandering cause's errors in the measured data because only a partial wake is measured as opposed to a full wake. Therefore, the CFD models should over-predict wake losses slightly.

## Results

In total, 14 different turbine arrangements were simulated, with three different velocities for each geometry (except in the altered wind direction cases) to provide data over a broad spectrum of operating conditions. The turbine row cases were designed to simulate what should be avoided when designing a wind farm with a particular average wind direction in mind. The first inlet velocity is set at the turbine rated speed of 12 m/s with a rotational speed of 15 RPM decreasing down the row. Figure 8 shows velocity contours along a cutting plane viewed from the side, indicating the size and intensity of the wake as it travels downstream. It is obvious from the contours that the wakes experienced by downstream turbines increase in size and intensity as air travels further downstream and encounters more turbines. This view of the contours provides an easy method of visualizing wake size and intensity. However, it does not provide specific information on the losses experienced by the downstream turbines due to the wakes.

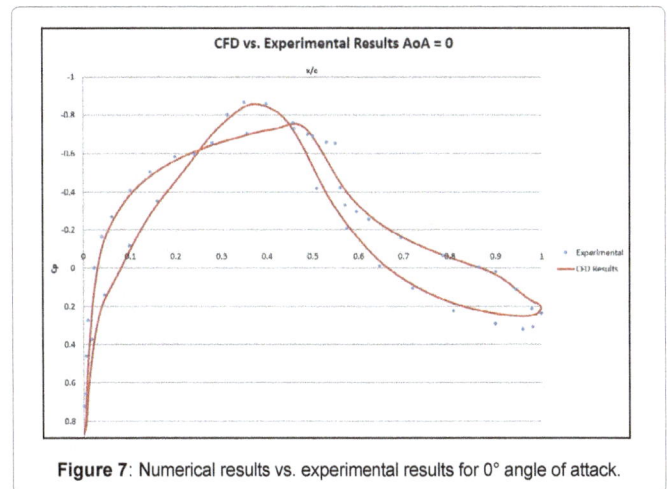

**Figure 7**: Numerical results vs. experimental results for 0° angle of attack.

| Rated Power Output | 1.5 MW |
|---|---|
| Type | Three Bladed, HAWT |
| Rotor Area | 4657 m² |
| Rotor Diameter | 77 m |
| Tower Height | 70 m |
| Rated Wind Speed, *V* | 12 m/s |
| Rated RPM, *ω* | 15 RPM |
| Maximum Wind Speed | 25 m/s |
| Cut-in Wind Speed | 3 m/s |

**Table 1:** Operating and design parameters for GE's 1.5 MW Turbine.

| Case | Measured Free-Stream Velocity (m/s) | Rotor Diameters Downstream | SODAR Relative Velocity Deficit (m/s) | CFD Relative Velocity Deficit (m/s) |
|---|---|---|---|---|
| 1 | 5.74 ± 0.20 | 2.8 | 2.28 | 2.45 |
| 2 | 6.37 ± 0.25 | 3.4 | 1.48 | 2.0 |
| 3 | 6.90 ± 0.59 | 1.7 | 4.26 | 3.1 |
| 4 | 7.54 ± 0.45 | 2.9 | 2.57 | 2.04 |
| 5 | 6.12 ± 0.74 | 7.4 | 0.61 | 1.5 |
| 6 | 8.19 ± 0.46 | 3.4 | 2.28 | 2.34 |

**Table 2:** Comparison of measured and calculated wake deficits at turbine hub height.

Figure 9 provides a more quantitative view of the wind velocity as the air travels downstream. The wind speed along the turbine group's centerline shows the velocity deficits after each turbine in detail. The first turbine experiences a wind speed of 12 m/s at the inlet, while the second and third turbines experience wind speeds of approximately 10.5 m/s and 9.7 m/s, respectively. The equation for available wind power is as follows,

Available wind power = $0.5 \times A \times v^3 \times \rho$

Due to the cubic relationship between available wind power and the wind velocity, a 1 m/s decrease in velocity from 10 m/s to 9 m/s can result in an available power decrease of 25%. When examining this simulation, it becomes increasingly clear why minimizing wake losses in wind farms is essential to maintaining the efficiency of the farm. Two additional simulations of this geometry were performed at wind speeds of 10 m/s and 8 m/s. The behavior at these lower speeds is almost identical to the behavior at 12 m/s. The observed velocities

from all three cases indicate that the two downstream turbines produce anywhere from 25 to 40% less power compared to the first turbine.

The first triangular geometry simulated was the 400 - 150 m case. This geometry is used as a base case for both the triangular and delta geometries simulated in this project. The cases were simulated at the same three velocities as the three turbine row case: 12 m/s, 10 m/s, and 8 m/s. In this case, there is almost no wake interaction between turbines. Each turbine should experience, at minimum, the inlet velocity of the first turbine. Simulations were performed for both these arrangements to determine if such arrangements might actually increase velocity at the inlets of the downstream turbines, either due to a cascading effect or the funneling of air flow.

Figure 10 shows the velocities at each turbine centerline. There is a small but distinct increase in velocity at both downstream turbines. Before the two turbines (green and blue lines), a 0.042 m/s increase in velocity is observed. While this increase in velocity may appear to be

**Figure 8**: Contours of velocity for the three turbines arranged in a row (12 m/s case).

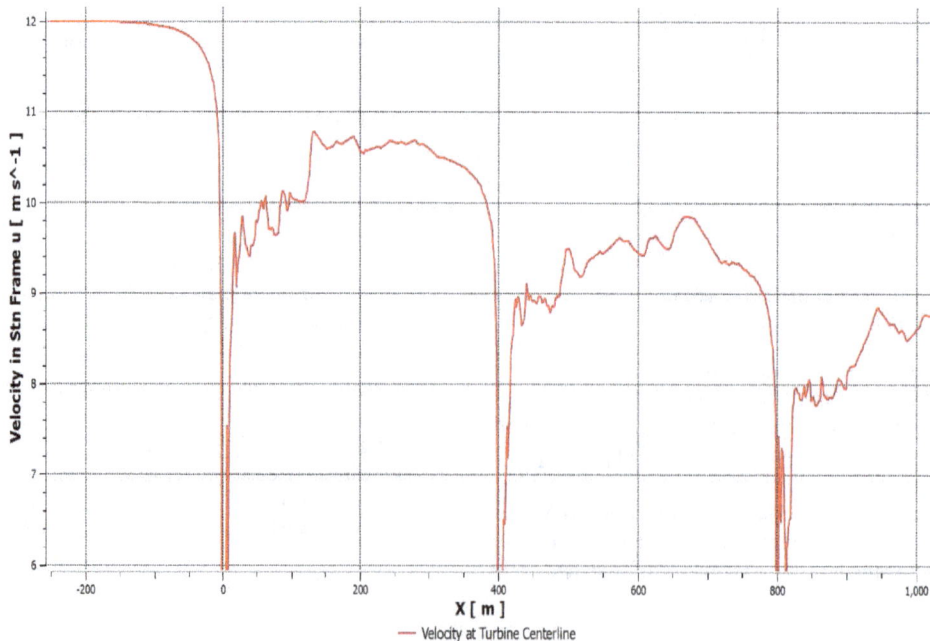

**Figure 9**: Velocity along the turbine row centerline for the 12 m/s case.

small, the cubic relationship between available wind power and wind speed shows that a 0.042 m/s increase in velocity at 12 m/s results in a 1.05% improvement in performance at each downstream turbine, or a 0.7% increase in energy production for the turbine group when compared to the three individual turbines operating at the given speed. For a group of three 1.5 MW turbines, a 0.7% performance increase results in an additional 32,000 W of power. The 10 m/s and 8 m/s cases result in similar, but smaller, velocity increases.

Figure 11, similar to Figure 10, shows the velocities at turbine centerlines, with the velocity increase at the center turbine visible. The delta cases were simulated to determine if the funneling of air between two wakes could increase energy production. In this case, the velocity increase occurs sooner than in the triangular case, and it is higher, at about 0.061 m/s. While this velocity increase is higher than the increase in speed due to the triangular arrangement, the percent increase in performance for the whole group is lower, at about 0.5%.

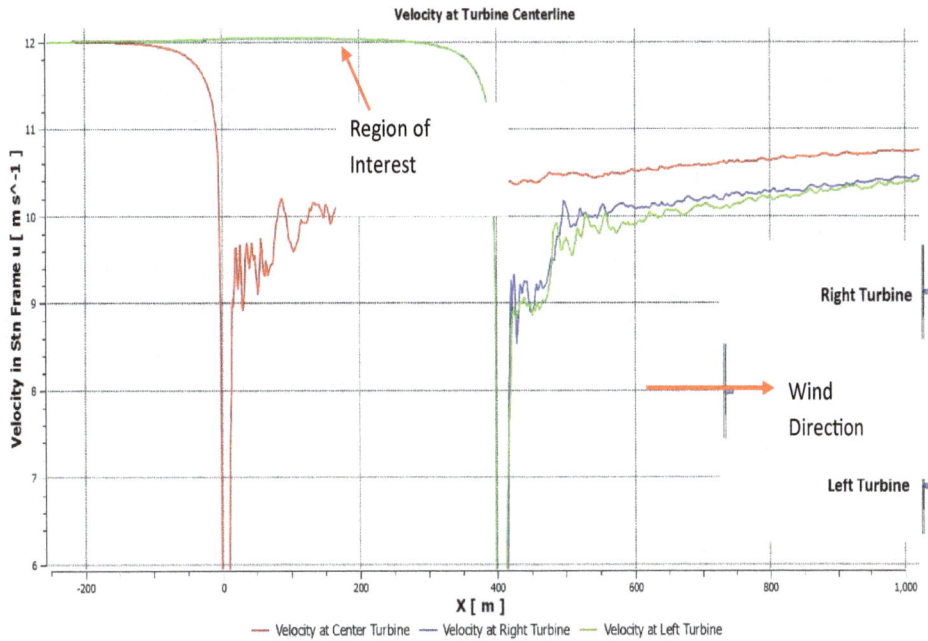

**Figure 10:** Turbine centerline velocities for the 400-150 m triangular case at 12 m/s.

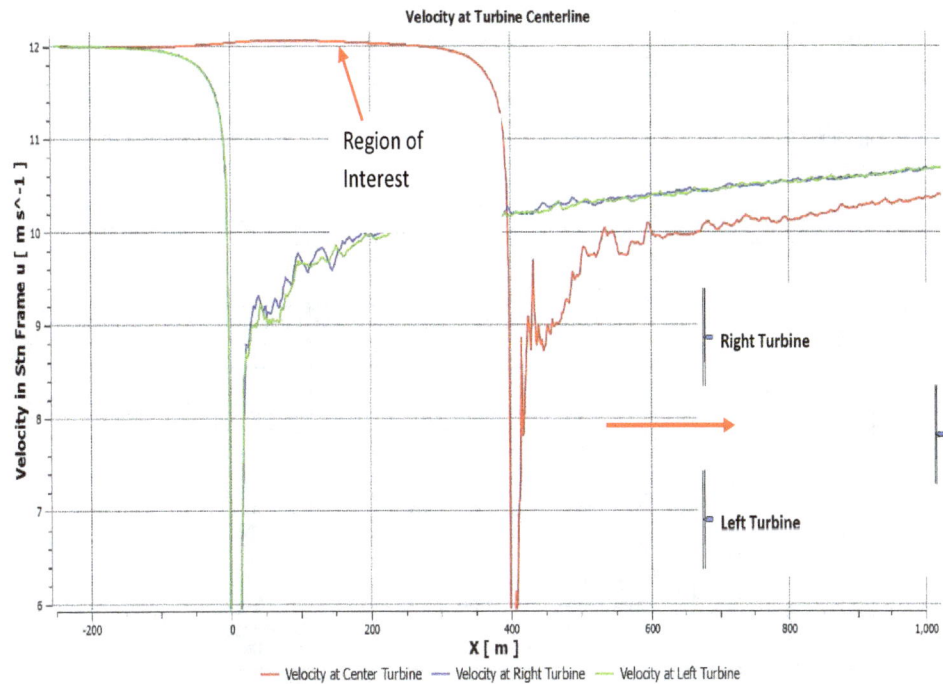

**Figure 11:** Turbine centerline velocities for the 400-150 m delta case at 12 m/s.

Similar results were observed for the delta cases when the wind speed was reduced.

When the wind direction is changed, wake interactions become much more significant. In fact, changes in wind direction completely negate any performance gains from the arrangement of the turbines. Therefore, in order to take maximum advantage of the arrangements presented in this study, the wind farm site should be located in a region with very little variability in wind direction. Figure 12 below shows an example of the velocity contours generated by one of the four simulated wind direction cases.

As seen in Figure 12, the left downstream turbine experiences a waked flow at half of its rotor inlet speed, which leads to both reduced power output from the turbine and additional fatigue stress due to the varying velocities at the turbine inlet.

From the simulations, it was determined that, in general, the triangular cases are more efficient than the delta arrangements at corresponding velocities. Also, the triangular 400 m by 100 m geometry produced the largest overall performance gains for the turbine group, with the triangular case remaining slightly more effective at increasing total power output than the delta case. The performance levels of the turbine group geometries are shown in Figure 13 and Table 3 below. The performance increases were determined by using the available

wind power equation to calculate the percent increase in available wind power at each turbine experiencing the velocity increase. The power increases were then averaged over the group of three turbines to determine the percent performance increase for the group as a whole, and by extension, for a wind farm using the given arrangement.

As mentioned before, the 400 m by 100 m simulations showed the largest performance gains compared to turbines experiencing the given speeds individually or without influence from a grouping, such as ones in a parallel row. The four simulations conducted with various wind directions indicated that the benefits of the triangular or delta arrangements are most pronounced when the wind travels directly downstream. If the wind direction changes roughly 10 to 15° at a given site, the 400 by 150 m cases can cause certain turbines to become waked, resulting in the negation of any performance improvements due to the geometry.

## Conclusion

This study proposes a possible method of generating more power from wind turbines by arranging turbines in either triangular or delta groups, arrangements designed to increase wind flow at rotor inlet planes by either cascading flow around a single leading turbine or by funneling it through the area between two turbine wakes. To investigate the appropriateness of each arrangement, several CFD simulations of

**Figure 12**: Velocity contours for a 400 – 150 m triangular arrangement with the wind direction at 10°.

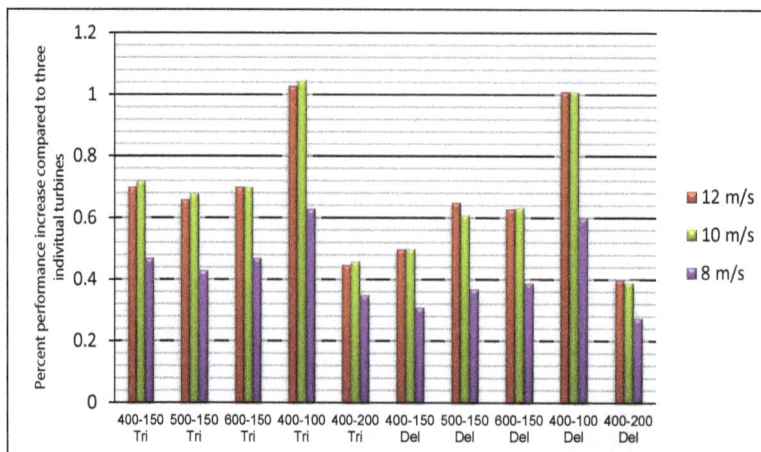

**Figure 13**: Graph of percent performance increase in each simulation geometry.

| Case | 12 m/s | 10 m/s | 8 m/s |
|---|---|---|---|
| 400 - 150 m Tri. | 0.7% | 0.72% | 0.47% |
| 500 - 150 m Tri. | 0.66% | 0.68% | 0.43% |
| 600 - 150 m Tri. | 0.7% | 0.7% | 0.47% |
| 400 - 100 m Tri. | 1.03% | 1.045% | 0.63% |
| 400 - 200 m Tri. | 0.45% | 0.46% | 0.35% |
| 400 - 150 m Del. | 0.5% | 0.5% | 0.31% |
| 500 - 150 m Del. | 0.65% | 0.61% | 0.37% |
| 600 - 150 m Del. | 0.63% | 0.634% | 0.389% |
| 400 - 100 m Del. | 1.01% | 1.01% | 0.6% |
| 400 - 200 m Del. | 0.4% | 0.39% | 0.275% |

**Table 3**: Recorded values of performance increase in simulation cases.

various turbine geometries were performed. Each geometric variation was simulated at three different velocities at the domain inlet (12 m/s, 10 m/s, and 8 m/s) to simulate the various wind speeds that might be encountered by turbines operating in a farm. In addition, four extra simulations were performed to determine the effects of wind direction.

The results of this study indicate that the triangular arrangements produce more energy than the delta arrangements. This is primarily due to the increase in velocity at the two rear turbines in the triangular arrangement, as opposed to an increase in velocity at only a single rear turbine in the delta cases. Of all the arrangements, the 400 m by 100 m triangular arrangement yielded the best results, with a performance increase of 1.05%, which could lead to an additional 47,000 W of power. Theoretically, in a utility scale wind farm capable of producing 200 MW of power, this performance increase applied to the entire farm would result in an increase in power generation about of 2.1 MW. If the given 200 MW wind farm has an installation cost of $100,000,000, the cost per MW of energy production capacity at installation is $500,000. Using the triangular groupings, the power production capacity would increase to 202.1 MW, resulting in a cost per MW value of about $494,070. Therefore, through the use of the triangular arrangement, the initial cost per MW of power generation capacity is reduced by approximately 1.1%. In the case described above, this translates to an overall savings of $6000 saved per MW of capacity installed.

However, this performance increase is not without certain drawbacks. Since the turbines are placed relatively close together, the possibility of intense turbulence from wake interactions is significantly higher. Also, in sites where the wind shifts direction frequently, the benefits of the triangular arrangement are significantly diminished. In fact, additional maintenance costs might be incurred if certain wind directions result in the upstream turbine directly waking one of the downstream turbines. Finally, all simulations were performed with the assumption that the turbines sat on completely level ground, such as a large field with few trees or an offshore farm. Therefore, the results may have little bearing on sites when terrain obstructions are present. In summary, these arrangements are recommended for use in utility scale farms only when the average annual wind direction is fairly consistent and the land is relatively level.

### Acknowledgements

This research was supported in part by US Department of Energy Grant DE-EE0003265 and US Department of Education Grant P116B100322. However, the contents do not necessarily represent the policy of the U.S. Department of Education, and endorsement by the Federal Government should not be assumed.

### References

1. NREL (2005) Wind Energy Myths, Report No. DOE/GO-102005-2137, U.S. Department of Energy.

2. Thomas BG, Urquhart J. (1996) Wind Energy for the 1990's and Beyond. J Energy Convers Manag 37: 1741.

3. Liu R, Ting, David K (2003) On the Aerodynamics of Wind Turbine, Proceedings of the 2003 International Joint Power Generation Conference, ASME, Atlanta, GA, USA.

4. Spera DA (2009) Wind Turbine Technology: Fundamental Concepts of Wind Turbine Engineering, ASME Press, New York, NY: 41 - 43, 83 - 87, 198 - 199.

5. Jha AR (2011) Wind turbine technology. CRC Press, Boca Raton, FL: 9 - 15.

6. Hansen AC, Butterfield CP (1993) Aerodynamics of Horizontal-Axis Wind Turbines. Ann Rev Fluid Mech 25: 115 - 149.

7. Vermeer LJ, Sørensen JN, Crespo A (2003) Wind Turbine Wake Aerodynamics. Prog Aerospace Sci 39: 467 - 510.

8. Kulunk, E, Yilmaz, N (2009) Aerodynamic Design and Performance Analysis of HAWT Blades, Proceedings of the ASME 2009 Fluids Engineering Division Summer Meeting, ASME, Vail, CO, USA.

9. Refan M, Hangan H (2010) Experimental and Theoretical Study of the Aerodynamic Performance of a Small Horizontal Axis Wind Turbine. ASME 2010 Power Conference, ASME, Chicago, IL, USA.

10. Haans W, Sant T, van Kuik G, van Bussel G (2006) Stall in Yawed Flow Conditions: A Correlation of Blade Element Momentum Predictions With Experiments. J Solar Energy Engg 128: 472 - 480.

11. Sørensen JN, Shen WZ (2002) Numerical Modeling of Wind Turbine Wakes. J Fluids Engg 124: 393 - 399.

12. Currin HD, Long J. (2009) Horizontal Axis Wind Turbine Free Wake Model for AeroDyn. Oregon Institute of Technology, Klamath Falls, OR, USA.

13. Currin HD, Coton FN, Wood B (2008) Dynamic Prescribed Vortex Wake Model for AERODYN/FAST. J Solar Energy Engg 130.

14. Currin HD, Coton FN (2007) Validation of a Dynamic Prescribed Vortex Wake Model for Horizontal Axis Wind Turbines. Proceedings of the 5th Joint ASME/JSME Fluids Engineering Conference, ASME, San Diego, CA, USA.

15. Sørensen JN, Michelsen JA, Schreck S (2002) Navier-Stokes Predictions of the NREL Phase VI Rotor in the NASA Ames 80 ft x 120 ft Wind Tunnel. Wind Energy 5: 151 - 169.

16. Anderson JD (1995) Computational Fluid Dynamics: The Basics with Applications. McGraw-Hill, Inc, New York, NY.

17. Amano RS, Malloy RJ (2008) Aerodynamic Comparison of Straight Edge and Swept Edge Wind Turbine Blade. ASME International Mechanical Engineering Congress and Exposition, ASME, Boston, MA, USA.

18. Palm M, Huijsmans R, Pourquie M, Sijtstra A (2010) Simple Wake Models for Tidal Turbines in Farm Arrangement. ASME 2010 29th International Conference on Ocean, Offshore and Arctic Engineering, ASME, Shanghai, China.

19. Wußow S, Stizki L, Hahm T (2007) 3D-Simulation of the Turbulent Wake Behind a Wind Turbine. J Physics 75.

20. De Bellis F, Catalano LA, Dadone A (2010) Fast CFD Simulation of Horizontal Axis Wind Turbines. ASME Turbo Expo 2010: Power for Land, Sea and Air, ASME, Glasgow, UK.

21. McStravick DM, Houchens BC, Garland DC, Davis KE (2010) Investigation of an Eppler 423 Style Wind Turbine Blade. ASME 2010 4th International Conference on Energy Sustainability, ASME, Phoenix, AZ, USA.

22. Wendt JF (2009) Computational Fluid Dynamics: An Introduction. Springer-Verlag Berlin Heidelberg.

23. Versteeg H, Malalasekra W. (2007) An Introductions to Computational Fluid Dynamics: The Finite Volume Method. McGraw-Hill, Inc, New York, USA.

24. www.en.wikipedia.org/wiki/GE_Wind_Energy#GE_1.5MW

25. http://wind.nrel.gov/airfoils/Shapes/S809_Shape.html

26. Wolfe WP, Ochs SS (1997) CFD Calculations of S809 Aerodynamic Characteristics. AIAA paper 97-0973.

27. Barthelmie RJ, Folkerts L, Larsen GC, Rados K, Pryor SC, et al. (2006) Comparison of Wake Model Simulations with Offshore Wind Turbine Wake Profiles Measured by Sodar. J Atmospheric Oceanic Technol 23: 888 – 901.

# Temporal Assessment of Wind Energy Resource in Algerian Desert Sites: Calculation and Modelling of Wind Noise

**Miloud Benmedjahed\* and Lahouaria Boudaoud**

*Research Unit Renewable Energy in Rural Sahara, URERMS, Renewable Energy Development Centre, CDER, BP 478 Route Reggane, Adrar, Algeria*

**Abstract**

Our study focuses on the assessment of wind resources of three desert sites in Algeria (Adrar, Ain Salah and Tindouf). The data used in this study span a period of 10 years. The parameters considered are the speed and direction of wind. For this purpose, the most energetic and frequent speed as well as the Weibull parameters to plot the wind rose were evaluated. The desert sites are favourable for large ZDE (Zone of Wind Development), why it was decided to investigate the possibility to set up a wind farm of 10 MW consisting of twelve wind turbine type WGT850 kW. Next, its noise was calculated and then modelled. The results obtained from the three sites gave annual mean speeds around 5 m/s; the West store is dominant for Tindouf, the East- North-East (ENE) store for Ain Salah and the East store. Our simulation of the noise propagation for wind farms shows that noise level is estimated around 45 dB at a distance of 300 m from the nearest turbine and 42 dB at a distance of 400 m. We can conclude that these noise levels have no effect on health and comply with the Algerian standard.

**Keywords:** Weibull parameters; Wind rose; Wind power; Wind farm; Noise; Algeria

## Introduction

In Algeria, the objectives established by the join-stock company NEAL (New Energy Algeria), focused on raising renewable energy production to 1400 MW in 2030 and 7500 MW at the beginning of 2050. Electrical power will be obtained from solar power plants, which are exclusively solar, or from hybrid solar plants, which also use other forms of renewable or conventional energy, preferably natural gas [1].

Harnessing the wind is one of the cleanest, most sustainable ways to generate electricity. Wind power produces no toxic emissions and none of the heat-trapping emissions that contribute to global warming. This, and the fact that wind power is one of the most abundant and increasingly cost-competitive energy resources, makes it a viable alternative to the fossil fuels that harm our health and threaten the environment. Wind energy is the fastest growing source of electricity in the world.

Many work indicated that Algeria was characterized by a competitive electricity generation cost per kW from Wind turbine; in particular, we can cite the Wind Potential Assessment of Three Coastal Sites in Algeria; Calculation and Modelling of Wind Turbine Noise using Matlab, the Wind Potential Assessment of Ain Salah in Algeria, Assessment of wind energy and energy cost in Algeria, Assessment of wind energy and energy cost in Algeria [2-4] and calculation of the Cost Energy the evaluation of electricity generation and energy cost of wind energy conversion systems in southern Algeria [5] .

The energy available in the wind varies as the cube of wind speed, so an understanting of the characteristics of the wind resource is critical to all aspects of wind energy exploitation, from the identification suitable sites to the prediction of the economic viability of wind farm project. The present study tries to determine various wind parameters and then focuses on the processing and simulation of their hourly data, collected during 10 years. Wind potential, its direction and frequency are assessed by plotting the wind rose, in order to select the appropriate site for future wind turbines. Finally, after the evaluation of wind power, the environmental impact of wind turbines was evaluated. For this purpose, the ISO 9613-2 calculation model is used in the case where octave data are available; otherwise some calculation formulas based on Matlab are developed.

## Site and Weather Data

In this study, the wind speed data were collected over a period of 10 years. The details of the sites are summarized in Table 1.

The meteorological measurements stations were made at 10 meters above ground level and registered every 3 hours. The geographical location of the metrological station is shown in Figure 1.

## Wind Potential

### Weibull distribution

The wind characteristics will determine the amount of energy that can be effectively extracted from the wind farm. In order to determine the properties of a site, measurements of the speed of wind and its direction are needed. This study was carried out over a period of ten years. However, previous studies in the field of wind energy showed that the most important and appropriate characteristic to exploit is the Weibull statistical distribution this is a probability function that can be expressed as [6-9]:

$$f(v) = \left(\frac{k}{C}\right)\left(\frac{v}{C}\right)^{k-1} \exp\left(-\left(\frac{v}{C}\right)^k\right) \tag{1}$$

k and C are the shape parameter (dimensionless) and the scale parameter (m/s), respectively. Usually, the shape parameter characterizes the symmetry of the distribution. The scale parameter is very close to the average speed of wind. The standard deviation method was chosen to determine both factors k and C. This method is based on the calculation of the standard deviation and the average speed [6].

**\*Corresponding author:** Benmedjahed M, Research Unit Renewable Energy in Rural Sahara, URERMS, Renewable Energy Development Centre, CDER, BP 478 Route Reggane, Adrar, Algeria
E-mail: benmedjahed_78@yahoo.fr

If the wind distribution is desired at some height other than the measurement height, the Weibull parameters can be adjusted to any desired height by the model of Justus [7].

## Wind energy

The power of the wind that flows at a speed v through the blade sweep area S can be expressed by the following equations [6,8,9]:

$$P(v) = \frac{1}{2}\rho \; s \; v^3 \qquad (2)$$

A wind turbine allows extracting the kinetic energy from the wind and converting it into and electric energy. The power curves of the wind turbines can be expressed by the following equations [9]:

$$P_e(v) = C_e P(v) \qquad (3)$$

Where, $C_e$ is the wind turbine efficiency. The efficiency of the wind turbines taken into consideration in this study are shown in Figure 2 and technical specifications of the selected wind turbines are listed in Table 2.

The histogram method is used to estimate the energy generated by a wind turbine. The superposition of the energy response curve (kW) and the frequency histogram give [8]:

$$E = \sum_{i=1}^{n} P_i(v_i) F_i(v_i) \times N \qquad (4)$$

## Wind farm planning

To produce a large amount of energy, a wind farm must be installed as followed. We use WGT 850 kW wind turbine model.

When several turbines are installed in block, the turbulence due to rotation of the turbine blades can affect other turbines nearby. To minimize this effect, the spacing of about 3 to 4 $D_T$ (with $D_T$ the

| Location | Latitude | Longitude | Altitude |
|---|---|---|---|
| Adrar | 27.88° N | 0.28° W | 263 m |
| Ain Salah | 27.25 ° N | 2.51° E | 269 m |
| Tindouf | 27.70° N | 8.17° W | 442 m |

Table 1 : Geographical coordinates of the data collection stations used in the study.

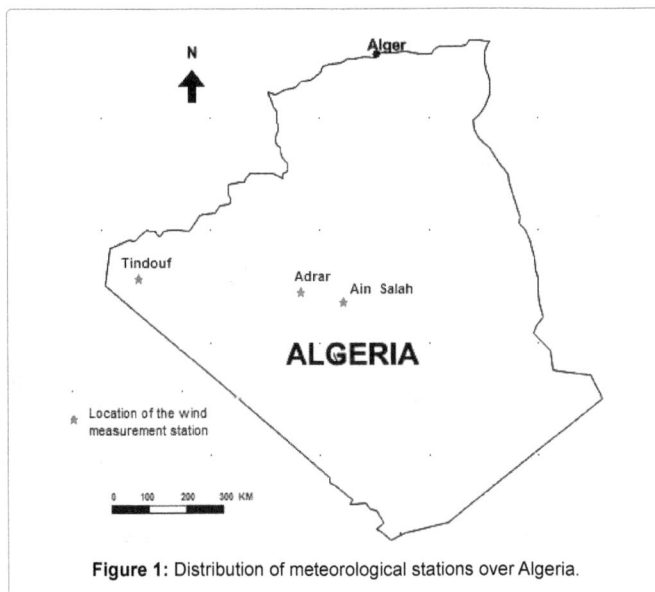

Figure 1: Distribution of meteorological stations over Algeria.

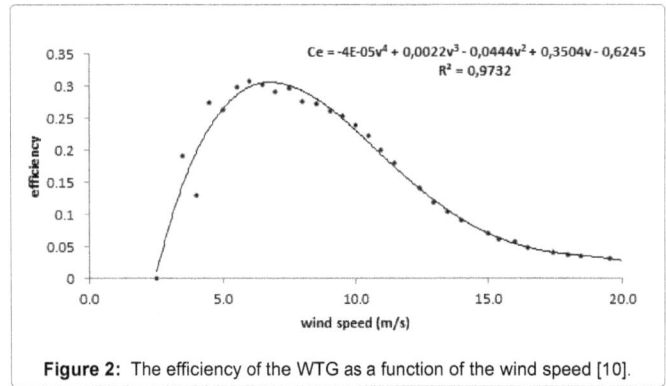

Figure 2: The efficiency of the WTG as a function of the wind speed [10].

| Characteristics | WGT 850 kW |
|---|---|
| Rated power | 850 kW |
| Rotor diameter | 52 m |
| Hub height | 55 m |
| Cut-in wind speed | 4 m/s |
| Rated wind speed | 16 m/s |
| Cut-out wind speed | 28 m/s |

Table 2: Technical specifications of the considered wind turbine [10].

diameter of the rotor) is provided inside the rows [6].

Similarly, the spacing between the rows may be of the order of 10 $D_T$, so that the air stream passing through a turbine is restored before its interaction with the following turbine. This spacing can be further increased for better performance but this implies more land used. In general, the energy loss due to the effect of park is about 5%.

## Noise Level

Usually, in order to measure the wind turbine noise, the level of the weighted acoustic power is calculated as an average level at 500 Hz. The impact of noise is calculated according to the international standard ISO 9613-2 [2,10,11] as follows:

$$L_{AT}(DW) = L_{WA} + D_C - A - C_{met} \qquad (5)$$

where $L_{AW}$ is the level of weighted acoustic power of the noise source. $D_C$ - $D_\Omega$ is the Correction made in order to take into consideration the directivity of the source (without directivity= 0 dB) and the reflection on the ground $D_\Omega$ which can be calculated as follows [2,11]:

$$D_\Omega = 10 Log \left( 1 + \frac{d_p^{\;2} - (h_s - h_r)^2}{d_p^{\;2} - (h_s + h_r)^2} \right) \qquad (6)$$

$h_s$ is the height of source above the ground (hub height), $h_r$ is height of point of noise impact (depending on the regulations but also adjustable when defining the calculation) and $d_p$ is the distance between noise source and point of impact, projected on the ground (m). The distance is calculated from the coordinates (x, y) of the source (index S) to the point of impact (index r) [2,11]:

$$d_p = \sqrt{(x_s - x_r)^2 + (y_s - y_r)^2} \qquad (7)$$

A is the attenuation during noise propagation between the source (the wind turbine nacelle) and the point of impact. The total attenuation is given by [2,11]:

$$A = A_{div} + A_{atm} + A_{sol} \qquad (8)$$

$A_{div}$ is the attenuation due to spatial propagation [2,11]:

$$A_{div} = 10 Log\left(d\right) + 11 \qquad (9)$$

d is the distance between source and point of impact (m) [2,11]:

$A_{atm}$ is the attenuation due to atmospheric absorption [2,11]:

$$A_{atm} = \frac{\alpha d}{1000} \qquad (10)$$

$\alpha_{500}$ is the absorption coefficient of air.

$A_{gr}$ is the attenuation of ground [2,11]:

$$A_{gr} = 4,8 - \left[\left(\frac{2h_m}{d_P}\right)\left(17 + \frac{300}{d_P}\right)\right] \geq 0 \qquad (11)$$

$h_m$ is the average height (m) of noise path above the ground. If no digital model of ground is found, then the average height can be calculated as follows:

$$h_m = \frac{\left(h_s + h_r\right)}{2} \qquad (12)$$

$C_{met}$ is the weather correction. This is done as follows

$$\begin{cases} c_{met} = 0 & if \quad d_p < 10\left(h_s + h_r\right) \\ c_{met} = c_0\left[1 - \frac{10\left(h_s + h_r\right)}{dp}\right] & if \quad d_p > 10\left(h_s + h_r\right) \end{cases} \qquad (13)$$

$C_0$ is a factor, in decibels, which depends on local meteorological statistics for wind speed and direction, and temperature gradients.

## Results and Interpretations

From the measured data for ten years, in the three weather stations (Adrar, Ain Salah and Tindouf) at a height of 10 m from the ground, the Weibull parameters could be calculated for the three sites (Table 2).

The annual shape parameter value range from 2.06 (Adrar) to 3.26 (Ain Salah), which means that winds are stable for all sites and the analysis of the annual scale parameter C shows that Adrar is the windiest site (7.4 m/s). Statistical data analysis allowed determining the wind rose which is the graphical representation of wind frequency as a function of direction in a polar reference. It is determined for ten years. The results obtained Figure 3 show that:

For Adrar the prevailing wind direction is East (E) with 15% and the predominant directions is East-North-East (ENE) with 14%. The prevailing Wind direction for Ain Salah is the East -North- East t (ENE) with 30% and predominant directions is East (E) with 14% and for Tindouf, the West (W) sector represents 30% of wind frequencies and the North western (NW) sector are predominant sectors with a percentage of around 25% (Figure 4).

Our choice fell on one rows of turbines, has twelve wind turbines. The distance between them 208 m, the wind turbines will be oriented from North to South for Adrar, West-North-west to East-South-East for Ain Salah and North to South for Tindouf. We estimated the power density and the energy produced by wind turbine WGT 8500 kW (Tables 3 and 4).

The power density of wind turbine WGT 850 kW varies from 84.21 W/m² (Tindouf) to 234.35 W/m² (Adrar), which means that Adrar has more important power density than Ain Salah and Tindouf .The annual energy we can produce for wind frame in Adrar, Ain Salah and Tindouf account respectively 49.70 GWh, 20.38 GWh, 17.86 GWh.

The noise emitted by a wind turbine constitutes the main impact on environment. Noise can be produced by any obstacle placed on an air flow trajectory. The tone of this noise depends on the shape and

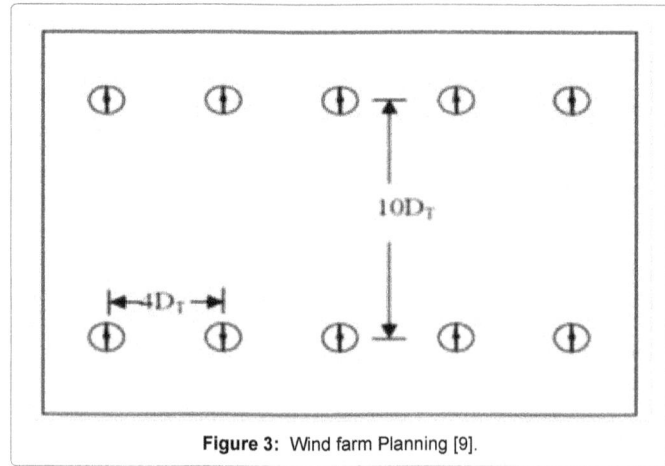

**Figure 3:** Wind farm Planning [9].

| Location | C (m/s) | k | v (m/s) |
|---|---|---|---|
| Adrar | 7.4 | 2.06 | 6.5 |
| Ain Salah | 6.0 | 2.48 | 5.4 |
| Tindouf | 5.9 | 2.27 | 5.2 |

**Table 3:** Annual mean wind speed and Weibull parameters at 10 m from the ground level.

| Location | Power density (W/m²) | Energy (MWh) | |
|---|---|---|---|
| | | Wind turbine | Wind farm |
| Adrar | 234.35 | 4.36 | 49.70 |
| Ain Salah | 96.11 | 1.79 | 20.38 |
| Tindouf | 84.21 | 1.57 | 17.86 |

**Table 4:** The power density and the energy produced.

dimensions of the obstacle as well as on the air flow speed, in addition to the mechanical noise from the operation of all components present in the enclosure. The main noise generating components are: the multiplier (except for some recent models), shafts, the generator, and auxiliary equipment (hydraulic systems, cooling units).

The estimation of wind turbine Noise of the WGT850, to be operated at the considered sites has been done under the following assumptions [2]:

- The level of weighted acoustic power of the noise source was considered to be 103 dB ± 1 dB/(m/s) at wind seeped 8 m/s .

- The absorption coefficient of air (α500) was taken as 1.9 dB/km.

- The attenuation due to a barrier and the attenuation due to miscellaneous other effects was considered Negligible.

The WGT 850 kW wind turbine was chosen considering its low noise power. We used Matlab software to calculate the noise generated by the wind turbine under the conditions that can be met by our three sites (flat ground). For stated sites, the results of our simulation of the propagation of noise by wind farm in Adrar, Ain Salah and Tindouf is a summarized in Table 5. The noise power given by the manufacturer is 103 dB for each wind turbine WGT 850 kW. According to our calculations, using the method (ISO 9613-2), the noise level is about 45 dB (A) at 350 m from the nearest turbine and at a distance of 400 m, the noise level will be about 42 dB for all wind farm.

The noise level of a wind turbine is 42 dB (A), which corresponds to the noise inside quiet house. Hence, these noise levels have no effect on health and are consistent with the national standard (Executive Decree No. 93-184 of July 27, 1993, regulating noise emission).

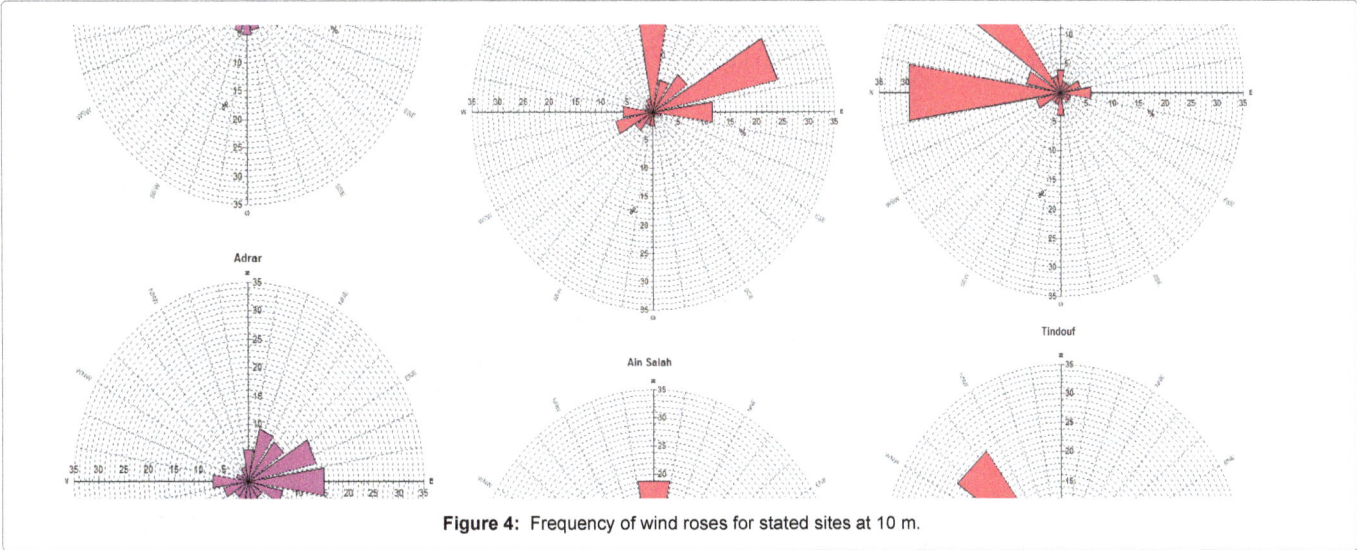

**Figure 4:** Frequency of wind roses for stated sites at 10 m.

| dp(m) | Noise (dB) | | |
|---|---|---|---|
| | Adrar | Ain Salah | Tindouf |
| 200 | 49.46 | 49.33 | 49.30 |
| 250 | 46.89 | 46.76 | 46.73 |
| 300 | 44.93 | 44.80 | 44.77 |
| 350 | 43.37 | 43.24 | 43.21 |
| 400 | 42.03 | 41.90 | 41.87 |
| dp(m) | Noise (dB) | | |
| | Adrar | Ain Salah | Tindouf |
| 200 | 49.46 | 49.33 | 49.30 |
| 250 | 46.89 | 46.76 | 46.73 |
| 300 | 44.93 | 44.80 | 44.77 |
| 350 | 43.37 | 43.24 | 43.21 |
| 400 | 42.03 | 41.90 | 41.87 |

**Table 5:** Level noise.

Disturbance from wind turbine noise during operation is low given the animal adaptability and the intermittent nature of noise emitted by wind.

## Conclusion

This study focused on the evaluation of wind potential of three Desert sites in Algeria (Adrar, Ain Salah and Tindouf), in order to use wind 850 kW turbines, based on wind speed measurements recorded during a ten year period. Wind resource analysis in the selected sites shows that South area in Algeria has a wind energy potential that can be effectively exploited. Indeed, statistical treatment of data allowed evaluating the characteristic speeds and wind potential for each site. The results obtained show that:

- The annual shape parameter value range from 2.06 (Adrar) to 3.26 (Ain Salah), which means that winds are stable for all sites.

- The analysis of the annual scale parameter C shows that Adrar is the windiest site (7.4 m/s).

- The West store is dominant for Tindouf, the East- North-East (ENE) store for Ain Salah and the East store.

- The power density of wind turbine WGT 850 kW varies from 84.21 W/m² (Tindouf) to 234.35 W/m² (Adrar).

- The annual energy we can produce for wind frame in Adrar, Ain Salah and Tindouf account respectively 49.70 GWh, 20.38 GWh, 17.86 GWh.

- The noise level is about 45 dB (A) at 350 m from the nearest turbine and at a distance of 400 m, the noise level will be about 42 dB for all wind farm, these noise levels have no effect on health and are consistent with the national standard (Executive Decree No. 93-184 of July 27, 1993, regulating noise emission).

## References

1. Hasni T (2006) Développement des Energies Renouvelables en Algérie, In National Conference on Renewable Energies and Sustainable Development. Laghouat, Algeria.

2. Benmedjahed M, Ghellai N, Benmansour A (2012) Wind Potential Assessment of Three Coastal Sites in Algeria; Calculation and Modeling of Wind Turbine Noise using Matlab . Int J Comp Appl 56: 20-25.

3. Benmedjahed M, Bouzid Z, Ghellai N (2015) Wind Potential Assessment of Ain Salah in Algeria; Calculation of the Cost Energy. Int J Energy Power Engg 4: 38-42.

4. Benmedjahed M, Ghellai N, Benmansour A, Boudai S M, Tabet Hellal MA (2014) Assessment of wind energy and energy cost in Algeria. Int J Renew Energy 9: 31-39.

5. Daif S, Belhamel M, Haddadi M, Louche A (2008) Technical and economic assessment of hybrid photovoltaic/wind system with battery storage in Corsica island. Energy policy 36: 743-754.

6. Sathyajith M (2006) Wind energy: fundamentals, resource analysis and economics. Springer-Verlag Berlin Heidelberg, UK.

7. Justus CG, Mikhail A (1976) Height variation of wind speed and wind distributions statistics. Geophy Res Letters 3: 261-264.

8. Yousef B, Aymeric G (2005) Collecte, organisation, traitement, analyse de mesures éoliques et modélisation énergétique. Doctoral Thesis, Université Tecnica Federico Santa Maria, UTFSM,Valparaso, Chili.

9. Arbaoui A (2006) Aide a la décision pour la définition d'un système éolien adéquation au site et a un réseau faible. Thèses doctorat, École Nationale Supérieure d'Arts et Métiers Centre de Bordeaux, France.

10. EMD (2012) the Wind Pro Software.

11. Judith L (1996) Industrial noise - Description of the calculation method. AR INTERIM-CM adaptation and revision of the interim noise computation methods for the purpose of strategic noise mapping. LABEIN Technological Centre.

# Application of Solar Energy Heating System in Some Oil Industry Units and its Economy

**A.M. Abd El Rahman [1\*], A.S. Nafey[2] and M.H.M.Hassanien[1]**

[1]Department of Petroleum Refining and Petrochemicals, Suez University, Egypt
[2]Department of Engineering Sciences, Suez University, Suez University, Egypt

## Abstract

Energy is one of the building blocks of modern society. Once an exporter of oil and gas, Egypt is now struggling to meet its own energy needs. In oil industry there is an energy problem due to fuel and electricity consumption and refinery losses in a way that reduces the net profit of the industry. There are also environmental problems due to carbon dioxide emissions which is a major source of the global warming problem. As Egypt is blessed with geographic location in the Sun Belt area with 325 days of sun in a year, solar energy can be used as a source of energy that reduces fuel consumption and $CO_2$ emissions. The current study presents solar energy heating system that can be used for heating applications in some oil industry units. The study has been divided in two parts, the first one concerned with choosing the most appropriate solar system that can be used in such applications. Four different mathematical models for prediction of optical efficiency and thermal losses for the chosen system have been analyzed and then computerized using excel sheet program. Numerical comparison and also practical validation of the selected model have been done. For that paper under title of "Evaluation of Mathematical Models for Solar Thermal System" was published in October, 2016, American Journal of Energy Science. Visual basic program is then done for the validated model for good and friendly user interface. For the second part of the study, this paper concerned with performing an economic evaluation for providing feasibility and reliability conception about using the proposed system in some oil industry applications in number of Egyptian companies as preheating of crude oil for desalting in oil production (Khalda petroleum company), preheating of viscous oil for transportation enhancing and preheating of boiler feed water (Cairo oil refining company). The results show that payback period for crude preheating before desalting is 20 years and for fuel oil and boiler feed water preheating are 7 years.

**Keywords:** Solar energy; Parabolic trough solar collector; Crude oil heating; Steam production

## Introduction

In Oil Industry there is an energy problem due to energy losses in a way that reduces the net profit of the industry. Energy Losses are due to fuel consumption, refinery losses and electricity consumption. For example Heating viscous oil for transportation enhancing and heating crude oil for preliminary treatment consume a great amount of fuel. In refinery also for 100 kbpd crude needs approximately 120 MW to be preheated up to 350°C [1]. A rule of thumb used by some refiners is that it takes 1 barrel of oil-equivalent energy to process 10 barrels of crude oil [2]. Petroleum refining in the United States is the largest in the world, refineries spend typically 50% of the cash operating costs [3]. Oil refining, petrochemicals, ammonia, paper, cement, and steel production consume about 18% of the primary energy in the European Union (EU).

So there is a great interest towards the technologies for increasing the energy efficiency by reduction of the energy consumption. The most productive energy-conserving measures appear to be in the areas of improved combustion, the recovery of low-grade heat, and the use of process modifications. Concerning these solutions Romulo, Lima S et al. made a comparison between energy efficiency in Brazilian and United States crude oil refinery and concluded that increasing the refinery complexity which means more heat integration inside the plant will lead to reduction of the energy consumption [4]. To meet the energy challenges faced by Chinese petroleum refiners. Liu X et al. indicated that upgrading process heaters is identified as apriority to enable short term energy optimization [5]. Refineries may be also able to use other sources of energy, and otherwise wasted heat, to reduce the combustion of gaseous and liquid fuels. So fuel substitution (such as the use of coal in refineries) is an important goal [6].

In addition to energy problems, there are also environmental problems as in 2008 around 81.3% of the world's primary energy was supplied from oil, gas, and coal products; resulting in around 29,381 million ton of $CO_2$ which is a major source of the global warming problem [7]. Renewable energies including solar, wind, hydropower and biomass are considered to be attractive alternatives that are highly abundant, sustainable and environmentally friendly resources, most countries have initiated programs to develop energy sources based on renewable resources.

To overcome economical and environment problems, solar energy is introduced in this paper for heating applications in oil industry. For the applications of solar energy in oil industry, researchers studied using solar thermal energy in heating viscous fuel oil to about 50°C and stored at that temperature [8,9], heating crude oil to maintain flow ability during transportation [10-13] and thermal treatment of crude oil by heating water in the collector to 85-90°C then heated water exchanges heat with crude which is heated to 55-60°C [14].

Reference to the first part of the study [15] parabolic trough collector (Figure 1) presents the most appropriate type of collector to be used in the discussed solar thermal system for usage in the oil

---

**\*Corresponding author:** A.M. Abd El Rahman, Department of Petroleum Refining and Petrochemicals, Faculty of Petroleum and Mining Engineering, Suez University, Suez, Egypt, E-mail: Ahmed_AbdElRahman10@yahoo.com

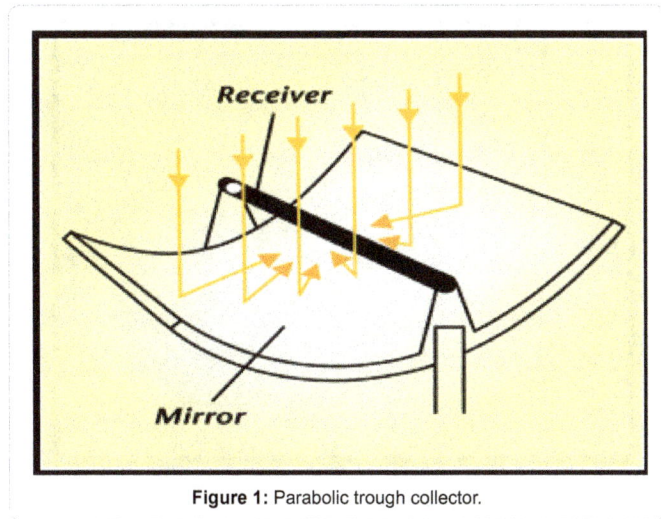

**Figure 1:** Parabolic trough collector.

industry.

For developing an economic study for any process unit, the process designer must be aware of selection of a basic process route, equipment used in the process and also details incorporated into the equipment. From a first draft flow sheet, a preliminary cost estimate can be prepared by the "factoring" or equivalent method. With more comprehensive and better information regarding the process, estimating engineers can prepare detailed estimates, which are often quite accurate, usually ±10% for the best [16].

## Methodology

### Cost estimation

For preliminary cost estimation of any process unit the following steps to be done:

1. Preparing a flow sheet for the process.
2. Preparing heat and material balances around each piece of equipment.
3. Sizing all of the equipment for cost estimation (the material cost of equipment often represents 20-40% of the total project cost for process plants).

Equipment cost prediction in the present study is done using the following methods:

a) Estimating Charts (Chemical Engineering Economics).

b) Matches (licensed Engineering Company) software.

c) Online cost estimator (McGraw-Hill Education).

d) Equipment cost V.1 software (University of Porto).

e) Actual commercial costs from Egyptian oil sector.

4. Analyzing the process carefully to determine what plant cost factors that should be used for calculation of (Elcctrical, Instrumentation, Utilities, Foundation, Installation, etc.) costs, and then calculation of total plant cost.
5. Finally investment decisions are taken often based upon several criteria such as payback period that is usually measured as the time from the start of production to recovery of the capital investment.

The above mentioned steps for cost estimation are done for each of the discussed applications (Figure 2).

### Proposed solar system

Heat transfer fluid (HTF) is designed to transfer heat from parabolic trough collector (PTC) to the fluid needed to be heated through heat exchanger. HTF must combines heat stability and low viscosity for efficient, dependable and uniform performance in a wide optimum use range of 120°C to 400°C.

**Basis of design:**

1. Configuration of the solar system is based on Kuraymat Plant (solar power project 100 km south of Cairo).
2. Design solar incident radiation=600 W/m$^2$.
3. Operating temperature range of HTF= (120-400°C).
4. Ambient temperature=20°C.
5. No over design is considered.

**Process description:** HTF is transferred by a transfer pump to be heated through PTC, and then directed to heat exchanger for heat transfer to the fluid to be heated. HTF is finally directed to surge tank to be pumped again (Figure 3).

**Scope of work:** This study scope of work in each application is to perform detailed engineering for the solar package in order to provide cost estimation for each project through the following steps:

a) Hysys simulation for the hot oil system to optimize the design of plant (Figure 4).

b) Hydraulic calculation for pump and piping is to provide piping size, system pressure drop (Figure 5) using HCALC (full pipe flow hydraulic calculator) and then pump required power.

Then commercial pump selection based on Tahoe Design Software's Pump Base (Figure 6).

c) Heat exchanger thermal design using HTRI (Figure 7) (Heat

**Plant Cost Estimating Factors**

| *Component* | *Plant Cost Factor, Fraction of Total Purchased Equipment Cost* | |
|---|---|---|
| Purchased equipment | 1.00 | |
| Piping | 0.15-0.70 | |
| Electrical | 0.10-0.15 | |
| Instrumentation | 0.10-0.35 | |
| Utilities | 0.30-0.75 | |
| Foundations | 0.07-0.12 | |
| Insulation | 0.02-0.08 | |
| Painting, fireproofing, safety | 0.02-0.10 | |
| Yard improvements | 0.05-0.15 | Optional Estimating |
| Environmental | 0.10-0.30 | Factors, Fraction |
| Buildings | 0.05-1.00 | of (Subtotal) |
| Land | 0-0.10 | Plant Costs |
| | Subtotal | 1.96-4.80 |
| Construction, engineering | 0.30-0.75 | 0.10-0.40 |
| Contractors fee | 0.10-0.45 | 0.03-0.10 |
| Contingency | 0.15-0.80 | 0.05-0.20 |
| Total | 2.51-6.80 | 2.31-8.16 |
| *Usual limits for total factor* | | *Lang Factor* |
| Minimum (solids processing) | 3.00 | 3.4 |
| Average (mixed processing) | 4.00 | |
| Maximum (fluid processing) | 5.50 | 5.2 |
| *Other capital requirements* | *Additional Factor, % of Total Plant Cost* | |
| Off-site facilities | 0-30 | |
| Plant start-up | 5-10 | |
| Working capital | 10-20, or 10-35% of mfg. cost | |

**Figure 2:** Plant cost estimating factor.

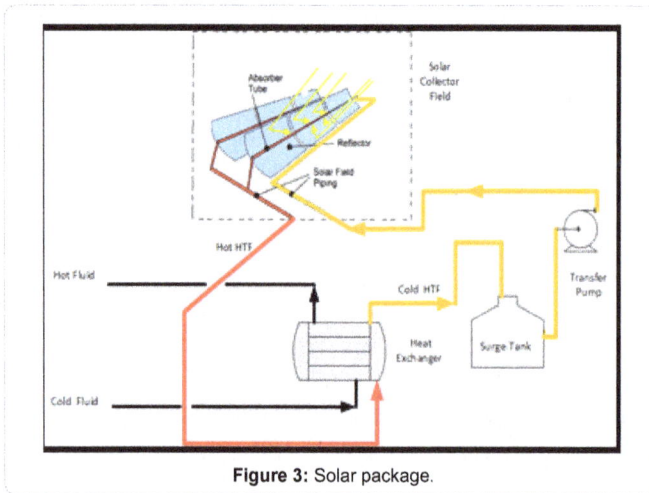

**Figure 3:** Solar package.

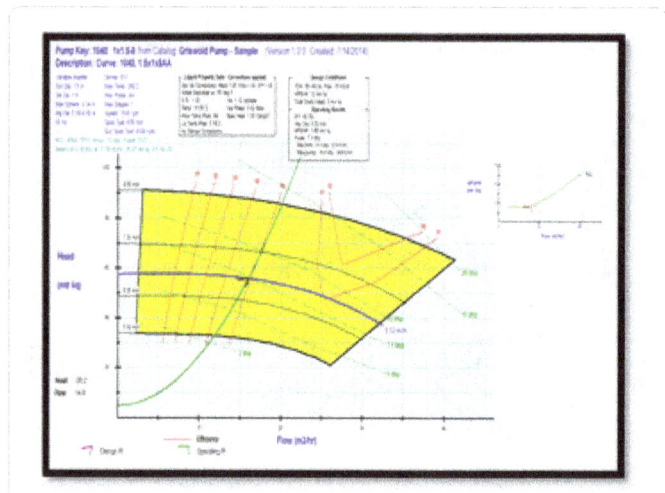

**Figure 4:** Hysys simulation for the proposed solar package.

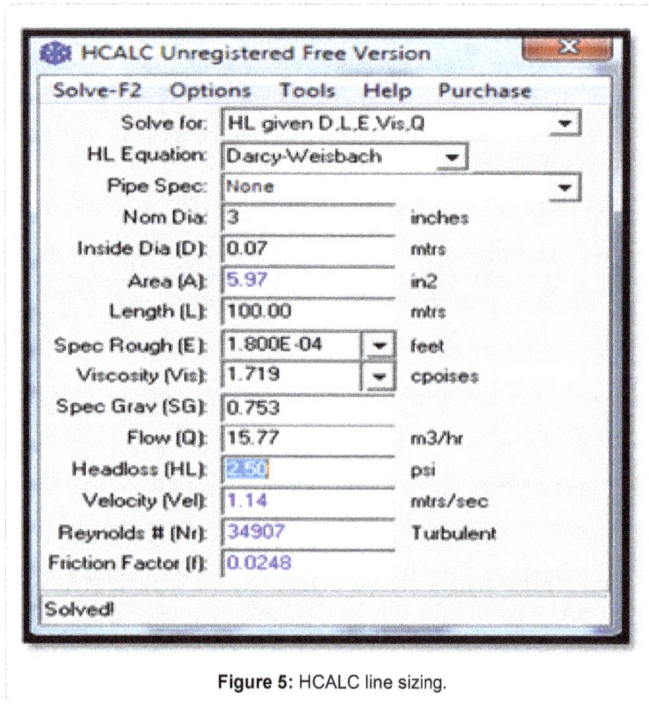

**Figure 5:** HCALC line sizing.

**Figure 6:** Q and H curve for Kalabsha.

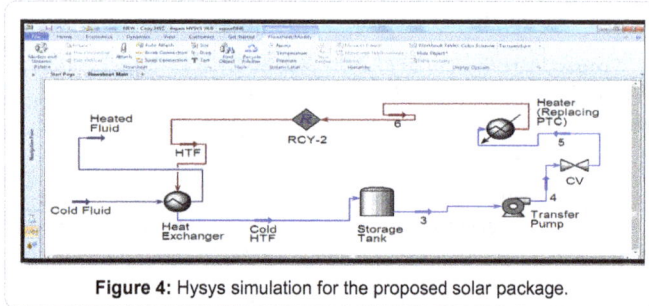

**Figure 7:** HTRI output summary.

One module area=12 × 5.76=69.12 m²

e)   Separator and tank sizing (if any).

f)   Cost estimation of the total solar package.

g)   Payback period calculation.

## Cases process description and study scope of work

### Case 1: Preheating of crude oil for desalting in oil production facilities KALABSHA (Khalda Petroleum Company):

*Process description:* The facility is located at KALABSHA field, Western Desert Egypt. Oil (Figure 9) is distributed to the six trains to separate the emulsion water in the oil to eventually achieve the product specifications through the following equipment for each train:

i.    Indirect forced heater.

ii.   Production separator.

iii.  Heater treater.

Transfer Research Institute) the global leader in process heat transfer to calculate heat transfer area and pressure drop across heat exchanger.

d)   Solar thermal system design using PTC calculation (Figure 8) (software developed in first part) to calculate solar collector required area.

**Figure 8:** PTC calculations.

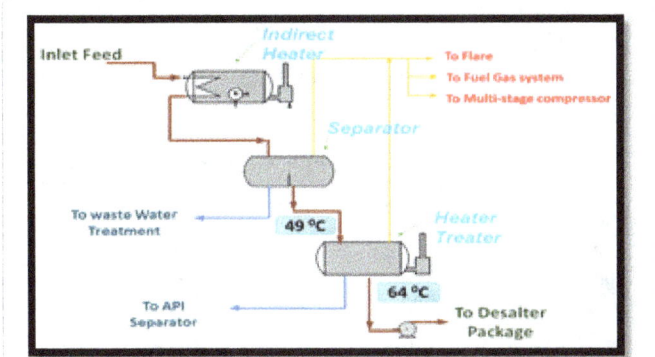

**Figure 9:** KALABSHA plant process description.

iv.      Oil feed pump.

Oil feed is heated in the indirect forced heater to 55°C to separate the emulsion water in the three phase separator. Flashed gases is directed to existing fuel gas system, separated water is directed to the waste water treatment facility and the oil is directed to the heater treater to further separate the emulsion water in oil. The outlet oil is pumped through oil feed pump to two desalter packages.

*Scope of work:* Produced crude oil is heated from about 49°C to 64°C in heater treater using fuel gas. Our scope of work is to bypass the heater (Figure 10) during day only (6 hours) with the proposed solar system and making a comparison between the existing and the proposed system.

### Case 2: Preheating of viscous oil for transportation enhancing Cairo Oil Refining Company (CORC):

*Process description:* The Facility is located at CORC (the biggest refinery in Egypt). Its capacity represents about 30% of Egypt's refining capacity. Fuel oil is produced from atmospheric distillation unit, stored at 50°C in storage tanks, and then heated in fired heater to about 90°C for transportation enhancing through overcoming pour point and viscosity problems (Figure 11).

*Scope of work (fired heater):* Produced fuel oil (Figure 12) is heated from about 50°C to 90°C in fired heater using diesel oil as heating fuel. Our scope of work is to bypass the heater during day hours (hours) with the proposed solar system and making a comparison between the existing and the proposed system.

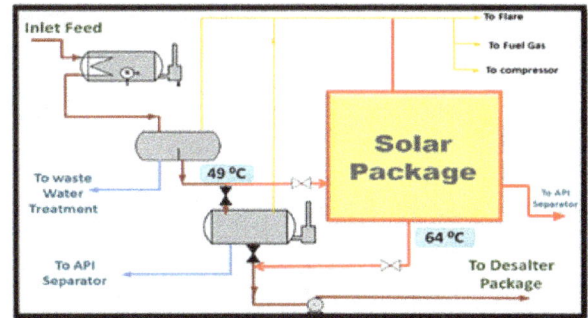

**Figure 10:** KALABSHA plant after introducing proposed solar package.

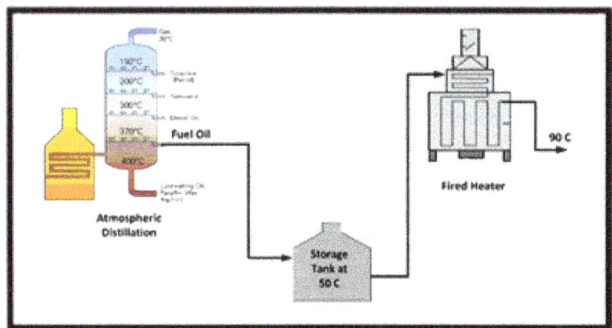

**Figure 11:** Fuel oil system process description.

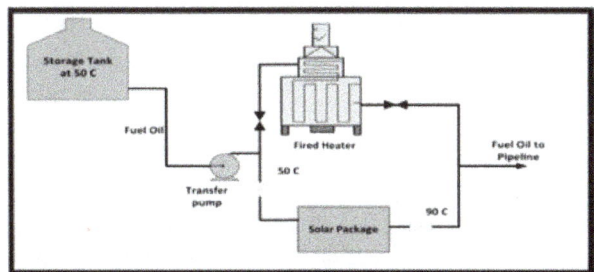

**Figure 12:** CORC fuel oil heating plant after introducing solar package.

### Case 3: Preheating of boiler feed Water Cairo Oil Refining Company (CORC):

*Process description:* The facility is located at CORC. In boiler (Figure 13), water is preheated firstly in the deaerator by steam. Burner mixes fuel and oxygen together and, with the assistance of an ignition device, provides a platform for combustion in the combustion chamber, and heat generated is then transferred to the water for steam production.

*Scope of work (deaerator):* Water is preheated in the deaerator from 25°C to about 110°C. Our scope of work is to preheat water before the deaerator from 25°C to 60°C in a way that the consumed steam in the deaerator is reduced. So for production of the same net amount of steam, the amount of consumed fuel in the boiler is reduced (Figure 14).

### Results and Discussion

•      **Summary of solar package design for the three cases** (Table 1).

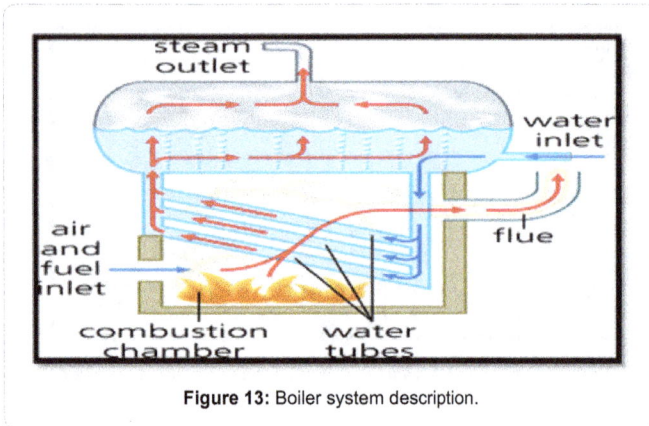

**Figure 13:** Boiler system description.

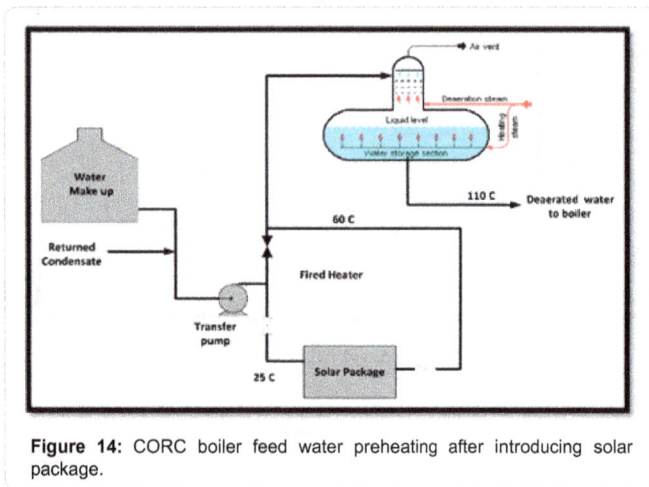

**Figure 14:** CORC boiler feed water preheating after introducing solar package.

- **Total Project Cost**

## HTF system equipment cost

For total project cost calculation firstly HTF system equipment cost is estimated using the above mentioned five methods. Then the total HTF system cost is calculated using plant cost estimating factor.

**Case 1: KALBSHA total project cost:** In Table 2, it is shown that costs from charts, excel sheet and commercial data are approximately equal. So the HTF system total equipment cost=115000$.

**Case 2: CORC fuel oil preheating:** In Table 3, it is shown that costs from charts and also data from actual commercial plant are approximately equal. So the HTF system total equipment cost=112000$.

**Case 3: CORC boiler feed water:** In Table 4, it is shown that costs from charts and commercial data are approximately equal. So the HTF system total equipment cost=29800$.

## PTC cost

PTC cost estimation is done (Table 5) according to updates to Solar Advisor Model (SAM) initiated by NREL (National Renewable energy Laboratory) which states that:

Solar plant Cost per m²=170$

Installation and Site Improvements per m²=200$

Total Cost of Solar Plant per m²=370$

|  | Case 1 | Case 2 | Case 3 |
|---|---|---|---|
| **Hydraulic Calculation** | | | |
| **Piping size    in** | 3 | 3 | 3.5 |
| **Line pressure drop bar** | 1.2 | 1.5 | 4.6 |
| **Heat exchanger pressure drop bar** | 0.5 | 0.5 | 0.5 |
| **Total pressure drop bar** | 1.7 | 2 | 5.1 |
| **Pump discharge pressure kg/cm²** | 8 | 8 | Booster pump 7 Main Pump    9 |
| **Heat Exchanger Thermal Design** | | | |
| **Heat exchanger surface area m²** | 15 | 83 | 71 |
| **PTC Design** | | | |
| **Number of required modules module** | 25 | 33 | 79 |
| **Total required area m²** | 1800 | 2281 | 5461 |

**Table 1:** Solar package design results.

| Equipment | Chart Cost $ | Matches software $ | Cost Estimator $ | Excel Sheet $ | commercial data $ |
|---|---|---|---|---|---|
| **Heat Exchanger** | 12180 | 16700 | 8200 | 12409 | 9000 |
| **Pump** | 31000 | 12900 | 12500 | 7488 | 17000 |
| **Surge Tank** | 1740 | 1236 | 10615 | 1284 | 12000 |
| **Vessel** | 69700 | 84872 | 21229 | 69749 | 40000 |
| **Total Cost $** | 114620 | 115708 | 52544 | 90930 | 78000 |

**Table 2:** Case 1 total HTF system equipment cost.

| Equipment | Chart Cost $ | Matches software $ | Cost Estimator $ | Excel Sheet $ | commercial data $ |
|---|---|---|---|---|---|
| **Heat Exchanger** | 40020 | 44200 | 18000 | 32000 | 50000 |
| **Pump** | 70000 | 44000 | 13000 | 30000 | 50000 |
| **Surge Tank** | 1740 | 1236 | 1500 | 1284 | 12000 |
| **Total Cost $** | 111760 | 89436 | 32500 | 63284 | 112000 |

**Table 3:** Case 2 Total HTF system equipment cost.

| Equipment | Chart Cost $ | Matches software $ | Cost Estimator $ | Excel Sheet $ | commercial data $ |
|---|---|---|---|---|---|
| **Heat Exchanger** | 34800 | 41000 | 16000 | 29700 | 42000 |
| **Booster Pump** | 87000 | 50000 | 14000 | 36000 | 122000 |
| **Main Pump** | 87000 | 50000 | 15000 | 36000 | 122000 |
| **Surge Tank** | 1740 | 1236 | 1500 | 1284 | 12000 |
| **Total Cost $** | 210540 | 142236 | 46500 | 102984 | 298000 |

**Table 4:** Case 3 total HTF system equipment cost.

**Total project cost summary:** After calculation of HTF system cost and PTC solar plant, the total project cost for the three cases is then calculated in Table 6.

## Comparison with the existing fuel consumption and payback period calculation

The purpose of this section is to calculate the payback period for each case and this can be done through the following steps:

|  | Case 1 | Case 2 | Case 3 |
|---|---|---|---|
| Required Area m² | 1800 | 2281 | 5461 |
| Cost $ | 666,000 | 843, 970 | 2, 020, 570 |

**Table 5:** PTC cost estimation for the three cases.

|  | Case 1 | Case 2 | Case 3 |
|---|---|---|---|
| HTF Equipment Cost $ | 115, 000 | 112, 000 | 298, 000 |
| Plant Estimating Factor | 1.95 | 1.95 | 1.95 |
| HTF System CPost $ | 224, 250 | 219, 000 | 581, 100 |
| PTC Cost $ | 666, 000 | 843, 970 | 2, 020, 570 |
| Total Project Cost with 10 % margin $ | 980, 0000 | 1, 170, 000 | 2, 862, 000 |

**Table 6:** Total project cost for the three cases.

|  | Case 1 | Case 2 | Case 3 |
|---|---|---|---|
| Heat consumption for the existing heaters | 3200 MMBTU/ HR | 1126 KW | 2640 KW |
| Existing fuel consumption / day | 0.5 MMSCFD | 2376 Lit | 5520 Lit |
| Amount of fuel Saved / 6 hrs. | 0.125 MMSCF | 591 Lit | 1385 Lit |
| Fuel Saved / year* | 41.25 MMSCF | 194809 | 456746 |
| $ Saved /year | 50, 000 | 176000 | 412000 |
| Total Project Cost $ | 980, 0000 | 1, 170, 000 | 2, 862, 000 |
| Payback Period year | 20 | 7 | 7 |

**Table 7:** Comparison with existing plant and payback period calculation.

*For Case 1 exiting heater uses natural gas as a fuel while for Cases 2 and 3 Diesel is used as fuel

1.  Calculation of the amount of fuel saved after implementing the new design.

2.  $$\text{Calculation of payback period} = \frac{\text{Total project cost}}{\text{Money saved due to fuel reduction per year}}$$

(Table 7).

Through (Table 7), the payback period for each application is calculated and after this period the annual profit for each case shall be the money saved for fuel consumption reduction. For example for Case 3 the payback period is 7 years and after these years the annual profit due to fuel saving shall be 412000$.

## Conclusion

The aim of this paper work is to provide economic study about using solar thermal system in some oil industry applications. For developing such study and through survey of different solar thermal systems, parabolic trough collector presents the most appropriate type to be used. Then evaluation of four available mathematical models for PTC was introduced for choosing the reasonable one that provides simulation of this collector under different operating conditions. For that paper under title of "Evaluation of Mathematical Models for Solar Thermal System" was published in October, 2016, American Journal of Energy Science. Finally steps and results of study using PTC solar heating system in some oil industry units are presented.

The study steps included design of all equipment and overall plant cost estimation for payback period prediction. The results showed that payback period for crude preheating before desalting was 20 years, fuel oil preheating was 7 years, water preheating 7 years. Fuel used in Case 1 was natural gas while for the other two cases was diesel fuel, and this is the most important factor that resulted in higher pay back period for Case 1.

### Recommendations

1.  Solar field presents about 50% of the total proposed system cost in each

application so optimization of solar system design affects greatly the total cost.

2.  Selection of PTC type with high efficiency and minor optical and thermal losses is essential for more economic application.

3.  Plant location is one of the main factors that affect collector performance so location with high direct radiation intensity is recommended.

4.  Heat exchanger optimum design is important as it presents about 25% of the HTF system cost.

5.  The oil industry application that the proposed system can replace in day hours is recommended to have fuel oil as a heating medium not fuel gas because of higher costs of fuel oil than fuel gas for generation of the same duty.

6.  The utilization of renewable energy in EGYPT should be increased, as there is a global concern that the developing nations could be faced with energy crisis and global warming. It is expected that more damage and pollution of the environment will continue. Parabolic trough solar collector is one of the options for renewable energy and this technology should be adopted in industries that utilize fossil fuel for heating and steam generation.

7.  The government has to support further research works in this area financially and allocating large area to test large grids of the parabolic trough system for heating applications and also power production.

### References

1.  Crude distillation unit (ADU VDU)-Alfa Laval is a world leader within the key technology areas of heat transfer, separation and fluid handling.

2.  No Authors Listed (1983) Industrial energy use workshop on the petroleum refining industry. Office of Technology Assessment, Washington DC.

3.  Worrell C, Galitsky E (2005) Energy efficiency improvement in the petroleum refining industry. 7th biennial ACEEE conference on Energy Efficiency.

4.  Romulo S, Lima D, Schaeffer R (2011) The energy efficiency of crude oil refining in Brazil: A Brazilian refinery plant case. Energy 36: 3101-3112.

5.  Liu X, Chen D (2013) An assessment of the energy-saving potential in China's petroleum refining industry from a technical perspective. Energy 59: 38-49.

6.  Petrick M, Pelligreno J (1999) The potential of reducing energy utilization in refinery. Argonne National Laboratory Report ANL/ESD/TM-158. US Department of Energy, Washington DC.

7.  International Energy Agency (2010) Key world energy statistics.

8.  Badran AA, Hamdan MA (1996) Utilization of solar energy for heating fuel oil. Mgmt 39: 105-111.

9.  Badran AA, Jubran BA (2000) Fuel oil heating by a trickle solar collector. Energy Conversion and Management 42: 1637-1645.

10. Xuesheng W, Ruzhu W (2004) Applied research of solar energy for heating crude oil of transportation. OGST 23: 41-45.

11. Xuesheng W, Ruzhu W, Wu J (2005) Experimental investigation of a new-style double-tube heat exchanger for heating crude oil using solar hot water. Applied Thermal Engineering 25: 1753-1763.

12. He Z, Shc C (2013) Application of solar heating system for raw petroleum during its piping transport. International conference on solar heating and cooling for buildings and industry, Freiburg, Germany.

13. Wang YX (2013) Solar thermal energy saving applications in oil and gas fields. Chinese scientific and technical journals.

14. Mamedov FF, Samedova UF, Salamov OM, Garibov AA (2008) Heat engineering calculation of a parabolo-cylindrical solar concentrator with tubular reactor for crude oil preparation for refining in the oil fields. Applied Solar Energy 44: 28-30.

15. Rahman AM, Nafey AS, Hassanien MHM (2016) Evaluation of mathematical models for solar thermal system. American Journal of Energy Science 3: 40-50.

16. Garrett DE (1989) Nostrand Reinhold softcover reprint of the hardcover. Chemical Engineering Economics Adjunct Professor, University of California, Santa Barbara.

# Exergetic Analysis of La Rumorosa-I Wind Farm

**Rafael Carlos Reynaga-López[1]\*, Alejandro Lambert[2], Oscar Jaramillo[3], Marlene Zamora[2] and Elia Leyva[2]**

[1]*Institute of Engineering, Autonomous University of Baja California, Mexicali, Mexico*
[2]*Faculty of Engineering, Autonomous University of Baja California, Mexicali, Mexico*
[3]*Institute of Renewable Energies, National Autonomous University of Mexico, Temixco, Mexico.*

### Abstract

Considering the produced power and the output wind velocity ($V_2$) for the year 2013 of five wind turbines in La Rumorosa I wind farm, located in the town of La Rumorosa, Baja California, Mexico; exergetic analysis was apply to determine relations between the efficiency variables, considering only the time the wind turbine is in use (worked hours H). Also, it was calculated the percentage of the entire year in which the turbines were producing energy. We found that relation between exergetic efficiency ($\varepsilon$) and power coefficient ($C_p$) is inversely proportional in all cases for the twelve months of the year 2013. In addition, we propose a new relation which shows the exergetic efficiency as a function of input wind velocity ($V_1$) and the power coefficient, for the mentioned period.

**Keywords:** Exergy; Wind energy; Efficiency; Power coefficient

## Introduction

### World scenario

Wind power is the world's fastest growing electricity generation technology. The year 2014 represented a record with the global installation of more than 50 GW, which added to the ones already operating make approximately a total of 369 GW. Nowadays, wind energy represents the 3% of the energy generated in the entire world. Some projections indicate that for 2019 the worldwide installed capacity will be over 660 GW. Wind turbine technology is moving fast, yet, there is the need to evaluate behavior of wind accurately [1].

### National scenario

Mexico has a great wind potential. Although, this resource has just started to be utilized in the recently, this sector shows high dynamism and competitiveness. Proof of this is the fact that more than 1,900 MW are in operation in both, independent production and self-supply as well as more than 5,000 MW are at different levels of development. Mexico has the commitment to decrease the fossil fuel electric generation from the actual 80% to a 65% for the year 2024, which implies installing more than 25,000 MW of clean technology in the next 10 years. Wind technology plays a fundamental role to achieve this goal, since wind power has been responsible of about two-thirds of the total objective in most countries with similar goals. The goal for 2020-2022 is to attain an installed capacity of at least 12,000 MW in the country, which is going to represent about the 40% of the national renewables target. This goal would have a cumulative impact on GDP of about 170,000 million pesos (approximately 9.5B USD) besides of the creation of more than 45,000 jobs [2].

### Exergy

Technically, exergy is defined as the maximum amount of work that can be produced by a system in non-equilibrium with its environment [3-5]. Exergy is a measure of the systems potential or flow to cause a change, as a consequence of not being completely in relative balance to a reference environment. Unlike energy, exergy is not linked to a conservation law. The exergy consumed during a process is proportional to the entropy created due to the irreversibility associated with that same process [6,7].

Exergetic analysis is a methodology based on the conservation of energy principle (first law of thermodynamics) along with the non-conservation of entropy one (second law of thermodynamics) for the analysis, design and improvement of energy and other systems. This analysis is useful to identify the causes, locations and magnitudes of inefficiencies in the processes. It recognizes that, although energy cannot be created or destroyed, it can be degraded in quality and eventually reached to a state in which it is in complete balance with its environment and therefore, without the ability to perform tasks [6]. This former process is currently used in innumerable fields involving energy transformation and optimization, such as the aforementioned sources of renewable energies (wind [8-12], geothermal [13], solar [14], biomass [15]), conventional sources (oil [16] and gas [17]), nuclear power [18], waste- water treatment [19] and even biology [20].

Regarding to energy efficiency of a wind turbine performance, the main factor to measure is the power coefficient $C_p$. Similarly, the main factor to be considered in performing the exergetic analysis is the exergetic efficiency $\varepsilon$. There are studies analyzing these relationships with interesting results [9,12]. In the present work, this type of analysis is carried out for the first time for five wind turbines at La Rumorosa I wind farm, pondering air density as a function of atmospheric pressure, relative humidity and temperature. Contrary to previous researches, this study only included data from the wind turbine (the output wind velocity $V_2$ and the electric power produced $W_{out}$), so it is necessary to represent the input wind velocity $V_1$, as a function of $V_2$ for the exergetic analysis.

## Overview of the Object of Study

### La Rumorosa I wind farm

La Rumorosa I wind farm is located within the town of its same name in the State of Baja California, Mexico; and it's operated under

---

**\*Corresponding author:** Rafael Carlos Reynaga-López, Institute of Engineering, Autonomous University of Baja California, Normal Street S/N and Blvd. Benito Juárez, Mexicali, Baja California 21100, Mexico, E-mail: rafael.reynaga@uabc.edu.mx

control of the State Energy Commission of Baja California (CEE BC). This area is the site, where wind resource is largely being exploited, in part, due to its excellent wind potential, to an acceptable transmission network of the Electricity Federal Commission (CFE for its acronym in Spanish) and to the fact of being border with the State of California, in the United States, which serves as the main customer. The park consists of five Gamesa G87-2.0 MW wind turbines with the following characteristics.

- Diameter: 87 m.

- Sweep area: 5945 m².

- Number of blades: 3 (Fiberglass pre-impregnated with epoxy resin).

- Cut-in speed: 4 m/s, cut-out speed: 25 m/s, rated speed: 16 m/s.

- Rotational speed 9.0-19.0 rpm.

- $\eta_{mec}$=0.98, $\eta_{el}$=0.95, $\eta$=0.93.

Figure 1 represents the power curve of the manufacturer; these results were obtained under the following conditions:

- Air density: 1,225 kg/m³ (at sea level)

- Intensity of turbulence: 10

- Rotor speed between 9.0-19.0 rpm

## La Rumorosas anemometric station (property of CONAGUA)

The meteorological data used for the exergetic analysis were provided by the National Water Commissions (CONAGUA for its acronym in Spanish) anemometric station located in the region of Agua Hechicera in the town of La Rumorosa, which is situated at approximately 27 km from the wind farm. The anemometer is positioned at a height of 10 meters above ground level and the altitude of the site is 1260 meters above sea level (Figure 2).

## Data

Data of a ten-minute measurement, for both generated electric power and wind output velocity, emitted by the five wind turbines were obtained during the 12 months of the year 2013. The monthly averages are presented in Tables 1-5. Here, $V_2$ is the wind output speed (in m/s), $H$ is the worked hours by the wind turbine, $W_{out}$ is the electric power produced (in kW), $E$ is the total electric energy produced in the month) and $\rho$ is the average monthly air density (in kg/m³). It is important to emphasize that the worked hours ($H$) represent the total number of hours the wind turbine was producing energy (that is, in operation and that the wind speed $V_1$ was higher than the cut-in speed); all average speed and output power take into account only the worked hours.

The behavior of the monthly electric power average and the total energy produced every month are presented in Figures 4 and 5, respectively.

## Theoretical Framework

### Energetic analysis

The kinetic energy of the wind can be written as

$$E = \frac{1}{2}\rho At V^3 \tag{1}$$

Therefore, the available power is,

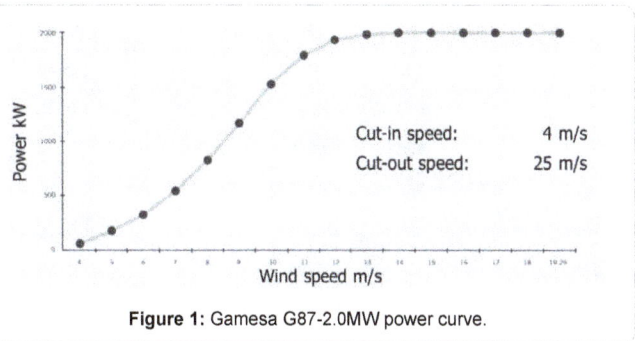

**Figure 1:** Gamesa G87-2.0MW power curve.

**Figure 2:** Anemometric tower in Agua Hechicera.

$$P = \frac{1}{2}\rho A V^3 \tag{2}$$

The power that absorbs the disk

$$P = m\ (V_1 - V_2)\ \overline{V} \tag{3}$$

Where,

$$\overline{V} = \frac{V_1 + V_2}{2}\ \text{or}$$

$$P = \rho A\ (V_1 - V_2)\ \overline{V}^2 \tag{4}$$

Making $\alpha = \dfrac{V_2}{V_1}$ we have

$$P = \frac{1}{4}\rho A\ (1-\alpha^2)\ (1+\alpha)\ V_1^3 \tag{5}$$

The maximum power will be for $\alpha = \dfrac{1}{3}$.

| Month | $V_2$ (m/s) | H (h) | $W_{out}$ (kW) | E (MWh) | $\rho$ (kg/m³) |
|---|---|---|---|---|---|
| January | 8.17 | 633.83 | 821.13 | 520.46 | 1.0941 |
| February | 8.29 | 529.00 | 825.20 | 436.53 | 1.0940 |
| March | 8.17 | 536.00 | 801.79 | 389.38 | 1.0729 |
| April | 8.78 | 587.00 | 936.52 | 549.74 | 1.1028 |
| May | 9.00 | 627.67 | 995.61 | 612.77 | 1.0536 |
| June | 7.73 | 637.33 | 669.82 | 426.90 | 1.0335 |
| July | 6.61 | 515.00 | 447.77 | 220.07 | 1.0250 |
| August | 6.39 | 526.83 | 417.68 | 220.05 | 1.0287 |
| September | 7.10 | 430.33 | 576.91 | 248.26 | 1.0325 |
| October | 8.29 | 533.50 | 790.63 | 421.80 | 1.0557 |
| November | 7.32 | 523.00 | 643.81 | 336.71 | 1.0739 |
| December | 8.75 | 641.33 | 891.78 | 571.93 | 1.0850 |

**Table 1:** Turbine 1.

| Month | $V_2$ (m/s) | H (h) | $W_{out}$ (kW) | E (MWh) | $\rho$ (kg/m³) |
|---|---|---|---|---|---|
| January | 8.01 | 634.17 | 779.32 | 494.22 | 1.094 |
| February | 8.13 | 538.17 | 793.30 | 426.93 | 1.094 |
| March | 7.91 | 557.83 | 742.79 | 375.31 | 1.073 |
| April | 8.53 | 581.00 | 889.22 | 516.63 | 1.103 |
| May | 8.68 | 627.83 | 936.10 | 576.56 | 1.054 |
| June | 7.42 | 636.50 | 617.11 | 392.79 | 1.034 |
| July | 6.48 | 501.83 | 442.66 | 213.18 | 1.025 |
| August | 6.34 | 396.17 | 405.94 | 160.82 | 1.029 |
| September | 7.26 | 338.83 | 608.03 | 206.02 | 1.033 |
| October | 8.00 | 532.67 | 737.38 | 392.78 | 1.056 |
| November | 6.99 | 522.17 | 602.75 | 314.74 | 1.074 |
| December | 8.56 | 627.83 | 817.17 | 513.05 | 1.085 |

**Table 2:** Turbine 2.

| Month | $V_2$ (m/s) | H (h) | $W_{out}$ (kW) | E (MWh) | $\rho$ (kg/m³) |
|---|---|---|---|---|---|
| January | 8.38 | 549.33 | 815.85 | 448.17 | 1.094 |
| February | 8.48 | 543.17 | 852.35 | 462.97 | 1.094 |
| March | 8.23 | 570.17 | 794.11 | 413.58 | 1.073 |
| April | 8.93 | 586.67 | 948.60 | 556.51 | 1.103 |
| May | 9.05 | 628.33 | 990.96 | 610.41 | 1.054 |
| June | 7.98 | 467.00 | 703.61 | 328.59 | 1.034 |
| July | 6.82 | 505.67 | 495.84 | 243.18 | 1.025 |
| August | 6.52 | 518.67 | 424.88 | 220.37 | 1.029 |
| September | 7.25 | 428.17 | 578.69 | 247.77 | 1.033 |
| October | 8.34 | 540.67 | 787.13 | 425.57 | 1.056 |
| November | 7.37 | 508.00 | 624.71 | 317.35 | 1.074 |
| December | 8.97 | 593.67 | 918.52 | 545.29 | 1.085 |

**Table 3:** Turbine 3.

| Month | $V_2$ (m/s) | H (h) | $W_{out}$ (kW) | E (MWh) | $\rho$ (kg/m³) |
|---|---|---|---|---|---|
| January | 8.48 | 635.83 | 841.27 | 534.91 | 1.094 |
| February | 8.54 | 535.67 | 860.23 | 460.79 | 1.094 |
| March | 8.22 | 568.83 | 781.57 | 406.47 | 1.073 |
| April | 8.89 | 586.33 | 940.29 | 551.32 | 1.103 |
| May | 9.07 | 611.83 | 985.58 | 591.38 | 1.054 |
| June | 7.83 | 624.67 | 670.98 | 419.14 | 1.034 |
| July | 6.71 | 503.17 | 477.14 | 229.02 | 1.025 |
| August | 6.43 | 520.50 | 408.93 | 212.85 | 1.029 |
| September | 6.64 | 381.67 | 457.99 | 174.80 | 1.033 |
| October | 8.63 | 538.67 | 792.79 | 427.05 | 1.056 |
| November | 7.82 | 516.17 | 656.19 | 338.70 | 1.074 |
| December | 9.05 | 647.50 | 912.47 | 590.83 | 1.085 |

**Table 4:** Turbine 4.

The energy efficiency of a wind turbine is characterized by its power coefficient [9].

$$C_p = \frac{W_{out}}{\frac{1}{2} \cdot \eta_{el} \cdot \eta_{mec} \cdot \rho \cdot \pi \cdot R^2 \cdot V_R^3} \tag{6}$$

Where,

$W_{out}$ is the electric power obtained, $\eta_{el}$ and $\eta_{mec}$ are the electrical and mechanical efficiencies of the turbine respectively; $\rho$ is the density of air, $R$ is the radius of the wind turbine and $V_r$ is the speed at the boundary of the disk.

Then

$$V_r = \frac{V_1 + V_2}{2} = \overline{V} \tag{7}$$

The kinetic exergy of the air.

$$ke_1 = \frac{V_r^2}{2} \tag{8}$$

The mass flow is the amount of matter per second that passes through the turbine (in units of kg/s):

$$\dot{m} = \rho \cdot A \cdot V_r = \rho \cdot \pi \cdot R^2 \cdot V_r \tag{9}$$

It can be obtained the input velocity of the mass flow ($V_1$) by means of the law of energy conservation.

$$\dot{m}ke_1 = C_p \cdot \dot{m}ke_1 + \dot{m}ke_2 \tag{10}$$

Substituting (8) and (9) into (10), leads to

| Month | $V_2$ (m/s) | H (h) | $W_{out}$ (kW) | E (MWh) | $\rho$ (kg/m³) |
|-------|-------------|-------|----------------|---------|----------------|
| January | 8.63 | 597.50 | 861.65 | 514.68 | 1.094 |
| February | 8.72 | 559.67 | 869.73 | 486.76 | 1.094 |
| March | 8.45 | 564.00 | 772.37 | 396.71 | 1.073 |
| April | 8.86 | 530.5 | 865.64 | 459.22 | 1.103 |
| May | 9.28 | 571.17 | 972.30 | 543.64 | 1.054 |
| June | 7.99 | 638.00 | 669.32 | 427.03 | 1.034 |
| July | 6.86 | 506.00 | 476.74 | 230.69 | 1.025 |
| August | 6.62 | 512.83 | 415.90 | 213.29 | 1.029 |
| September | 7.34 | 431.33 | 572.23 | 246.82 | 1.033 |
| October | 8.42 | 549.83 | 779.90 | 428.81 | 1.056 |
| November | 7.47 | 448.00 | 608.26 | 272.50 | 1.074 |
| December | 8.82 | 632.67 | 873.44 | 552.60 | 1.085 |

**Table 5:** Turbine 5.

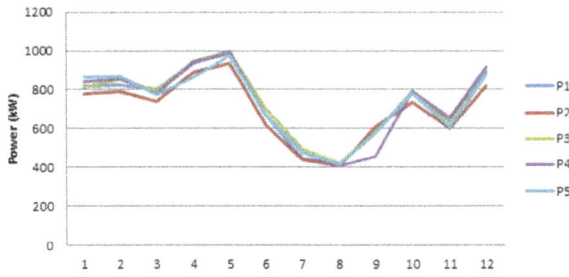

**Figure 3:** Monthly electric power average for 5 wind turbines.

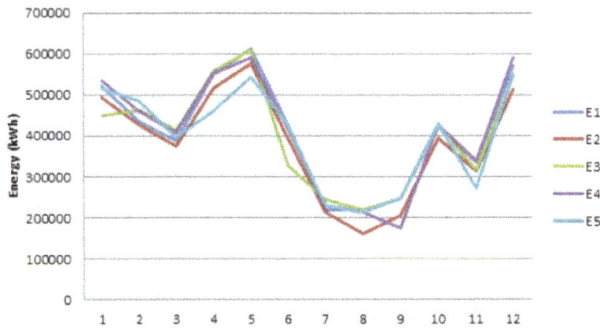

**Figure 4:** Total energy produced each month for the 5 wind turbines.

$$V_1 = \frac{V_2}{\sqrt{1-C_P}} \qquad (11)$$

Energy efficiency is defined as

$$\eta = \frac{W_{out}}{W_{wind}} \qquad (12)$$

Combining Equations (12) and (6) and then

$$\eta = \eta_{el} \cdot \eta_{mec} \cdot C_p \qquad (13)$$

The total efficiency of the turbine is a function of both the rotor power coefficient and the mechanical and electrical efficiencies [21].

**Exergetic analysis**

Exergetic efficiency is defined as

$$\varepsilon = \frac{W_{out}}{W_u} \qquad (14)$$

Where,

$W_u$ is the useful power.

Neglecting the change in temperature of the mass flow when transferring the rotor, the useful power depends only on the change of pressure:

$$W_u = (p_1 - p_2)\frac{m}{\rho} \qquad (15)$$

Similarly, the loss of exergy (I) is defined as

$$I = W_u - W_{out} \qquad (16)$$

Substituting the expression for the useful work (Equation 15) into exergetic efficiency (Equation 14), in addition to the equation for pressure as a function of velocity,

$$p_{1,2} = p_{at} + \frac{\rho}{2}V_{1,2}^2 \qquad (17)$$

Is obtained then

$$\varepsilon = \frac{W_{out}}{W_u} = \frac{W_{out}}{\frac{\rho}{2}(V_1^2 - V_2^2)AV_r} \qquad (18)$$

Replacing now the expressions for $V_r$ (Equation 7) and $V_1$ (Equation 11):

$$\varepsilon = \frac{W_{out}}{\frac{\rho}{2}\left(\frac{V_2^2}{(1-C_p)}-V_2^2\right)A\frac{V_2}{2}\left(\frac{1}{\sqrt{1-C_p}}+1\right)} = \frac{4W_{out}}{\rho A\left(\frac{1}{\sqrt{1-C_p}}-1\right)\left(\frac{1}{\sqrt{1-C_p}}+1\right)^2 V_2^3} \qquad (19)$$

Equation 19 relates the exergetic efficiency directly to the power coefficient for fixed values of $V_2$ and $W_{out}$.

Likewise, it can be find an expression for the loss of exergy:

$$I = W_u - W_{out} = \frac{\rho A}{4}\left(\frac{1}{\sqrt{1-C_p}}-1\right)\left(\frac{1}{\sqrt{1-C_p}}+1\right)^2 V_2^3 - W_{out} \qquad (20)$$

## Results and Discussions

Using Equation [19] for monthly average values of $V_2$ and $W_{out}$ (Table 1), it can be found the relation between the efficiency variables $\varepsilon$ and $C_p$. Figure 5 shows that as the power coefficient increased, the exergy efficiency decreased. The curve with the lowest values for the power coefficient in the allowed range {$0<\varepsilon<1$ and $0<C_p<0.59$} were presented in the month of May, when the average speeds are higher, (alike the $W_{out}$ output power, as shown in Figure 3). In the extreme case, the curve that shows the maximum values of power coefficient is the one that represents the month of August, which, conversely, refers to the lowest speeds and average output power. Figures 6-10 represent the relationship between the exergetic efficiency and the power coefficient for the months of May (blue) and August (red) for the five wind turbines.

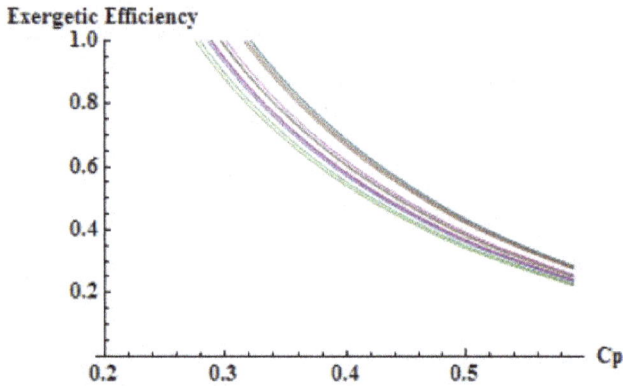

**Figure 5:** Relation between the power coefficient and exergetic efficiency for the 12 months of 2013 (Turbine 1).

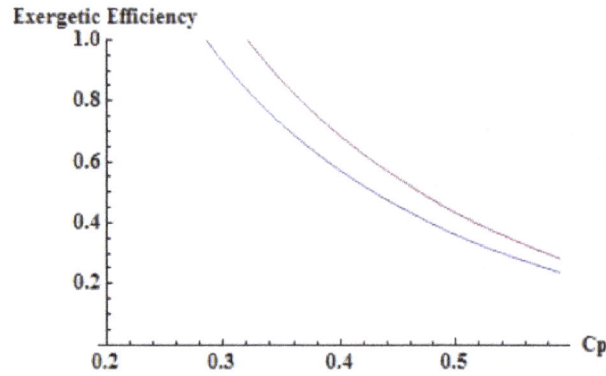

**Figure 6:** Relation between the power coefficient and exergetic efficiency for May (blue) and August (red) 2013 (Turbine 1).

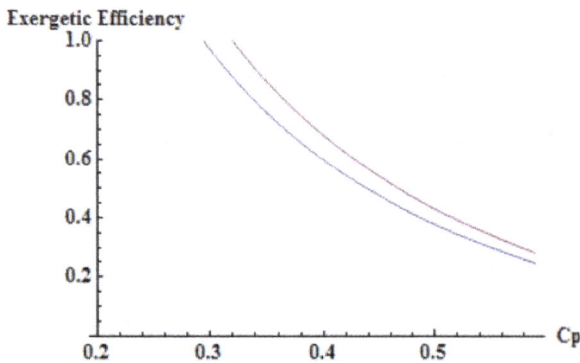

**Figure 7:** Relation between the power coefficient and exergetic efficiency for May (blue) and August (red) 2013 (Turbine 2).

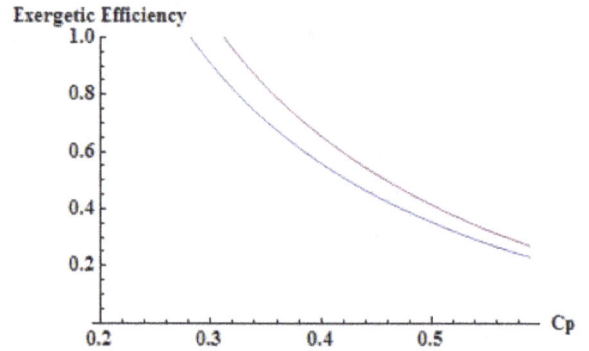

**Figure 8:** Relation between the power coefficient and exergetic efficiency for May (blue) and August (red) 2013 (Turbine 3).

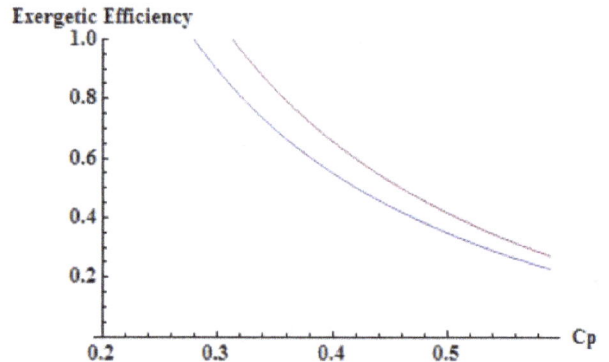

**Figure 9:** Relation between the power coefficient and exergetic efficiency for May (blue) and August (red) 2013 (Turbine 4).

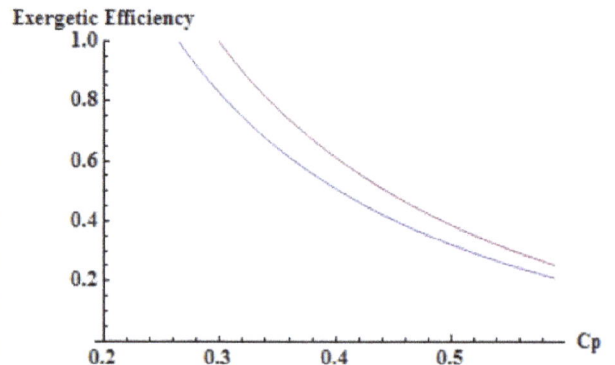

**Figure 10:** Relation between the power coefficient and exergetic efficiency for May (blue) and August (red) 2013 (Turbine 5).

In Figure 11 the curves for the month of May of the five wind turbines are compared.

Figure 12 is obtained using Equation 20 and shows the relationship between the loss of exergy and the power coefficient for the five turbines in the month of May 2013. Expectedly, the turbine 5, when showing the lowest effciency between 5, presents the greatest exergetic loss.

Using Equation 18

$$\varepsilon = \frac{W_{out}}{W_u} = \frac{W_{out}}{\dfrac{\rho}{2}(V_1^2 - V_2^2)AV_r}$$

It can be directly replace $V_1 = \dfrac{V_2}{\sqrt{1 - C_p}}$ and $V_r = \dfrac{V_1 + V_2}{2}$ to get to

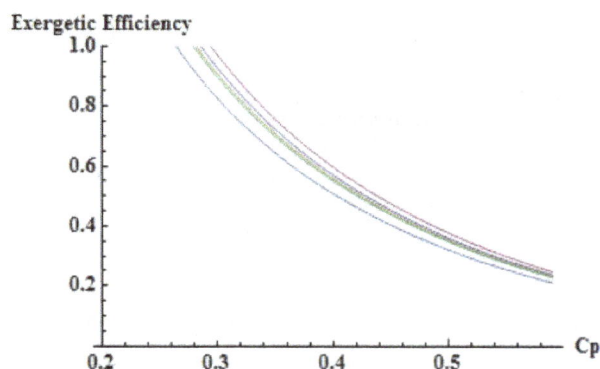

**Figure 11:** Relation between power coefficient and exergetic efficiency for the 5 turbines in May 2013.

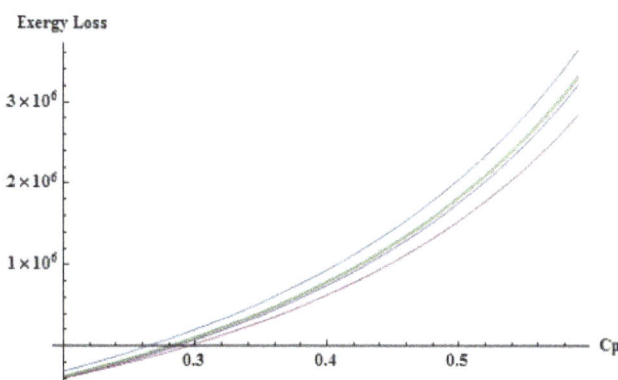

**Figure 12:** Relation between the power coefficient and exergy loss (in W) for the 5 turbines in May 2013.

$$\varepsilon = \frac{4 \cdot W_{out}}{\rho \cdot A \cdot \left(\dfrac{1}{1-C_p}-1\right)(V_1+V_2) \cdot V_2^2} \qquad (21)$$

This equation was plotted for turbine 1 in May 2013, the validity intervals of the variables are $\{0<\varepsilon<1\}$, $\{0<C_p<0.59\}$ and $\{4<V_1<25\}$. Analyzing equation 19, it could be find a relation between the exergetic efficiency and the power coefficient. The equation became characteristic for each turbine (defining the relationship between the two important performance variables in this analysis), by setting the values of $V_2$ and $W_{out}$ with measured data.

The relationship given in Equation 21 shows how exergy efficiency, power coefficient and input velocity $V_1$ varied each another. Figure 13 ensures (when reviewing the slopes) that exergetic efficiency decreases to a greater extent with respect to the power coefficient than with regard to the velocity in the bounded intervals of definition.

## Conclusion

In the present study, the behavior of five wind turbines of the same model in the La Rumorosa I wind farm was analyzed. Even though, they are the same kind of turbine, when comparing their individual performance, it can be found that there are notable differences in both, their exergetic efficiency and the power coefficient. Differing the way exergy studies have previously been performed on wind turbines. This method was only based on using parameters obtained from the wind

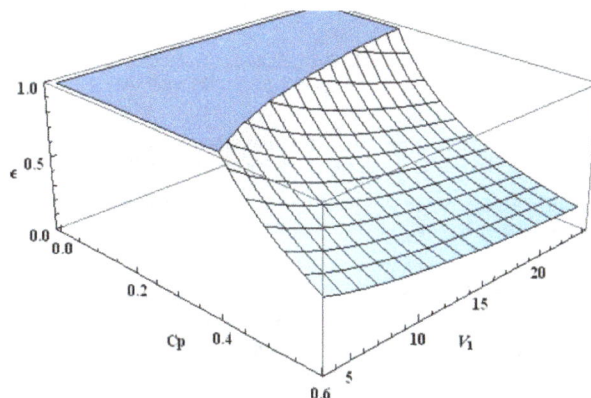

**Figure 13:** Relation between the power coefficient, exergetic efficiency and input velocity $V_1$, for turbine 1 in May 2013.

turbine ($V_2$, $W_{out}$). In addition, it only measured the time the wind turbine was producing energy (worked hours $H$).

The percentages of the annual time (2013) that the wind turbines produced energy (worked hours) were of 76.72%, 74.14%, 73.51%, 76.15% and 74.67%, for turbines 1 to 5 respectively.

Equation 19 demonstrate the characteristic exergy performance of the wind turbine by exposing all the parameters ($V_1$, $V_2$, $\rho$, $W_{out}$ and $C_p$) that are involved in the conversion of wind energy to electric energy.

The graphs demonstrate that the exergetic efficiency decreased as the power coefficient increased, contrary to previous studies results, in which a linear relationship between the parameters has been reported. On the other hand, when we consider the value of $C_p$ constant, we confirmed that the exergetic efficiency is greater for months with average $V_2$ approximately 6-7 m/s, like July and August. Furthermore, for months with an average $V_2$ of about 8-9 m/s the exergetic efficiency is smaller, as in April and May. This corresponds to the normal behavior of the wind speed ($V_1$) in the region, being lower in the summer and higher in spring and autumn [22,23].

According to the axial induction factor, the aforementioned data is an indication that the Cp has a maximum corresponding to wind speeds ($V_1$) between 9 and 11 m/s for the wind turbine analyzed.

### Acknowledgments

The authors would like to thank the National Council for Science and Technology of Mexico (CONACyT) for the scholarship awarded. In addition to the State Energy Commission of Baja California (CEE BC) and the National Water Commission (CONAGUA) for the information provided.

### References

1. http://www.gwec.net/

2. http://www.amdee.org/

3. Szargut J (1980) International progress in second law analysis. Energy 5: 709-718.

4. Moran MJ (1989) Availability analysis: A guide to efficient energy use. Science. p. 260.

5. Kotas TJ (1995) The exergy method of thermal plant analysis. (1stedn), The Exergy Method of Thermal Plant Analysis. p. 320.

6. Rosen MA, Dincer I, Kanoglu M (2008) Role of exergy in increasing efficiency and sustainability and reducing enviromental impact. Energy Policy 36: 128-137.

7. Dincer I (2002) The role of exergy in energy policy marking. Energy Policy 30: 137-149.

8. Baskut O, Ozgener O, Ozgener L (2010) Effects of meteorological variables on exergetic efficiency of wind turbine power plants. Renew Sustain Energy Rev 14: 3237-3241.

9. Baskut O, Ozgener O, Ozgener L (2011) Second law analysis of wind tur- bine power plants: Cesme, Izmir example. Fuel Energy 36: 2535-2542.

10. Sahin AD, Dincer I, Rosen MA (2006) New spatio- temporal wind exergy maps. J Energy Resour Technol 128: 194-202.

11. Sahin AD, Dincer I, Rosen MA (2006) Thermodynamic analysis of wind energy. Int J Energy Res 30: 553-566.

12. Redha AM, Dincer I, Gadalla M (2011) Thermodynamic performance assessment of wind energy systems: An application. Fuel Energy 36: 4002-4010.

13. Koroneos C, Rovas D (2013) Exergy analysis of geothermal electricity using the Kalina cycle. Int J Exergy 12: 54-69.

14. Karakilcik M, Bozkurt I, Dincer I (2013) Dynamic exergetic performance assessment of an integrated solar pond. Int J Exergy 12: 70-86.

15. Pellegrini LF, Oliveira SD (2007) Exergy analysis of sugarcane bagasse gasification. Energy 32: 314-327.

16. Taniguchi H, Mouri K, Nakahara T, Arai N (2005) Exergy analysis on combustion and energy conversion processes. Energy 30: 111-117.

17. Gungor A, Bayrak M, Beylergil B (2013) In view of sustainable future energetic-exergetic and economic analysis of a natural gas cogeneration plant. Int J Exergy 12: 109.

18. Durmayaz A, Yavuz H (2001) Exergy analysis of a pressurized-water reactor nuclear-power plant. Appl Energy 69: 39-57.

19. Khosravi S, Panjeshahi MH, Ataei A (2013) Application of exergy analysis for quantification and optimisation of the environmental performance in wastewater treatment plants. Int J Exergy 12: 119-138.

20. Borgert JA, Moura LM (2013) Exergetic analysis of glucose metabolism. Int J Exergy 12: 31-53.

21. Manwell JF, McGowan JG, Rogers AL (2009) Wind energy explained: Theory, design and application (2nd edn). Wiley, USA. p. 704.

22. Zamora M, Lambert A, Montero G (2014) Effect of some meteo rological phenomena on the wind potential of Baja California. Energy Procedia 57: 1327-1336.

23. Machado MZ, Sanchez EL, Lambert Arista AA (2003) Wind resource in Baja California.

# Tilted Wick Solar Still with Flat Plate Bottom Reflector: Numerical Analysis for a Case with a Gap Between Them

Hiroshi Tanaka*

*Department of Mechanical Engineering, National Institute of Technology, Kurume College, Komorino, Kurume, Japan*

## Abstract

A tilted wick solar still with a flat plate bottom reflector was analyzed theoretically when there is a gap between the still and reflector at 30°N latitude. A mirror-symmetric plane of the wick relative to the reflector was introduced to calculate the amount of solar radiation reflected from the reflector and absorbed on the wick. Heat and mass transfer in the still were also analyzed to determine the temperature in the still and the distillate production rate of the still. The inclinations of both the still and the reflector should be adjusted adequately for each month and for the gap length in order to increase the distillate productivity. The optimum inclinations of both the still and the reflector throughout the year were determined. The effect of the reflector on distillate productivity decreases with an increase in gap length. However, the distillate productivity can be increased by the reflector even if the gap length is equivalent to that of the still and reflector. The sum of the daily amount of distillate on each month throughout the year was predicted to be increased about 28, 19 and 14% by the reflector when the gap length is 0, 0.5 and 1 m.

**Keywords**: Solar desalination; Solar still; Tilted wick; Bottom reflector; Gap

## Nomenclature

$A$: area, m²

$G$: solar radiation on a horizontal surface, W/m²

$l$: length, m

$mc_p$: heat capacity, J/K

$Q_c$: convective heat transfer rate, W

$Q_d$: conductive heat transfer rate, W

$Q_e$: heat transfer rate by mass transfer, W

$Q_f$: enthalpy increase, W

$Q_r$: radiative heat transfer rate, W

$Q_{sun}$: absorption of solar radiation, W

$T$: temperature, K

$t$: time, s

$w$: width, m

$\alpha$: absorptance

$\beta$: incident angle of direct solar radiation on glass cover

$\beta'$: incident angle of reflected solar radiation on glass cover

$\varphi$, $\phi$: azimuth and altitude angle of the sun

$\theta$: inclination angle

$\rho$: reflectance of reflector

$\tau$: transmittance of glass cover

## Subscripts

$a$: ambient air

$df$: diffuse radiation

$dr$: direct radiation

$g$: glass cover

$gp$: gap

$m$: reflector

$re$: reflected radiation

$s$: still

$w$: wick

## Introduction

A tilted wick solar still consisting of a transparent cover and a wick as absorber/evaporator is one of the simplest types of solar stills. Compared with a basin type still consisting of a transparent cover and a basin liner, the advantages of the tilted wick still are as follows:

1) Heat capacity of water flowing in the wick is considerably smaller than that of water in a basin liner.

2) The inclination angle of the tilted wick still can be adjusted according to seasons and locations to increase the solar radiation absorbed on the wick, while the basin liner should be set horizontally.

3) Size of the still is smaller.

4) The tilted wick still can be installed on a slope.

Due to these advantages, tilted wick stills have been studied numerically and experimentally [1-21], and also reviewed by

*Corresponding author: Tanaka H, Department of Mechanical Engineering, National Institute of Technology, Kurume College, Komorino, Kurume, Fukuoka 830-8555, Japan
E-mail: tanakad@kurume-nct.ac.jp

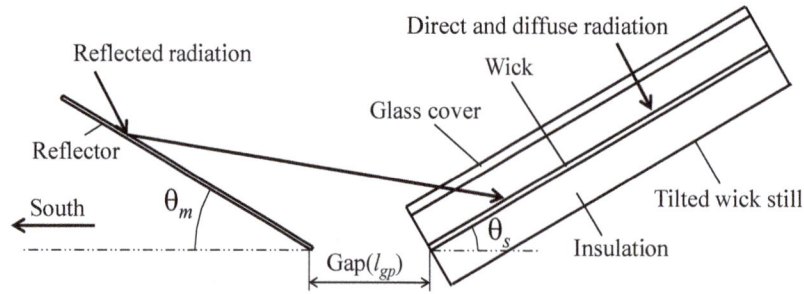

Figure 1: Schematic diagram of a tilted wick solar still with a flat plate bottom reflector with a gap between the still and reflector.

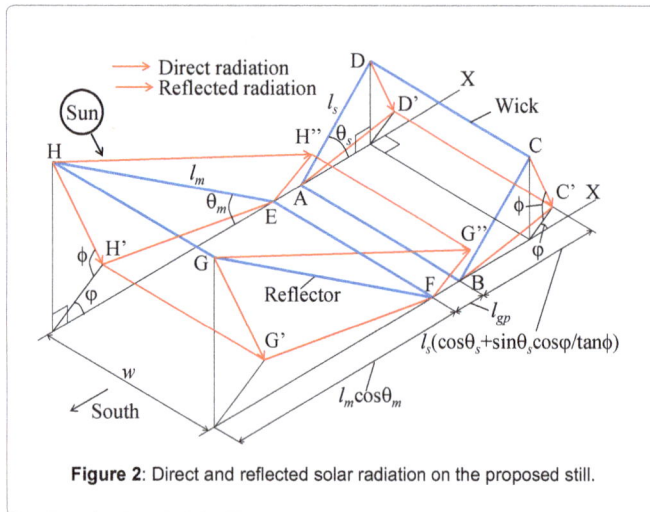

Figure 2: Direct and reflected solar radiation on the proposed still.

Manikandan et al. [22] and Murugavel et al. [23].

A flat plate reflector is an inexpensive and simple modification to increase the distillate productivity of tilted wick stills. However, there had been few reports about tilted wick stills with a flat plate reflector [3,10], and a quantitative analysis of the effect of a flat plate reflector on tilted wick stills had not been done. Tanaka and Nakatake [24-26] have presented geometrical models to predict the solar radiation reflected from a top reflector extending from the upper edge of the still, and then absorbed onto the wick. The top reflector is assumed to be set vertically in spring and autumn [24], and inclined forwards slightly in winter [25] and backwards slightly in summer [26]. Tanaka [27] also predicted the optimum inclinations of both the still and the top reflector throughout the year using these geometrical models. After that, Tanaka presented another geometrical model for a bottom reflector extending from the lower edge of the still [28], and theoretically analyzed the optimum inclinations of both the still and the bottom reflector throughout the year [29]. Through these studies, it was found that a top reflector as well as a bottom reflector can be used to increase distillate productivity of the still by adjusting the inclinations of both the still and the reflector appropriately according to the seasons.

In all the geometrical models mentioned above, it was assumed that the still and the reflector touch each other and there is no gap between them. However, in practical cases, especially when the reflector would be installed on the already mounted solar still, there would be many instances in which the reflector should be installed apart from the solar still due to limitations of the installing site. But the geometrical models which have been proposed cannot be applied if there would be any gap between the still and the reflector. Recently, Tanaka numerically analyzed a solar thermal collector and a flat plate bottom reflector with a gap between them [30], and presented a new geometrical model to predict the amount of solar radiation reflected from a bottom reflector and then absorbed on the collector in cases where there is a gap between collector and reflector. The geometrical model for a system with a gap is similar to those for stills without a gap, but more complicated. The collector/reflector system with a gap analyzed by Tanaka [30] consists of two flat plates; therefore, the geometrical model for a collector/reflector system with a gap can be applied for a tilted wick still with a flat plate bottom reflector with a gap since they are also two flat plate systems. Therefore, in this paper, the geometrical model for a collector/reflector system with a gap is applied to a tilted wick solar still with a bottom reflector with a gap between the still and reflector, and the optimum inclinations of both the still and reflector as well as an increase in distillate productivity of the still using a bottom reflector with a gap are predicted for each month of the year at 30°N.

## Theoretical Analysis

### Amount of solar radiation absorbed on the wick

Figure 1 shows a tilted wick solar still with a flat plate bottom reflector and a gap between the still and reflector. A tilted wick still consists of a glass cover and wick as an absorber/evaporator. The bottom of the still is insulated. The still is facing south, and the bottom reflector is placed parallel to the still at the south side of the still with a gap. It is assumed that the lower edge of the reflector and the wick of the still are at same level. The inclinations of the still and the reflector are assumed to be adjustable.

Direct and reflected solar radiation to the proposed still is shown in Figure 2. ABCD is the wick of the still and EFGH is the reflector, respectively. The walls of the still are neglected in this study since the height of the still (0.01 m) is negligible in relation to the still's length (1 m) and width (1 m). The length of the still and the reflector are shown as $l_s$ and $l_m$, and the inclinations of the still and the reflector from horizontal are shown as $\theta_s$ and $\theta_m$, respectively. The length of a gap between the still and the reflector is shown as $l_{gp}$. Solid arrows (CC', DD', GG' and HH') show direct radiation and dashed arrows (GG" and HH") show reflected radiation. The azimuth and altitude angle of the sun are shown as $\phi$ and $\varphi$, respectively.

The amount of direct solar radiation absorbed on the wick, $Q_{sun,dr}$, can be calculated by determining the area of the shadow of the wick on a horizontal surface (a trapezoid ABC'D') as

$$Q_{sun,dr} = G_{dr} \tau_g (\beta) \alpha_w \times A_{dr} \qquad (1)$$

$$A_{dr} = w l_s (\cos \theta_s + \sin \theta_s \cos \phi / \tan \varphi) \qquad (2)$$

$$\cos \beta = \sin \phi \cos \theta_s + \cos \phi \sin \theta_s \cos \varphi \qquad (3)$$

where $G_{dr}$ is direct solar radiation on a horizontal surface, $\tau_g$ is transmittance of the glass cover, $\beta$ is the incident angle of direct solar radiation on the glass cover, $\alpha_w$ is absorptance of the wick and $A_{dr}$ is the area of the wick's shadow (ABC'D'). When the sun moves north in the early morning and late evening in the months of April to August, the shadow of the wick would be shorter than $l_s \cos \theta_s$ and $A_{dr}$ should be determined as

$$A_{dr} = w l_s (\cos \theta_s - \sin \theta_s \cos \phi / \tan \varphi) \qquad (4)$$

Diffuse solar radiation absorbed on the wick, $Q_{sun,df}$, can be determined assuming that diffuse radiation comes uniformly from all directions in the sky dome, and this can be expressed as

$$Q_{sun,df} = G_{df} (\tau_g)_{df} \alpha_w \times w l_s \qquad (5)$$

where $G_{df}$ is the diffuse solar radiation on a horizontal surface, and $(\tau_g)_{df}$ which is a function of the inclination of the still, $\theta_s$, can be calculated by integrating the transmittance of the glass cover for diffuse radiation from all directions in the sky dome [24].

The reflected projection from the bottom reflector is shown as EFG"H". Not all of the reflected radiation from the bottom reflector can reach the wick, and part or all of the reflected radiation would escape to the ground without hitting the wick. To calculate the amount of solar radiation reflected from the bottom reflector and then absorbed on the wick, a mirror-symmetric plane of the wick relative to the reflector is introduced. A side view of the wick, reflector and mirror-symmetric plane is shown in Figure 3. BC and FG show the wick and reflector, and JK shows the mirror-symmetric plane. As shown, the position of the upper edge of the mirror-symmetric plane (point J) can be determined using lengths $l_1$ and $l_2$.

$$l_1 = 2 l_{gp} \sin^2 \theta_m \qquad (6)$$

$$l_2 = 2 l_{gp} \sin \theta_m \cos \theta_m = l_{gp} \sin 2\theta_m \qquad (7)$$

The incident point and incident angle of the reflected radiation on the wick are exactly the same as those of the direct radiation which goes through the reflector and incidents on the mirror-symmetric plane. Therefore, the amount of the reflected radiation absorbed on the wick can be calculated by determining the amount of direct radiation which goes through the reflector and then is absorbed on the mirror-symmetric plane. Here, $\omega_1$ is the inclination from vertical of the mirror-symmetric plane determined as

$$\omega_1 = 2\theta_m + \theta_s - \pi / 2 \qquad (8)$$

Figure 3 shows only case in which $\omega_1$ is positive. If $\omega_1$ is negative,

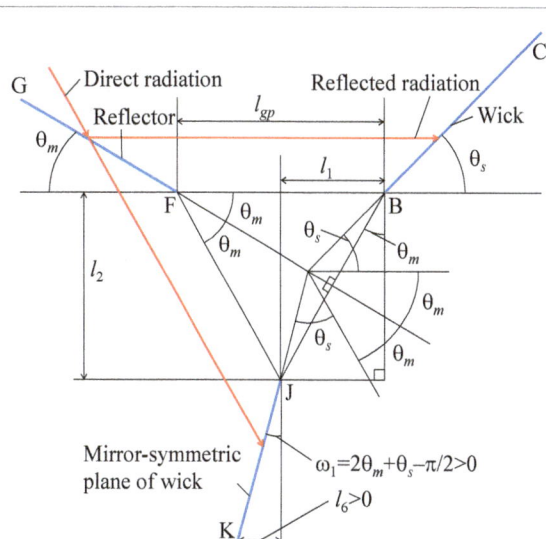

Figure 3: A side view of the wick, reflector, and a mirror-symmetric plane of the wick relative to the reflector.

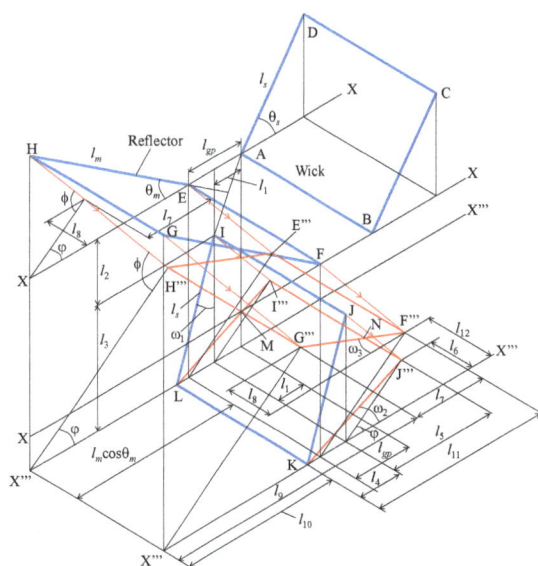

Figure 4: Geometrical model to determine the amount of solar radiation which goes through the reflector and is absorbed on a mirror-symmetric plane.

length $l_6$ mentioned below will also be negative, however, the following calculations are valid even if $\omega_1$ and $l_6$ have negative values.

A geometrical model to calculate the amount of direct radiation which goes through the reflector and is absorbed on the mirror-symmetric plane is shown in Figure 4. The wick of the still (shown as ABCD) and the reflector (shown as EFGH) are exactly the same as those shown in Figure 2. A mirror-symmetric plane shown as IJKL is placed on a virtual horizontal surface X''' which is $l_2 + l_3$ below the horizontal surface X on which the still and reflector are placed. The shadows of the reflector (EFGH) and the mirror-symmetric plane (IJKL) caused by direct radiation on a virtual horizontal surface X'' are shown as E'''F'''G'''H''' and I'''J'''KL, respectively. Therefore, the amount of direct radiation which goes through the reflector and is absorbed on the mirror-symmetric plane, which is exactly the same as that reflected from the reflector and then absorbed on the wick, $Q_{sun,re}$, can be calculated by determining the area of the overlapping part of these shadows ($A_{re}$) shown as I'''NG'''M. This can be expressed as

$$Q_{sun,re} = G_{dr}\rho_m\tau_g(\beta')\alpha_w \times A_{re} \tag{9}$$

$$\cos\beta' = \sin\phi\cos\omega_1 + \cos\phi\sin\omega_1\cos\varphi \tag{10}$$

where $\rho_m$ is reflectance of the reflector and $\beta'$ is incident angle of the reflected radiation on the glass cover. Here, lengths $l_3$ to $l_{12}$ and angles $\omega_2$ and $\omega_3$ shown in Figure 4 to calculate the overlapping area can be determined as follows:

$$l_3 = l_s\cos\omega_1 \tag{11}$$

$$l_4 = l_s\sin\omega_1 \tag{12}$$

$$l_5 = l_3\cos\varphi/\tan\phi \tag{13}$$

$$l_6 = l_3\sin|\varphi|/\tan\phi \tag{14}$$

$$l_7 = l_m(\cos\theta_m - \sin\theta_m\cos\varphi/\tan\phi) \tag{15}$$

$$l_8 = l_m\sin\theta_m\sin|\varphi|/\tan\phi \tag{16}$$

$$l_9 = (l_m\sin\theta_m + l_2 + l_3)\cos\varphi/\tan\phi \tag{17}$$

$$l_{10} = l_m\cos\theta_m + l_{gp} - l_1 - l_4 \tag{18}$$

$$l_{11} = l_9 + l_7 - l_{10} \tag{19}$$

$$l_{12} = (l_2 + l_3)\sin|\phi|/\tan\varphi \tag{20}$$

$$\tan\omega_2 = l_6/(l_4 + l_5) \tag{21}$$

$$\tan\omega_3 = l_8/l_7 \tag{22}$$

The overlapping area $A_{re}$ should be calculated according to the conditions of the positions of shadows of both the reflector and mirror-symmetric plane, and the calculating methods were described in detail in a previous paper [30].

### Effect of shadow of reflector on wick

When the solar altitude angle $\varphi$ is small in the early morning and late evening, the reflector will shade the wick and this will cause decrease in direct radiation absorbed on the wick, especially in the case where the inclination of the reflector $\theta_m$ is large. In the calculations, the effect of the shadow is taken into consideration. The way to estimate the effect of the shadow was described in a previous paper in detail [30]. Here, the effect of the shadow is also affected by a gap length $l_{gp}$ and decreases with an increase in gap length $l_{gp}$ as mentioned in Results.

### Heat and mass transfer in the still

The wick of the still absorbs solar radiation ($Q_{sun,w}$). A part of the energy escapes to the surroundings through the bottom insulation by

conduction ($Q_d$), and is consumed to heat up saline water fed to wick ($Q_f$). The remaining energy transfers to the glass cover by radiation ($Q_r$), conduction and mass transfer ($Q_e$). Therefore, energy balance for the wick can be expressed as

$$Q_{sun,w} = Q_{(r+d+e),w-g} + Q_{d,w-a} + Q_f \tag{23}$$

where subscripts $w$, $g$ and $a$ show the wick, glass and ambient air, respectively. Subscripts of $Q$, e.g. $w-g$, show the heat transfer rate from the wick ($w$) to the glass cover ($g$).

The glass cover of the still absorbs solar radiation ($Q_{sun,g}$) as well as heat transfer from the wick, and the energy is transferred to the surroundings by radiation and convection ($Q_c$). Therefore, energy balance for the glass cover can be expressed as

$$Q_{sun,g} + Q_{(r+d+e),w-g} = Q_{(r+c),g-a} + (mc_p)_g\frac{dT_g}{dt} \tag{24}$$

Htin the calculations, and $mc_p$ is heat capacity, $T$ is temperature and $t$ is time. Each heat transfer rate in Eqs. (23) and (24) was described in a previous paper in detail [28].

Solar radiation absorbed on the wick, $Q_{sun,w}$, and the glass cover, $Q_{sun,g}$, can be expressed as

$$Q_{sun,w} = Q_{sun,dr} + Q_{sun,df} + Q_{sun,re} \tag{25}$$

$$Q_{sun,g} = \left(Q_{sun,dr}/\tau_g(\beta) + Q_{sun,df}/(\tau_g)_{df} + Q_{sun,re}/\tau_g(\beta')\right) \times \alpha_g/\alpha_w \tag{26}$$

where $\alpha_g$ is absorptance of the glass cover.

Eqs. (1) to (26) and relating equations were solved together to find the solar radiation absorbed on the wick and the glass cover, temperatures of the wick and the glass cover and distillate production rate of the still with 600 s time steps. Temperatures of the wick and the glass cover were set to be equal to ambient air temperature at $t = 0$ just before sunrise as the initial condition. The weather and design conditions and physical properties employed in the calculations are listed in Table 1.

### Results

Figure 5 shows theoretical predictions of hourly variations in (a) global solar radiation on a 1 m² horizontal surface and distillate production rate of a tilted wick still and (b) direct ($Q_{sun,dr}$), diffuse ($Q_{sun,df}$) and reflected ($Q_{sun,re}$) solar radiation absorbed on the wick on a spring equinox day. The daily global solar radiation on this day is about 23.3 MJ/m²day. Here, the inclination of the still $\theta_s$ is determined as 35° which maximizes the distillate production rate of the still when there is no gap between the still and reflector and the reflector's length $l_m$ is the same as still's length $l_s$ on that day [29]. In Figure 5, the results for a still with a reflector (RS), in which the reflector's inclination $\theta_m$ is 30° and 40° and gap length $l_{gp}$ is 0, 0.5 and 1 m as well as a still without a reflector (NS) are shown. The distillate production rate as well as each solar radiation absorbed on the wick varies similarly with the change in global solar radiation. The distillate production rate and reflected solar radiation absorbed on the wick for RS is larger for $\theta_m = 30°$ than for $\theta_m = 40°$, and decreases with an increase in gap length $l_{gp}$ for both inclinations $\theta_m$. When the reflector inclination $\theta_m$ is 40° and gap length $l_{gp}$ is 1 m, the wick cannot absorb the reflected radiation and the distillate

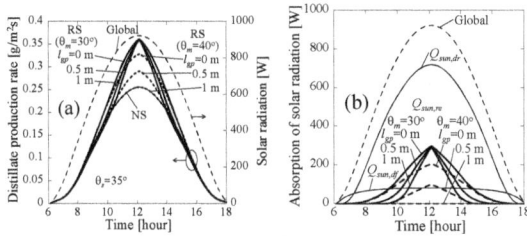

**Figure 5**: Time variations of (a) the distillate production rate of the still (RS is a still with a reflector and NS is one without a reflector) and global solar radiation on a horizontal surface and (b) absorption of solar radiation on the wick at $\theta_s$ = 35° on a spring equinox day.

production rate is the same as NS throughout the day. This indicates that the inclinations of both the still and reflector should be determined carefully when the gap is large. When the reflector's inclination $\theta_m$ = 30°, the distillate production rate as well as reflected radiation absorbed on the wick decreases with an increase in gap length $l_{gp}$, however, at noon, these figures are almost the same at any gap length $l_{gp}$. The reason is as follows: the altitude angle of the sun $\varphi$ at noon on this day is about 60°, and reflected radiation from the reflector with an inclination of 30° goes nearly horizontal. Therefore, the effect of the gap length would be very little or negligible at noon.

The global solar radiation on a 1 m² horizontal surface and distillate production rate on the summer solstice (daily global solar radiation is about 30.3 MJ/m²day) and winter solstice (daily global solar radiation is 12.6 MJ/m²day) are shown in Figures 6 and 7. The still's inclination $\theta_s$ is determined as 10° or 65° on the summer and winter solstices, respectively, due to the same reason as on the spring equinox. On both days, the distillate production rate can be increased by using a flat plate bottom reflector even if the gap length is equivalent to the lengths of both the still and reflector (1 m). These results indicate that a flat plate bottom reflector can be used effectively even if there is a gap between the still and reflector throughout the year by adjusting both the inclinations of the still and the reflector.

Isometric diagrams of the daily amount of distillate produced by the still (kg/m²day) varying with both the inclinations of the still and reflector on the spring equinox and summer and winter solstices are shown in Figure 8. The daily amount of distillate of each figure in Figure 8 was calculated at 1° steps for each inclination $\theta_s$ and $\theta_m$. The daily amount of distillate of NS with $\theta_s$ = 30° on each day is also shown in Figure 8. The inclinations $\theta_s$ and $\theta_m$ which maximize the daily amount of distillate vary considerably with the seasons, and are slightly affected by gap length $l_{gp}$. The inclination of the still $\theta_s$ which results in the maximum daily amount of distillate is largest in winter and smallest in summer, while the inclination of the reflector $\theta_m$ which results in the maximum daily amount of distillate is largest in summer and smallest in winter, since the solar altitude angle is highest in summer and lowest in winter. The range of the reflector's inclination $\theta_m$ which can increase the daily amount of distillate decreases with an increase in gap length $l_{gp}$ from about $\pm$ 20° at $l_{gp}$ = 0 m to about $\pm$ 10° at $l_{gp}$ = 1 m. Therefore, the reflector's inclination $\theta_m$ should be adjusted to the proper angle more carefully when the gap length is long. When the gap length $l_{gp}$ = 0 m, a considerable decrease in daily amount of distillate is observed when the reflector's inclination is large, especially in spring and winter. This is due to the shadow of the reflector on the wick. However, the effect of the shadow of the reflector decreases with an increase in gap length $l_{gp}$ since the shadow of the reflector does not reach the wick when the gap

is large.

The combinations of the optimum inclinations of the still ($\theta_s$) and the reflector ($\theta_m$) which result in the maximum daily amount of distillate on each month throughout the year are shown in Figure 9. The optimum inclination of the still without a reflector (NS) is also shown. The optimum inclinations of both $\theta_s$ and $\theta_m$ are approximately symmetrical with the months of June and December.

In winter, the optimum inclination of the still $\theta_s$ slightly decreases and of the reflector $\theta_m$ slightly increases with an increase in gap length $l_{gp}$. The reason is as follows: the altitude angle of the sun at noon on the winter solstice is about 36°, and the reflected radiation from the reflector inclined at $\theta_m$ is around 10° goes upwards. And the angle from horizontal of the reflected radiation can be decreased by increasing the reflector's inclination. Therefore, the reflector's inclination should be increased with an increase in gap length $l_{gp}$ to hit the reflected radiation on the wick. When the gap length $l_{gp}$ is short, an increase in absorption of reflected radiation overcomes a decrease in direct radiation absorbed on the wick by increasing the still's inclination $\theta_s$ from the optimum one of NS. However, this effect decreases with an increase in gap length $l_{gp}$. Therefore, the optimum inclination of the still slightly decreases with an increase in gap length $l_{gp}$ and comes close to that of NS.

In summer, especially in June, the optimum inclination of the still $\theta_s$ is not affected by the gap length $l_{gp}$, while the optimum inclination of the reflector $\theta_m$ slightly decreases with an increase in gap length $l_{gp}$. The reason is as follows: the altitude angle of the sun at noon on a summer solstice day is about 83°, and the reflected radiation from the reflector with inclination $\theta_m$ at around 50° goes downwards. The angle from horizontal of the reflected radiation can be decreased by decreasing the reflector's inclination. Therefore, the reflector's inclination should be decreased with an increase in gap length $l_{gp}$ to hit the reflected radiation on the wick.

The optimum inclinations of both the still $\theta_s$ and the reflector $\theta_m$, which are determined in 5° steps for ease of operation, are listed in Table 2. Here, the deviation of the daily amount of distillate obtained with $\theta_s$ and $\theta_m$ listed in Table 2 compared with that obtained with $\theta_s$ and $\theta_m$ shown in Figure 9 is less than 1 %. Therefore, the approximate maximum daily amount of distillate can be obtained with inclinations $\theta_s$ and $\theta_m$ listed in Table 2.

The variations of the daily amount of distillate throughout the year and the cumulative productivity (sum of the daily amount of distillate on each month in Figure 10) with gap length $l_{gp}$ are shown in Figures 10 and 11. Here, NS shows the results of a still without a reflector in which the still's inclination is fixed at 30° throughout the year, and NS* shows the results for one in which the still's inclination is set to the optimum listed in Table 2 according to month. The daily amount of distillate can be increased by the flat plate bottom reflector throughout the year even if there is a gap between the still and the reflector, and the cumulative productivity of the still with reflector is predicted to be about 81.2, 75.6 and 72.4 kg/m², and about 28, 19 and 14 % more than that of NS (63.3 kg/m²) when the gap length $l_{gp}$ is 0, 0.5 and 1 m. However, the daily amount of distillate and the range of the inclination of the reflector that can increase the daily amount of distillate decreases with an increase in gap length $l_{gp}$. Therefore, it is better to set the reflector nearer to the still.

## Conclusions

The effect of a flat plate bottom reflector on the distillate production rate of a tilted wick solar still was analyzed theoretically for instances when there is a gap between the still and reflector. The results of this

$G_{dr}$, $G_{df}$: Bouger's and Berlage's equations [31] with transmittance of atmosphere of 0.7, solar radiation incident on the atmosphere of 1370 W/m² at 30ºN latitude.

Ambient air temperature: 25 °C (Feb., March and April), 33 °C (May, June and July), 30 °C (Aug., Sep. and Oct.) and 20 °C (Nov., Dec. and Jan.).

Wind velocity of ambient air = 1 m/s

$l_s = l_m = w = 1$ m, $\rho_m = 0.85$, $\alpha_w = 0.9$, $\alpha_g = 0.08$

Diffusion gap between the wick and the glass cover = 10 mm.

Thickness and thermal conductivity of bottom insulation = 50 mm and 0.04 W/mK.

Transmittance of glass cover [32]:

$$\tau_g(\beta) = 2.642\cos\beta - 2.163\cos^2\beta - 0.320\cos^3\beta + 0.719\cos^4\beta \cdot$$

Feeding rate of saline water to the wick: Twice as large as the steady-state evaporation rate from the wick calculated on the assumption that solar radiation is kept constant at its peak value throughout the local day time.

**Table 1**: The weather and design conditions and physical properties.

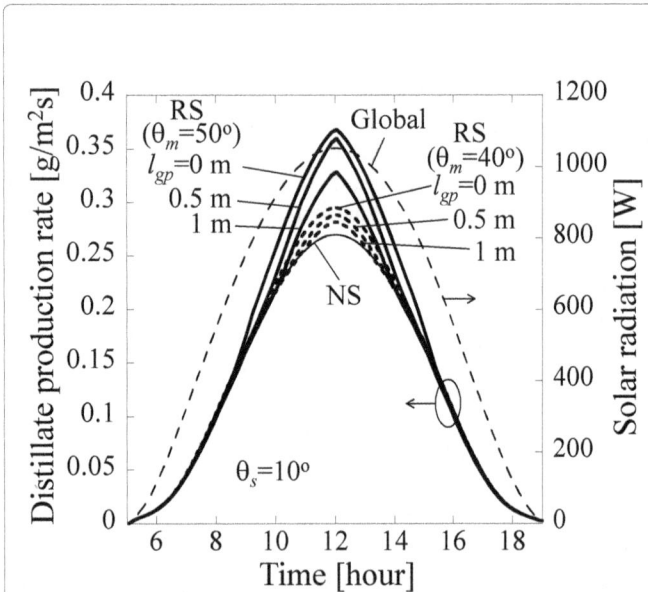

**Figure 6**: Time variations of the distillate production rate of the still and global solar radiation on a horizontal surface at $\theta_s = 10°$ on a summer solstice day.

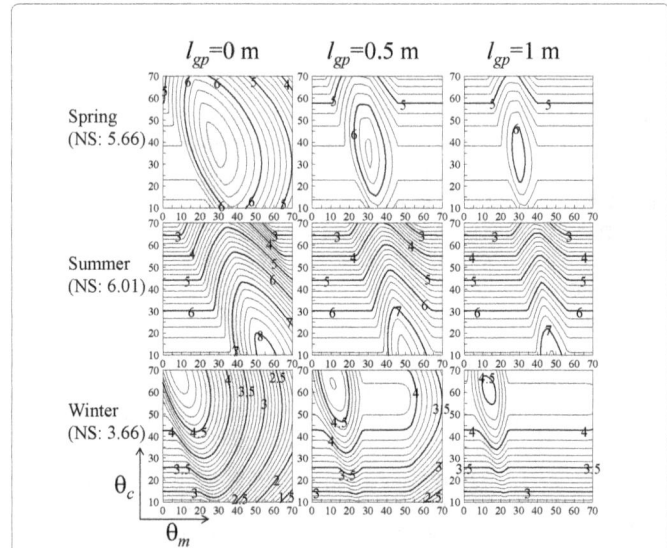

**Figure 8**: Isometric diagrams of the daily amount of distillate (kg/m²day) varying with both the inclinations of the still $\theta_s$ and reflector $\theta_m$ at $l_{gp} = 0$, 0.5 and 1 m on three typical days (spring equinox and summer and winter solstices).

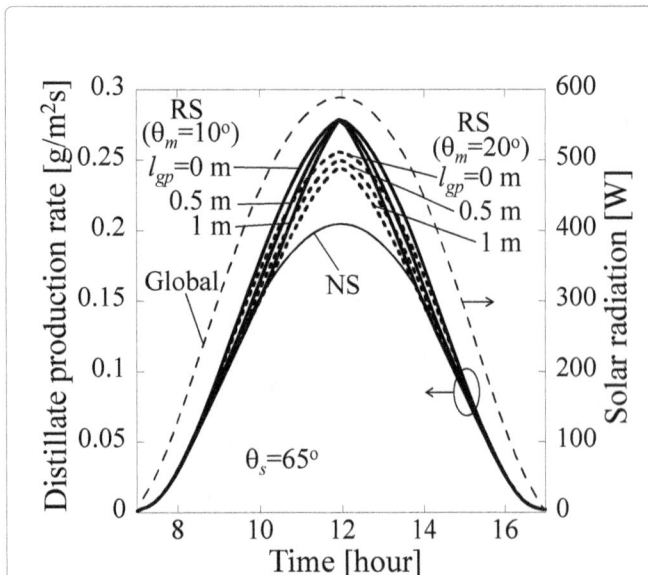

**Figure 7**: Time variations of the distillate production rate of the still and global solar radiation on a horizontal surface at $\theta_s = 65°$ on a winter solstice day.

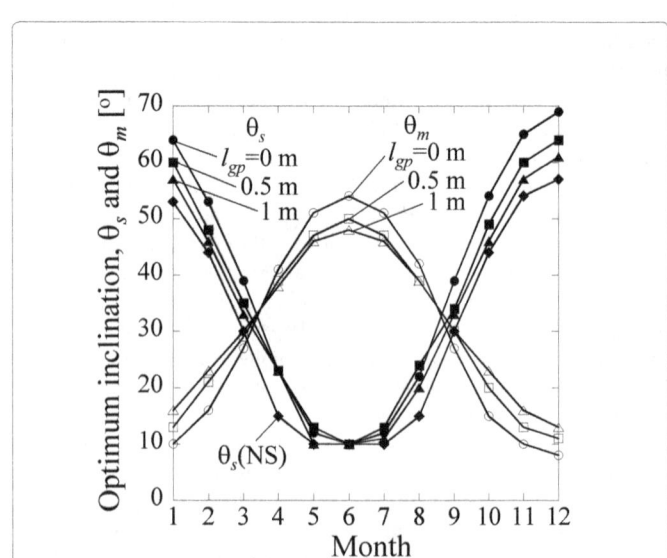

**Figure 9**: Combinations of the optimum inclinations of the still and reflector which maximize the daily amount of distillate throughout the year.

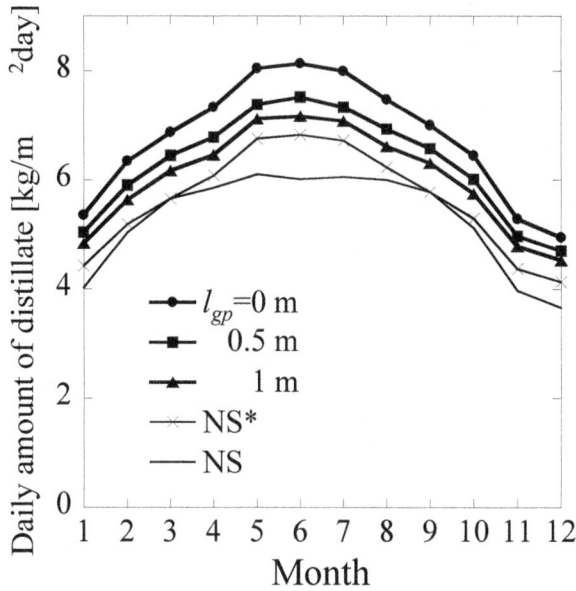

**Figure 10**: Daily amount of distillate varying with the gap length $l_{gp}$ throughout the year.

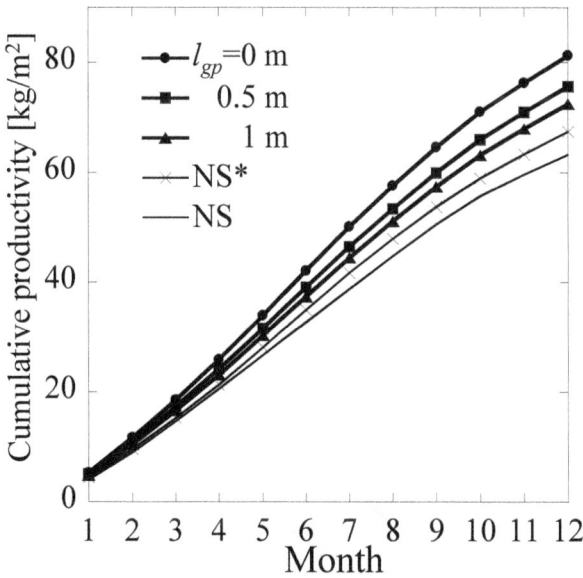

**Figure 11**: Cumulative daily amount of distillate on each month varying with the gap length $l_{gp}$.

| t | NS | RS ($l_{gp}$ [m]) | | | | | |
|---|---|---|---|---|---|---|---|
| | | 0 | | 0.5 | | 1 | |
| | $\theta_s$ | $\theta_s$ | $\theta_m$ | $\theta_s$ | $\theta_m$ | $\theta_s$ | $\theta_m$ |
| Dec. | 55 | 65 | 10 | 65 | 10 | 60 | 15 |
| Jan., Nov. | 55 | 65 | 10 | 60 | 15 | 55 | 15 |
| Feb., Oct. | 45 | 55 | 15 | 50 | 20 | 50 | 20 |
| Mar., Sep. | 30 | 35 | 30 | 35 | 30 | 35 | 30 |
| Apr., Aug. | 15 | 25 | 40 | 20 | 40 | 20 | 40 |
| May, July | 10 | 15 | 50 | 15 | 45 | 10 | 45 |
| June | 10 | 10 | 55 | 10 | 50 | 10 | 45 |

**Table 2**: Optimum inclinations of still $\theta_s$ and reflector $\theta_m$ throughout the year at 5° steps.

lowest in summer, while the optimum inclination of the reflector is lowest in winter and highest in summer.

(4) Both the inclinations of the still and reflector should be adequately adjusted according the seasons, especially when the gap is large.

(5) The distillate productivity of the still decreases with an increase in gap length even though both the inclinations of the still and reflectors are set to the proper angle.

(6) The sum of the daily amount of distillate of the still on each month throughout the year was predicted to increase about 28, 19 and 14% due to the reflector when the gap length is 0, 0.5 and 1 m, respectively.

### References

1.  Tanaka K, Yamashita A, Watanabe K (1981) Experimental and analytical study of the tilted wick type solar still. International Solar Energy Society Congress, Brighton: 1087-1091.

2.  Sodha MS, Kumar A, Tiwari GN, Tyagi RC (1981) Simple multiple wick solar still: Analysis and performance. Solar Energy 26: 127-131.

3.  Malik MAS, Tiwari GN, Kumar A, Sodha MS (1982) Solar Distillation. Pergamon Press, UK.

4.  Gandhidasan P (1983) Theoretical study of tilted solar still as a regenerator for liquid desiccants. Energy Convers Manag 23: 97-101.

5.  Tiwari GN, Sharma SB, Sodha MS (1984) Performance of a double condensing multiple wick solar still. Energy Convers Manag 24: 155-159.

6.  Tiwari GN (1984) Demonstration plant of multi wick solar still. Energy Convers Manag 24: 313-316.

7.  Yeh HM, Chen LC (1986) The effect of climatic, design and operational parameters on the performance of wick-type solar distillers. Energy Convers Manag 26: 175-180.

8.  Mahdi JT, Smith BE (1994) Solar distillation of water using a V-trough solar concentrator with a wick-type solar still. Renew Energy 5: 520-523.

9.  Minasian AN, Al-Karaghouli AA (1995) An improved solar still: The wick-basin type. Energy Convers Manag 36: 213-217.

10. Al-Karaghouli AA, Minasian AN (1995) A floating-wick type solar still. Renew Energy 6: 77-79.

11. Hongfei Z, Xinshi G (2002) Steady-state experimental study of a closed recycle solar still with enhanced falling film evaporation and regeneration. Renew Energy 26: 295-308.

12. Aybar HS, Egelioglu F, Atikol U (2005) An experimental study on an inclined solar water distillation system. Desalination 180: 285-289.

13. Janarthanan B, Chandrasekaran J, Kumar S (2005) Evaporative heat loss and heat transfer for open- and closed-cycle systems of a floating tilted wick solar still. Desalination 180: 291-305.

14. Boukar M, Harmim A (2005) Performance evaluation of a one-sided vertical

work are summarized as follows:

(1) The distillate productivity of the still can be increased by a flat plate bottom reflector throughout the year even if the gap length is equivalent to the length of the still and reflector.

(2) The optimum inclinations of both the still and reflector are considerably affected by seasons, and slightly affected by the gap lengths.

(3) The optimum inclination of the still is highest in winter and

solar still tested in the desert of Algeria. Desalination 183: 113-126.

15. Shukla SK, Sorayan VPS (2005) Thermal modeling of solar stills: An experimental validation. Renew Energy 30: 683-699.

16. Eltawil MA, Zhengming Z (2009) Wind turbine-inclined still collector integration with solar still for brackish water desalination. Desalination 249: 490-497.

17. Mahdi JT, Smith BE, Sharif AO (2011) An experimental wick-type solar still system: Design and construction. Desalination 267: 233-238.

18. Omara ZM, Eltawil MA, ElNashar EA (2013) A new hybrid desalination system using wicks/solar still and evacuated solar water heater. Desalination 325: 56-64.

19. Alvarado-Juarez R, Alvarez G, Xaman J, Hernandez-Lopez I (2013) Numerical study of conjugate heat and mass transfer in a solar still device. Desalination 325: 84-94.

20. Zerrouki M, Settou N, Marif Y, Belhadj MM (2014) Simulation study of a capillary film solar still coupled with a conventional solar still in south Algeria. Energy Convers Manag 85: 112-119.

21. Hansen RS, Narayanan CS, Murugavel KK (2015) Performance analysis on inclined solar still with different new wick materials and wire mesh. Desalination 358: 1-8.

22. Manikandan V, Shanmugasundaram K, Shanmugan S, Janarthanan B, Chandrasekaran J (2013) Wick type solar stills: A review. Renew Sustain Energy Rev 20: 322-335.

23. Murugavel KK, Anburaj P, Hanson RS, Elango T (2013) Progresses in inclined type solar stills. Renew Sustain Energy Rev 20: 364-377.

24. Tanaka H, Nakatake Y (2007) Improvement of the tilted wick solar still by using a flat plate reflector. Desalination 216: 139-146.

25. Tanaka H, Nakatake Y (2009) Increase in distillate productivity by inclining the flat plate external reflector of a tilted-wick solar still in winter. Solar Energy 83: 785-789.

26. Tanaka H (2011) Increase in distillate productivity by inclining the flat plate reflector of a tilted-wick solar still in summer. Energy Sci Technol 1: 11-20.

27. Tanaka H (2009) Tilted wick solar still with external flat plate reflector: Optimum inclination of still and reflector. Desalination 249: 411-415.

28. Tanaka H (2011) Tilted wick solar still with flat plate bottom reflector. Desalination 273: 405-413.

29. Tanaka H (2013) Optimum inclination of still and bottom reflector for tilted wick solar still with flat plate bottom reflector. Desalin Water Treat 51: 6482-6489.

30. Tanaka H (2015) Theoretical analysis of solar thermal collector and flat plate bottom reflector with a gap between them. Energy Reports 1: 80-88.

31. Japan Solar Energy Society (1985) Solar Energy Utilization Handbook, Onkodo Press, Tokyo.

32. Tanaka H, Nosoko T, Nagata T (2000) A highly productive basin-type – multiple-effect coupled solar still. Desalination 130: 279-293.

# Application on Solar, Wind and Hydrogen Energy - A Feasibility Review for an Optimised Hybrid Renewable Energy System

Subhashish Banerjee*, Md. Nor. Musa, Dato' IR Abu. Bakar Jaafar and Azrin Arrifin

*Department of Renewable Energy, Universiti Teknologi Malaysia, Malaysia*

### Abstract

Present status of development of the energy systems, like solar, wind and of Hydrogen use as renewable energy sources, have been examined in ascertaining the scope of their application from techno-economic viability.

Efficiency rating of different generations of PV cells with cost component and scope of their use, including practical application and economic fall outs thereby, were also spelt out from case study. Cost of PV module being shown to be the main contributor, deciding the economy of PV – based power generation scheme, the methodology of determining its total requirement could also be shown. Economy evaluation of solar PV and solar heating (solar cooker, solar pond, etc.) were also assessed stating their merits and demerits.

Different aspects on evaluation of economy for off-shore and on-shore wind energy for developing a wind farm at a particular site were assessed. The scope of availing the wind energy value /sq. m. at a concerned site for a particular wind speed (measuring it at anemometer height and converting to its value at the hub height), could also be ascertained.

Necessary study on the economy of hydrogen production by splitting water (electrolysis) was made and was noted that the OTEC (ocean thermal energy conversion system) generated electricity would be the best cost-effective method, mainly because of its prospect of earning huge royalty from its different by-products. On examination of $H_2$-fuel cell combine as transport fuel, it could be shown that a single 100 MW OTEC can cater to 30 hydrogen refuelling stations, each with 250 vehicle movement/day.

Feasibility study on optimised hybridization of the combination of PV-wind and $H_2$, for uninterrupted power supply with methodology on resource assessment at concerned sites, particularly for PV and wind energy with economic fall outs, could be ascertained. Grey areas of research in exploring acoustic energy is also discussed.

**Keywords:** Solar energy; PV module; Solar pond; Wind energy; Wind speed; Off-shore wind energy; Hydrogen; Electrolysis; OTEC; Hybrid system; Acoustic energy; Piezoelectric effect

## Introduction

Sustainable development would require transition from fossil fuel based procurement of energy to renewable energy (RE) development. But total transition from fossil fuel to renewable energy is becoming difficult, mainly because of the economic constraint coming in the way of their large scale application. However, considering the problems faced on fossil fuel use, from the depletion of this very resource itself, as well as of environmental degradation from carbon equivalent gases emission inducing global warming, RE systems may emerge to be competitive with fossil fuels, if the social cost of the latter are taken into account.

In this context, Solar and Wind energy systems are the two most important source of renewable energy which has become very popular in recent times. They are being tried to develop with constant improvement of cost component from R and D studies and trials with necessary subsidies. In fact, in order to assess any energy system's commercial acceptability, its economic evaluation is very important for its further development as also for availing the research funding. For several decades extensive research work are being done on these two renewable energy systems for advancement of their technology. As a result of these R and D efforts their application cost have also fallen down. Another reason for their decreasing trend of cost is the increment of their volume of use.

The other RE energy which is being considered to be the most favoured energy by the turn of the century, particularly as the transport fuel, is the Hydrogen/ Fuel cell system. Hydrogen is important not only because it is an easily transportable clean energy. But it can be used to store electricity, producing it by electrolysis using electricity, and thereafter it can be used to produce electricity through fuel cell.

With this in view it is proposed to study the present status of development of these energy systems from the following perspectives. They are:

- The theoretical basis of their operation on electricity production.

- The brief review of their technology with emergence of cheaper and more acceptable 2nd and 3rd generation system.

- Suitable site of their implementation.

- Their efficiency and scope of use with economic evaluation

*Corresponding author: Subhashish Banerjee, Department of Renewable Energy, Ocean Thermal Energy Centre, Universiti Teknologi Malaysia, Malaysia, E mail: wave.banerjee@gmail.com

• Environmental fall outs including LCA studies on GHG emission etc.

• Limitations.

• Scope of further improvement with hybridization etc.

• Grey area of research

A brief review on above aspects on energy systems, Solar and Wind and Hydrogen type fuel, have been outlined below with feasibility study of their hybridization.

## Solar Energy

Most of the energy available--including the fossil fuels, owe their origin directly or indirectly, from the sun. The upper atmosphere of the earth receives from the sun more than $1.5 \times 10^{21}$ watt- hour of solar radiation annually, which is more than 23,000 times the energy used by human population globally [1]. Even if much less than 0.1% of this energy could be used effectively, it would meet the entire global demand of energy many times over.

In fact, direct tapping of solar radiation may be made, either by the generation of electricity from photovoltaic effect of incident solar radiation – using solar cells, or utilizing the heating effect of the solar insolation.

## Solar cells

**Theoretical basis of electricity production from solar PV cells:** Solar PV cells are basically semiconductor materials. It conducts electricity only when light falls upon the solar PV cells; otherwise it behaves as a semiconductor. Hence, they are termed photo-voltaic cell or, PV cell. Over 95% of the solar cells that are produced are Silicon (Si) based. This is because Si is abundantly found in nature. At the same time for processing, it does not put much burden to the environment. For making the solar cells, which is a kind of semiconductor, it involves doping them. Doping means that it is to be contaminated with certain elements that will allow flow of electricity. Such doping with selective chemical elements, produces either excessive positive types of charge carriers (p-type semiconductor), or negative types of charge carriers known as (n-type semiconductor). If both these type of semiconductors are combined, then p-n junction occurs at the boundary of the layers. As a result an interior electric field is built up which leads to the separation of the charge carriers. When light falls on them, then they are released, and on making metallic contacts, generation of electricity from electric charge flow, can be achieved. But the current build up thus made, is DC. A typical such type of PV cell, showing flow of DC current is shown below in (Figure 1) [2].

**Emergence of improved types of solar cells:** There are three categories of Solar PV Cells, mainly classified as the three generation cells.

• The first generation cells are single as well as polycrystalline silicon cells.

• Second generation cells are polycrystalline thin film crystal structure cells; amorphous Si: H cells, etc.

• Third generation cells are high efficiency multi -junction concentrator solar cells; like, dye sensitized cells, organic cells, polymeric cells, nanostructured cells including multi carrier photon cells, quantum dot and quantum confined cells. They are all of the third generation PV cells [3].

The 1st generation single crystalline Si cells occupy 31% of the market and its efficiency is as high as 24.7%. However it is expensive and require very pure silicon, like 99.999999999% pure silicon. Its processing requires long time and high temperature [3].

Second generation polycrystalline cells are fastest growing technology and have efficiency a little less. In case of second generation technology, it requires lower material use, fewer processing steps and simpler manufacturing technology. Hence it has cost advantages over the 1st generation crystalline Si: PV cells. The major systems in this 2nd generation of flat plate thin film PV cells are: amorphous Silicon (a: Si), Cadmium Telluride (Cd: Te), Copper Indium di-Selenide (CIS). Though they are cheaper than the 1st generation PV cells, but their efficiency is lower, around 13%. They occupied 15%-20% of the market in the year 2010 [3].

The efficiency percent of different types of 2nd generation Silicon based cells are shown below in (Table 1) [3].

On the other side, the third generation PV cells are high efficiency concentrator cells consisting of Gallium Arsenide substrate (Ga: As). They are twin junction cells with Indium Gallium phosphide made on Gallium Arsenide wafers. Their laboratory scale efficiency is around 40%. Dye sensitized PV cells, Organic PV cells and Nano structural solar cells are included in this category of 3rd generation PV cells. They are however, in the R and D stage and not yet realized experimentally [3].

**Application of Solar PV cells:** P-N junction of PV cell produces 0.5 V/cell. Agglomerate of a number of cells constitute a solar module, and multiple modules make array of PV cells, as shown in (Figure 2) given below [4].

As early as in 2000, at an inaccessible small island, in Sundarbans, India, (Gayenbazar in Sagar Islands, Sundarbans, WB, India) a PV array containing 320 modules of PV cells, with 36 solar cells in each module was installed, which provided 100 watts of power to 93 consumers. It required coverage of 300 m² area for their installation [5]. The power generated being DC had to be converted to AC using inverters and were stored in battery for night supply to the consumers.

Schematic diagram of a solar power plant with its different outfits is shown below in (Figure 3) [6].

These additional outfits (battery and inverter) escalates the cost of PV based power generation still further, cost of which in 2000,

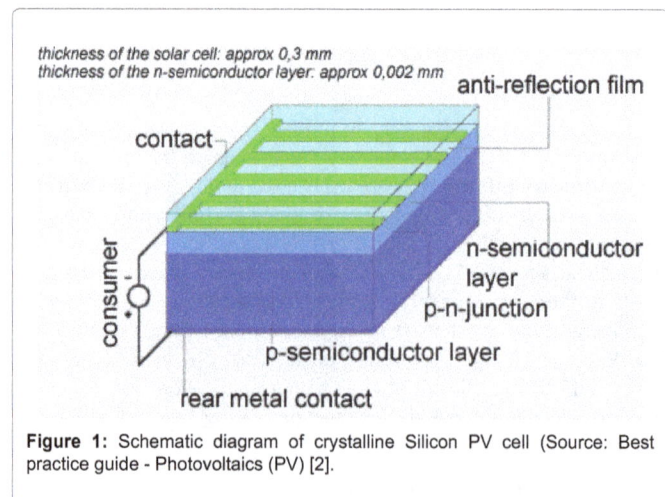

thickness of the solar cell: approx 0,3 mm
thickness of the n-semiconductor layer: approx 0,002 mm

anti-reflection film

contact

consumer

n-semiconductor layer
p-n-junction
p-semiconductor layer

rear metal contact

**Figure 1:** Schematic diagram of crystalline Silicon PV cell (Source: Best practice guide - Photovoltaics (PV) [2].

| Type/Material of PV cell | Level of efficiency in Laboratory (%) | Level of efficiency in Production (%) |
|---|---|---|
| Mono crystalline Silicon based | 24 (Approximate ) | 14-17 |
| Polycrystalline Silicon based | 18 (Approximate) | 13-15 |
| Amorphous Silicon based | 13 (Approximate) | 5-7 |

**Table 1:** Efficiency percent of different types of Silicon based PV cells [3].

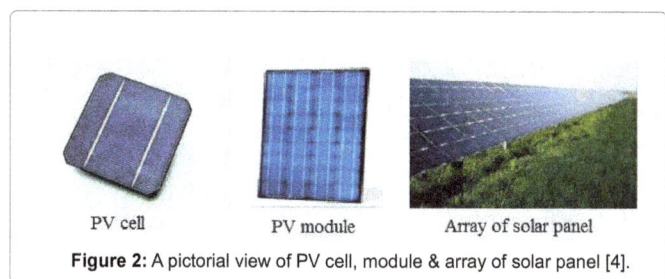

PV cell          PV module          Array of solar panel

**Figure 2:** A pictorial view of PV cell, module & array of solar panel [4].

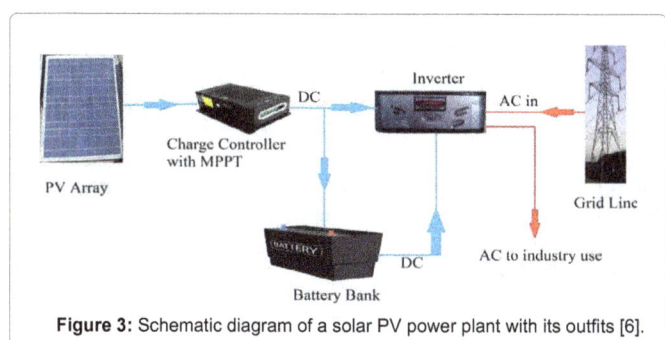

**Figure 3:** Schematic diagram of a solar PV power plant with its outfits [6].

was $ 4-6/W$_{peak}$ [6]. Such installation though apparently uneconomic, but it helped to improve the economy of that inaccessible place, at Sundarbans, Sagar island, India, from the availability of power supply. They are now being used more widely with Government subsidies in many such inaccessible places.

Such installations subsequently made could be the hybrid system along with other power grid supply line or, only stand-alone type -as per the suitability of installation site. In fact, such hybridization can also be made with wind energy supply, if available.

It may be added that solar PV installation cost has improved a lot since then, and have become much cheaper, not only from the availability of rather cheaper 2$^{nd}$ generation PV cells, but also because of their increased volume of use.

It may be relevant to add that the efficiency of crystalline Silicon PV products ranges from 15-20% with requirement of space 100 sq. ft./kW; whence for thin film PV, average efficiency is 7-15% with space requirement of 200 sq. ft. /kW [4].

PV array site is determined from the availability of space; normally 1 kW$_p$ requiring 8 m$^2$ of roof/space, that faces south with slope of around 30-40˙ [7]. Since PV functions from the incident solar light, shades of trees etc. would lower the efficiency.

PV systems are fixed to the roof using stainless steel "roof hooks". They robustly get attached to the roof. Fixing these hooks to concrete tiled roof is a rather easier process. But in case they are required to be

hooked in slate and clay tiles, adequate measures are needed for their fixing/removal etc., increasing the cost from these extra measure with extra labour cost of the same [7]. In the following figure is shown the mode of fixing solar module over roof surface showing 30˙ tilt angle, facing the preferred southern side for availing longer hours of sunlight (Figure 4) [4].

## Economic assessment of PV cells

The economic assessment of a PV power plant is mainly decided from the cost of the PV modules required for the concerned plant. Of course, other costs including inverter, battery assembly, labour cost in fixing the modules and installation cost, O and M cost, land cost etc.-are also to be taken into account. The sizing of the inverter, battery assembly, land area requirement etc. are all dependent on the sizing/ number of the PV modules required. The relationship deciding the number of PV modules are hence important, which can be estimated from the following equations that determine the panel generation factor (PGF) and total watt peak rating of the PV modules, as shown below [6].

**Panel generation factor** = Solar irradiance x sunshine hours / Standard test condition irradiance; (where, solar irradiance (kWh/m$^2$), depends on site concerned and sunshine hours, which is normally considered to be 9-10 hours and standard test condition irradiation is considered 1000 kWh/m$^2$) .. [1]

**Total watt peak rating** = Total energy requirement from the PV module (kWh/day) / Panel generation factor (PGF); (where, PGF is determined from eqn. 1 and total energy requirement is case specific, depending on the requirement). [2]

**Number of PV modules required** = Total watt peak rating / PV modules peak rated output; (where, Total watt peak rating is decided from eqn. 2 above and PV modules peak rated output depends on the module type chosen for installation). [3]

It may be relevant to cite a case study as made for a PV plant of 2.5 MW at Jaipur city India. It required 714.1 million INR excluding the land cost, out of which PV module cost was 587.87 million INR [6]; which is more than 82% of the total cost of installation.

Of course, capacity factor of PV plant, which is decided from annual kWh generated for each KW AC peak/8760; is quite decisive in PV economy. In Jaipur case, it showed around 35% [6]. However, it would be case specific and site specific.

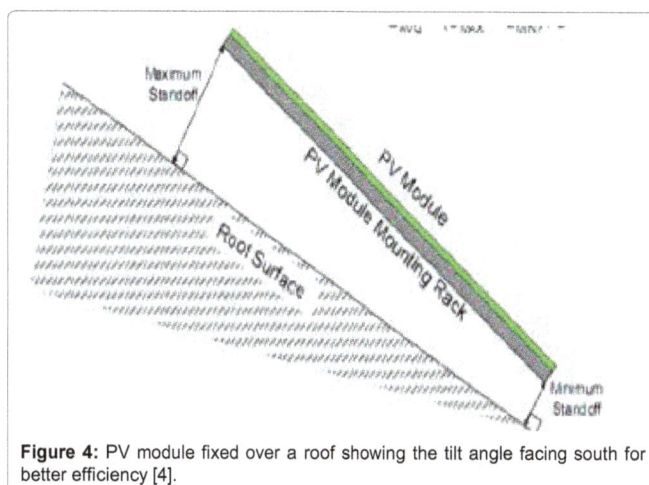

**Figure 4:** PV module fixed over a roof showing the tilt angle facing south for better efficiency [4].

Thereby is determined LCOE (levelised cost of energy), giving the average price the consumer has to pay the investor, for the capital cost, O and M cost etc. incurred with a rate of return which is equal to the discount rate.

In Jaipur case of 2.5 MW PV plant, India, with solar irradiance 617, and capacity factor a little more than 35%, discount rate at 10% and life considered 25 year, LCOE (excluding land cost) showed a value 11.40 INR/kWh, which is = $ 0.016ʹ11.4 = $ 0.18/kWh. (1 INR being $ 0.016 dt 20.8.2015).

## Scope of hybridization

It may be relevant to add that the assembly of batteries may be eliminated if necessary hybridization with Wind energy is made - both the PV and Wind being time specific. The hybridization of one may help the other for 24 x 7 power supply. In addition Hydrogen generation may be tried in for storing electricity during scope of availability of excess supply, through $H_2$/Fuel cell route.

## Utilization of sunlight's heating effect

Different techniques have been improvised to trap the radiant energy of the sun and utilize its heating effect fruitfully both for domestic use as well as in industry -- without taking recourse to conventional energy resources. A few of them briefed below--include Solar Cooker, Solar Pond, and generation of electricity concentrating the heating effect of the sun to derive clean Energy. A brief outline of them are given below.

**Solar-cooker:** A report from IREDA, India, states that an amount of 6.7 million tonne of $CO_2$ emission could be avoided if 3% of Indians switch over to using Solar Cooking [8]. It could be demonstrated building a prototype of cheaper quality solar cooker (life expectancy of 5-6 years ) that 2-3 hours of clear sunshine in tropical country like India, could be enough to cook 2-4 items --- making it a viable proposition. A sample survey made in India, amongst poorer sections of the society, proved that this is cheaper than using conventional fuels used for their cooking [9].

Various types of Solar Cookers are already available in the market - which however needs proper information dissemination, besides R and D on cost reduction (without compromising the quality) for achieving wider acceptability.

**Solar pond:** Solar pond entraps heat energy from sun-light in a novel way. The bottom layer of a water pool is artificially made denser with adequate addition of common salt. Convection current of the heavier bottom layer does not move upwards to shed its heat to the upper layer of the pond and thereafter to the atmosphere as in a normal pond. Thus, heat energy received from sunlight remain stored in the bottom zone. Therefore the heat received from the sun is virtually entrapped in the bottom layer of such ponds, termed Solar Pond.

In Solar Ponds the bottom layer temperature may go up as high as 70-80°C, when the upper surface layer temperature would virtually equalize with the atmospheric temperature Since the salinity gradient increases with depth, the middle layer with salinity/density in between, would have a temperature in between and virtually act as the insulation between the cooler surface layer and the hot bottom layer. This heated water in the bottom layer of the Solar Pond can have industrial use providing heat energy [10].

The Bhuj Solar Pond in India, perhaps one of the largest solar ponds - with dimension of 100 m × 80 m × 3.5 m, required 4000 tonnes of common salt and is functioning successfully supplying heated water

required for an adjacent large dairy farm [11].

It is obvious that such solar ponds can only be successful where adequate supply of sea water/cheaper salt supply, and high radiant energy of the sun is assured.

**Electricity generation from solar heating:** It is a power tower, which is a system for trapping solar energy from a large field of mirrors and converting it to heat at high temperature for efficient generation of electricity. All the mirrors track the sun and the heat is focused on a single boiler thermal system. The purpose is to cover the midday load as experienced by utilities. To counter the effect of passing cloud, there is a thermal storage capability filled with oil.

## Merits and demerits of solar energy

Solar Power is rather decentralized by its very nature and the advantage of Solar Energy is that it can supply energy, in inaccessible places as well.

The generation of liquid and solid wastes during wafer slicing, cleaning, processing and assembling of solar PV cells may create health hazards to workers of manufacturing units. This can however be reduced by recycling and from use of suitable traps. Wastes arising from disposal of array of batteries used for solar power stations are also a problem that needs to be addressed. Environmental life cycle assessment made at Utrecht University on multi crystalline solar cells reports emission of pollutants like, fluorine, chlorine, nitrate, iso-propanol, $SO_2$, $CO_2$, silica particles etc. during the production stage of the PV cells as well as from the mining and refining of silica (Utrecht Univ. Report). Module life time has also been suggested in their study as to not exceed 30 years. Recycling of materials in the PV cell manufacture has been advocated in their study for cost reduction.

Solar Energy is by its very nature site and time specific. It has good potential particularly in inaccessible areas, though further R and D studies are in progress for its wider application with economic viability.

## Wind Energy

Wind Energy was in use since long mainly for grinding purposes in wind mills. Generation of Electricity with Wind Turbines, utilizing the aerodynamic lift of the wind, is the recent trend-with global stress upon Renewable Energy Resources. Wind turbines capture the wind's energy with two or three propeller-like blades, which are mounted on a rotor, to generate electricity. The turbines sit atop high towers, taking advantage of the stronger and less turbulent wind at 30 meters or, more above ground.

In fact, the availability of wind is the most important criterion that would determine the deployment of wind turbine in a certain place. By availability it means the wind speed due to which it will rotate the wind turbine so that mechanical energy of the wind is converted into electrical energy through generator. The wind speed decides the efficiency and economy of wind energy application for creating wind farms.

### Theoretical basis of wind energy formation and its scope of use

Wind movement is known to be formed due to the uneven heating of the earth from solar insolation, irregularities of the earth's terrain, and also from the rotation of the earth. The kinetic energy possessed by wind movement can be converted into mechanical energy, which rotates the blade and spin a shaft of the turbine, which produces electricity through generator. The wind turbines are of two types. One of them is horizontal axis wind turbine and the other is vertical axis.

Horizontal axis based wind turbines are rather more common [12].

Since the power that a wind turbine generates is a function of the cube of the average wind speed of the site concerned; hence small differences in wind speed would cause large differences in productivity and thereby of electricity cost. Also, the swept area of a turbine rotor is a function of the square of the blade length. Hence a modest increase in blade length would enhance energy capture, and thus of the cost component on power generation [13].

Since wind speed is higher at high altitude and also less turbulent, so wind turbines are to be placed at higher hub heights. Thus it becomes important to determine the relationship between the wind speed measured using anemometers at the anemometer measurement height, and the wind speed at high hub height around 25-30 m atop. It is known to maintain a logarithmic relationship between the wind speed at anemometer height (where wind speed is measured) and at the hub height as below [13]:

$$V_h = \ln (H/z)/[\ln(A_h/z)/Va].. \qquad [4]$$

where, $V_h$ is the wind velocity of hub height at the concerned site, $A_h$ is the anemometer measurement height measuring the wind speed, $V_a$ is velocity of wind at the anemometer measurement height, z is the surface roughness index expressed in length and H is the hub height.

Hub height of wind turbine (at which there is rotor) may vary from 25 m for smaller wind turbines like below 50 kW, to as high as even 100m for large multi - megawatt wind turbines. The surface roughness index varies from 0.008 m to 3 m depending on site characteristics [Surface roughness length for terrain with lawn grass it is 0.008 m, fallow field is 0.03 m, with associated few trees 0.01 m, sites with many trees and few buildings it is 0.25 m, for forest it is 0.5 m, suburbs is 1.5 m, but in city centre with tall buildings it is considered to be 3.0m] [13].

Wind turbines are constructed as per the designated power production capability of the concerned site. It may have three situations, like operating with cut in speed, which is the minimum speed at which the wind turbine can give useable power (3-5 m/s); or, rated speed at which the wind turbine will make designated rated power (8-15 m/s), or the cut out speed at which it will cease to function giving power, like in a cyclone.

A typical diagram showing the relationship of power generation from different wind speed from use of a FL100 brand turbine is shown below in (Figure 5) [13].

It is needed to first determine the wind speed at the concerned hub height of the wind turbine. This is done using logarithmic equation shown above, from measurement of wind speed by the anemometer at the anemometer height. Then to apply it to the turbine power curve as per the above figure to calculate the power output under standard condition of temperature and pressure. Thereafter to multiply with the air density corrections for getting the real condition.

## Power production from wind turbine

Cluster of Wind turbines constituting the wind-farm is an expanding industry. These Wind farms may be stand- alone system or connected with utility power grids. They are fruitfully in operation in many countries - including India (Tamil Nadu, Gujarat) and European countries. It has been reported that global scenario of tapping energy from Wind power increased from 2500MW in 1998 to 10,000 MW in 2003 [8]. The global wind industry now expands at 44 percent year-on-year growth, with a total now (2014) at 369.553 GW. In 2014, the

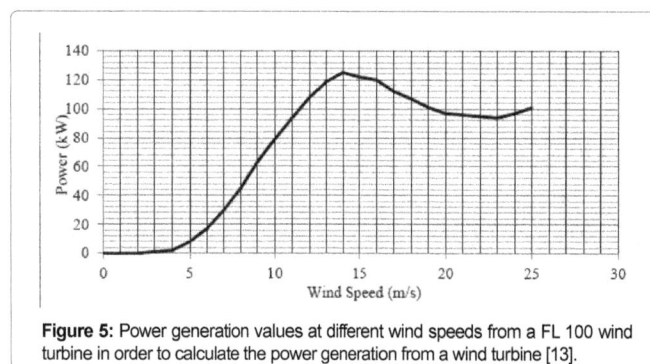

**Figure 5:** Power generation values at different wind speeds from a FL 100 wind turbine in order to calculate the power generation from a wind turbine [13].

United States represented 17.8% of the world's total installed wind energy capacity, second only to China, which is followed by Germany, Spain, and India [14].

The power generation from a wind farm at a certain average wind speed may be calculated from the following case study made at Malaysia, considering the wind speed at hub height to be 3 m/s and with a turbine of 25 m dia blades, which can be operational in low wind speed as well [15].

The energy per sq.m of an area = Ea = 0.5˙air density in $kg/m^{3*}$ wind speed [5].

Thus considering air density to be 1.3 $kg/m^3$, and at wind speed of 3 m/s, Ea =17.55 W [15]. The circular area swept by the turbine of 25m diameter=π˙$(25/2)^2$= 491 $m^2$.

Thus, total power grenerated from a single wind mill of above operational data would be = 491˙17.55 =8617 W;

which after correcting the efficiency at 50% = 4.309 kW...... 　[6]

In order to estimate the scope of power generation of a wind farm, covering a particular land area, it is needed to estimate the optimum number of wind mills that can be set up. Too close placement of wind mills will curtail the available wind speed; too far placement would cause uneccesary land pressure. As a rule of thumb, the optimum distance margin between two wind mills is said to be kept around five times the turbine diameter.

Thus per square meter land area the scope of tapping power, from above type of turbine with above wind speed would be=4.309 kW / $(5˙25)^2$ = 0.28 W/sq.m land area. ..........　[7]

But it is also important to know the capacity factor or load factor of the wind mill, to determine the period when it may be non-functional, from lowering of wind speed below the cut in speed (said to be between 3-5 m/s).The wind speed profile is thus important to know the capacity factor of the concerned wind mill. In Malaysia's context the wind speed profile as determined is shown below in (Figure 6) [15].

It would be obvious that day time wind speed is higher than during night.

But off shore wind speed is higher than on-shore wind speed (of only around 2 m at anemometer height and 3 m at hub height of 30 m, at Malaysia as noted before). Besides off-shore wind is also less turbulent and thus generates more electricity.

In Malaysia, 16 number of such off-shore wind sites could be located, as shown below in (Figure 7). At these sites wind speed reached more than 5 m/s; but during north-west monsoon season only [16].

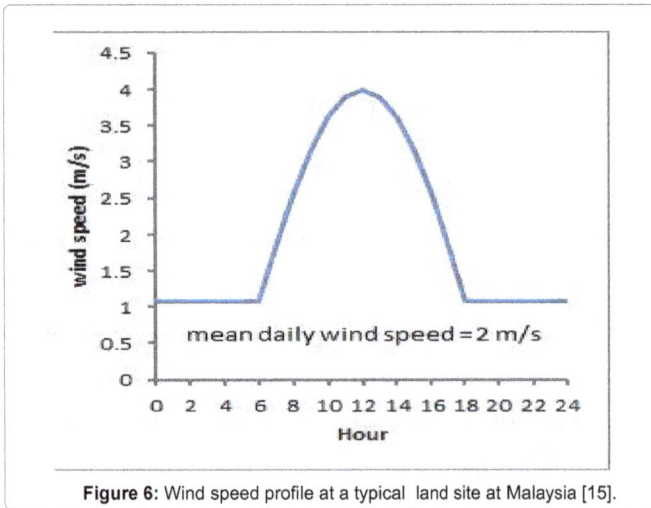

**Figure 6:** Wind speed profile at a typical land site at Malaysia [15].

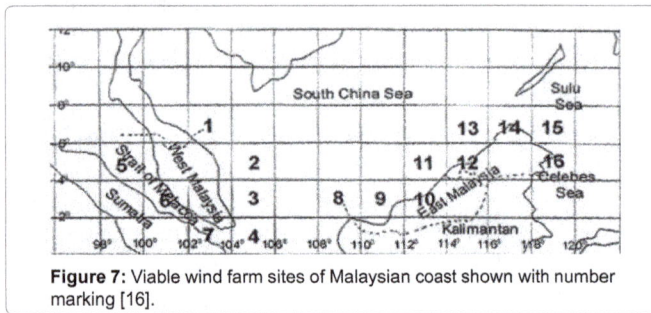

**Figure 7:** Viable wind farm sites of Malaysian coast shown with number marking [16].

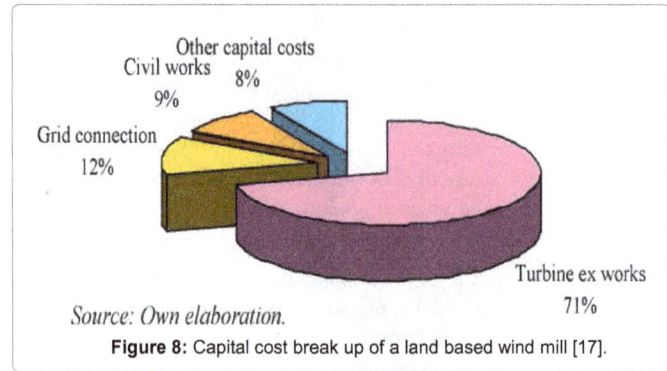

*Source: Own elaboration.*

**Figure 8:** Capital cost break up of a land based wind mill [17].

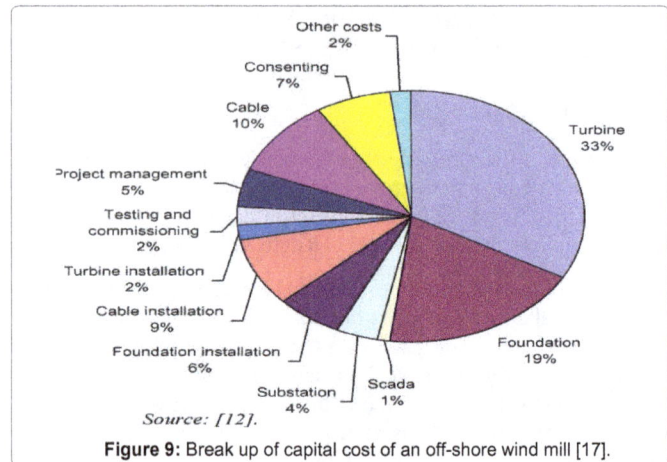

*Source: [12].*

**Figure 9:** Break up of capital cost of an off-shore wind mill [17].

Though normally wind speed at Malaysia is about 2 m/s, which can reach 3 m/s at hub height around 30 m; but at selected sites, particulatrly in south west monsoon it can reach as high as 7-15 m/s [15].

### Economy of wind farm

The cost of the turbine and grid line connecton cost together constitute 80% of the capital cost involvement of a wind mill. Other costs like, construction cost for land – based mills though much less than off-shore wind mills, but is also appreciable. It involves around 10% of the total capital cost. The break up of capital cost involvemengt of a land based wind mill is shown below in (Figure 8) [17].

But in case of off-shore wind mill, the turbine cost together with the foundation cost costitute nrearly 50% of the total capital cost involvement. Cable cost including foundation of cables involve nearly 20% of the total capital cost. The break up capital cost of an off-shore wind mill is shown below in (Figure 9) [17].

It may be added that though the power generation cost of wind energy is the cheapest amongst all the RE systems, competitive to fossil fuel; but its installation is not cheap. Including all costs, even for a good cut in wind speed site, it involves around $1 million per megawatt (MW) of generating capacity installed [14]. In order to make economically viable, a wind farm should have the minimum power availability of 20 MW or more. Presuming each wind mill having production capability of 750 kW, there should be installation of at least 26 turbines, to generate 20 MW involving $ 20, 000, 000 [14].

Cost of wind energy/kWh is therefore not only dependent on wind speed but also the grid line distance and also whether it is off-shore or, on-shore wind. It had been estimated that with very good wind speed

(10 m/sec), it would be around $ 0.04/kWh, whereas $ 0.06-09/kWh for moderate wind speed (6 m/sec) [8].

### Merits and demerits

Life time emission of Carbon equivalent gas is minimum in Wind energy systems; 12 g/kWh (Post note 2004).

Though initial investment needed for wind farm is higher than that of the fossil fuel based power plants, but its life cycle cost requires no extra fuel cost etc. with limited O and M cost only. This makes wind energy an economically viable proposition compatible with fossil fuel power stations at suitable sites, having higher wind velocity for prolonged period.

But its problem is that, it is intermittent depending on wind flow with the cut in speed. Thus stand-alone wind system cannot provide power 24 x 7, round the year. Hence its hybridization with solar PV is stressed upon. Also at times it may create land pressure, if off-shore wind is not harnessed. But off-shore wind though have huge resource availability possibility, is rather costlier from increased grid line availability distance and also foundation cost increase.

In addition it should ensure not to affect species (Eagle etc. birds) and should not block the air traffic route. It is also not encouraged for placement in populated areas from local protests due to noise pollution hazards from it.

### Hydrogen

Hydrogen is considered to be the future energy in the next millennium, being an absolute clean energy. It can be used as the fuel source, that can be availed by splitting of water and also leaves behind nothing but water on its liberation of the energy. Hence it has been identified to be the future energy source in the new millennium, for the energy supply and security, climate stewardship with ensured sustainability, particularly in the transport sector [18].

It may be relevant to add in this context that Hydrogen though an environmentally clean fuel, which leaves behind only water on its combustion/liberation of energy, but it is available only in the combined form with water/hydrocarbons, etc. Hence though its resource is almost inexhaustible; but it would always require energy input, either conventional fossil fuels or, renewable energy etc. for its production. Since it can be produced passing electric current in water, and can also be used to generate electricity through fuel cells; hence it is considered to be instrumental in the storage of electricity, which unlike battery need not require periodic charging with power to derive power from it.

Hydrogen and Oxygen can be obtained by splitting water in Proton Exchange Membrane Electrolyser (PEME), using in acidic medium [19].

In view of the fact that Hydrogen is a clean fuel for sustainable growth, the Internal Energy Agency (IEA), as early as in 1977 started to promote Hydrogen, with measures to meet the different challenges faced on its production from carbon free sources, as also to sort out its storage problems and explore its scope of use as a clean fuel source [18]. But it has not yet taken off, excepting small scale trials as transport fuel, which still remains in experimental stage only and has not yet picked up commercial viability.

## Economy evaluation

More than 90% of the present day Hydrogen preparation, around 500billion $m^3$/year is produced from steam reforming of fossil fuels (Natural gas, Naphtha, or coal) and only 4% from splitting of water, by electrolysis routes etc. the latter being uneconomic (Isao A. Anon.). Though it can be produced from bio-mass decomposition (gasification and fermentation) and also splitting of water by electrolysis, or by Photo electrochemical/ photo-biological etc routes. The former, bio-mass decomposition is not yet economically viable and the latter is still in the R and D stage [20].

Efforts are now being made to make $H_2$/fuel cell combine vehicles replacing the fossil fuel. Presently attempted alternate energy like, hydrogen/fuel cell route for vehicle movement are: [21]:

• Battery charged vehicle (BEV) using different types of RE systems.

• Fuel cell based electrical vehicles (FCEV).

• Plug-in-hybrid electrical vehicle (PHEV), which combines battery with fuel cell in operating the internal combustion engine (ICE).

But they all are yet to take off [20]. Despite the availability of the technology for availing such vehicles excluding use fossil fuel altogether, lack of facilities like $H_2$ refuelling stations etc. affecting the volume of their use, limit the economic viability of vehicles with $H_2$- fuel cell combine. The scenario may however, change if the social cost (like GHG emission cost and fossil fuel depletion escalating cost) is included, making future Hydrogen economy a viable and better alternative.

It has been said that the cost of production of electricity decides the economy of Hydrogen production by electrolysis, as shown from the figure 10 given below (Figure 10) [22].

In this context it may be added that electricity cost from ocean thermal energy conversion system (OTEC) may be drastically reduced, from the prospect of huge royalty as may be earned from its different by products availed free in course of power production [23]. Particular mention may be made of the improved thermodynamic cycle like Uehara cycle with increased power advantage generation efficiency and with higher by product availability from OTEC (like potable water, sea food, and also mineral water from cold water feed, etc.) [24]. Thus OTEC driven power supply may prove to be economically viable for hydrogen production by electrolysis, when considered such scope of advancement of OTEC technology.

## Merits and demerits

It has been projected by UAE that production of $H_2$ from electrolysis (PEM electrolyte etc.) and thereafter electricity generation through fuel cell, would cost 90% of its electricity production by the turn of the century [25]. As stated before OTEC route of $H_2$ production has the potential to produce cheaper electricity.

But such economic viability of $H_2$ production from OTEC plants are achieved only for larger sized plants, above 50 MW [26]. It has been noted that for use as transport fuel, $H_2$ production of 1500 kg/day is needed in refuelling station feeding 250 cars/day [22]. It can be thereby be estimated that a single 100 MW OTEC plant (net power) producing 2160 MWh/day (running at 90% capacity factor) with production/day of 44, 497 kg $H_2$ by electrolysis, can cater to around 30 such refuelling stations [27], 4.33 kWh is reported to produce 1 $NM^3$ of hydrogen, which is equivalent to 48.5 kWh/kg).

Of course the cost component is required to be improved upon if compared with present day cost of

fossil fuels. This can be achieved from performance improvement and on availing the by-product royalty from OTEC plants, over which it has immense scope.

## Scope of Feasibility Study on Use of Hybrid System (PV, Wind and $H_2$)

Feasibility on scope of application of an energy system would need to identify quantitatively the resource potential of them which mainly decides the economy.

In case of solar PV the

• measurement of solar irradiance at site concerned and

• studies on availability of sunlit hours at site and

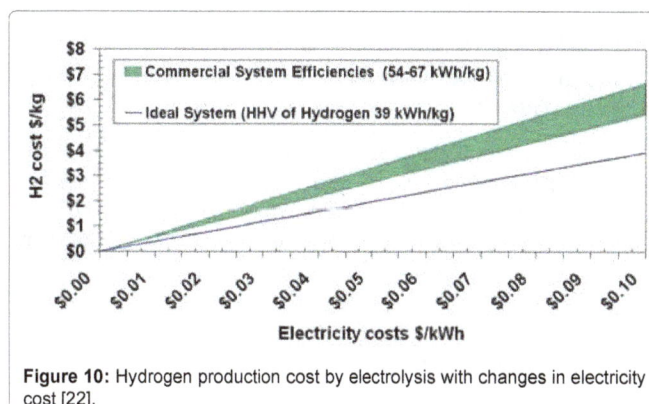

**Figure 10:** Hydrogen production cost by electrolysis with changes in electricity cost [22].

• the scope of area availability for placement

are the three data which are to be availed. They can be availed either from actual site study and/or search from meteorological department data of the concerned zone.

Thereafter based on equation 1, 2 and 3, can be decided number of PV module requirement, which decides the economy, being more than 80% contributing factor of the total cost.

In case of wind energy the

• Measurement of wind speed using anemometer at the concerned site and

• Studies on period/ hours that the cut in wind speed is available at the concerned site and

• Convert the said wind speed at the hub height from the measurement made at anemometer height using logarithmic relationship shown in equation 4 and knowing the terrain data (surface roughness index) and

• Determining the swept area from the diameter of the blades of wind turbine being used.

Thereby can be determined the value electricity availability/sq. m. of the concerned zone using equations 5, 6, 7 -shown before.

It may be added that the turbine cost together with the gridline connecting cost constitute 80% of the total cost of a land based wind farm. In case of off-shore wind farm turbine cost together with gridline cable cost constitute 70% of the total cost plus foundation cost constitute 20% of the total cost.

In case of $H_2$, particularly OTEC-driven $H_2$ production by electrolysis, it is needed to assess:

• scope of temperature differential available between surface warm water (SOW) with deep ocean water (DOW) and

• size of the OTEC plant with net power availability and

• the scope of by products available from the concerned plant, which is an important criterion deciding the economy and

• electrolysis efficiency which varies from a little above 50- 70% (the $H_2$ production estimations as made in the text, as per the data from Nilhous and Vega [27] conforms to 61% efficiency).

• Based on the above data, the cost component of $H_2$ availability may be availed.

In case of $H_2$, it seems further RandD studies are needed on fuel cell technology for $H_2$ production /electricity generation which yet remains an economic challenge.

The combined Hybridization of the above three RE systems are shown in (Figure 11) [28].

The optimized Hybrid model consisting of PV/Wind and $H_2$ system is optimized on sizes of concerned RE systems, using the tool HOMER (Hybrid Optimized Model for Electric Renewable), based on meteorological data and data generation at the concerned site, along with necessary economy evaluation of them [28].

## Exploring Grey Areas of Research

Discussions on RE remained confined to the energy resources like, Solar, Wind and Hydrogen energy and also of their suitable hybridization, depending on the concerned site with scope of resource availability. Besides these conventional RE systems, hitherto unused energy resource that is abundant in our environment, is the low frequency acoustic energy, harvesting of which ensures sustainable development [29]. Such acoustic energy can advantageously be used in the rapidly increasing industries like, electronic devices with low power consumption, such as electronic communication devices, wireless sensor network etc, replacing the electrochemical batteries having limited life span [30].

The acoustic energy may be harnessed to generate electricity on introduction of a quarter wave length straight tube acoustic resonator using a PVDF (poly vinylidene fluoride) piezoelectric transducer. The acoustic pressure gradient causing vibratory motion of the PVDF piezoelectric beams generate electricity due to the direct piezoelectric effect [31]. It may be added that a quarter-wavelength resonator is the preferred choice; showing three times more efficiency than even half wave length resonator with similar diameter and frequency [31]. Using lead zirconate titanate (PZT) piezoelectric cantilever plates it could be noted producing power output of 30 mW from an incident sound pressure level of around 160 db [30]. Of course output power can be increased using multiple piezoelectric plates along a quarter wave length straight resonator tube; a schematic diagram of which is shown below in (Figure 12) [29].

With incident sound energy of 100 db, 4.06 V at 189 Hz could be generated using 5 PZT plates; and 0.37 mW at 190 Hz using 4 plates [29].

It however remains a grey area of research for developing various types of piezoelectric transducers and sound tracking resonators to explore with better harvesting of hitherto unused acoustic energy hugely available in our environment.

## Conclusion

A method could be developed from assessment of the present status of development of solar PV, wind energy and $H_2$ –fuel cell combine, for effective hybridization of these renewable energy systems for availing uninterrupted power supply with better economy and making optimized hybrid model of them.

The parameters needed for resource assessment of wind energy farm installation, by estimating the energy density /sq. m of the site as per the available wind speed; and also of the requirement of PV modules for a given power generation from PV modules, based on solar irradiation at the concerned site – could also be well defined. These

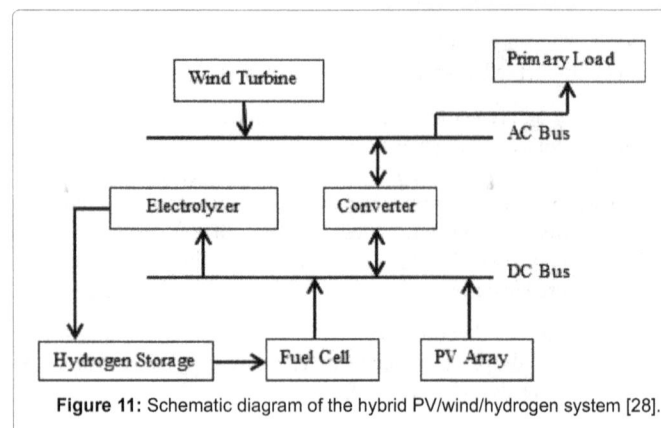

**Figure 11:** Schematic diagram of the hybrid PV/wind/hydrogen system [28].

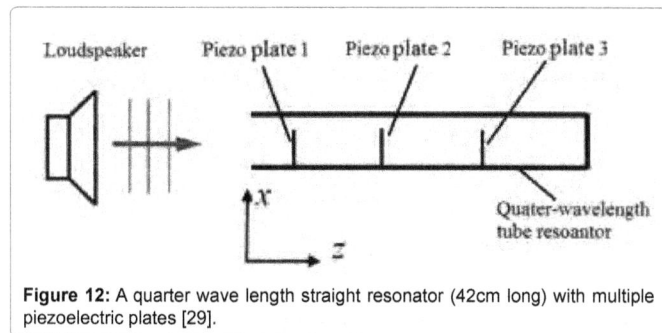

**Figure 12:** A quarter wave length straight resonator (42cm long) with multiple piezoelectric plates [29].

measurements were noted to be important in economy assessment of these two RE systems, Wind/PV combine, where speed of wind at hub height in case of wind energy and number of PV module in case of PV systems, mainly decides their cost effectiveness.

Hydrogen production by electrolysis route, for availing sustainable energy system, though noted to be much costlier than its production through steam reforming of fossil fuels; but OTEC-driven power supply using higher capacity advanced OTEC types, were found to be the solution. The scope of huge royalty earning from different by-products of OTEC makes its electricity production cost cheaper and thus economically viable on availing transport fuel from $H_2$ fuel cell combine.

Making of use of hybrid optimised model for electric renewable (HOMER) tool has been suggested for deciding the extent of application of the RE systems, Solar PV, wind and $H_2$/fuel cell combine, to achieve better economy.

Methodology of data generation at site on determining feasibility for application of HOMER tool, for developing optimised hybrid model of all the three energy systems (PV, wind and $H_2$) could be spelt out.

A grey area of research has been suggested in harvesting low frequency acoustic energy remaining unused, though abundant in the environment. It could be noted that a quarter wave length straight resonator with multiple piezoelectric plates, show better result in electricity production.

## References

1. McGraw (2002) Hill Encyclopaedia. 9th Edition, 16: 693.

2. http://www.seai.ie/Publications/Renewables_Publications_/Solar_Power/Best_Practice_Guide_for_PV.pdf

3. http://www.cie.unam.mx/lifycs/ITaller2011/PV-Tutorial2011/Lecture3-AH.pdf

4. http://www.pge.com/solar

5. Banerjee S (2000) Project report on Studies on Power generation from solar energy.

6. Chandel M, Agrawal GD, Mathur S, Mathur A (2014) Techno-economic analysis of solar photovoltaic power plant for garment zone of Jaipur city. Case Studies in Thermal Engineering 2: 1-7.

7. http://www.goodenergy.co.uk

8. IREDA News (1007) New Delhi, India EWEA Report - 2004; an analysis of Wind Energy - EU-25-20.

9. Banerjee S (1999) Towards commercialization of a low cost solar cooker. Proc. National Convention of Renewable convention 481-484.

10. Saifullah AZA, Shahed Iqubal AM, Saha A (2012) Solar pond and its application to desalination. Asian Transactions on Science & Technology 2: 1-25.

11. http://edugreen.teri.res.in/explore/renew/pond.htm

12. http://windeis.anl.gov/guide/basics/

13. Mohammadi S, Vries de B, Schaefer W (2014) Modeling the allocation and economic evaluation of PV panels and wind turbines in urban areas. Procedia Environmental Sciences 22: 333-351.

14. American Wind Energy Association, Wind Energy Fact Sheet, 10 steps in building wind farms.

15. Aziz AA (2011) Feasibility Study on Development of a Wind Turbine Energy Generation System for Community Requirement of Pulau Banggi Sabah. A report from Mechanical Engineering Department, UTM, Malaysia.

16. Mekhilef S, Chandrasegaran D (2011) Assessment of Off-Shore Wind farms at Malaysis.

17. Blanco MI (2009) The economics of wind energy. Renewable and Sustainable Energy Reviews 13: 1372-1382.

18. Elam CC, Padro GP, Sandrock G, Luzzi A, Lindblad, P, et al. (2003) Realizing the hydrogen future: the International Energy Agency's efforts to advance hydrogen energy technologies. International Journal of Hydrogen Energy 28: 601-607.

19. Symes MD, Chronin L (2013) Decoupling hydrogen and oxygen evolution during electrolytic water splitting using an electron-coupled-proton buffer, Nature Chemistry 5: 403-409.

20. Abbasi T, Abbasi SA (2011) Renewable hydrogen: Prospects and challenges. Renewable and Sustainable Energy Reviews 15: 3034-3040.

21. Ball M, Weeda M (2015) The hydrogen economy - Vision or reality? International Journal of Hydrogen Energy 40: 7903-7919.

22. Levin JI, Mann MK, Margolis R, Milbrandt A (2005) An Analysis of Hydrogen Production from Renewable Electricity Sources.

23. Banerjee S, Duckers L, Blanchard RE (2015) A case study of a hypothetical 100MW OTEC plant analysing the prospect of OTEC technology 1: 98-129.

24. Kobayashi H (2002) 'Water' from the ocean with OTEC. Forum on Desalination using Renewable Energy.

25. Kazim A (2010) Strategy for a sustaniable development in the UAE through hydrogen energy. Renewable Energy 35: 2257-2269.

26. Vega LA (2012) Ocean Thermal Energy Conversion; Encyclopaedia of sustainability science and technology. Springer 2012: 7296-7328.

27. Nilhous GC, Vega L (1993) Design of a 100MW OTEC plant ship. Marine structures 6: 207-221.

28. Kalinci S, Hepbasi A, Dincer I (2015) Techno-economic analysis of a stand-alone hybrid renewable energy system with hydrogen production and storage options. International Journal of Hydrogen Energy 40: 7652-7664.

29. Bin L, You HJ (2015) Simulation of Acoustic Energy Harvesting Using Piezoelectric Plates in a Quarter-wavelength Straight-tube Resonator, Conference paper.

30. Hassan HF, Idris S, Hassan S, Rahim RA (2014) Acoustic Energy Harvesting Using Piezoelectric Generator for Low Frequency Sound Waves Energy Conversion. IJET 5: 4702-4708.

31. Bin L, Laviage AJ, You JH, Kim Y (2013) Harvesting low-frequency acoustic energy using multiple PVDF beam arrays in quarter-wavelength acoustic resonator. Applied Acoustics 74: 1271-1278.

# Small Wind Power Energy Output Prediction in a Complex Zone upon Five Years Experimental Data

**Ba MM, Ramenah Harry\* and Tanougast C**

*Department of Electronics, University of Lorraine, France*

## Abstract

In this paper, we investigate the performance of a micro-wind turbine in a complex through the power output prediction. The purpose is to show that due to long time period and very subtle onsite measurements the ideal position for the wind turbine can be determined considering experimental data under real conditions. More precisely, from well measured data (wind speed), the power output at one particular location can be approximated by the Weibull function. The considered model is tested and validated at an urban landscape location in Metz city, France, where anemometry is positioned at adjacent to the turbine and the instrumentation is specific to its surrounding location including record wind turbine data thanks to real time wireless communications. Technical data including wind speed and output power were analyzed and reported allowing to provide a reliable estimation of the wind energy potential in an urban location upon five years experimental data.

**Keywords:** Wind energy; Weibull distribution function; Wind speed; Urban wind powers

## Introduction

Small and micro-wind turbines are designed to work in an urban environment and can meet the electricity needs of individual homes, farms, small businesses and villages or small communities which can be as small as 0.2 kW. Therefore, several micro-wind installations were carried out and demonstrate benefits and possibility of producing an adequate amount of power required to a rural area even at lower wind speed with most cost effective way [1-7]. Micro-wind turbines can play a very important role in urban electrification schemes in mini-grid applications and off-grid application for rural applications and even be complement to solar photovoltaic systems in off-grid systems or mini-grids. Architects are now incorporating wind turbines in their new build designs.

Major countries in the European Union (EU) have developed strategies to promote the growth of RES-E but French renewable output has lagged that of neighbor countries. The recent France's energy transition bill is encouraging householders to use micro renewable generation through financial incentives. Sitting urban wind power needs preliminary resources assessment such as wind characteristics and wind profile, topography of the terrain with respect to the roughness class and near-by obstacles. These parameters of a specific location are essentials for power output prediction and estimation to reduce payback time on capital investment. However, most research works used numerical models to predict power output of urban micro wind power.

Some researchers have ascertain the potential of building mounted turbines by providing the on-site measured data for design and assessment of micro-wind turbines installed in building blocks [8-13]. Similarly, the specific technology and design issues in the use of wind energy in buildings have been described by Mertens [14]. In considering where these technologies are likely to be installed, little is known of the wind resource in these environments and due to the very rough and heterogeneous landscapes, turbines close to the urban surface will experience site-specific [15]. Consequently, the wind fields undergo significant changes in urban areas compared to rural areas due to the channeling effect of the urban buildings. Therefore, most research works used numerical models needed to predict power output of urban micro

wind power and assess this particular effect. Hence, some researchers have employed computational fluid dynamic modeling to indicate that turbines installed in urban environments are subject to wind particular effect. These works demonstrates the significance of turbine position and mounting height facing the building, such that small changes in location can have dramatic effects on the power generated. Accordingly, these installations appear to underperform when compared to installations in wind field undergo or rural environments.

Another approach is based on an appreciation/ quantification of how turbulence affects the productivity of a wind turbine which are required for the installation locations [16]. However, such analysis requires intensive computation resources and validation of results is very difficult to achieve owing to the requirement of the turbulence intensity modelization. Therefore, a genetic algorithms to wind farm performance evaluation and optimization for wind turbine placement has been applied [17,18]. Furthermore, an algorithm simulating the power output from the wind turbine based on wind average speed, the electrical load and the power curve has been developed [13]. A numerical wind speed data to estimate energy yields as well as analysis of financial payback periods under various scenarios applicable to micro-wind devices and the urban environment has been considered in Bahaj et al. [13]. Wind distribution functions and power evaluation models for optimization of wind farm configurations by genetic algorithms have been used in Wan et al. [19].

The goal of this paper is to investigate the performance of a micro-wind turbine in a complex urban area and show that due to long time period and very subtle onsite measurements the ideal position for the

\*Corresponding author: Ramenah Harry, Department of Electronics, University of Lorraine, 34 Cours Léopold, 54000 Nancy, France
E-mail: harry.ramenah@univ-lorraine.fr

wind turbine can be determined. The originality of our study is that the model arising from the well mounting turbines in such urban areas may provide the energy output prediction and estimation of payback time on capital investment.

In contrast with previous works, this paper focuses on the decision pertaining to installation so that optimal performance can be achieved considering the hub height with respect to proximity/influence of adjacent buildings/obstructions. Our predicting turbine productivity in the urban environment based on evaluation of roughness coefficient and Weibull distribution.

This paper presents a methodology to evaluate the real power output of a micro wind turbine based on experimental results under a long time period 2012 to 2016 thanks to very subtle onsite measurements to determine the ideal position for the wind turbine in an urban environment. More precisely, it is possible to define a reliable model for a particular complex zone. The originality of the considered method is to adapt the theoretical Weibull predictable model with the real experimental data production based only on the wind profile of the installation area.

The following Section 2 describes the GREEN platform and micro wind turbine as well as wind power parameters and data location. In this section, roughness coefficient is determined proofing that the experimental site correspond to a turbine urban sitting. Section 3 gives the wind real measurements of the considered location. The wind power output for the urban sitting is presented in the Section 4 which briefly describes the considered Weibull Model estimation and its application for the considered study case where the measurement power output performance during five years are detailed. The modeling wind speed and power output methodology for the case study is presented in the Section 5, where experimental data have been compared with simulation results to validate a micro wind turbine output power prediction model and proving the reliability micro wind power estimation in a specific urban sitting. Section 6 gives a discussion to justify the considered approach achieving the goal attempt. Finally, Section 7 draws appropriate conclusions.

## Micro Wind Turbine Power Parameters and Location

### Wind observation location

For our study, the experimental data are provided from the GREEN platform where several renewable energy technologies are implemented for modeling, managing and optimization of energy consumption. All technologies are monitored, including real weather conditions data are recorded and processed for prediction analysis (Figure 1). The platform GREEN is equipped with a residential Skystream three blades 2.6 kWatts horizontal axis wind turbine, which is at 12 m above the ground level. Installing a domestic micro wind turbine in France is usually subject to planning permission and total height must not exceed 12 m. The blades are constructed from two halves of compression molded fiberglass. The curve of the blade helps to more efficiently capture the energy in the wind and to reduce the sound of the blades as they move through the air. The turbine is a downwind design where the blades of the turbine are downstream from the nacelle, which is quieter and inherently better at finding the wind direction than upwind designs. The associated 3 blades Wind power Skystream in an urban area is represented in the Figure 2.

The inverter constantly monitors the turbine and the electrical connection to ensure that the electric energy generated by the turbine is synchronized with the frequency and voltage of the building's electrical system. Table 1 indicates the Skystrean technical specifications. The inverter actually draws about 5-7 Watts to operate the monitoring system. Consequently, the turbine will not generate electricity when the electrical grid to the building is down.

Wind speed is measured by a 3-cups rotor anemometer at the same height as the wind power. The anemometer is part of a meteorological weather station and wireless computer interface allows direct communication with the anemometer. It displays the current weather station data in a real-time report on the computer. Figure 3A presents the block diagram of the considered micro-wind turbine system while the Figure 3B gives a snapshot of the power output monitoring. Thus, a wireless wind monitors allows to save large quantities of data for download (2 Megabytes of internal memory) and use a ZigBee wireless link to computer for data acquisition. Indeed, the wind turbine has a built in 2.4 GHz wireless radio that sends performance data to a desktop computer in the GREEN platform monitoring with a wireless receiver. More precisely, the wireless wind monitor measures wind speed, direction data every minute and stores wind statistics once per minute. Figure 3B shows the Skyview software track the generation of the turbine displaying the data wind turbine in a defined sample time [20]. Thus, from data acquisition we can calculate histogram data.

The GREEN platform is located in building at the University of Lorraine, Metz, France. The investigation site location is given in the Table 2.

Figure 4 gives the climate of localization platform. More precisely, Figure 4A provides a map specifying isobaric and wind curves, temperature and nebulosity. Figure 4B gives the wind direction in percentage of the urban localization of the considered micro-wind

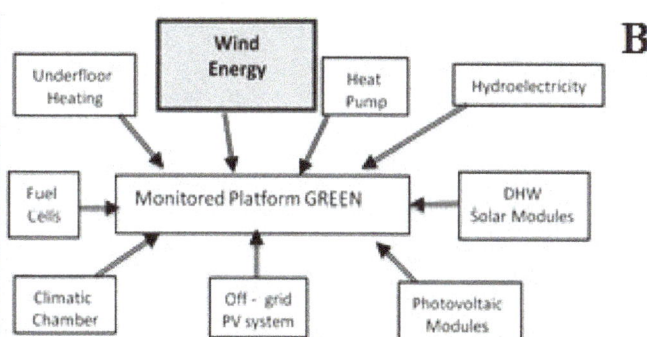

Figure 1: **A.** Monitoring room; **B.** Synoptic of the GREEN platform.

**Figure 2**: The 3 blades wind power skystream of the GREEN platform.

| Technical Specifications | |
|---|---|
| Rated capacity | 2.4 KW |
| Rotor diameter | 3.72 m |
| Swept area | 10.87 m² |
| Rated speed | 50-330 rpm |
| Cut-in Wind speed | 3.5 m/s |
| Rate Wind speed | 13 m/s |

**Table 1:** Skystream technical specifications.

turbine which mainly directed to south-west. Figure 4C details the real measured wind direction of the used micro-wind turbine for our experimental which is in concordance to local wind forecast where the maximal wind field is at the direction East and South-East with 5.5 m/s.

## Wind power function

A micro-wind turbine is characterized by the wind power function which gives the power output, in kilowatts (W), for a given wind speed V given in m/s. The wind turbine converts wind power into mechanical power. The mechanical power generated by the wind turbine at the shaft is given by the Equation 1 [21].

$$P(W) = \frac{1}{2} C_p \rho S V^3 \qquad (1)$$

Where $\rho$ is the air density which depends on altitude, air pressure and temperature. $C_p$ is the wind-turbine power coefficient (dimensionless), S is the swept area of the rotor blades (in m²) and V is the wind speed (in m/s). Figure 5 shows the measured power curve for the Skystream of the experimental GREEN platform site for an experimental day. It can be noted that the cut in speed is below the value indicated in the technical specifications as mentioned in the Table 1.

## Micro-wind turbine coefficient

The output power from the wind is given by the Equation 1, where the typical value of mean air density ($\rho$) used in this work is 1.22 kg/m³. From the Equation 1, it can be concluded that each parameter has an effect on the output power. $C_p$ is the wind-turbine power coefficient. The theoretical maximum power efficiency of any design of wind turbine is 0.59 (i.e., no more than 59% of the energy carried by the wind can be extracted by a wind turbine). This is also called the Betz Limit. Wind turbines cannot operate at this maximum limit. The $C_p$ value is unique to each turbine type and is a function of wind speed applying to operating turbine. For this experimental GREEN platform site, we determined the power coefficient from measurements as represented in the Figure 6 which gives the $C_p$ value given as $C_p$=0.34.

## Roughness coefficient

The roughness coefficient depends on the variability of wind speed at the site due the height above the ground and the rough- ness of the terrain which is a function of the wind direction. Several researchers have investigated wind speed profiles at different turbine heights and various expressions have been established to determine wind profiles while estimating the increase in wind speed with height [22]. In this study, we assumed the change in speed is less pronounced and we simply used the power law exponent relation [9] which is given by the following Equation 2.

$$\frac{V_0(h_0)}{V_0(h_0)} = \left(\frac{h_0}{h_1}\right)^\alpha \qquad (2)$$

Where $V_0(h_0)$ (m/s) and $V_1(h_1)$ (m/s) are the measured mean wind speeds at the reference height $h_0$ (m) and new height $h_1$ (m) at which the wind speed is predicted, respectively. $\alpha$ is the roughness coefficient. From Equation 2, the roughness coefficient is given as follows:

$$\alpha = \frac{\ln V_0(h_0) - \ln V_1(h_1)}{\ln(h_0) - \ln(h_1)} \qquad (3)$$

Table 3 indicates some roughness's for different types of topography.

The roughness coefficient for this experimental site is determined using reference height ($h_1$) and the corresponding mean wind speed ($V_1$) data from the military airport only 5 km away from the Green platform. The other values are measured from the investigation site given in the Table 4.

The roughness coefficient calculated from Equation 3. Therefore, we obtain $\alpha$=0,389 $\approx$ 0.4. As indicated in the Table 3 and showing in the Figure 2, the used micro-wind turbine is close to trees and buildings. Indeed, the considered micro-wind turbine of the Green platform is situated in an urban zone.

## Urban Zone Versus Rural Zone

The output power of a wind turbine is strongly influenced by the mean wind speed to which it is subjected. In urban areas, the effect of buildings, tall trees, may not only reduce the mean wind speed but may also increase the standard deviation of the wind fluctuations. The experimental site as indicated in section 3 is an urban case. In the Table 5, we determine the daily, monthly mean speed for the five years period (2012, 2013, 2014, 2015 and 2016) and the corresponding standard deviation. Figure 7 represents monthly mean speed of the five years analysis. The bar charts illustration shows good monthly similarity.

The standard deviation $\sigma$ is determined by using the Equation 4 [23].

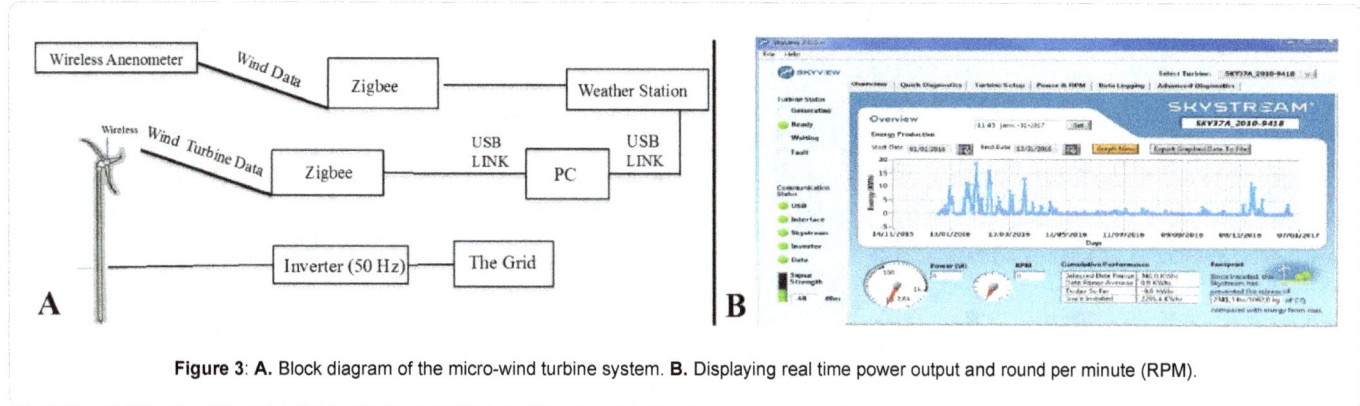

**Figure 3**: **A.** Block diagram of the micro-wind turbine system. **B.** Displaying real time power output and round per minute (RPM).

| Location | Latitude | Longitude | Elevation |
|----------|----------|-----------|-----------|
| Metz | 49°05'N | 6°13'E | 182 m above sea level |

**Table 2:** GPS coordinates of the GREEN platform.

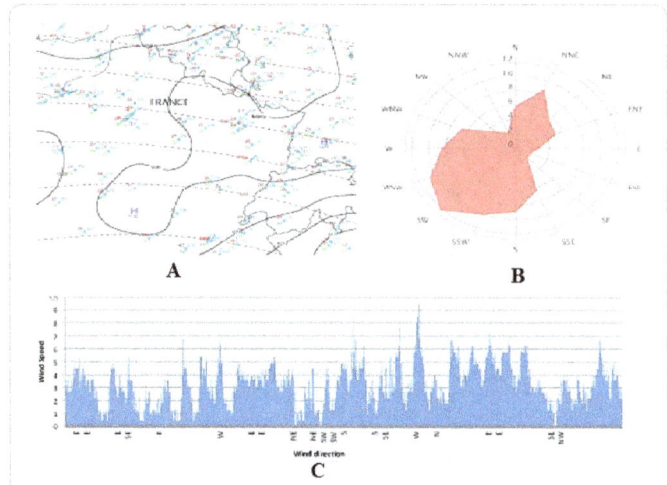

**Figure 4: A.** France climate map; **B.** Wind direction in percentage of the urban localization test site; **C.** Wind direction of the GREEN platform experimental zone.

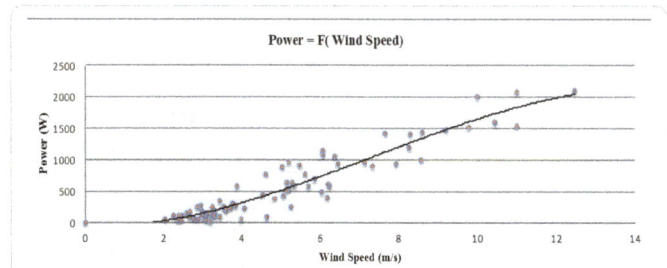

**Figure 5:** Power output as a function wind speed for an experimental day.

$$\sigma = \left[ \frac{1}{N-1} \sum_{i=1}^{n} \left( V - \overline{V} \right)^2 \right]^{\frac{1}{2}} \qquad (4)$$

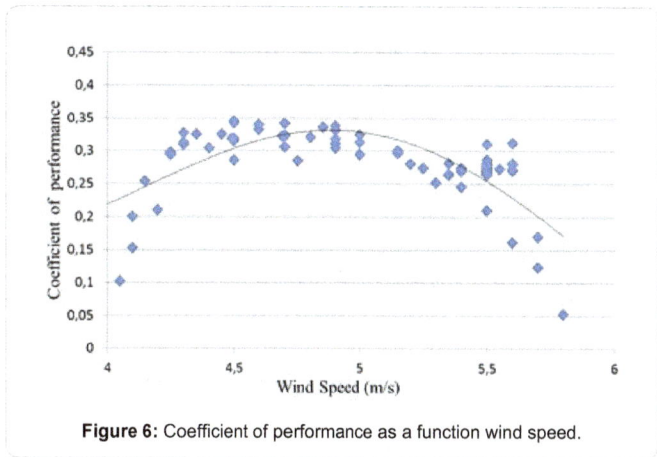

**Figure 6:** Coefficient of performance as a function wind speed.

| Wind Shear Coefficients | |
|-------------------------|--------------------------------------|
| **α** | **Description** |
| 0.1 | Perfectly smooth |
| 0.2 | Flat grassland or low |
| 0.3 | Trees or hills, bulding in area |
| 0.4 | Close to trees or buildings |
| 0.5 | Very close to trees or buildings |
| 0.6 | Surrounded by tall trees or buldings |

**Table 3:** Some roughness coefficients.

| Reference values | $h_1$=192 m | $V_1$=5 m/s |
|------------------|-------------|-------------|
| **Predicted values** | $h_0$=12 m | $V_0$=1.7 m/s |

**Table 4:** Experimental data of wind speed versus height.

$\overline{V}$ is the mean wind speed (m/s), N is the number of wind speed data. The calculated values of the standard deviation of the five years are given as follows:

$$\sigma_{2012}=0.4137 \text{ (m/s)} \qquad (5A)$$

$$\sigma_{2013}=0.3494 \text{ (m/s)} \qquad (5B)$$

$$\sigma_{2014}=0.3826 \text{ (m/s)} \qquad (5C)$$

$$\sigma_{2015}=0.3254 \text{ (m/s)} \qquad (5D)$$

$$\sigma_{2016}=0.4568 \text{ (m/s)} \qquad (5E)$$

As the standard deviation of the wind speed fluctuations relative to

| Month | 2012 | 2013 | 2014 | 2015 | 2016 |
|---|---|---|---|---|---|
| January | 2.10 | 1.83 | 1.93 | 2.17 | 2.01 |
| February | 1.40 | 1.42 | 1.32 | 1.87 | 1.90 |
| March | 1.42 | 1.20 | 1.47 | 2.08 | 2.07 |
| April | 2.05 | 2.19 | 1.38 | 1.61 | 1.63 |
| May | 1.56 | 1.24 | 1.75 | 1.49 | 1.32 |
| June | 1.33 | 1.35 | 1.44 | 1.54 | 1.08 |
| July | 1.41 | 1.49 | 1.17 | 1.62 | 1.07 |
| August | 1.30 | 1.05 | 1.24 | 1.40 | 1.03 |
| September | 1.29 | 1.31 | 1.01 | 1.60 | 0.87 |
| October | 1.40 | 1.55 | 0.92 | 1.00 | 1.30 |
| November | 1.03 | 1.87 | 1.09 | 2.02 | 1.65 |
| December | 2.44 | 1.96 | 2.20 | 1.59 | 0.79 |
| Mean speed (m/s) | 1.56 | 1.54 | 1.41 | 1.67 | 1.39 |
| σ (m/s) | 0.4137 | 0.3494 | 0.3826 | 0.3254 | 0.4568 |

**Table 5:** Yearly mean wind speed.

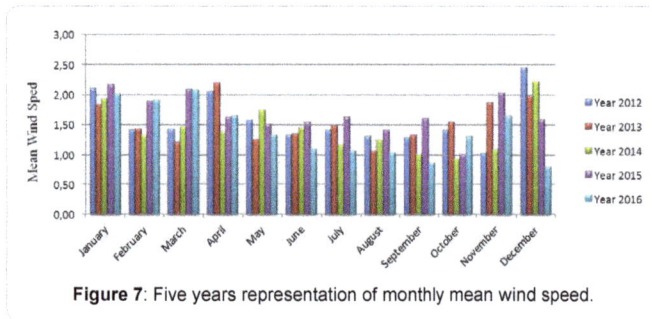

**Figure 7**: Five years representation of monthly mean wind speed.

| Year | 2012 | 2013 | 2014 | 2015 | 2016 |
|---|---|---|---|---|---|
| Month | Power (W) | | | | |
| January | 84.83 | 74.89 | 61.56 | 95.27 | 80.24 |
| February | 48.84 | 44.21 | 35.67 | 53.90 | 50.91 |
| March | 45.64 | 24.44 | 42.87 | 37.81 | 71.2 |
| April | 90.94 | 82.13 | 80.23 | 95.81 | 92.67 |
| May | 31.15 | 22.71 | 34.66 | 30.73 | 21.78 |
| June | 30.95 | 23.75 | 46.34 | 27.98 | 18.56 |
| July | 40.22 | 5.63 | 40.29 | 27.16 | 16.45 |
| August | 28.15 | 20.22 | 42.53 | 28.32 | 15.34 |
| September | 30.19 | 25.13 | 32.31 | 20.38 | 10.12 |
| October | 58.80 | 21.32 | 40.77 | 22.18 | 55.67 |
| November | 65.18 | 72.19 | 16.24 | 50.98 | 76.23 |
| December | 187.13 | 110.28 | 125.97 | 166.6 | 67.78 |

**Table 6:** Five years monthly mean power.

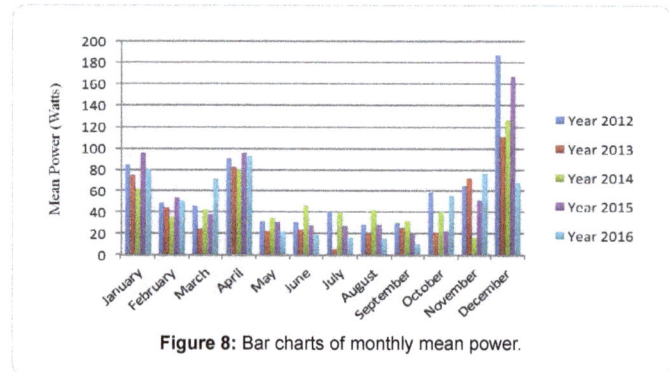

**Figure 8:** Bar charts of monthly mean power.

the mean wind speed in rural areas is likely to be reasonably constant. The standard deviations from Equation (5A) and Equation (5E) reflect a more rural area than urban area for the GREEN platform experimental site. The mean standard deviation upon the five years is 0.3855. As the urban environment not only influences the mean wind speed, it also affects the standard deviation and consequently the mean power output. In the Table 6, we determined the monthly mean power output for the five experimental years.

Figure 8 gives comparison of five years power output. We can see close similarity of the monthly power output for different years confirming a rural zone profile for an urban installation. Indeed, this is due to a long time period and very subtle onsite measurements to determine the ideal position for the wind turbine. The mean power output from a wind turbine as a function of mean wind speeds, is determined by a probability density distribution. The Weibull function is commonly used for fitting measured wind speed probability distribution. This point is discussed in the next section.

## Weibull Function and Wind Speed

The most commonly observed distribution providing small errors in the calculating of densities and better experimental matching is the Weibull [24,25] probability distribution function (pdf) which is given by the three-parameter Weibull distribution as given by the Equation 6.

$$F(T) = \frac{k}{\eta} \cdot \left(\frac{T-\gamma}{\eta}\right)^{k-1} \exp -\left(\frac{T-\gamma}{\eta}\right)^{k} \tag{6}$$

Where, F(T) ≥ 0, T ≥ 0 or k, γ> 0, η> 0,

- ∞<γ<∞ and:

- k is the shape parameter, also known as the Weibull slope

- η is the scale parameter

- γ is the location parameter

Frequently, the location parameter is not used and the value for this parameter can be set to zero. When this is the case, the pdf equation reduces to that of the two-parameter Weibull distribution given by the Equation 7 as follows:

$$F(T) = \frac{k}{\eta} \cdot \left(\frac{T}{\eta}\right)^{k-1} \exp -\left(\frac{T}{\eta}\right)^{k} \tag{7}$$

Usually, when climate has no zero values and then keep the wind turbine on run, the location parameter must be determined. Frequently, the location parameter is not used when the data base of the mean climate have very low or zeros measured values. Thereby, the location parameter may be set to zero. In our case study, measurement shows that the wind speed is too low or null during at least 850 h/year and then under the cut-in wind speed lower than 3.5 m/s, see the Table 1 to set micro wind turbine on. Consequently, in our modeling the location parameter is set to zero. Therefore, the pdf equation reduces to that of the two parameters Weibull distribution given by the Equation 8.

$$F(V) = \frac{k}{A}\left(\frac{V}{A}\right)^{k-1} \exp\left(-\left(\frac{V}{A}\right)^{k}\right) \tag{8}$$

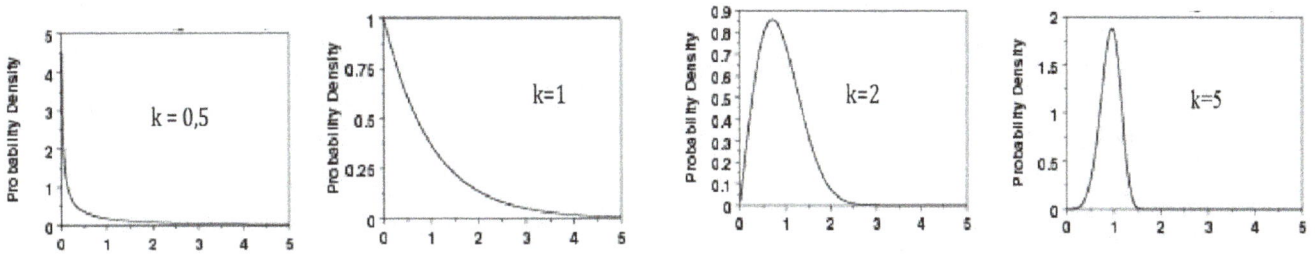

**Figure 9:** Weibull curves for some k values.

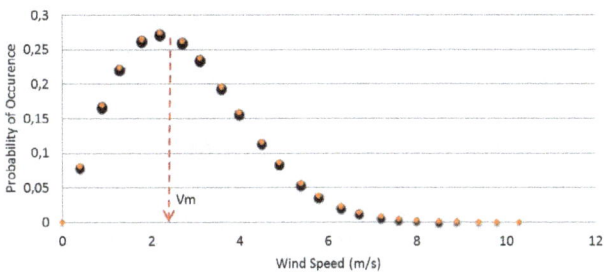

**Figure 10:** Probability of occurrence as a function of wind speed for a particular day.

- where, η=A=Weibull scale parameter in (m/s),

- k is s the unitless shape parameter,

- T=V=wind speed (m/s).

Figure 9 gives a representation of pdf for different k values while keeping η constant. There are several methods that are used for estimating the Weibull parameters A and k, depending on which wind statistics are available. We used the mean wind speed and standard deviation (σ) method as suggested in Hussein et al. [26]. Thus, the calculated value of k specifying the GREEN platform conditions can be obtained by using the mean wind speed and the standard deviation as follows by the Equation 9.

$$k = \left(\frac{\sigma}{\overline{V}}\right)^{-1.0983} \tag{9}$$

Thereby, considering the measured mean wind speed (Table 5) and the mean standard deviation during the year 2015, the calculated value of k is 2.37. The resulting equation is transformed into Equation 10.

$$F(V) = \frac{2.37}{A}\left(\frac{V}{A}\right)^{1.37} \exp\left(-\left(\frac{V}{A}\right)^{2.37}\right) \tag{10}$$

Figure 10 is an example of the experimental image of the probability density distribution calculated from the Weibull function, for a particular day. The particular wind speed $V_m$ is nearly 2,5 m/s and probability of occurrence is more that 25%.

Using the Weibull statistical method for evaluation of local wind probabilities for five consecutive years (2012, 2013, 2014, 2015 and 2016) of the GREEN platform, wind power site are represented in Figure 11. The curve shape is similar to the Figure 9 with k=2. From this figure, the comparison between the probability distribution function

calculated from the Weibull function and the wind speed distribution based on data for the studied urban location, indicates that the most probable and corresponding wind speed upon 5 years are close and lies between 1.3 and 1.4 m/s.

Mean speed over each period of Figure 11 is nearly the same and defined as the total area under the curve F (V) - V integrated between

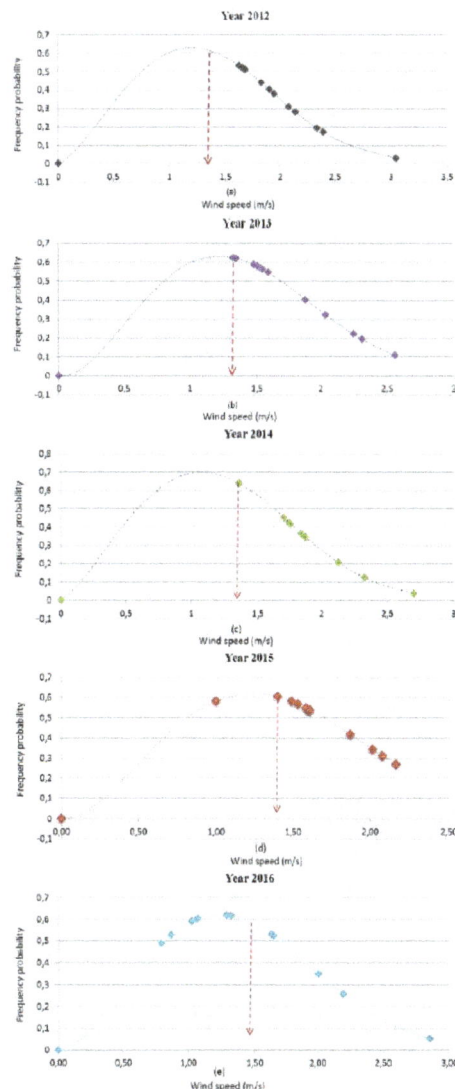

**Figure 11:** Weibull function versus wind speed upon 5 years.

windless days (V=0 m/s) to a very windy day (V=∝ m/s) divided by the total number of hours in the year which is nearly the same for each year as deduced from the Figure 11. The annual mean speed [27] is therefore the weighted average speed and is given by Equation 10:

$$V_{mean} = \frac{1}{H} \int_0^\infty F(V).\ V\ dV \tag{11}$$

H is the number of operating hours and in this work, H=3600 h per year for this particular site. The integral expression of $V_{mean}$ of Equation 11 can be approximated to the Gamma function as expressed by the Equation 12:

$$V_{mean} = A \cdot \Gamma\left(1 + \frac{1}{k}\right) \tag{12}$$

where Γ is the Gamma function which is obtained from the Equation 13:

$$\Gamma(X) = \int_0^\infty \exp(-t)\ t^{X-1}\ dt \tag{13}$$

Table 7 indicates the experimental site results for the four years experimental data. Each resulting mean speed from the Table 7 is a similar trend to the area under the corresponding curve of the Figure 11.

## Weibull Function and Power Output

### Power output

Output power from the wind is proportional to various parameters as given section 2.2. The primary parameter is wind speed, where the output power is proportional to the cube of the speed and the speed varies with height according to power law exponent relation. We should state all effected parameters on the power output and concentrated on how to find the average annual wind speed by using Weibull probability distribution. From this average annual wind speed, we can determine the expected output power as discussed later. The output power from the wind is given by the Equation 14 [28]:

$$P_{out} = \frac{1}{2}\ \rho S C_P V_{rmc}^3 \tag{14}$$

Where $V_{rmc}$ is the root mean cube speed given as follows by the Equation 15:

$$V_{rmc}^3 = A^3 \cdot \Gamma\left(1 + \frac{3}{k}\right) \tag{15}$$

Replacing the Equation 15 into the Equation 14, we deduice the following expression:

$$P_{out} = \frac{1}{2}\ \rho S C_P A^3 \Gamma\left(1 + \frac{3}{k}\right) \tag{16}$$

The monthly calculated energy output using the Weibull-data, $E_{Weibull}$ (kWh) is given by the Equation 17 as follows:

$$E_{Weibull}\ (kWh) = \sum_i^n P_{i,output}\ F_i(V) \cdot T_i \tag{17}$$

Where Ti is the number of hours of wind power operation and the number of days in a particular month. Figure 12 shows the relationship between Weibull function and mean power output upon 5 years measured data. The most yearly probable output for this site lies in the range of 45 W to 70 W and for an operating time of 5300 hours per year. The estimating output per year lies between 240 kWh and 371 kWh.

| Year | k | A (m/s) | $V_{mean}$ (m/s) |
|------|------|---------|------------------|
| 2012 | 2.37 | 2.03 | 3.59 |
| 2013 | 2.37 | 1.54 | 2.72 |
| 2014 | 2.37 | 1.41 | 2.50 |
| 2015 | 2.37 | 1.67 | 2.81 |
| 2016 | 2.37 | 1.48 | 2.62 |

**Table 7:** Mean speed from experimental data.

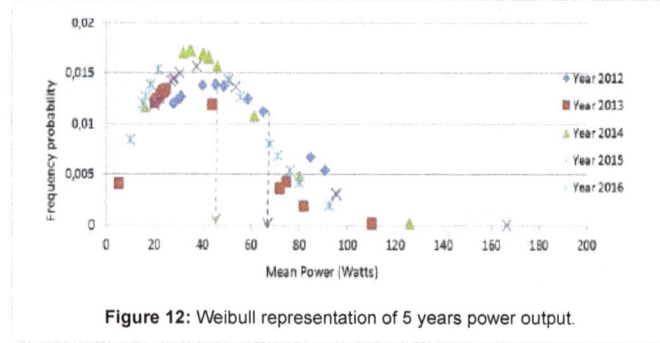

**Figure 12:** Weibull representation of 5 years power output.

| Year | k | A (m/s) | Energy (kWh) (Weibull) | Measured Energy (kWh) |
|------|------|---------|------------------------|------------------------|
| 2012 | 2.37 | 2.03 | 372.77 | 279.83 |
| 2013 | 2.37 | 1.54 | 329.92 | 346.94 |
| 2014 | 2.37 | 1.41 | 311.36 | 297.14 |
| 2015 | 2.37 | 1.67 | 348.12 | 361.16 |
| 2016 | 2.37 | 1.48 | 314.16 | 295.17 |

**Table 8:** Comparing 5 years energy output.

Table 8 gives the five years of energy calculated from Weibull function and measured ones. The measured energy is the raw data recorded using data loggers. We deduce that the Weibull-representative energy outputs is nearly of good agreement with the recorded data energy outputs from the experimental platform site. The root mean square error (RMSE) between the estimated Weibull distribution from mean wind and the Weibull distribution from measured power is given in the Table 9. We calculated the RMSE for 2014 corresponding to the middle year between 2012 until 2016.

The results of measurements and power calculation shows a strong correlation with low RMSE value of around 0.5%, obtained from the Weibull model and the measured energy data can be found using the Equation 18 [25]:

$$Error\ (\%) = \left| \frac{E_{(Weibull)} - E_{(Measured)}}{E_{(Measured)}} \right| \tag{18}$$

Table 10 gives the errors for the five years between 2012 until 2016. We observe from this table that for year 2012, we obtained a higher error rate due to mainly some lost wireless zigbee communication occured over experimental measurements and leading the lack of data during the data acquisition. Also few days between the 20th and 31st are missing for December 2016 due to technical problem.

## Discussion

To investigate the potential performance of micro-wind turbines, we have considered a real well-studied siting case of a micro wind turbine in complex urban areas. We used the quality of wind speed assessment and Weibull probability density function for describing the measured wind speed frequency distribution upon five years real experimental data. Our goal is to provide the energy output prediction of specific locate urban places for possible installation sites of a wind

| Year 2014 | |
|---|---|
| Calculated Weibull value (Mean wind) | Weibull distribution value (Measured power) |
| 0.011299 | 0.010805 |
| 0.004239 | 0.017171 |
| 0.015104 | 0.016452 |
| 0.015541 | 0.004874 |
| 0.012982 | 0.017167 |
| 0.015253 | 0.015709 |
| 0.015957 | 0.01685 |
| 0.015916 | 0.016512 |
| 0.015616 | 0.017045 |
| 0.015137 | 0.016786 |
| 0.015863 | 0.01171 |
| 0.008608 | 0.000175 |
| RMS | 0.005738 |

**Table 9:** RMSE between the estimated Weibull distribution from mean wind and Weibull distribution from measured power for the year 2014.

| Year | Error (%) |
|---|---|
| 2012 | 0.2493 |
| 2013 | 0.0516 |
| 2014 | 0.0457 |
| 2015 | 0.0434 |
| 2016 | 0.0551 |

**Table 10:** Errors for the 5 years.

turbine and which are essential before any installation or modeling of the expected energy in urban configurations. To meet our goal, our approach performs an initial investigation which is required to know the wind speed and turbulence characteristics at the corresponding turbine height for specific location before installing micro wind turbine in an urban site. In summary, from real experimental results and although siting micro wind turbines in an urban environment is not generally considered as finest locations, we can deduce when local climate conditions can be taken into consideration to deduce good locations generating useful amounts of electricity at a reasonable cost and can be a worthwhile investment.

## Conclusion

The paper investigates the potential performance of micro-wind turbines in complex urban areas upon five years real experimental data. The main contribution is to provide a pragmatic approach providing the energy output prediction and estimation of payback time on capital investment. More precisely, our study is to provide the ideal position for the wind turbine allowing optimal performance can be achieved in function of the hub height with respect to proximity/influence of adjacent buildings/obstructions in the urban zones through very subtle onsite measurements.

In our approach, we used the quality of wind speed assessment and Weibull probability density function for describing the measured wind speed frequency distribution. The root mean cube speed is useful in quickly estimating the annual energy potential of the site. Thereby, by considering the proposed modeling with real conditions only depending of the speed and wind direction statistics, we can estimate the power production potential for a specific micro wind turbine installation in a particular urban area site. The originality of our study upon many years data is to provide the energy output prediction and estimation of payback time on capital investment can be determined. The consequently, our approach allows to know the potential wind

speed which can be exploited as an installing micro wind turbine in a complex urban site.

### Acknowledgments

The authors wish to acknowledge the GREEN platform industrial partners. The work described in this paper has been supported by UEM French local electricity company and also funded by the Tous Chercheurs project of the Bettencourt Foundation.

### References

1. Chvez-Ramrez A, Cruz J, Espinosa-Lumbreras R, Ledesma-Garca J, Durn-Torres SM, et al. (2013) Design and set up of a hybrid power system (PV-WT-URFC) for a stand-alone application in Mexico. Int J Hydrogen Energy 38: 12623-12633.

2. Andaloro A, Salomone R, Andaloro L, Briguglio N, Sparacia S (2012) Alternative energy scenarios for small islands: A case study from Salina Island (Aeolian Islands, Southern Italy). Renew Energy J 47: 135-146.

3. Islam MR, Saidur R, Rahim NA (2011) Assessment of wind energy potentiality at Kudatand Labuan, Malaysia using Weibull distribution function. Energy 36: 985-992.

4. Cabello M, Orza JAG (2010) Wind speed analysis in the province of Alicante, Spain. Potential for small-scale wind turbines. Renew Sustain Energy Rev 14: 3185-3191.

5. Fyrippis I, Axaopoulos PJ, Panayiotou G (2010) Wind energy potential assessment in Naxos Island, Greece. Appl Energy 87: 577-586.

6. Jowder FAL (2009) Wind power analysis and site matching of wind turbine generators in Kingdom of Bahrain. Appl Energy 86: 538-545.

7. Dharmakeerthi CH, Atputharajah A, Ekanayake J (2008) Field experience with an islanded micro wind power plant. IEEE Int Conf Sustain Energy Technol, pp: 966-971.

8. Li D, Cheung KL, Chan WWH, Cheng CCK, Wong TCH (2014) Analysis of wind energy potential for micro wind turbine in Hong Kong. Build Serv Eng Res Technol 35: 268-279.

9. Lu L, Sun K (2014) Wind power evaluation and utilization over a reference high-rise building in urban area. Energ Build 68: 339-350.

10. Li QS, Chen FB, Li YG, Lee YY (2013) Implementing wind turbines in at all building for power generation: A study of wind loads and wind speed amplifications. J Wind Eng Ind Aerodyn 116: 70-82.

11. Mithraratne N (2009) Roof-top wind turbines for microgeneration in urban houses in New Zealand. Energ Build 41: 1013-1018.

12. Lu L, Ip KY (2009) Investigation on the feasibilty and enhancement methods of wind power utilization in high-rise buildings of Hong Kong. Renew Sustain Energy Rev 13: 450-461.

13. Bahaj AS, Myers LE, James PAB (2007) Urban energy generation: Influence of micro-wind turbine output on electricity consumption in buildings. Energ Build 39: 154-165.

14. Mertens S (2002) Wind energy in urban areas. Renew Energy focus, pp: 22-24.

15. Ayhan D, Salam S (2012) A technical review of building-mounted wind power systems and a sample simulation model. Renew Sustain Energy Rev 16: 1040-1049.

16. Fleury A, Arteiro F, Brasil D, Franceschi A (2012) Integration of wind power plants into the electric system-The Brazilian experience. IEEE/PES Transmission and Distribution: Latin America Conference and Exposition 6: 1-6.

17. Mosetti G, Poloni C, Diviacco B (1994) Optimization of wind turbine positioning in large wind farms by means of a genetic algorithm. Wind Eng Indust Aerodyn 51: 105-116.

18. Grady SA, Hussaini MY, Abdullah MM (2005) Placement of wind turbines using genetic algorithms. Renew Energy 30: 259-270.

19. Wan C, Wang J, Yang G, Li X, Zhang X (2009) Optimal micro-siting of wind turbines by genetic algorithms based on improved wind and turbine models. IEEE Conf Decision Control.

20. Skyview (2014) Software for skystream wind turbines.

21. Burton T, Sharpe D, Jenkins N, Bossanyi E (2001) Wind energy handbook

(2ndedn). John Wiley Sons, New York.

22. Justus CG, Mikhail A (1976) Height variation of wind speed and wind distributions statistics. Geophys Res Lett 3: 261-264.

23. Oner Y, Ozcira S, Bekiroglu N, Senol I (2013) A comparative analysis of wind power density prediction methods for anakkale, Intepe region, Turkey. Renew Sust Energy Rev 23: 491-502.

24. Conradsen K, Nielsen LB, Prahm LP (1984) Review of weibull statistics for estimation of wind speed distributions. J Appl Meteorol 23: 1173-1183.

25. Celik AN (2003) A statistical analysis of wind power density based on the weibull and rayleigh models at the southern region of turkey. Renew Energy 29: 593-604.

26. Elkinton MR, Rogers AL, McGowan JG (2006) An investigation of wind-shear.

27. Hussein M, Mousa M, Abdel-Akher M, Orabi M, Ahmed ME, et al. (2010) Studying of the available wind and photovoltaic energy resources in Egypt. 14th International Middle East Power Systems Conference, pp: 657-662.

28. Patel MR (1999) Wind and solar power systems. U.S. Merchant Marine Academy Kings Point, New York.

# PERMISSIONS

# LIST OF CONTRIBUTORS

**Sagar M Kande**
Energy Technology, Department of Technology, Shivaji University, Kolhapur, India

**Wagh MM**
Energy Technology, Shivaji University, India

**Ghane SG**
Department of Botany, Shivaji University, India

**Shinde NN**
Energy Technology, Shivaji University, India

**Patil PS**
School of Nano Science and Technology, Shivaji University, India

**Aadesh Rajkrishna**
Department of Engineering, University of Petroleum and Energy Studies, India

**Bardsley A, Whitty JPM, Howe J and Francis J**
Energy and Power Management Research Group, School of Computing, Engineering and Physical Sciences, University of Central Lancashire, Preston, PR1 2HE, UK

**Vivek Kumar Singh, Lakshman Ravi Teja and Jitendra Tiwari**
University of Coimbra, MIT Portugal Program, Coimbra, Portugal

**Abdul-Kareem Mahdi Salih**
Department of Physics, College of Science, Thi-Qar University, Iraq

**Abdullh Saiwan Majli**
Department of Electronics, College of Engineering, Thi-Qar University, Iraq

**Mei Gong**
School of Business and Engineering, Halmstad University, PO Box 823, SE-30118 Halmstad, Sweden

**Göran Wall**
Oxbo gard, SE-43892 Härryda, Sweden

**Meryem Oudda and Abdeldjebar Hazzab**
Department of Electric and Electronics Engineering, Tahri Mohamed Bechar University, Algeria

**Ruben M. Mouangue**
Department of Energetic Engineering, UIT, University of Ngaoundere, Cameroon

**Alexis Kuitche**
Departments of GEEA, PAI, ENSAI, University of Ngaoundere, Cameroon

**Myrin Y. Kazet**
Department of Energetic Engineering, UIT, University of Ngaoundere, Cameroon

Departments of GEEA, PAI, ENSAI, University of Ngaoundere, Cameroon

**Daniel Lissouck**
Department of Renewable Energy, HTTTC, Kumba, Cameroon

**J.M. Ndjaka**
Department of Physics, Faculty of Sciences, University of Yaounde 1, Cameroon

**Ali Razmjoo**
Department of Energy Systems Engineering, Islamic Azad University, South Tehran Branch, Tehran, Iran

**S. Mohammadreza Heibati**
Department of Mechanical Engineering, Islamic Azad University, Pardis Branch, Iran

**Mohammad Ghadimi**
Department of Mechanical Engineering, Islamic Azad University, Roudehen Branch, Iran

**Mojtaba Qolipour**
Department of Industrial Engineering, Yazd University, Iran

**Javad Rezaei Nasab**
Islamic Azad University, Bushehr Branch, Iran

**Samira Ben Abdallah, Nader Frikha and Slimane Gabsi**
Samira Ben Abdallah, Research Unit of Environment, Catalysis and Analysis Processes, University of Gabes, National Engineering school of Gabès, Street Omar Ibn ElKhattab, 6029 Gabès, Tunisia

**Domènec Jolis and Natalie Sierra**
San Francisco Public Utilities Commission 750 Phelps Street, San Francisco, California, USA

**Harwinder Singh, Aftab Anjum, Mohit Gupta, Aadish Jain and Amrik Singh**
Department of Mechanical Engineering, Delhi Technological University, India

**Abouelfadl S**
Architectural department- College of Engineering, Assiut University, Asyut, Egypt

**Ouda K**
Department of Geology- College of Science, Assiut University, Asyut, Egypt

**Atia A, AL-AMIR N, Ali M, Mahmoud S, Said H and Ahmed A**
Architect, Asyut, Egypt

**Oyedepo SO**
Mechanical Engineering Department, Covenant University, Ota, Nigeria

**Agbetuyi AF**
Electrical Engineering Department, Covenant University, Ota, Nigeria

**Odunfa MK**
Mechanical Engineering Department, Covenant University, Ota, Nigeria
Mechanical Engineering Department, University of Ibadan, Nigeria

**Hossein Sheykhlou**
Department of Mechanical Engineering, Urmia University, Iran

**Vijayamohanan Pillai N**
Centre for Development Studies, Prasanth Nagar, Ulloor, Kerala, India

**Banafsheh Abolpour**
Department of Civil Engineering, Science and Research Branch, Islamic Azad University, Iran

**Mohsen Yaghobi**
Department of Chemical Engineering, Shahid Bahonar University of Kerman, Iran

**Bahador Abolpour**
Department of Chemical Engineering, Shahid Bahonar University of Kerman, Iran
Department of Aerospace Engineering, Payame Noor University, Iran
Department of Computer Engineering, Payame Noor University, Iran

**Hosein Bakhshi**
Department of Civil Engineering, Hakim Sabzevari University, Iran

**Akintayo T Abolude and Stefano C Sarris**
School of Energy and Environment, City University of Hong Kong, Hong SAR, China

**Akintomide Afolayan Akinsanola**
School of Energy and Environment, City University of Hong Kong, Hong SAR, China
Department of Meteorology and Climate Science, Federal University of Technology Akure, Nigeria

**Kehinde Olufunso Ogunjobi and Kehinde O Ladipo**
Department of Meteorology and Climate Science, Federal University of Technology Akure, Nigeria

**Harwinder Singh and Pushpendra Singh**
Department of Mechanical, Production and Industrial Engineering, Delhi Technological University, India

**Xingxing Zhang, Jingchun Shen, Llewellyn Tang, Tong Yang, Liang Xia, Zehui Hong, Luying Wang and Yong Shi**
Department of Architecture and Built Environment, University of Nottingham, Ningbo, China

**Yupeng Wu**
Department of Architecture and Built Environment, University of Nottingham, UK

**Peng Xu**
Beijing Key Lab of Heating, Gas Supply, Ventilating and Air Conditioning Engineering, Beijing University of Civil Engineering and Architecture, China

**Shengchun Liu**
Key Laboratory of Refrigeration Technology, Tianjin University of Commerce, China

**Elmehdi Karami, Amine Haibaoui and Abderraouf Ridah**
Department of Physics, LIMAT Laboratory, Ben M'sick, Morocco

**Mohamed Rafi and Bouchaib Hartiti**
Mohammedia Faculty of Science and Technology, MAC & PAM Laboratory, ANEPMAER Group, Morocco

**Philippe Thevenin**
Laboratory Optical Materials, Photonics and Systems, University of Lorraine, Metz, France

**Samer Yassin Alsadi and Yasser Fathi Nassar**
Department of Electrical Engineering, Palestine Technical University, Palestine

**Aragón-Aguilar Alfonso, Izquierdo-Montalvo Georgina and López-Blanco Siomara**
Instituto de Investigaciones Eléctricas, Gerencia de Geotermia, Morelos, México

**Gómez-Mendoza Rafael**
Instituto Mexicano de Tecnología del Agua, Morelos, México

**Tyamo Okosun and Chenn Q Zhou**
Department of Mechanical Engineering, Purdue University, Center for Innovation through Visualization and Simulation, Purdue University Calumet, USA

**Miloud Benmedjahed and Lahouaria Boudaoud**
Research Unit Renewable Energy in Rural Sahara, URERMS, Renewable Energy Development Centre, CDER, BP 478 Route Reggane, Adrar, Algeria

**A.M. Abd El Rahman and M.H.M.Hassanien**
Department of Petroleum Refining and Petrochemicals, Suez University, Egypt

**A.S. Nafey**
Department of Engineering Sciences, Suez University, Suez University, Egypt

**Rafael Carlos Reynaga-López**
Institute of Engineering, Autonomous University of Baja California, Mexicali, Mexico

**Alejandro Lambert, Marlene Zamora and Elia Leyva**
Faculty of Engineering, Autonomous University of Baja California, Mexicali, Mexico

**Oscar Jaramillo**
Institute of Renewable Energies, National Autonomous University of Mexico, Temixco, Mexico

**Hiroshi Tanaka**
Department of Mechanical Engineering, National Institute of Technology, Kurume College, Komorino, Kurume, Japan

**Subhashish Banerjee, Md. Nor. Musa, Dato' IR Abu. Bakar Jaafar and Azrin Arrifin**
Department of Renewable Energy, Universiti Teknologi Malaysia, Malaysia

**Ba MM, Ramenah Harry and Tanougast C**
Department of Electronics, University of Lorraine, France

# Index

**A**
Aerodynamic Geometries, 15
Aerofoil, 15, 17, 19-20

**B**
Beneficial Reuse, 67
Biofuels, 22, 37, 87
Biosolids, 67-71
Boost Converter, 41-42
Buck Converter, 41

**C**
Carbon Footprint, 22
Concentrated Solar Power, 7, 9, 14
Converter Topology, 41
Cumulative Distribution Function, 48, 52, 113

**D**
Dc-dc Converter, 41-42, 46
Decentralized Rural Energy, 22
Desalination, 59, 65, 195, 201-202, 211
Direct Current, 22-23, 73
Dry Air Density, 28
Drying Technology, 67-68, 71
Duty Cycle, 41-42, 44

**E**
Economic Growth, 55, 95
Electric Power Network, 23, 88
Electric Pump, 48, 50, 52-53, 124
Electrical Devices, 72, 77
Energy Audit, 72-73, 77
Energy Flow Diagram, 34
Entrepreneurial Activity, 22
Environmental Audit, 72
Evapotranspiration, 1
Exergy Analysis, 33-36, 39-40, 102, 194
Exergy Destruction, 34, 97, 101-102
Exergy Payback Time, 33, 35, 38-39

**F**
Flat Plate Collector, 7, 14, 77, 123-124, 126, 135, 159
Fossil Energy, 55, 89
Fresnel Lens, 7-12, 14

**G**
Gamma Function, 48, 218

Grid Electricity, 24

**H**
Higher Heating Value, 35
Hollow Fiber Module, 59, 64-65
Humidity, 27-31, 67-71, 139, 156, 188

**I**
Instantaneous Efficiency, 1, 10, 12, 14, 126

**L**
Life Cycle Analysis, 33, 35, 87
Linear Fresnel Reflector (LFR), 9

**M**
Mat Lab, 55
Mathematical Modeling, 15, 149
Maximum Deflexion, 19-21
Mechanical Equipment, 72, 131
Micro Grid, 22-25
Modeling, 15, 17, 19, 55, 58-59, 62-63, 65, 67, 115-116, 149, 162, 168, 177, 181, 202, 211-213, 216, 219
Moist Density, 28

**N**
Ngaoundere, 48, 50, 52-53

**O**
Off-grid Electrification, 22-23, 25

**P**
Panel Cooling, 5
Photovoltaic System, 14, 23, 38-39, 41, 44, 46, 150, 155
Photovoltaics, 6-7, 23, 89, 96, 204
Piston Pump, 48, 51-53
Polycrystalline Silicon, 1, 204-205
Poor Electrification, 22
Primary Lighting Fuel, 22
Probability Density Function, 48, 50, 117, 218-219
Pyranometer, 2

**R**
Rayleigh Distribution, 48
Recycled Exergy, 37-38
Regression Coefficients, 55-57
Renewable Energy, 6, 15, 22-23, 26-27, 31, 33-34, 37, 39-41, 48, 54-55, 58-59, 78-81, 84, 86-92, 95-97, 102, 122, 128-129, 140, 150, 160, 168, 178, 187, 203, 206, 209-211, 213

Renewable Energy Sources, 6, 22, 86-87, 89, 91, 95-97, 129, 160, 203

Roto-dynamic Pump, 48, 52-53

Rural Energy Infrastructure, 22

**S**

Semiconductors, 41, 91, 204

Solar Drying, 67-68, 71

Solar Energy, 6, 11, 14, 23, 26, 33, 39, 53, 55, 58-59, 65, 68, 71, 77, 87, 90, 92, 95, 97, 123, 128-130, 134, 136, 140, 144, 146, 148-151, 155-159, 177, 182, 187, 201-204, 206, 211

Solar Photovoltaic, 6, 22-24, 33, 38-39, 55-56, 58, 77, 89, 148, 155, 211-212

Solar Pond, 59, 62-63, 65-66, 194, 203, 206, 211

Solar Radiation, 1, 14, 33-35, 42, 55-59, 62-64, 67-69, 130, 136-137, 139, 154, 156-159, 195-200, 204

Solar Thermal, 7-11, 33-35, 37-40, 59, 67-69, 77, 86-87, 123, 129-130, 133-134, 136-139, 143, 145, 147-149, 159, 182, 184, 187, 196, 202

Specific Humidity, 27

Structural Stability, 15

Sun Radiation, 55

Sun Tracking System, 8, 14

Sustainable Development, 26, 33, 40, 48, 72, 77-78, 84, 86, 95, 122, 181, 203, 210

**T**

Tidal Energy, 33

**U**

Utilization Efficiency, 41

**V**

Vegetation Cover, 1

Ventilation Flux, 67

Volume Decrement Effect, 30

**W**

Water Conservation, 72, 77

Water Pumping, 47-48, 52, 91-92

Weibull Distribution, 48, 50, 52-53, 113-115, 122, 178, 212-213, 216, 218-219

Wind Energy, 6, 21, 27, 31-32, 48, 52-54, 58, 79, 87, 89, 95-97, 103, 116-117, 121-122, 169, 171, 177-179, 181, 188, 193-194, 203, 205-206, 208, 210-212, 219

Wind Power, 21, 27-28, 31-32, 47-48, 53, 58, 87-89, 91-93, 95, 98, 102, 113, 117, 121-122, 169, 174-176, 178, 188, 207, 212-214, 217-220

Wind Turbine Blades, 15, 19, 21, 171-173